Innovation and Entrepreneurial Opportunities in Community Tourism

Jakson Renner Rodrigues Soares
Universidade da Coruña, Spain & Universidade Estadual do Ceará, Brazil

IGI Global
PUBLISHER of TIMELY KNOWLEDGE

A volume in the Advances in Hospitality, Tourism, and the Services Industry (AHTSI) Book Series

Published in the United States of America by
IGI Global
Business Science Reference (an imprint of IGI Global)
701 E. Chocolate Avenue
Hershey PA, USA 17033
Tel: 717-533-8845
Fax: 717-533-8661
E-mail: cust@igi-global.com
Web site: http://www.igi-global.com

Library of Congress Cataloging-in-Publication Data

Names: Soares, Jakson Renner Rodrigues, 1980- editor.
Title: Innovation and entrepreneurial opportunities in community tourism / Jakson Renner Rodrigues Soares, editor.
Description: Hershey, PA : Business Science Reference, [2021] | Includes bibliographical references and index. | Summary: "This book provides relevant theoretical frameworks and the latest empirical research findings in the community tourism for professionals who want to improve their understanding of the innovation in the community tourism network and its contribution to the development of the territory. The relevance of the theme revolves around the dynamic capacities related to small tourism entrepreneurs in the coastal space"-- Provided by publisher.
Identifiers: LCCN 2020020523 (print) | LCCN 2020020524 (ebook) | ISBN 9781799848554 (hardcover) | ISBN 9781799858911 (paperback) | ISBN 9781799848561 (ebook)
Subjects: LCSH: Tourism--Management--Case studies. | Tourism--Planning--Case studies. | Community life--Case studies. | Community development--Case studies. | Entrepreneurship--Case studies.
Classification: LCC G155.A1 I464 2021 (print) | LCC G155.A1 (ebook) | DDC 910.68/4--dc23
LC record available at https://lccn.loc.gov/2020020523
LC ebook record available at https://lccn.loc.gov/2020020524

This book is published in the IGI Global book series Advances in Hospitality, Tourism, and the Services Industry (AHTSI) (ISSN: 2475-6547; eISSN: 2475-6555)

British Cataloguing in Publication Data
A Cataloguing in Publication record for this book is available from the British Library.

For electronic access to this publication, please contact: eresources@igi-global.com.

Advances in Hospitality, Tourism, and the Services Industry (AHTSI) Book Series

Maximiliano Korstanje
University of Palermo, Argentina

ISSN:2475-6547
EISSN:2475-6555

Mission

Globally, the hospitality, travel, tourism, and services industries generate a significant percentage of revenue and represent a large portion of the business world. Even in tough economic times, these industries thrive as individuals continue to spend on leisure and recreation activities as well as services.

The Advances in Hospitality, Tourism, and the Services Industry (AHTSI) book series offers diverse publications relating to the management, promotion, and profitability of the leisure, recreation, and services industries. Highlighting current research pertaining to various topics within the realm of hospitality, travel, tourism, and services management, the titles found within the AHTSI book series are pertinent to the research and professional needs of managers, business practitioners, researchers, and upper-level students studying in the field.

Coverage

- Leisure & Business Travel
- Sustainable Tourism
- Customer Service Issues
- Service Management
- Service Training
- Food and Beverage Management
- Health and Wellness Tourism
- Hotel Management
- Tourism and the Environment
- Casino Management

IGI Global is currently accepting manuscripts for publication within this series. To submit a proposal for a volume in this series, please contact our Acquisition Editors at Acquisitions@igi-global.com or visit: http://www.igi-global.com/publish/.

The Advances in Hospitality, Tourism, and the Services Industry (AHTSI) Book Series (ISSN 2475-6547) is published by IGI Global, 701 E. Chocolate Avenue, Hershey, PA 17033-1240, USA, www.igi-global.com. This series is composed of titles available for purchase individually; each title is edited to be contextually exclusive from any other title within the series. For pricing and ordering information please visit http://www.igi-global.com/book-series/advances-hospitality-tourism-services-industry/121014. Postmaster: Send all address changes to above address.

Titles in this Series

For a list of additional titles in this series, please visit:
http://www.igi-global.com/book-series/advances-hospitality-tourism-services-industry/121014

Role and Impact of Tourism in Peacebuilding and Conflict Transformation
Jorge Tavares da Silva (GOVCOPP, University of Aveiro, Portugal) Zélia Breda (GOVCOPP, University of Aveiro, Portugal) and Fabio Carbone (Centre for Trust, Peace, and Social Relations, Coventry University, UK & GOVCOPP, University of Aveiro, Portugal)
Business Science Reference • © 2021 • 402pp • H/C (ISBN: 9781799850533) • US $195.00

Strategies for Promoting Sustainable Hospitality and Tourism Services
Maximiliano Emanuel Korstanje (University of Palermo, Argentina) Babu George (Fort Hays State University, USA) and Alexandru-Mircea Nedelea (Stefan cel Mare University of Suceava, Romania)
Business Science Reference • © 2020 • 243pp • H/C (ISBN: 9781799843306) • US $195.00

Handbook of Research on Resident and Tourist Perspectives on Travel Destinations
Patrícia Pinto (Faculty of Economics and Research Centre for Tourism, Sustainability, and Well-Being (CinTurs), University of Algarve, Portugal) and Manuela Guerreiro (Faculty of Economics and Research Centre for Tourism, Sustainability, and Well-Being (CinTurs), University of Algarve, Portugal)
Business Science Reference • © 2020 • 529pp • H/C (ISBN: 9781799831563) • US $285.00

Global Niche Tourism and Opportunities for the Medical Tourism Industry
Manjeet Singh (Central University of Jammu, India) and Subbaraman Kumaran (Alliance University Bangalore, India)
Business Science Reference • © 2020 • 300pp • H/C (ISBN: 9781799834274) • US $215.00

Cases on Tour Guide Practices for Alternative Tourism
Gulsun Yildirim (Recep Tayyip Erdogan University, Turkey) Ozlem Ozbek (Bandirma Onyedi Eylul University, Turkey) Ceyhun Caglar Kilinc (Akdeniz University, Turkey) and Abdullah Tarinc (Akdeniz University, Turkey)
Business Science Reference • © 2020 • 351pp • H/C (ISBN: 9781799837251) • US $245.00

Accelerating Knowledge Sharing, Creativity, and Innovation Through Business Tourism
Muhammad Waseem Bari (Government College University, Faisalabad, Pakistan) Sadia Shaheen (Government College University, Faisalabad, Pakistan) and Meng Fanchen (Beijing Institute of Technology, China)
Business Science Reference • © 2020 • 351pp • H/C (ISBN: 9781799831426) • US $225.00

Contemporary Management Approaches to the Global Hospitality and Tourism Industry
Abraham Pius (Arden University, UK) Husam H. Alharahsheh (University of Wales Trinity Saint David, UK) and Adenike A. Adesanmi (University of East London, UK)
Business Science Reference • © 2020 • 290pp • H/C (ISBN: 9781799822042) • US $225.00

IGI Global
PUBLISHER of TIMELY KNOWLEDGE

701 East Chocolate Avenue, Hershey, PA 17033, USA
Tel: 717-533-8845 x100 • Fax: 717-533-8661
E-Mail: cust@igi-global.com • www.igi-global.com

Editorial Advisory Board

Table of Contents

Detailed Table of Contents

This work derives from postdoctoral research funded by Funcap-CAPES in Brazil. It investigates innovation in community-based tourism in coastal spaces. The competitiveness of tourist companies has a great relationship with innovation and it with the sustainability of the environment. The objective of this chapter is to present the innovations implemented that resulted in being practices derived from concepts of corporate social responsibility and to present the innovations that the coastal communities of Ceará have implemented in recent years. The innovations were identified in relation to community management for the benefit of collective rights, innovative activities in accommodation, intermediation, and use of space, among others. This work is expected to improve the ability to learn and multiply innovation in dynamic environments such as tourism and being able to add knowledge about the innovation capacity of Ceará tourism companies, collaborating for a discussion of their learning, and the multiplication of their knowledge.

Small tourism enterprises play a key role in generating income and local development of a community in rural areas through community tourism (TC) ideas, services, processes, and products. These have been generated in the communities of San Pedro, Valdivia, Dos Mangas, and Sacachún from the province of Santa Elena, which deserve to be registered through empirical research. The results show that the population is interested in working towards the sustainable development of its natural and cultural resources. This interest is transforming the tourist activity in their responsible daily lives, and they are committed to caring for the environment. Therefore, the people's own lifestyles are highlighted in a scene of tranquility and harmony.

The State of Ceará, located in the Brazilian Northeast, has a high tourist potential due to the beauty of its beaches and the great geomorphological differences within the territory. In addition, there are warm winds that allow the practice of wind sports throughout the state seacoast. This chapter presents three categories of analysis here interrelated with planning: wind sports, the use of social projects to improve the tourist experience, and the participation of the local community in social projects. The main objective of this chapter is to present innovations that might be implemented with the creation of a tourist route based on existing structures in the state of Ceará. The focus is to use the attractiveness of extreme wind sports linked to social responsibility (NGOs) and the strengthening of the tourist image. This chapter presents the social projects that relate tourism, social projects, and planning. Thru this job, the authors hope to give light to other destinations, which might perform the practice of benchmarking to face similar problems.

All over the world, institutions and organizations that recognize the social responsibility of companies are identified. They work on social innovation, which is companies' capacities to influence problems, generating synergies between the various sectors of society. In this research, a qualitative methodology was applied to a sample of companies from Puebla (Mexico) and the region to identify innovative practices of corporate social responsibility in the tourism sector and related companies. The central question that guides this research is, through which strategies or actions are companies in the tourism sector socially responsible and innovative? Therefore, the purpose is to show the good practices of different companies in the tourism sector in Puebla (Mexico) as well as their areas of opportunity and therefore strategies to strengthen responsibility and social innovation in the sector.

which drives the strengthening of partnerships; collaboration and cooperation between stackholders; the appreciation and diffusion of its heritage and culture; encouraging increased visitor numbers, consumption of goods and services, and decreasing seasonality of tourism. The attraction of its waters for tourism and leisure activities represents a resource for the strengthening of the economic activities within the region with concrete effects on trade, hotels, and local restaurants, in view of the practices involved, the natural and cultural contexts that the territory offers, and the opportunities for new activities.

In this chapter, the authors present a part of an analysis of the organizational culture of micro and small tourist companies in the state of Colima, Mexico. It is the result of mixed research that belongs to an exploratory study carried out from a systemic perspective of tourism organizations. The purpose of the study was to identify elements capable of triggering the productivity, competitiveness, and innovation of Mexican micro and small tourism companies based on their language and organizational structure. The results obtained revealed that the "Clan" type culture is predominant in this type of company in the state of Colima. Hence, a transformation of the mental structure of senior management is necessary and urgent, as it requires a change in an organizational culture focused on the "Market" to achieve its permanence in the sector.

Tourism, with a high economic and social significance for the economy of many countries, is one of the fastest developing fields. A new challenge of transition to an innovative way of developing tourism industry is strongly proposed with the result of existing conditions of fierce rivalry in the field, economic and political instability in many regions, including the Republic of Uzbekistan, the fast-growing needs and demands of tourists. Therefore, destinations, importantly, have to determine their competitiveness to figure out their strengths and weaknesses and thereby develop their future conditions. The authors discuss the issues related to the innovation in tourism destinations and present its main directions in the sector, as well as factors affecting the innovative development of the fieldin Khorezm region of Uzbekistan. Moreover, the chapter illustrates the principles and basic requirements for the system of managing innovative activities in tourism destinations at macro, mesa, and microeconomic levels.

This chapter performs a bibliometric analysis on innovation in hotels field. More specifically, it maps the current research front on innovation in hotels field based on 287 papers published in refereed journals

Foreword

The last months of global history have been characterized by the expansion of an unexpected pandemic, the one caused by the SARS-COVID19 virus. Several million people have been infected worldwide and the dead number in the tens of thousands. Apart from the tragic data of this disease, COVID19 has changed our way of life. We have stopped going far, practicing tourism in remote destinations and we calculate with a very prudent perspective how many days we can take a vacation or enjoy a weekend of well-deserved rest. The pandemic has hit with an unusual force the tourist sector, especially the mass destinations, linked to the enjoyment of the party, the sun and the beach. As some authors point out, tourism has been domesticated, emerging as alternatives the quiet places of rural or nearby coastline where to stroll, contemplate nature and benefit from a simple offer, in which to live the experiences of contact with local communities. At this time, there is not enough perspective to predict whether the world of tourism has changed forever or simply a brief pause in a runaway growth process in recent years. What does coincide in affirming is that the tourist experience based on the values of the territory, not overcrowded and that supposes a unique experience will have a future in any of the post-pandemic scenarios that we imagine.

It is evident that the classic capitalist development model is coming to an end, either due to its exhaustion or because it is inherently tending towards crisis. For this reason, the uncomplicated offers, built from the communities and the sustainable management of the resources of the territory, constitute an obvious alternative to any of the scenarios that the future offers us. We are immersed in a green transition of the economy, in a search for new systems of social governance and in a return to the small, to the local, as a counterpoint to the excesses of globalization. Therefore, community tourism and the expansion of the scope of Corporate Social Responsibility will be components with an increasing role in the activity. Tourism must be reorganized in these times of turbulence, and the collective, the responsible and the egalitarian must gain strength in a sector that is not going to decline. . Tourism is not going to decline because urban workers subjected to very heavy work and life routines need and will need to get away more or less traveled, resting and living other experiences that separate them from their daily lives.

The only question about the future of community and responsible tourism developed in peripheral and southern countries resides in knowing if the movements that maintain it will continue to be long-range or if, on the contrary, it will be based on attracting visitors from the vicinity, from the country itself or from the same region. This question is of great importance, since classical tourism, the one we met until a few months ago, involved a large number of journeys from societies from the North to the South, towards those called by some peripheries of pleasure, where the activity had grown and it had become remarkably sophisticated in recent years. However, in the current situation it is difficult to predict whether long journeys will continue in the direction of Eastern Europe, Latin America or some regions of Asia and Africa. For all this, the undeniable capacity for resistance and the future that must be predicted for social, community and responsible tourism will be conditioned by the intensity of the visitor flows themselves and by their origin. This will always be important information to approximate the average cost that the activity will generate in the near future. These types of tourism undoubtedly have a tomorrow, but their influence on the development of the towns and regions where they are promoted will depend on the origin and consumer markets that they manage to attract after the impact generated by the COVID-19 crisis.

Given the foreseeable emergence of simple forms of tourism, based on the community and the responsible, this book can become a first-rate reference for the academic world. This powerful statement answers two reasons. The first, the thematic diversity that it presents and, the second, the transcontinental scale that it uses to include diverse and complementary analyzes. The different parts and chapters of the book cover community tourism, small-scale entrepreneurship, social responsibility, innovation, sustainability or the role of women in these new ways of conceiving the tourism business. Very specific products are also studied, such as adventure sports, gastronomy or everything related to the enjoyment of water. In summary, it is about building a multiple approach to community and responsible tourism that is booming during this period of health uncertainty. On the other hand, and as we have already indicated, it is worth noting the enormous wealth of places and countries that constitute the case studies of this work, Mexico, Puerto Rico, Brazil and Ecuador in Latin America, Portugal and different references to the EU in different texts., Albania or Uzbekistan, on the eastern periphery of the European imaginary. Without a doubt, a wide panorama of case studies to confirm the global vocation of this work.

To finish, we must make a reference to the promoter of this book, Dr. Jakson Renner, PhD professor at the University of Santiago de Compostela (Spain) and the State of Ceará (Brazil). First of all, this condition of professor and researcher in the field of tourism on both sides of the Atlantic gives us an account of Jakson's international dimension. A dimension reflected, both in the origin of the studies and in the will to build a great work on community and social tourism of very broad influence. Jakson Renner has cultivated the scientific study of tourism activity from different disciplinary perspectives, ranging from Business Sciences to Geography through Sociology. This interdisciplinarity is combined with a growing commitment by the societies of the South, which find tourism a factor of development and economic growth, which in any case must be fair and equal, managed from the common and the public. With this objective, the young coordinator has worked very intensely in recent years, trying to give an international dimension to his work that, in no case, renounces the commitment to fragile and threatened societies in Latin America and the rural and peripheral world of Europe.

Rubén C. Lois-González
University of Santiago de Compostela, Spain

Preface

Innovation and Entrepreneurial Opportunities in Community Tourism is an edited collection of papers that provides conceptual and empirical insights to innovation, entrepreneurship and tourism. The book is suitable for students, researchers and practitioners in any of those fields. The chapters of this volume provide current and practical examples that can be used in various university disciplines addressing reality and claiming your attention for model cases.

Tourism stakeholders, every so often, seek to achieve ventures capable of attracting tourists and maintaining the sustainable destination development. Positive and negative human interventions related to tourism generate very different effects, both positive and negative. This has a particular influence to the economic, social, cultural and environmental aspects of societies at a global level. However, there are territories that are vulnerable due to their culture and the environment in which they operate. Hence, there is a need for extensive research, conceptual and empirical studies to address these concerns and provide the tourism sector with valuable insights that can improve the stakeholder relationships. At times, innovation can come from public policies and public actors through various schemes and incentives. In addition, social innovations often emerge to provide opportunities for alternative forms of tourism that seek, among other reasons, to tackle the negative impacts of the conventional model of mass tourism. Thus, it was in this context that the theme of community-based tourism has emerged as the guiding theme of this book.

This edited collection focuses on innovation, tourism and entrepreneurship and provides various perspectives to the extent to which they contribute to the community development. In this context, the book involves everything from epistemological issues in those fields to the specificities of innovative practices, which contribute to the development and ethical maturity of practices, processes and studies in the field of tourism. The main objective of the volume is to provide a comprehensive investigation aligned to the contemporary complex processes and demands of community development.

The founding idea of this collective work involving researchers from various parts of the world initially came from the project "Innovation in the community tourism network and its contribution to the development of the territory (2018-2020)" conducted at the State University of Ceará, Brazil. It was financed by the Cearense Foundation for Support for Scientific and Technological Development (FUNCAP) in partnership with the Coordination for the Improvement of Higher Education Personnel (Capes). During the fieldwork stage, it was established that it would be relevant to expand the scope of the study and publish a material in which the theme to revolve around the dynamic capacities related to small tourism entrepreneurs in the coastal areas. Given the context of the study it was decided that the most suitable theme of the material is community tourism, including various approaches such as social innovation, entrepreneurship and tourism. This is largely because, even though unprepared, without theoretical and technical knowledge, this sectors, and tourism in particular sector, are very creative and often provide many unique approaches to tackle community development problems and issues. In this context, community-based tourism is considered as an arena for innovative approaches for environmental conservation, local development and improvement of the social well-being of the host communities.

The studies part of this collection are related to the innovative approaches that have emerged in the context of community-based tourism. It is understood that they have an impact on tourism as an economic activity in regional social and economic development. The book includes studies on the economy of sustainable tourism, the relationship between tourism, local development and sustainability, debates around social responsibility and tourism, the communication and image of community-based tourist destinations. In addition, there are is a focus on gender studies in tourism and, finally, heritage and culture in a community-based tourism context.

In Chapter 1, Jakson Renner Rodrigues Soares, explores a number of innovations that the coastal communities of Ceará have implemented in the recent years. The innovations were identified in relation to community management for the benefit of collective rights in accommodation, intermediation and use of space, among others. In Chapter 2, Myriam Yolanda Sarabia Molina, Lupe Cecilia García Espinoza and Arnaldo Efrén Mendoza Tarabó explore small tourism enterprises and the how they play a key role in generating income and local development for the community in rural areas. Their results show that the population is interested in working towards the sustainable development and this interest is about transforming the tourist activity. In the same vein, in Chapter 3, Luciano Barbosa Almeida and Jakson Renner Rodrigues Soares presents cases that discuss coastal tourism development and social innovation projects. in doing so, they hope to give light to other destinations, which might perform the practice of benchmarking to face similar situations. In Chapter 4, Maria del Carmen De la Luz Lanzagorta, Edith Sarai Lozada Sánchez, Jessica Abigail Cortés González and Concepción Nancy De Cristobal González study cases in community tourism. The main purpose is to show the good practices of different companies in the tourism sector in Puebla (Mexico) as well as their opportunity for further development and strategies how to strengthen responsibility and social innovation in the sector. In Chapter 5, the concluding one to this section, Maria Sâmia de Oliveira and Gil Célio de Castro Cardoso explore the relationship of residents with community-based tourism practiced in the coastal area of the extreme west of the state of Ceará, located in Northeast Brazil. This is a very important contribution that considers the perspective of residents on the importance of local tourism and its relationship with elements of social innovation.

The book also covers the topic of entrepreneurship in tourism. In Chapter 6, Nick Naumov and Gheorghia Costandachi explore the role of gastronomy for the development of tourism. This chapter seeks to address the gap in academic literature focusing on how food tourism can drive community development, stimulate local businesses and inspire creativity and innovation. In a similar vein, in Chapter 7, Aida Ciro and Merita Toska present case studies of community tourism development. In their chapter, they focus on the role entrepreneurial eco-systems rooted in local communities can play in the development of sustainable tourism models in Albania.

The last part of the book focuses on resilience and community development In Chapter 8, Maria Francisca Casado-Claro analyzes the efforts of the Foundation for Puerto to help communities around the island transition from relief stage to recovery stage after the devastating impact of Maria hurricane. The lessons extracted from this case could be useful for destinations that are exposed to natural disasters, or those who wish to retain their population by creating tourism-related businesses and local employment opportunities.

From a different perspective, Chapters 9, 10, and 11 study the concept of innovation from a gender perspective. Chapter 9 presented by Karla Juliana Pinheiro Melo, Eliana Coelho da Silva, Raquel Santiago Romo and Williany Loren Gaia Ferreira Austrégesilo, stresses that, over the years, women have positioned themselves to transform the reality of their labor occupations. They leave low hierarchical level positions, which have little value to position themselves in business management jobs, which, consequently, contributes towards the transformation of social structures. In Chapter 10, Iria Caamaño-Franco, Antonia Pérez-García and María Andrade-Suárez explore the notion of women entrepreneurship. The aim of this chapter is to analyze women's entrepreneurship in the frame of marine tourism as an innovative and territorial development dynamics that collaborate with other sectors (for example, fishing-tourism, gastronomy, canning) promoting in turn a sustainable economy based on blue growth and development of indigenous resources. Finally, Chapter 11, written by Luana Moreira de Araújo, Cintia Martins and André Riani Costa Perinotto, discusses about the innovation in communities for women who travel. This study addresses innovation in the community of women who travel alone by analyzing a platform where women share lodging facilities. This chapter focuses on the applicability of this innovation to the accommodation market.

The next three chapters relate to destination development and destination. In Chapter 12, Gonçalo Poeta Fernandes and Hugo Gomes explore the importance of water resources in tourism and leisure activities. In order to to establish a framework of activities in inland waters and their meaning in the territory of the Serra da Estrela, the authors investigate the valorization of differentiated tourism products. In Chapter 13, Vicente Alejandro Jiménez Olivera and Irma Magaña Carrillo seek to identify elements capable of triggering the productivity, competitiveness, and innovation of Mexican micro and small tourism companies, based on their language and organizational structure. The results obtained revealed that a transformation of the mental structure of senior management is necessary and urgent, as it requires a change in organizational culture focused on the "Market" to achieve its full potential permanence in the sector. Finally, in Chapter 14, Dilmurad Bekjanov and Bunyod Matyusupov discuss the issues related to the innovation in tourism destinations and present its main directions in the sector, as well as factors affecting the innovative development of the field in Khorezm region of Uzbekistan. Moreover, the chapter illustrates the principles and basic requirements for the system of managing innovative activities in tourism destinations at macro, mesa and microeconomic levels.

The book also include two chapters, that provide a critical bibliographic review on innovation in tourism. In Chapter 15, Catarina Fernandes and Rui Alexandre R. Pires carry out a bibliometric analysis on innovation in the hotel sector. More specifically, their study maps the current research front on innovation in hotels field based on 287 papers published in refereed journals indexed in the Scopus database. Lastly, Chapter 16 seeks to explore the knowledge produced by the scientific community about tourism, innovation and community tourism. Eunice Ramos Lopes, Joao Tomaz Simoes and Maria Rita Nunes notice that publications on community tourism and innovation are limited, which demonstrates the lack of research within the two areas.

Overall, the chapters of this volume recognize that sustainability of destinations that choose to offer experiences in a community tourism context needs to be considered in a dialogical way. It is important to consider the necessity to have a debate from different perspectives, and this volume contributes to this need by providing different views and approaches to destination development. Thus, *Innovation and Entrepreneurial Opportunities in Community Tourism* addresses these issues of innovativeness and social technologies to broaden models and understand innovation in tourism from new perspectives.

Acknowledgment

I would like to thank the collaboration of the researchers who are part of the Editorial Advisory Board. Thank you for the work committed to the quality of the analyzed material.

Furthermore, I would like to thank all the authors for the chapters presented in this book. Community Tourism has been studied all over the globe and it is a pleasure to present in a single work experiences from such different places such as Brazil, Ecuador, Mexico, Puerto Rico, Spain, Portugal, Albania or Uzbekistan.

It is important to note that we received a total number of 29 proposals and for various reasons we had to select the ones that best fit the purpose of this manual. The process was long, but I believe the result is worth it.

Lastly, the author is thankful for the support from the *Fundação Cearense de Apoio ao Desenvolvimento Científico e Tecnológico* (FUNCAP) in partnership with the *Coordenação de Aperfeiçoamento de Pessoal de Nível Superior* (Capes) [Poyect Number 88887.162342/2017-00].

Chapter 1
Community-Based Tourism in Ceará's Coastal Space:
Innovative Socially Responsible Activities

Jakson Renner Rodrigues Soares
Universidade da Coruña, Spain & Universidade Estadual do Ceará, Brazil

ABSTRACT

This work derives from postdoctoral research funded by Funcap-CAPES in Brazil. It investigates innovation in community-based tourism in coastal spaces. The competitiveness of tourist companies has a great relationship with innovation and it with the sustainability of the environment. The objective of this chapter is to present the innovations implemented that resulted in being practices derived from concepts of corporate social responsibility and to present the innovations that the coastal communities of Ceará have implemented in recent years. The innovations were identified in relation to community management for the benefit of collective rights, innovative activities in accommodation, intermediation, and use of space, among others. This work is expected to improve the ability to learn and multiply innovation in dynamic environments such as tourism and being able to add knowledge about the innovation capacity of Ceará tourism companies, collaborating for a discussion of their learning, and the multiplication of their knowledge.

INTRODUCTION

The State of Ceará is located in the Northeast of Brazil and offers numerous tourist activities. However, most are concentrated on the coast, as it has a tropical climate that allows sun and beach activities throughout the year. Regardless of the level of professionalism in the tourism sector, even if it is not tourism with professionals with regular training, on the coast of Ceará this economic activity was able to develop alternative tourism, community-based tourism. This tourism seeks to offer visitors preserved environments, respecting local ways of life and environments. This is because nowadays it is perceived that tourist movements are related to environmental values. These community groups plan and promote exchanges by building a form of tourism that values cultural diversity and strengthens traditional activities.

DOI: 10.4018/978-1-7998-4855-4.ch001

This chapter derives from postdoctoral research funded by Funcap-CAPES in Brazil. It investigates innovation in community-based tourism in coastal spaces. The sun and beach tourist today seeks to consume, in addition to aspects related to the sea, authentic experiences that respect the territory and the environment. In this way, traditional, widespread tourism leaves space for projects that allow less risk to the place and cause less impact on the environment and the local community.

The competitiveness of tourist companies has a strong relationship with innovation. And this one, with environmental sustainability. However, despite the existence of important efforts and investments in clean technologies by organizations, the environmental impact continues to increase. In this context, the need for this study arose, which identified the innovative activities related to the Corporate Social Responsibility of businesses in Ceará's coastal tourism. The objective of this chapter is to present the innovations implemented that resulted in being practices derived from concepts of corporate social responsibility and not only that, to present the innovations that the coastal communities of Ceará have implemented in recent years. Understanding the transformations of Ceará's coastal space is very useful to understand the dynamics of tourism development and the metamorphosis suffered by business models (sometimes informal) in the context of the work performed by the local community.

Soares and López (2015) point out that tourism is an activity that when poorly planned, its attractions can deteriorate in a way that will have very negative consequences for the place and, consequently, for its residents and indigenous people. Even so, it is necessary to highlight the need to seek new orientations and productive diversification based on taking advantage of the potential of the natural environment and coordination by the community that offers the service, so that the bases of sustainable tourism are really followed and they are balanced. In this context, there is a need for a paradigm shift. It must shifts from a tourism where profit matters to a non-hegemonic tourism, which is more human and seeks a cordial relationship with the environment, understanding that the environment is a lot more than just nature, or environmental environment, it goes beyond and it involves the society, the people who make up the place. Therefore, community-based tourism is capable of organizing small companies (sometimes they are not even conformed to a legal figure), economic, political and intellectual agents to develop and control management, avoiding speculation by other interested sectors.

The chapter is organized into V sessions. In addition to this first, where the study is presented, there is a second session where the territory to be studied is presented. In the third part, tourism on the coastal space is discussed, from sun and beach tourism to tourism on the coastal space. It is in the fourth session that innovations in Ceará community tourism are presented: Constant work, shared management and innovations in Ceará coastal tourism?; Joint marketing and brand positioning; or, Social technologies and concern for sustainability.

BACKGROUND: THE TERRITORY IN QUESTION

The State of Ceará is the third richest in the Northeast Region. Its economy is diverse; with a strong presence of the tertiary sector of commerce and services, with a special emphasis on tourism. In this sector, the biggest tourist attraction is related to its coastline and tropical climate. Temperatures average 28° throughout the year, with little variation in temperatures. The consistency of the trade winds on this coast both serves as a natural regulator for the temperature not to be so high, and favors the appearance of wind sports.

The Litoral do Ceará is the region of focus for this chapter and the study area, focusing on innovative activities developed in community-based tourism. The coast of Ceará is known nationally and internationally for the beauty of its beaches, making this one of the main tourist attractions of the State. The study area is limited to the east with the State of Rio Grande do Norte and to the west with the State of Piauí. The coastal region has 573km of varied beaches, with fine sand, cliffs or mangroves. It is formed by 21 municipalities and interconnected by two roads (freeways) that connect the entire coastline from east to west from the capital, Fortaleza: *Costa do Sol Nascente* and *Costa do Sol Poente*.

It is worth mentioning that part of Ceará's coastline belongs to the group of sixty-five destinations that induce tourism, according to the classification done by the Brazilian Ministry of Tourism, which is carried out since 2003. They should be a source of investments and improvements for the product "Brasil" to be marketed with excellence, both nationally and internationally. The coast of Ceará is an important tourist attraction and the tourist activities developed generate employment and income for the State.

Along the entire coastal strip there are basic infrastructure to receive tourism. From road infrastructures such as those mentioned above, to the recent creation of airports at the eastern and western ends

Figure 1. Map of the Ceará coast (Own elaboration with images from Dreamstime and Revista Fácil)
Source: Dreamstime and Revista Fácil, 2018.

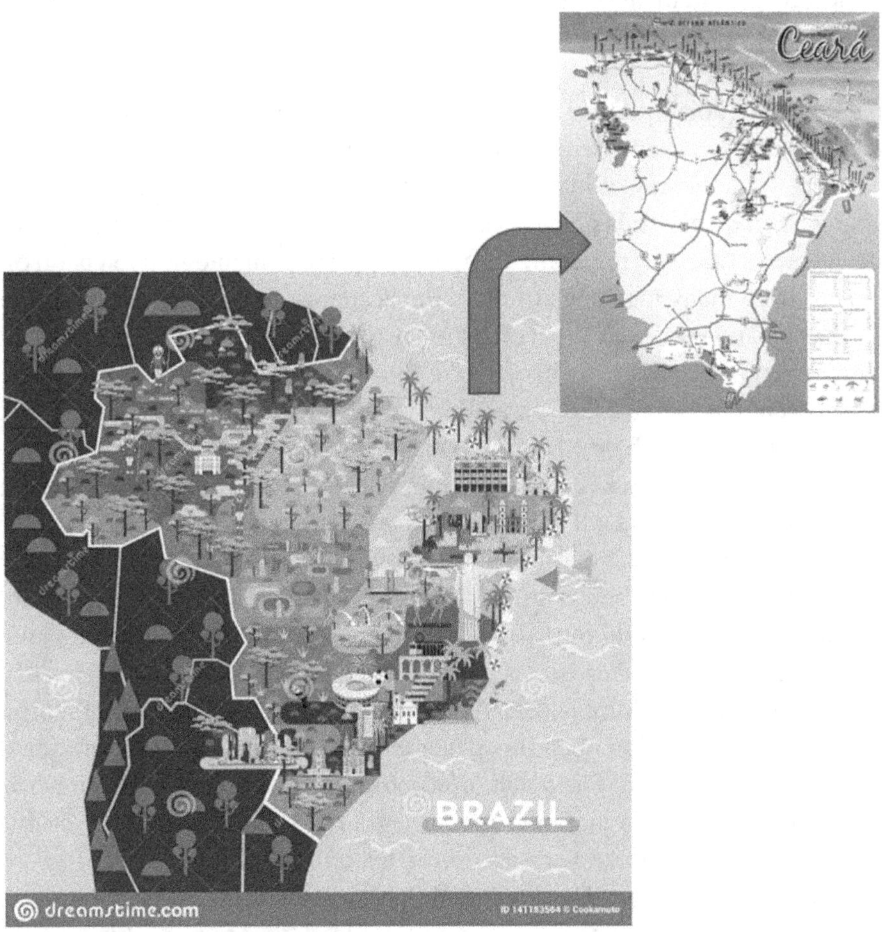

of the state, in addition to the capital's airport, Fortaleza. Hospital infrastructure, traditional means of accommodation of various levels, as well as numerous food services and complementary services. In this territory there is a variety of landscapes, quite different from place to another, ranging from mangroves to completely urbanized or deserted beaches, cliffs, colonial architecture...

The state government has spent decades betting on the professionalization of tourism in the region. However, even understanding that this institutional support is necessary, given the capacity of tourism to develop territories and reduce inequalities, it is also known that, for capitalist reasons, these results are not always found. That is why alternative destinations have emerged in this destination to this traditional tourism, which aims at profit rather than development. That is, these alternatives were able to take advantage of the benefits of these policies to create different proposals, controlled by the community itself, and with a real focus on aspects related to the sustainability of the environment, with the reduction of inequalities and with the involvement of all those involved in the tourism design and management.

FROM SUN AND BEACH TOURISM TO TOURISM IN THE COASTAL AREA

As was well identified in the previous session, Ceará´s coast was and is a basic element of tourism in this territory. Fernandes, Soares and Coriolano (2020) confirm that sun and beach tourism predominate in Ceará, but also indicate that in the territory there is also the practice of sports linked to the sea and the wind, such as wind surfing and kite surfing, given the conditions from the coast of Ceará (Fernandes, 2014).

As Yepes and Cardona (2000) pointed out at the beginning of this century, fine sandy beaches bathed by clean waters on sunny coasts are (were and continue to be) fundamental elements in the selection of the destination for the summer stay of many families. However, the transformations that have been occurring in the uses and characteristics of the coast lead to discuss this metamorphosis from a traditional approach to sun and beach tourism (regarding the mass models) to an integrative vision of tourism in the coastal space. There were many meanings used by the academy and marketing professionals to define tourism that has the sun and the sea as inputs for the tourism product. However, it should be noted that they all deal with common elements to characterize this segment. That is, tourist activities related to recreation, entertainment or rest have the sun and the sea as their integrating axis.

In the early days of sun and beach destinations, from the moment that man "discovered" natural places with exuberant beauties like its beaches, this environment started to undergo transformations to adapt the infrastructure (urbanize) and make it more conducive to enjoy others (accommodation, restaurants, complementary services). It was like that in countless world destinations, now consolidated, some in decline and others in the process of reinvention. This added value to the place, valued the now destiny, but generated changes in the lifestyle of the natives, changes in the environment, socio-economic restructuring that still exists today. Likewise, as highlighted by the Brazilian Ministry of Tourism in its study on Sun and Beach Tourism (MTUR, 2010), this segment is associated with the number of annual hours of sunshine in a given location. And this generates a large concentration of tourists (national or international) in these destinations at times that offer more hours of sunshine. It is also worth mentioning the seasonality factor of tourism that influences the tourist influx, since these destinations depend mostly on the summer seasons (holidays or vacations) that tourists enjoy.

Regarding the tourist *Sol e Praia, Da Silva* (2014) highlights that this segment is associated with mass tourism, due to the enormous concentration in the place. In addition, this tourist is sensitive to the variation of his income. On the one hand, the increase in income may mean an increase in demand for this type of place. However, it is observed that in recent years, one of the factors that tourists value when choosing their destination is the possibility of carrying out complementary activities that offer a more complete experience during their visit to the destination.

INNOVATIONS IN CEARÁ'S COMMUNITY TOURISM

Discussing Innovation

Tourism management models have been affected in recent decades by the emergence of new factors worldwide that have had major impacts on the traditional tourism model. Tourists more aware and concerned about the footprint they leave in the environment. Residents unhappy with the negative impacts of tourism or with the non-positive results of this activity in their lives. Businesses that have to reinvent themselves to compete in a competitive market such as tourism, where competitors are no longer considered only the closest destinations, but often the main competitor is in a destination thousands of kilometers away. And as not, institutions that regulate and encourage the use of activities that respect the local communities and the environment where tourism takes place. Therefore, innovating is nothing more than an obligation of the company that wants to remain in a market as competitive as tourism.

According to Valls (1996), the tourist destination is characterized as a geographical space with its own characteristics of climate, roots, infrastructure and services, which have a certain administrative capacity capable of developing common planning instruments. In addition, this author also considers that the attraction of tourists must be made through products perfectly structured and adapted to the satisfaction sought by these subjects. However, the actors are not always organized to achieve ventures capable of attracting and maintaining, in one side, there is a satisfied tourist mass, and in the other side, the sustainability of the destination.

We could understand innovation as the activity which the result is to come up with new products, processes or significant improvements of the existing ones. That is, it is not just about technology. As Fundación Cotec (2015) points out, innovation is any change that is not only technological, it is not based only scientific knowledge, but which brings not only economic value.

The theme when related with the activities developed in the coastal tourism environment of Ceará, it can be said that they carried out several technological, process, product innovations ... In fact, in the community-based tourism in Ceará it is noted that there is a propensity to develop new products, what Garcia and Calantone (2002) relate to the capacity for innovation. In this destination there are innovations related to new markets (changes in target), new business models (joint marketing), new processes (electricity production) and also new sources of supply (proximity purchase). Not only that, also referring to product innovations (housing construction method, recycling and reuse of materials that would previously become waste) or management process (they used joint intermediation and marketing methods to strengthen the network image of community tourism in the region). Although this chapter deals with community tourism businesses, it is understood that innovation in these companies is very important to create competitive advantages over competitors and, due to the community tourism business model itself, to generate sustainability of activities over time and a preservation of the environment where tourist activities are inserted. Finally, as highlighted by Shoham et al. (2012) the creative capacity of the ventures, understanding it as a process of connecting ideas to generate new ones capable of adding value to the existing one.

Speaking of innovation in tourism usually meant dealing with major tourist developments. Something contrary to the object of study in this book, where community tourism is formed by small enterprises formed by the local community itself. In these communities, what Stahel (1997) relates to the functional economy is identified, since it optimizes the use (or function) of goods and services and thus, the management of existing wealth (goods, knowledge and nature), aiming to create the highest possible use value for the longest possible time, using the least amount of material resources and energy.

This chapter, like the entire book, discusses innovation in small businesses, often-rudimentary businesses, which in some cases are unaware of the concepts that we academics give to things, but that use them effectively. This benefits both the operation of the enterprise and it is capable of generating a positive impact from tourism onto the community. Finally, according to Menguc and Auh (2006), for a company to be innovative, it needs to adopt a new mentality or new attitude, which theses new values must be shared and they must be disseminated throughout the organization in order to be effective. These common aspects are present in community-structured tourism, where the whole community acts as a basic management entity. It should also be noted that innovation involves offering new solutions to old problems. As Oliveira (2019) points out, innovating is more about adapting tools, systems, equipment or behaviors, it is a broad concept and in tourism, it goes far beyond just having cutting edge and innovative technology. Thus, from now on, the innovative activities developed in community-based tourism on the coast of Ceará will be discussed.

Community-based Tourism on the Coast of Ceará?

Before presenting in detail the innovations developed in the territory, it is worth highlighting the process that led to the emergence of alternative tourism in the State of Ceará. This has been defined as community-based tourism. The coast of Ceará was previously composed of small fishing villages that, with the implementation of policies for the development of tourism, provided a new economic process. The use of the territory for tourism has generated a change in the economic matrix, leading many people who previously depended on traditional jobs to relocate themselves in this new activity. In this context, some communities did not see themselves in the embroilment of this transformation and decided to protect themselves and the environment in which they operate

Therefore, a momentum was found where few and large enterprises dominated the tourism production chain, relegating the local community to a minor role. Real estate speculation was present in this space that was growing within a market orbit and large consolidated networks started to control the development of activity in the region. However, it was from the awareness of some communities that they were resistant to the exclusive process of touristization of the territory, seeking well-being from the tourist activity, which, in itself, was not considered negative. Therefore, there was a need for a remodeling in the way of implementing this activity, so that tourism could really be a tool for local development.

As a consequence, community-based tourism appears in the State, and even since its foundation, we can say that innovation is present. That is, the very way of thinking about tourism and organizing decisions is an aspect that can be considered innovative, where the local community has the power to manage the territory.

Community-based tourism proposes a different use of the territory and its resources, considering both service providers and its users. It implies practices of environmental valuation, recognition of the cultural heritage of communities and their interaction with them (Villavicencio, Zamora and Pardo, 2016). As the authors point out, it is, therefore, a more conscious and integrated tourism capable of generating economic and social benefits at the local level.

Likewise, as highlighted by De Moraes, Irving and Mendonça (2018), this type of tourism is gaining more and more visibility, as forms of resistance outlined based on locally rooted cultural and political strategies, against the conventional patterns of mass tourism. Likewise, the authors insist that it refers to strategies developed by organized groups and social movements seeking to guarantee and preserve the local culture and the territories they traditionally occupied.

Constant Work, Shared Management Ideas and Innovations in Ceará's Coastal Tourism?

This section of the chapter presents innovative activities identified in Ceará's coastal tourism. They are consequences of the model of managing tourism in these communities. Furthermore, one must start from the idea that tourism is a major driver of the economy on a global scale. This leads businesses / destinations into constantly adapt to meet the concerns of all stakeholders in the sector. However, in the case of community tourism, considering mainly the satisfaction of the needs of the local community and the changes in the motivations and behavior of the new tourist, increasingly demanding with quality and respect for the environment.

Regarding innovation, Zahra and George (2002) relate it to the market's ability to learn. The authors state that it would be the ability to identify, assimilate, transform and apply new knowledge obtained outside the company, which makes it possible to formulate strategies capable of improving the performance of tourist organizations.

Joint marketing or brand positioning

We could, therefore, understand that the cooperative organization of these communities alone is a social innovation that allows an innovative and creative approach to these productive tourism arrangements.

This administrative model allows us to join efforts to generate solutions in production and in the relationship between producers and consumers. Thus, it presents an important form of circular economy and sustainability because it allows saving resources, energy and diverse inputs. In community tourism communities, cooperative initiatives play an important role in innovation for sustainability through the promotion of ecological production, services for better environmental performance (servitization), production and consumption in the vicinity (local short chains), etc. That is, in the case of tourism, the current tourist seeks to have experiences that are more authentic, these experiences are also related to the knowledge of the quality of the product offered to him. Even so, if we only think about a basic aspect of the authenticity of tourism, such as gastronomy, we can identify that the inputs needed to offer a gastronomic experience based on the local area will need, primarily, local food, produced and distributed locally. In this sense, even the transport chain will have less impact on the environment. In the same way, if tourism has impacts on the life of the native of the receiving region, it is best that they are not only negative. Therefore, if the resident realizes that not only the problems arising from tourism will affect him, but as well the benefits and in this case the economic gain, they are more likely to bet on the proper functioning of the activity.

Therefore, the first innovation identified in this community-oriented tourism is the self-organization of the local community. For example, in the State of Ceará they organized themselves in a network called Tucum. It aims to strengthen and give visibility to Community Based Tourism (TBC) initiatives in the state of Ceará. Common principles include solidarity, equitable income distribution, valuing production, local culture, the local identity and socio-environmental sustainability (Rede Tucum, 2020). It should be noted that the Tucum Network understands tourism as an opportunity for local communities to control its development, being responsible for the planning and management of tourist activities, for the control of the proposed tourist structures and services. According to the Code of Conduct created by the Tucum Network itself (Rede Tucum, 2013), it calls itself an autonomous articulation that must seek its financial sustainability and constitute fundraising strategies according to collective principles and criteria.

Figure 2. Tucum Network web presentation
Source: Tucum Network, 2020

For community tourism, this is an innovation because they work in a single image, projecting a quality image to attract tourists. Thus, they achieve a change in the captured tourist, improving the user's profile, finding a less conflicting and more sensitive consumer in relation to the environment than the mass tourist. Finally, organizing in a network allows service providers to innovate to generate greater attractiveness in the market. Not competing with each other, but adding value to the product they offer. Finally, Tucum works with a brand that is used to position this TBC in the tourist market.

Social Technologies and Concern for Sustainability

Other innovations were identified in the TBC Ceará. We can say that these are very related to technology, more specifically to social technologies, as they seek to improve people's quality of life (Medeiros et al., 2017), solving social and environmental problems. According to this group of authors, social technologies present simple, low-cost and easily applicable solutions that generate social impact in fields as diverse as education, agriculture, health, the environment and leisure. They seek to transform society and could solve problems such as inequality, poverty, access to energy, education or health, among many others.

A strong presence in terms of recycling such as reuse, ecological construction, second life cycle of materials and products used in the communities was identified on the coast of Ceará. The communities, regardless of whether they are linked to the Tucum Network or not, carry out environmental awareness campaigns to implement selective garbage collection or encourage the use of the three R's in the locality, for example. Regarding these three R's, it is expected that recycling, reusing and reducing to the maximum will be a way to protect and not to explore the environment where they are inserted.

Thus, another example that can be used in this chapter are the countless activities performed by Eco Aldeia Flecha da Mata. A permacultural station located in the municipality of Aracati, 5km from the world known Canoa Quebrada Beach, where community experimentation is practiced, with permaculture, sustainability and self-knowledge as the guiding lines (Flecha da Mata, 2018). In the community, in addition to seeking harmony with nature through the management of agroforestry gardens, the use of photovoltaic energy, bioconstructions. At Flecha da Mata, the treatment of organic and solid waste is done through recycling, composting, treatment and reuse of water from ecological pits, visitors and the local community participate in experiences in the area of permaculture, bioconstructions, holistic therapies, shamanism, music, dance and yoga. It is noted that in the TBC in Ceará the communities are in constant learning, therefore relating to innovative activities in the destination. Relating to what was indicated by Jantunen (2005), that innovation in a dynamic environment renews the knowledge base of organizations.

At this venture, all electrical power that is consumed is generated through solar panels that were built by the villagers themselves. Not only that, the village is committed to disseminating this practice in permaculture courses and theoretical and practical workshops that they teach. Regarding the three R's, the organic and solid waste from the village is recycled and turned into compost, earthworm and pet bottle bricks (filled with non-recyclable waste). Here is an interesting example to be followed by other destinations. The use of plastic bottles filled with litter, such procedure transforms them into a base structure where further construction can be built, this method is a clever a way to reduce the amount of plastic that cannot be stored.

Figure 3. Solar panel, solar oven and plastic bottle bricks
Source: Eco Aldeia Flecha da Mata, 2020

Most of the solid waste is disposed of and they became compost, they also reuse materials, which would otherwise be litter in other occasions, though this action they give a second life to plastics and automobile tires. At Flecha da Mata, they also make ecological construction, such constrictions are carried out using the Superadobe technique, polypropylene bags filled with sand and clay. Quick to build and little contaminants, where little heating or cooling is needed (Mannise, 2012).

Figure 4. Ecoconstruction with superadobe technique
Source: Eco Aldeia Flecha da Mata, 2020

There are initiatives to improve the village´s water consumption, such as reuse and decrease its quantity, there also is a water treatment network called the evapotranspiration basin through banana circles. It is used to treat black water to prevent contamination of the water table. In addition, they use a dry toilet where there is no water discharge and sawdust is used instead of water to clean the toilet. According to Daterra (2016), the banana circle is an appropriate technology for the destination of gray waters from the kitchen, laundry and bathing. It consists of a basin excavated around bananas trees and other plants with high evapotranspiration rates. This is an ideal system for wastewater reuse exactly at the place where this sewage is originated.

Finally, at Eco Aldeia da Flecha da Mata they carry out eco-pedagogical experiences, activities which aim at elementary, high school, higher education students and at any community interested in developing practices of reconnecting with nature. Through the numerous actions identified, it is perceived in community tourism in Ceará, a perspective of strengthening the relationship between society, culture and nature, with environmental justice, characterized by a presence of innovations in the most diverse areas that directly influence tourism. This is positive both for the environment and for the local community, who benefit from a more natural environment and who suffer less negative impacts than normally occur in traditional tourism destinations.

Figure 5. Círculo de bananeiras para tratamento de águas
Source: Daterra, 2016

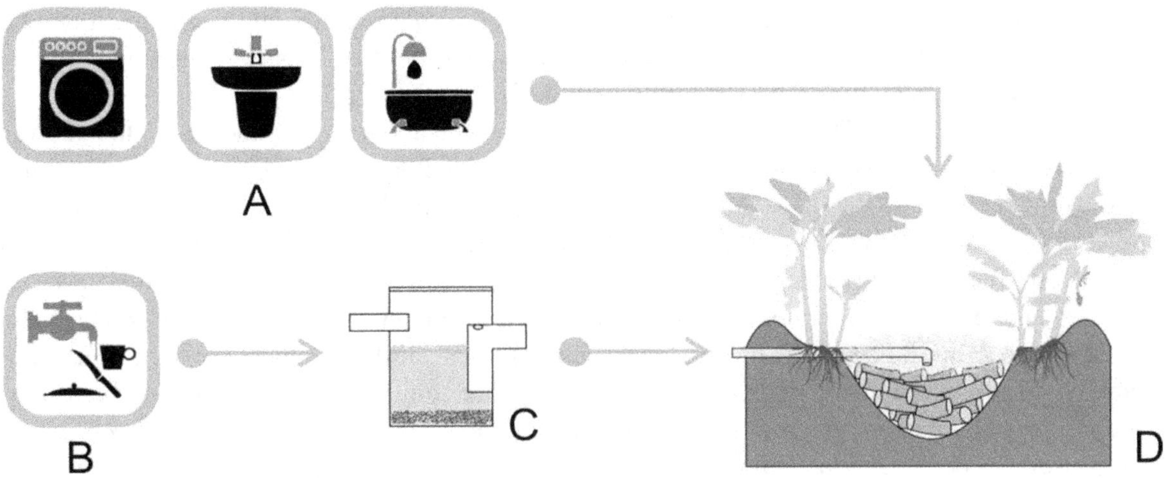

CONCLUSION

From the study conducted, it is possible to identify the existence of innumerable innovations present in community-based tourism in Ceará that it can clearly be exported to other places and different types of destinations which work with the same concern for the environment, with aspects of sustainability, for respect for the community and the collective. Although it can be certain that, these innovations are not exclusive to this territory, due to their very essence.

However, it can be said that one of the most significant aspects of the innovation observed is the strategy of joining a single brand. The Ones who know the saying *"Together we are strong"* would clearly fit this case. People working in community-based tourism in Ceará understood that they should not compete in isolation to monopolize the market. This type of actions would be better related to other cultures, more capitalist, than really to the positioning of a society that thinks in collective. Thus, they innovated by understanding that cooperation would lead to equity and satisfaction for the local population and, consequently, make the destination much more attractive. Mainly because they created a label capable of identifying not only its attractions, but also the sense of community and generating a greater sense of security in the customer. Joint marketing under the Tucum seal is an innovation in community tourism because, as in traditional sectors, such as the accommodation sector, creating a brand places the product / service on the market, identifies and represents the community, and helps to expand the exchange with the consumer.

Another conclusion reached in relation to the innovation present in this territory is the great presence of aspects related to social responsibility in the activities performed. That is, understanding that social responsibility is related to the ethics and morals of individuals, based on the impacts of decisions on society, the experiences that work in community-based tourism are closely related to this concept. The implemented innovations promote an inclusion of indigenous individuals in tourist activities in an equitable manner, respecting the particularities of the activities developed by each one and, at the same time, define the profile of the tourist they wish to attract, thus reducing the negative effects of mass sun and beach tourism.

In addition, it should be noted that the innovations implemented in the territory, from a socioeconomic point of view, generates quality in the offer of products / services, competitiveness and increasing the presence of the destination in the international market. It also generates equity, qualification of the communities and greater access to income. Consequently, it has been an important weapon in the fight against sex tourism, a negative impact of traditional tourism on the territory.

Despite the fact that the TBC on the coast of Ceará has been able to foster and strengthen the value chains and productive networks that involve the activities of the region. It is optimizing the resources offered by the territory so that its benefits are increasingly reflected in the real conditions of the communities' lives. However, there are some other problems contrasting with this type of sustainable tourism, which persist. Such as, the problem regarding the occupation and inadequate use of public space, this is still present and requires solutions capable of generating sustainable tourism, which it is increasingly noticed less impact and that is able to involve local communities in the development of effective initiatives.

Finally, it is observed that with innovation, there are more complementary offers and opportunities for the visitor to participate in activities such as: fishing, handicrafts, visual arts workshops, regional dance classes, sweets production, among others that enable a greater contact with local life. Although it is certain to affirm that all of the benefits from community-based tourism derives from sun and beach tourism. Thus, in Ceará, it is noticed that there is distancing from mass tourism, and it is betting on the use of the coastal space to attract a more tourist profile sensitive to the environment and interested in a destination that offers much more than the sun as its main activity.

ACKNOWLEDGMENT

I am grateful for the collaboration of all the people who helped in the achievement of the data of this research, mainly Dr. Conceição Malveira, who introduced me to some of the tourist initiatives. Also, Dr. Larissa Gabriel, who was always willing to revise the writing to improve consistency in the text.

REFERENCES

Da Silva, R. G. O. (2014). *O papel dos serviços prestados pelos bugueiros na composição e no diferencial do produto turístico potiguar de sol e praia: um estudo na APCBA/RN* (Unpublished master's thesis). Universidade Federal do Rio Grande do Norte, Natal, RN.

Daterra. (2016). *Círculo de bananeiras*. Retrieved January 5, 2020, from http://www.ambientaldaterra.com.br/circulo-de-bananeiras/

De Moraes, E. M., Irving, M. A., & Mendonça, T. C. M. (2018). Turismo de base comunitária na América Latina: uma estratégia em rede. *Revista Tur., visão e ação, 20*(2), 249-265.

Dreamstime. (2020). *Señales, lugares de visita turístico de excursión en el mapa de Suramérica (141183564)*. Academic Press.

Fernandes, L. M. M. (2014). O Ceará turístico: Política de regionalização e governança nos destinos indutores. *Revista GeoUECE, 3*(5), 213–214.

Fernandes, L. M. M., Soares, J. R. R., & Coriolano, L. N. M. T. (2020. (in press). Governance in the regionalization tourism policy of the state of Ceará/Brazil. *Revista Portuguesa de Estudos Regionais.*

Flecha da Mata. (2018). *A permacultura em vivência comunitária.* Retrieved January 5, 2020, from http://flechadamata.com/#rt-utility

Fundación Cotec. (2015). *Innovación.* Retrieved February 2, 2020, from https://cotec.es/quienes-somos/presentacion/

Garcia, R., & Calantone, R. (2002). A critical look at technological innovation typology and innovativeness terminology: A literature review. *Journal of Product Innovation Management, 19*(2), 110–132. doi:10.1111/1540-5885.1920110

Jantunen, A. (2005). Knowledge-processing capabilities and innovative performance: A empirical study. *European Journal of Innovation Management, 8*(3), 336–349. doi:10.1108/14601060510610199

Mannise, R. (2012). *Casas de "superadobe", ¿Qué son y como se hacen?* Retrieved January 5, 2020, from https://ecocosas.com/construccion/casas-de-superadobe-como-hacen/?cn-reloaded=1

Medeiros, C. P. (2017). Inovação social além da tecnologia social constructos em discussão. *Race: revista de administração, contabilidade e economia, 16*(3), 957-982.

Menguc, B., & Auh, S. (2006). Creating a firm-level dynamic capability through capitalizing on market orientation and innovativeness. *Journal of the Academy of Marketing Science, 34*(1), 63–73. doi:10.1177/0092070305281090

MTUR. (2010). *Turismo de Sol e Praia: Orientações Básicas.* Retrieved November 11, 2019, from http://www.turismo.gov.br

Oliveira, R. (2019). *Turismo e inovação, 5 questões para pensar a inovação em turismo.* Retrieved December 19, 2019, from http://turismoeinovacao.com/conceito/5-questoes-para-pensar-a-inovacao-em-turismo/

Rede Tucum. (2013). *Caderno de normas e procedimentos internos Turismo Comunitário.*

Rede Tucum. (2020). *Quem somos: Tucum – Rede Cearense de Turismo Comunitário.* Retrieved January 5, 2020, from http://www.redetucum.org.br/rede-tucum/quem-somos/

Revista Fácil. (2018). *Tourist Map of Ceará.* Retrieved January 15, 2020, from http://www.revistafacil.net/2017/09/mapa-do-turismo-do-ceara-cresce-e.html?m=1

Shoham, A., Vigoda-Gadot, E., Ruvio, A., & Schwabsky, N. (2012). Testing an organizational innovativeness integrative model across cultures. *Journal of Engineering and Technology Management, 29*(2), 226–240. doi:10.1016/j.jengtecman.2012.01.002

Soares, J. R. R., & López, M. P. V. (2015). O uso do etnoconhecimento para o desenvolvimento do turismo sustentável. *Tourism and Hospitality International Journal,* (4), 19-28.

Stahel, W. R. (1997). The service economy: 'wealth without resource consumption'? *The Royal Society, 35*(1728), 1309-1319.

Valls, J. F. (1996). *Las claves del mercado turístico. Cómo competir en el nuevo entorno*. Deusto.

Vieira, J. M. (2011). Inovação no Turismo: Um processo catalisador de desenvolvimento. *Journal of Tourism Studies*, *4*, 133–143.

Villavicencio, B. P., Zamora, J. G., & Pardo, G. L. (2016). El turismo comunitario en la Sierra Norte de Oaxaca: Perspectiva desde las instituciones y la gobernanza en territorios indígenas. *El Periplo Sustentable*, *30*, 6–37.

Yepes, V., & Cardona, A. (2000). Proyecto e instalación de infraestructuras higiénicas en las playas turísticas. *Cimbra*, *333*, 56–59.

Zahra, S., & George, G. (2002). Absorptive Capacity: A review, reconceptualization, and extension. *Academy of Management Review*, *27*(2), 185–203. doi:10.5465/amr.2002.6587995

KEY TERMS AND DEFINITIONS

Circular Economy: Model that strives for the use of resources and the reduction of materials seeking effective use of resources.

Coastal Tourism: Segment that uses activities in the coastal space as the greatest attraction for the configuration of the tourist product.

Cooperative Strategies: Social innovation that allows an innovative and creative approach to local productive arrangements.

Ethnocognition: Ancestral knowledge, knowledge, and traditions (culture) passed from generation to generation in traditional communities, which must be taken, into account in the sustainable management of territories.

Innovative Capacity: Element that represents the company's orientation, proactivity or inclination towards innovation.

Local Community: According to the stakeholder theory, it is one of the factors that must be considered in order to achieve sustainable tourism, concerned with the environment and responsible for reducing the impacts of tourism activity.

Servicizing: Term that arises in the context of growing concern for the environmental impact of economic activities, in this case, tourism.

Social Technology: Technologies that present simple, low-cost and easily applicable solutions that generate social impact in fields as diverse as education, agriculture, health, the environment and leisure.

Chapter 2
Small Businesses Like a Sign of Innovation and Sustainable Development in the Community Tourism Centers in Santa Elena

Myriam Yolanda Sarabia Molina
Universidad Estatal Península de Santa Elena, Ecuador

Lupe Cecilia García Espinoza
Universidad de Guayaquil, Ecuador

Arnaldo Efrén Mendoza Tarabó
Universidad Estatal Península de Santa Elena, Ecuador

ABSTRACT

Small tourism enterprises play a key role in generating income and local development of a community in rural areas through community tourism (TC) ideas, services, processes, and products. These have been generated in the communities of San Pedro, Valdivia, Dos Mangas, and Sacachún from the province of Santa Elena, which deserve to be registered through empirical research. The results show that the population is interested in working towards the sustainable development of its natural and cultural resources. This interest is transforming the tourist activity in their responsible daily lives, and they are committed to caring for the environment. Therefore, the people's own lifestyles are highlighted in a scene of tranquility and harmony.

DOI: 10.4018/978-1-7998-4855-4.ch002

INTRODUCTION

Small tourism businesses in rural destinations play an important role in supporting rural development (Lane & Kastenholz, 2015: Thomas, Shaw, & Page, 2011). Ecuador is a country that has advanced in its vision and development of the Tourist Offer by Base Organizations and Communities that have become fully aware of their ownership of natural and community resources and they have gone from being an object to being a subject of development. Acquiring a collective commitment to ensure their control and participation (García et al, 2017), in the province of Santa Elena, community tourism practices support community work for the benefit of the population, through local management for the sustainability of the themselves (Sarabia et al, 2019)

Community tourism (TC) understood as a tool that provides communities with opportunities and decision-making capacity in conservation and development to achieve direct benefits to the entire community Giampiccoli & Mtapuri (2014), which it should be considered as the main objective of this kind of tourism, which is linked to economic development, cultural conservation, women empowerment, poverty alleviation, and income generation. The main objective of this kind of tourism is to include and benefit the population of local communities, under the central characteristic of community empowerment as subjects increase more than objects of development, Nunkoo, Ramkisson and Gursoy (2012).

Local communities and organizations have their strengths and weaknesses, but there seems to be a consensus among those working in the fields of local development, regarding the need to promote participatory democracy and gender equality, within organizations and their links with others from the public and private sectors that surround them, that is, social capital. In addition, increasingly, international organizations that work in favor of local development through projects, these organizations seek to make grassroots organizations the conductors of local development processes; therefore, they strengthen their leadership, and promote the construction of capital that allows them to have control of the resources.

The search for new ways of undertaking in the TC is reflected in ideas, services, processes, products generated along the coastal profile in the Santa Elena province. The search for these new ways deserve to be registered through empirical research, so it is necessary to return to look at the communities located in the rural area that are interested in working towards a sustainable development of their natural and cultural resources, making tourist activity a responsible daily life and committed to caring for the environment. The inhabitants in a setting of tranquility and harmony highlight their own lifestyles.

These are the reasons that motivate the small entrepreneurs of the Community Tourism Centers (CTC) to stay in their territories and generate their permanent employment opportunity. It is expected that in the medium term, innovative TC processes will be consolidated that encourage the participation of its members within a framework of equal opportunities and adequate distribution of profits for the benefit of the entire population in search of the desired well-being.

Job creation is closely related to the range of problems identified under social sustainability, and which should be included as an important thematic domain of indicators. Employment is related to both the number and quality of jobs created, suggesting a shift from the quantitative effects of tourism (i.e. the number of jobs) to the inclusion of the quality of jobs work created, which is little emphasized by tourism companies or tourism plans (Liu and Wall, 2006). Since the number of jobs created does not reflect other relevant issues surrounding the nature of the jobs (i.e. salary, working conditions and opportunity for advancement), several academics have advocated the inclusion of job quality for an understanding of the effects of tourism on employment. In essence, the economic indicator of job creation should be evaluated in terms of quantity and quality of jobs created.

Certification is important for the communities since that on one hand they formalize the presence of the State and on the other hand it gives the community a status in front of the private offer, which encourages the system of norms and agreements with which they have been operating.

The current four CTC Sacachún, San Pedro, Dos Mangas and Valdivia located in Santa Elena are examples of endogenous ventures that they are in the initial stage of tourism development, where the population has experimented with sources of work, to provide solutions to the economic and social needs of the population located in rural areas of the province. The consolidation of the TC in the studied CTCs is expected to give answers to the contextualized needs in each center and it might be the way to achieve the desired well-being of the population.

This chapter argues approaches to sustainable TC ventures at the local level, the contents of tourist ventures come from a direct analysis through interviews and surveys of the leaders, tourists and residents of the four CTCs under study, which allowed identifying activities that contribute to generating ideas, services, processes and products created in the community within the practice of TC. Currently, the tourist activity in the province of Santa Elena is more committed to the management of natural and cultural resources to be able to preserve and conserve responsibly, and it takes care of the environment with the own initiatives of the local people involved and interested in protecting their resources of recognized good practices in the communities.

It also documents personal experiences and practices of tourist entrepreneurs who were generated from the arrival of tourists to their communities because of the certification as CTC. Therefore, it organizes the opportunity to create a job that allows them to save, distribute their earnings for the family, for their well-being, and this way they stay in their community doing what they know how to do well from their own knowledge. That said, the study will be useful information for tourism research and it will bring visibility to the work that is being developed and gestating in the coastal communities of the province.

BACKGROUND

This development implies a process that raises the standard of living of people, relating not only to increasing the wealth of individuals but also to changing behaviors, aspirations and the different ways in which people view the world around them (Bartelmus, 1986;Hall, 2019). Therefore, development is not only concerned with institutional and economic changes but it also involves broader concerns such as quality of life, poverty reduction and prosperity (Hall, 2019). the term has sustainable development influenced the ways in which tourism research is carried out and how governments and other organizations plan for and manage tourism development (Nunkoo, 2017).

Innovation in the research is seen as key to improving the beneficial outcomes of sustainable tourism. As Bramwell and Lane (2011) argue, "sustainable tourism is becoming more innovative at identifying ways to secure positive benefits, in addition to the established approaches of regulation and development control" (p. 1). Like Bramwell and Lane (2011), we argue that for sustainable tourism research to be impactful, innovation in their search process is vital innovation in sustainable tourism. Schumpeter (1934), the father of the economic theory of innovation, innovation is the critical dimension of economic change and "a creative destruction." He describes an innovation as:

An application and implementation of a new idea or a new application of an existing idea that results either in a new kind of product, or a new and better process for producing an existing product. (Divisekera and Nguyen, 2018, p. 158).

Since the diffusion of the principles of sustainable development in tourism studies, the term has influenced the ways in which tourism research is carried out and how governments and other organizations plan for and manage tourism development (Nunkoo, 2017), such that it is now considered as a dominant paradigm in tourism research (Bramwell and Lane, 2011; Ruhanen et al., 2015). Systematic reviews, journal editorials and commentaries on this topic suggest that sustainable tourism research covers a broad range of topics such as those related to climate change, ecotourism and other alternative forms of tourism, social responsibility and ethics, sustainable mass tourism, climate change, community involvement, tourism impacts, visitor behaviors and attitudes, responsible tourism, rural tourism, heritage and cultural tourism and travel and transport (Bramwell and Lane, 2011; Gross and Grimm, 2018; Hashemkhani Zolfani et al.,2015; Holden, 2018; Lu and Nepal, 2009; Ruhanen et al., 2015; Ruhanen et al., 2018). Sustainable tourism research also relies on a variety of approaches such as case study, empirical study, critical and systematic reviews, theory building and testing, development of new concepts and historical perspectives to advance knowledge (Font et al., 2019; Ruhanen et al., 2015).

Residents' attitudes to tourism is among the most studied topics and is perhaps one that has undergone considerable innovation in terms of problem and analytical framing, theory, methodology, and context (Muler Gonzalez et al.,2018; Nunkooet al., 2013). Innovation is not limited to creativity; it extends to implementation that requires social interaction. Therefore, collaboration becomes an important element of social innovation, in the sense that social innovation always takes a cooperative approach (Ziegler, 2017). Stakeholders partnering together to fight a social problem offers a holistic solution to a problem. In this sense, collaboration between universities and the industry is essential for innovation to take place, as it facilitates knowledge sharing between institutions (Hjalager, 2002, 2010; Martínez-Pérez et al., 2019).

Inherent to social innovation in sustainable tourism is the change occurring at a systemic level. Systemic level change challenges the status quo and brings about transformation by introducing new products, processes, or services. Because the change operates at the institutional level, building legitimacy becomes important. Third party audits and certifications have been recognized as legitimacy building tools. Having its origin in the manufacturing sector, environmental certification is regarded as an important indicator of sustainable tourism (Font, 2002; Klein and Dodds, 2018; Rivera, 2002)

Building local capacity serves as a mechanism for social innovation, as it targets working with the marginalized to enable them to help themselves (Alvord et al., 2004). This approach is based on the participatory principle and takes place at the community level. From a sustainability perspective, slum tourism should aim at improving the life of slum residents (that essentially amounts to building local capacity), as evidenced by the various journal articles and books published on this topic (Frenzel, and Koens, 2012; Frenzel et al., 2015; Meschkank, 2011; Von Schuckmann et al., 2018)

The literature on small businesses mentions that owner-managers with non-economic motivations dominate such business and they underpin their lifestyle goals. They enjoy a lifestyle supported by the natural environment and its associated opportunities is frequently one of the reasons why participants in this sector become involved in rural tourism (Ateljevic and Doorne, 2000; Hall, 2009a). These non-economic motivations are often integrated into the various aspects of meaning and representation that small business owners and managers attribute to the rural destination in which their tourism business is located and to which they are attached (Carlsenet al., 2008). .

Despite the importance of entrepreneurs for the development of the destination, the various functions they perform require research to understand how social objectives are achieved. Perhaps an obvious role is that their efforts, through product development, can attract tourists to destinations. However, the contributions of multiple entrepreneurs are less obvious and they can have a combined effect, which is often greater than the impact on their own individual companies and it is exhibited through the formation of networks that contribute to the development of the destination (for example, Beritelli, 2011; Pavlovich, 2003; Tinsley & Lynch, 2007). Russell and Faulkner (2004) examine tourism entrepreneurship within the framework of chaos theory and claim that entrepreneurs "play a major role both as creators of chaos and initiators of adaptive responses to chaos induced by external events" (2004, p. 556). Entrepreneurs also act as leaders within the destination, often developing a long-term vision (Ryan et al., 2012), being innovative (Weiermair, Keller, Pechlaner, & Go, 2010) and identifying opportunities (Johns & Mattson, 2005). It is also notable that local social situations and environments can affect business strategies (Bosworth & Farrell, 2011; Kline & Milburn, 2010; Mottiar, 2015). As such, entrepreneurs tend to be deeply connected to the rural economy and community, what leads us to further consider the connection of the business community and, in particular, the role of tourism ventures (Mottiar et al., 2018).

COMMUNITY DEVELOPMENT VERSUS COMMUNITY TOURISM

The concept of community development lacks a direct definition. Academics have defined the concept in relation to specific groups and specific problems (Goodness, 2016). According to Theodori (2008), community development is "the process ... that depends largely on the intentional actions of people in a locality, so, they unite and interact for the purpose of solving their local problems, and by improving their quality of life and shaping their future well-being "(p. 64). In the definition above, local people work together to solve their shared challenges with the goal of improving their quality of life.

Community development can be oriented towards the improvement of negative economic situations such as poverty reduction and income generation. Another aspect is human development where community members strive to be healthy and educated people, stronger social ties, and the ability to unite to pursue determined collective action.

Consequently, community development is a process of building and strengthening the community as a social characteristic of the local population. It is primarily based on place-based communities as "natural will" occurs predominantly, if not exclusively, in direct and continuous contacts between people living in the same place (Tönnies 1988; Wilkinson 1991; Summers 1992; Theodori 2005; Muthuri 2007).

Another aspect is human development where community members strive for healthy and educated individuals, stronger social ties, and the ability to come together to pursue purposeful collective action. Finally, community development can be achieved through infrastructure development (Richards and Dalbey, 2006, p. 20).

Regarding the TC, the origins referring to this type of tourism are heterogeneous, since they can derive from a specific or strategic practice of duly planned processes, it can also develop spontaneously from individual or family initiatives that, over time, make them become in structured practice (Ruiz Ballesteros & Cáceres Feria, 2016). In Murphy's (1985) contributions, he already considered community tourism initiatives as the center of tourism planning and sought to establish the community as a key focus for decision-making and approval by tourism planners.

Along the same lines, (Ruíz et al., 2007) sustains that the TC must be understood fundamentally from the social economies, the popular and solidarity economies, before the tourism sector itself, since its reason for being is its community organization. Likewise, for (López-Guzmán & Sánchez-Cañizares, 2009a) TC is an activity that crystallizes in the creation of tourist products under the basic principle of the necessary participation of the local community. For Tolkach and King (2015) the TC refers to maximizing the positive impacts of tourism and minimizing the negative ones, thus attending to the needs of hosts and guests without compromising the well-being of future generations and the physical environment.

Community-based tourism (CBT) has been widely identified for its ability to improve local economies, and it has been introduced in many countries (e.g., Dodds, Ali, & Galaski, 2018; Lee, 2009b, 2013; Lepp, 2007). Consequently, traditional communities, such as rural communities (e.g., Wang, Cater, & Low, 2016), fishing communities (e.g., Thompson, Johnson, & Hanes, 2016), small islands (e.g., Teh & Cabanban, 2007), and aboriginal communities (e.g., Reggers, Grabowski, Wearing, Chatterton, & Schweinsberg, 2016), could develop CBT to improve their economic status.

TOURIST ENTERPRISES IN RURAL AREAS

The roles that social entrepreneurs play in the progress of rural destinations has developed a new theoretical framework in this area (Jaafar & Rasoolimanesh 2015). Social enterprises can influence the development of community tourism (TC); however, there is a lack of literature that documents the roles they play leaving a raising a gap in the understanding of the ways in which TC develops. Examining the roles of entrepreneurship in tourist destinations sheds light on the commitment of entrepreneurs who are motivated not only by profits, but also by social goals (Mottiar et al., 2018)

It is evident that the tourism activity is mainly made up of small companies (Lashley & Rowson, 2010), the exploration of its impact is absent in the literature (Power, Di Domenico and Miller, 2017). Specifically, small businesses can cultivate strong social capital, as they can act as a link between the local community and tourism activity (Shaw and Williams, 1998). However, the tourism literature, according to Koh and Hatten (2002), has "overlooked" the tourism entrepreneur as an actor in tourism development. Recently, some academics point out the important role that entrepreneurs can play within destinations. For example, Komppula (2014, p. 361) concludes that "without committed and risk-taking innovative entrepreneurs, no destiny will flourish." Ryan, Mottiar and Quinn (2012) demonstrate that entrepreneurial influence can continue long after the original entrepreneur engages through the development of an environment in which entrepreneurship is fostered and supported.

Despite the importance of entrepreneurs for the development of the destination, the various functions they perform require research to shed light on how social objectives are achieved. Perhaps an obvious role is that their efforts, through product development, can attract tourists to destinations. However, the contributions of multiple entrepreneurs are less obvious and can have a combined effect, which is often greater than the impact on their own individual companies and is exhibited through the formation of networks that contribute to the development of the destination (for example, Beritelli, 2011; Pavlovich, 2003; Tinsley and Lynch, 2007). Russell and Faulkner (2004) examine tourism entrepreneurship within the framework of chaos theory and claim that entrepreneurs "play a major role both as creators of chaos and initiators of adaptive responses to chaos induced by external events" (2004, p. 556). Entrepreneurs also act as leaders within the destination, often developing a long-term vision (Ryan et al., 2012), being innovative (Weiermair, Keller, Pechlaner, & Go, 2010) and identifying opportunities (Johns & Mattson,

2005). It is also notable that local social situations and environments can affect business strategies (Bosworth and Farrell, 2011; Kline and Milburn, 2010; Mottiar, 2015). As such, entrepreneurs tend to be deeply connected to the rural economy and community, leading us to further consider the connection of the business community and, in particular, the role of tourism endeavors.

RESIDENTS' PERCEPTIONS OF SUSTAINABILITY

Residents' perceptions are crucial for supporting the sustainable development of CBT (Lee, 2013; Nicholas, Thapa, & Ko, 2009). Based on the social exchange theory (SET; Ap, 1992), residents may support CBT according to their positive or negative perceptions (Lee, 2013). Moreover, during CBT development, residents are the critical stakeholders who may affect tourism planning, development, and support based on their perceptions of the impacts of tourism (Lundberg, 2015). Based on the perception of tourism impacts, resident attitudes toward tourism development may change from positive to negative (Lee, 2013; Woo, Uysal, & Sirgy, 2018); specifically, residents' perceptions of CBT may change over time according to the level of tourism development (Butler, 1980; Diedrich & García-Buades, 2009; Hunt & Stronza, 2014; Lundberg, 2015).

Residents' perceptions of tourism impacts are an important issue that has garnered considerable attention in the tourism literature (Almeida-García, Peláez-Fernández, Balbuena-Vázquez, & Cortés-Macias, 2016; Nunkoo & Gursoy, 2012). Residents' perceptions can be influenced by demographic factors (e.g., age, gender, education, and years of residence; Huh & Vogt, 2008; Vargas-Sánchez, Plaza-Mejia, & Porras-Bueno, 2009), community attachment (Lee, 2013), tourism planning (Choi & Murray, 2010), environmental sustainability (Choi & Murray, 2010), the state of the local economy (Gursoy, Jurowski, & Uysal, 2002), and the stage of tourism development (Diedrich & García-Buades, 2009; Hunt & Stronza, 2014; Long, Perdue, & Allen, 1990; Lundberg, 2015).

As mentioned above, residents seem to be the most important stakeholders in CT development because they interact with tourists directly and provide unforgettable experiences for tourists; thus, residents with positive perceptions will become major stakeholders in tourism planning and management (Davis, Allen, & Cosenza, 1988; Lee & Hsieh, 2016). Since the perspectives of stakeholders are different across different stages (Ngo, Lohmann, & Hales, 2018), destination managers may identify the stage of tourism development to plan for and manage destination resources to increase the attractiveness of communities (Lundberg, 2015).

Scholars have introduced life satisfaction as an indicator to measure residents' perceptions of CBT (Long et al., 1990). For example, Kim, Uysal, and Sirgy (2013) proposed the perception of the residents' wellbeing under tourism, which may influence the direction of tourism policy and planning. Several scholars have also used well-being as an indicator to examine residents' perceptions for supporting tourism de- velopment (Kim et al., 2013; Vargas-Sánchez et al., 2009). Additionally, Woo, Kim, and Uysal (2015) indicated that residents' perceptions of community life satisfaction is a crucial indicator for supporting sustainable tourism development. Thus, this study includes residents' perceptions regarding the impacts of tourism on the sustainability of the economy, socio-cultural factors, environment, and life satisfaction as effective predictors of support for further CT development.

COMMUNITY TOURISM CENTERS IN ECUADOR

In 2009 (MINTUR) [1] issued the Instructive for Registration of Community Tourism Centers (CTC), and in 2010, the Regulations for CTC considering that "it is urgent to support communities seeking tourism development as a tool for fight against poverty" (MINTUR, 2010).

The offer of CTCs according to MINTUR in the last ten years (2010-2020) has carried out an information dissemination program to encourage registration in its Official CTC Registry. This program has the collaboration of public entities such as MAGAP[2], MIES[3], IEPS[4], together with communities of seven zonal coordination, the main objective was to publicize fundamentally the requirements and procedures to carry out the registration of communities in MINTUR as CTC.

MINTUR currently maintains the CTC program, which is a certification for organizations that are active in the sector, accrediting them for the exercise of their operations; which also allows allocating funds for training and infrastructure. It is important to point out that this Certification on the one hand formalizes the presence of the State and on the other hand gives the community a status in front of the private offer which encourages the system of norms and agreements with which they have been operating (García et al., 2017).

Despite all these advances, the registration of CTCs is still incipient. Thus we have 20 CTC in 2014 and 38 CTC in 2018. The 38 are registered CTC and they have 2940 accommodation places, which is equivalent to an average of 77 places in each of them, the bar chart in FIgure 1 presents CTC accommodation places registered by MINTUR (2014-2018)

Figure 1. CTC accommodation places registered by MINTUR (2014-2018)
Source: Own elaboration

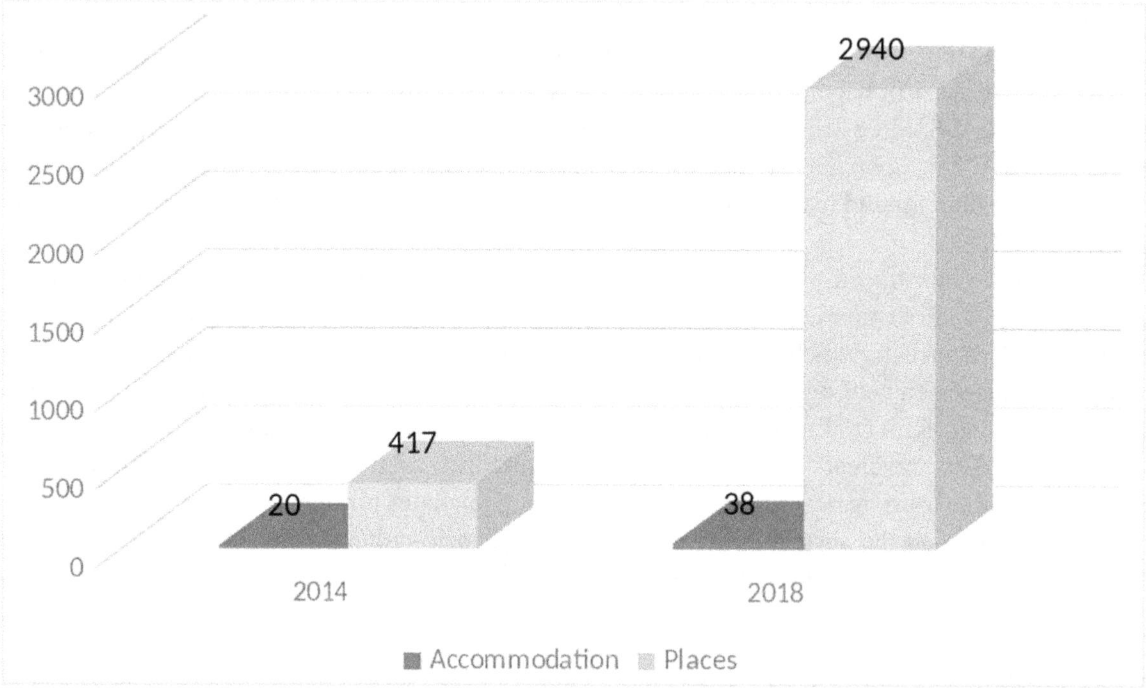

In the country, a significant number of organizations with experiences of autonomy and empowerment in the management of natural resources have been registered since its inception in the 1950s. They are strong organizations that have come to consolidate a tourist offer under the control of grassroots organizations. They faced long processes of struggle for land, for water, and in defense of their natural resources, which has allowed them accumulate an organizational experience that today constitutes the main base that supports tourism activity, as shown in table 1: Community Tourism Offer in Ecuador.

Table 1. Community tourism offer in Ecuador

Community	Geographic location	Tourist project	Start date
Yunguilla	Pichincha/Quito	Accommodation services / guided visits to the forest	1999
Íntag	Imbabura/Cotacachi	Accommodation / forest tour services	1997
Ishkayaku	Pastaza	Accommodation Services	2003
Isla Corazon	Manabí/San Vicente	Accommodation Services	2000
Salango	Manabí/Puerto López	Services Accommodation and guides	2004
Agua Blanca	Manabí/Puerto López	Accommodation and guides	1986

Source: Ecuadorian community tourism, environmental conservation and defense of the territories (Roux, F. 2013)

To understand the successful TC projects in Ecuador, it is necessary to mention Cabanilla's doctoral thesis (2017), which it describes successful experiences managed by the community that made the right decisions, and it knew how to make the most of the help they received from international organizations. Understanding that tourism is a medium and long-term activity and that with work, effort, innovation, incentives, diversification and strategic alliances with the private sector, it can become so profitable until it becomes the first source of income in a community. Thus, there are six cases: Kapawi Eco Lodge and Reserve, Alta Florencia, Sacha Ñampi trail, Saraguro community, Salinas de Guaranda community, Shandia community and Yunguilla community.

Community Tourism Centers in the Province of Santa Elena

Tourism practices in this province are relatively new and a recent addition to tourism activities, since Santa Elena has resources and attractions that are at an early stage of their life cycle, counting with immense potential to be developed, this is alone reason enough to be studied. The decision to travel to the province in the last three years compared to 2015 has increased significantly (MINTUR 2017). The trips are motivated by vacations, recreation and leisure; tourist destinations in the area are concentrated in the canton of Santa Elena.

Yumisaca & Mendoza (2016), ensure that 61% of the provincial tourist plant are located in Salinas and Montañita. Firstly, it is possible to say that 56% are distributed for the restaurant sector, followed by 40% for servers' accommodation and lastly only 4% for intermediaries, the operators and travel agencies represent this last sector. While, in Montañita 58% corresponds to accommodation, 34% to catering and 8% for tourism operators, Montañita registers arrivals from nearby provinces such as Guayas, Azuay, Bolívar and a small number of tourists from the Pichincha province in recent years.

Considering that development can be conceived as the result of a deliberate political strategy or public action, where actors, resources and means intervene to achieve the set of previously established goals, and it also as the result of the operation of an institutional and social structure generated by the population itself (Hernández, 2015). The purpose of this chapter is to study the contributions of the Community Tourism Centers to the development of the province of Santa Elena. The Tourist Offer in the province involves the following decision-makers: MINTUR with national competence, (Emuturismo)[5] with local competence, the Municipality with local competence and finally the Community that through the Cabildo[6] approves and regulates all activity in the community space. Currently from 2012 to 2017 MINTUR has certified the communes: San Pedro, Dos Mangas Sachachún and Valdivia respectively as CTC (Table 2), accrediting them for the exercise of their operations: accommodation places and food and drink places, which it also allows allocating funds for training and infrastructure. Generate direct employment to 105 tourism enterprises.

Table 2. Tourist offer of the CTC in Santa Elena, according to (MINTUR, 2019)

Name	Parish	Status of establishment	Men	Women	total employees	Rooms	table places	bed places
Comuna San Pedro	Manglaralto	Open	16	15	31	0	448	112
Dos Mangas			38	17	55	11	64	42
Valdivia			9	3	12	0	24	0
Sacachún	Simón Bolívar (Julio Moreno)		4	3	7	0	60	0

Source: Own elaboration

Fieldwork carried out during the months of November and December 2019, some in-depth interviews were applied to the leaders and entrepreneurs of the four CTCs at the province under study, finding illustrated bellow. 13 enterprises generated where neighborhood stores are the ones that found the highest percentage of activities, followed by restaurants, cottege-dinings rooms and handicrafts. The entrepreneurship activity in the CTC is indicated in Figure 2.

In this way, our research uses the methodology with the technique based on a series of qualitative interviews with a semi-structured script, where we try to understand the phenomenon studied according to the perception of the resident subjects, tourists and entrepreneurs of the communities under study. Therefore, in this section of the investigation, the qualitative analysis of the data was determined, to achieve more effective results, in accordance with the objective proposed in this chapter, detailed below.

CTC San Pedro

The organization they maintain is through the communal council, who is in charge of ensuring the well-being and development of the commune. It is mentioned locality that it began as CTC from 2015 offering the paragliding service as an adventure sport that today is one of the consolidated services within the province, another service as a start is gastronomy, within the information gathered The following is documented in the field work. The governing body of the tourist activity in Ecuador MINTUR has

Figure 2. Entrepreneurship activity in the CTC
Source: own elaboration

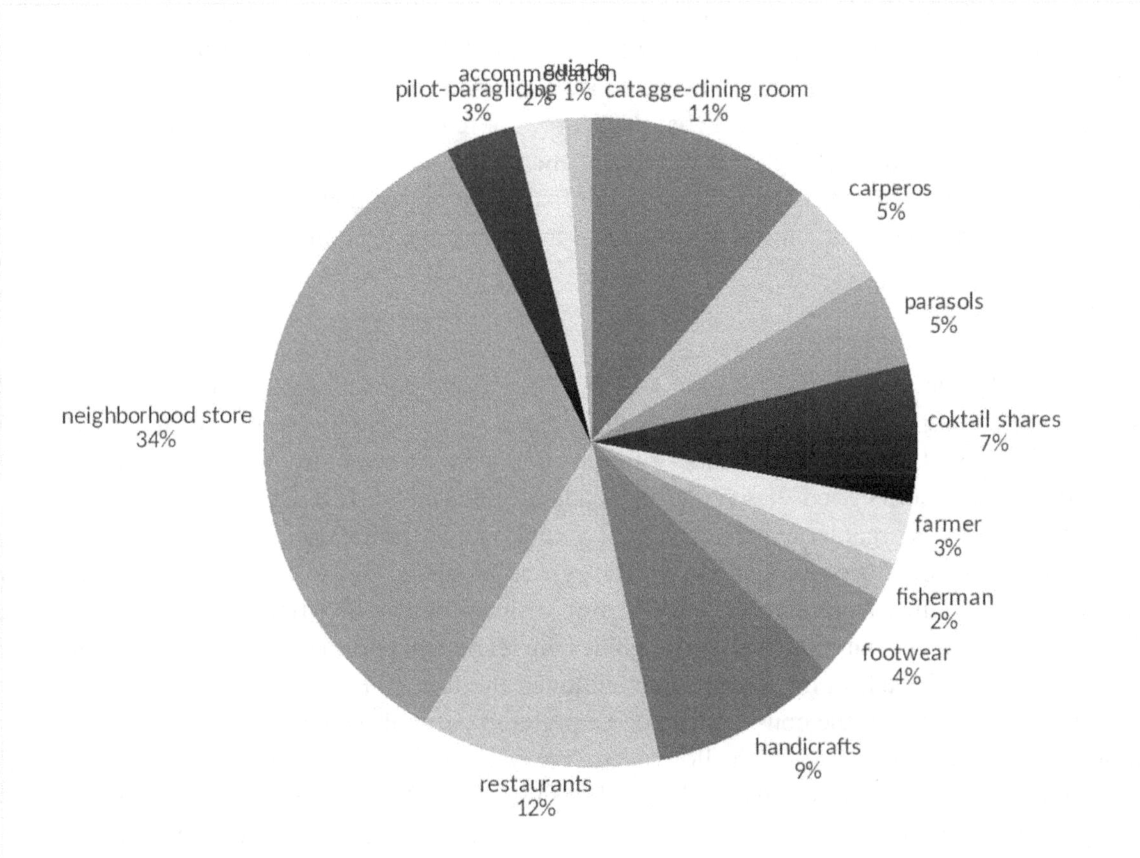

worked since the beginning of its certification in programs and projects such as: provision of services, accommodation, food and beverages. These services allow operators to provide quality products to develop the tourist activity; however, despite the work done, it is determined that San Pedro is a community that has not adequately developed as a CTC in the gastronomy service, due to the fact that they have not been able to organize the cabañeros as a group.

This also illustrate the perception of the population regarding the benefits achieved through certification as CTC. Not alone saying that a group of 12 professional guides who practice paragliding live well in their locality and they generate a 5% contribution to the community. Thus, they contribute to the local development of this place and at the same time to the practice of TC through its participation as a clear form of organization. The words said by one of the interviewed guides were "the work here is good, we organize ourselves by monthly shifts, we are all certified by the Ministry of Sport and we look for tourists to achieve an optimal experience with all safety standards". Another interviewee mentioned words of one of the women who is dedicated to gastronomy in the cabins, her were "we offer food service with fresh seafood here, passing thru tourists come and visit us, we want them to always come".

It is also evident that the population in general has started, since 2019, in greater numbers, the generation of enterprises in activities such as dining rooms, shops, pharmacies, restaurants, cabins.

Community leaders believe that it is possible to live well through tourism ventures (Graph 2), which are already evidence of the local ideas and processes that have natural and cultural resources that must be used. In which, the residents are starting their practices of small tourist enterprises and they are still encouraged and convinced that the constancy of visitors will lead to positive results.

When we consulted the residents, the response was immediate, the level of importance is high), because it is the sustenance of daily life that they carry out in their communities and try to protect their resources, they comment "fishing is an activity of the males" is a natural resource that must be cared for and "ask the sea for permission to remove the fish", they agree that it must be maintained and be responsible to conserve the resources, customs and traditions of each community.

CTC Dos Mangas

This community has a privileged location and the TC activity consolidated at the provincial level, it has a tourist offer organized and controlled by the indigenous population. Its management model, is the result of a process of almost three decades of collective work, it began in the tourist activity with the preparation of handicrafts goods based on (tagua and toquilla straw). It is a result of the CTC certification the community is organized with a group of young people leading tours of the tropical forest that they have done the tours over the last 5 years. The group increased to a number of 25 guides, in which 14 are currently waiting to receive the certification of local guides by MINTUR, the 14 new comers will allow them to carry out more work. The guides are responsible for natural resources owned by one of the guides interviewed; in his speech, he mentioned the following: "The community members are increasingly committed to the conservation of natural and cultural resource and especially the guides who work here. This is our job, which allows us to live well, but that implies that we must be constantly training ourselves to improve our work. At another moment in the interview, the guide mentioned "at the beginning of the tourist activity in our community, not all people believed that it would benefit us. Our own families were the ones who helped to make tourism visible, they vibrated as we did our work and by this way, new people join TC every year. This is the activity that allows us to develop locally ".

Also within the fieldwork, it was documented that most of the residents believe in the tourist activity, because it has allowed them to generate their ventures with their own employment places and to acquire profit. The earnings are distributed and invested in the development of capacities that contribute to quality of life and well-being.

The greatest commitment is with the adequate management of natural and cultural resources, this good management is the main reason why tourists come to this community and the number of visits is increasing. With the active participation of the majority of the residents, and the practice of the principle of gender equality and the distribution of local benefits, the locals make this TC an example of innovation and sustainable development of the community.

Likewise, the perception of tourism guides is that they are convinced that TC is a solidarity tourism practice for the benefit of the community, where residents become aware that it is appropriate and precise to undertake within their locality, where organization and training are taking strength to find the sustainable development of their territory.

CTC Valdivia

It began its tourist activity as CTC from 2017 with the administration of the Valdivia Marine Park. At that time the locals were nor very interested in participate; this perception has been changing for the better by the improvements in cultural, economic and political reality of the Santa Elena province. The organization that this community, like the rest, is done through the communal council, what differentiates it from the rest is the creation of a tourism commission here. The commission has the purpose of making tourism activity viable, such as the ventures generated in the last two years. (2018-2020)

The management carried out by its leaders since 2019 was positive for sustainable development and it can be translated in a more direct participation in the organization of the small tourist ventures generated among them such as: cabañeros, parasoleros, carperos, cocktail shakers, artisans, dining rooms, shops and restaurants. Residents' perception is encouraging and visible towards the local development that the population aspires to by implementing their ideas. They implement their own initiatives as evidence of better living conditions; they are taking advantage of their natural and cultural resources. The motivations that lead them to undertake are having their own job, increasing the number of services and the possibility to reach more customers, sale and income continuously, speed of collection (at the moment) and work with members of the family.

Among the documented speeches, an entrepreneur commented the following: "the arrival of tourists to our community is a reason to create tourism service ventures, we are trained to do and we have the support of the leaders, we are started work from 2019 to offer our services to more tourists".

The words of a tourist were: "I came to this community for its tranquility, because you can enjoy family vacations, above all the children they like freedom, they enjoy the sand, the water so much and we take advantage of the proximity to be able to vacation".

Thus it was also documented that the ventures are along the boardwalk of Valdivia, where other young entrepreneur mentioned in his speech "Valdivia commune is small and almost all of us here are dedicating ourselves to a job created by ourselves. In addition, this is the reason we are just realizing that it is a benefit to be CTC certified. Tourists come due to the certification; they want the information and recommendation.

CTC Sacachún

Sacachún received CTC certification since 2012 as the San Biritute monolith returned to the community, accompanied by an advertising campaign. The community and the Government decided to return the emblematic figure of the commune to its place of origin! San Biritute is a 2.35 m high stone structure to which the inhabitants attribute the grant of women fertility. Since its arrival, the gastronomy service was certified but it has not been developed. There are 3 local guides, they carry out advanced guide services to the cultural attractions as well as the preparation of the gastronomy service.

The commune has not managed to offer services associated with a tourist offer, despite having fifteen potential tourist attractions, distributed among natural sites and cultural events.

The community is not interested in generating tourist ventures since they feel they have everything there; they are motivated by the cool weather due to its location, their lifestyle is leisured without noise nor stress. The locals are friendly, respectful, solidarity, and concerned with the care of their community, they maintain its cleanliness and order.

In one of the interviews, a mature woman with a lot of life experience commented "here, the people are mostly older adults who live quietly in this place, young people work in agriculture and they should start tourism so that they can live happy in this place".

Another woman who was interviewed, works in the gastronomy service, mentions that: "If you work with the gastronomy service, the "*seco de chivo*" for groups who request by phone after they arrive here! We don't have tourists every day! If they always came! Well, there would be work for us and we would be willing to attend them, the constant flow of tourists"

As for the responsible activities carried out by tourists who come to the communities of the Santa Elena canton to visit, they mention: that they do not seek information because they stay for a short time and go only for specific activities such as "adventure sports, rest, gastronomy, photography" and We note that the community is not fully prepared to receive tourists, despite having enough tourist attractions. We respect their "customs and traditions" but we expect more tourist offer to stay longer in rural communities to get to know the community and support local development.

Once the actions carried out in the 4 CTCs in the province of Santa Elena have been analyzed, the reasons why the communities make the decision to undertake are detailed below, what are their objectives, motivations, and benefits that they find in tourism enterprises, what are established community commitments and the obstacles they face to decide to undertake. This information was taken in the field where tourists and local residents themselves participated and is classified according to the most homogeneous criteria of those interviewed, the actions carried out in the CTC is indicated in Table 3.

Table 3. Actions carried out in the CTC Santa Elena province

Reasons to undertake (invest)	I include myself in decision making and local participation. I seek access to good health, education, employment, fair wages. I facilitate self-respect, security, happiness. The acquired profits are distributed for the development of capacities that contribute to the quality of life and well-being I recommend tourist activities that tourists can do.
Benefits	Secure family income Allows me to save I create jobs for my children, residents of the community I am my own boss and I organize my way I live as I like calm and happy
Objectives and motivations	Guarantee family income (ethics and morals) and the organization. Have my job. Increase the number of services and reach more clients. Clients who visit CTCs request our services. The services we offer are the image of the community organization. Sales and income continuously (weekly, monthly, etc.). Collection speed (at the time) Personal relationships with clients. I work with my family.
Obstacles	Residents are not convinced of community tourism. Not everyone's support in the community. Residents expect full support from central government and do not seek their own means. Residents are unsure to undertake. Lack of credit lines for community ventures.
Community commitment	Solidarity tourist activity. Active participation of the community. Adequate management of natural and cultural heritage. Valuation of the natural and cultural heritage. Principle of gender equality. Distribution of local benefits.

Source: Own elaboration

CONCLUSION

The small tourist ventures in the Santa Elena province are generated from the personal experiences and the activities of the local inhabitants that started their ventures as the tourists arrived in their communities. Tour communities were certified as CTC from 2012 to 2017: San Pedro, Dos Mangas, Valdivia and Sacachún. The certification ensured the opportunity to create a permanent job them to save, distribute their earnings with the family, for their well-being, there are able stay in their communities and they continue doing what they know how to do well from their own knowledge.

This study is useful information for empirical tourism research in order to make visible the work that is being developed and it is gestating in the coastal communities of the province of Santa Elena. There are very particular endogenous realities, at these four CTC; despite, being communities all next to other. This confirms the level of innovation and sustainable development that each CTC can develop with the organized tourist activity committed to the protection of their natural and cultural resources. It should be noted that only the case of the Dos Mangas commune has a consolidated tourist activity. In the rest of the 3 studied communities, it is found traces of community participation; however, they are communities which have the resources and capacities to generate larger entrepreneurship from the certification achieved as CTC. The development of the enterprises in the 4 CTC is indicated in Figure 3.

By studying the local sustainable development of the Community Tourism Centers, it is confirmed that they are an example of innovation in tourist ventures. Certification is important for communities. On one hand, it formalizes the presence of the State and on the other; it gives the community a status ahead of private sector. It encourages the system of norms and agreements with which they have been operating and considering the motivations and obstacles that arise.

SOLUTIONS AND RECOMMENDATIONS

The contributions of the 4 CTCs studied in this chapter are evident to local development, as mentioned (Bartelmus, 1986; Hall, 2019) this development implies a process that raises the standard of living of people, relating not only to increasing the wealth of individuals but also to changing behaviors, aspirations and the different ways in which people view the world around them.

Small tourism businesses in the province of Santa Elena are a sample of new ideas that generate positive results for the local population through simple processes that are appropriate to the context, as mentioned (Divisekera y Nguyen, 2018), the research of sustainable tourism In this chapter, he based himself on the empirical study of four communities to advance knowledge (Font et al., 2019; Ruhanen et al., 2015) of the province of Santa Elena.

The perceptions of both internal and external actors are important in the study of community tourism, giving a wide space to residents since they are responsible for the natural and cultural resources in their communities; and they are responsible for making their jobs sustainable for future generations. Moreover, during CBT development, residents are the critical stakeholders who may affect tourism planning, development, and support based on their perceptions of the impacts of tourism (Lundberg, 2015).

Empirical studies are recommended to small entrepreneurs on ideas, services, processes and products created in the community within the practice of TC, taking into account the criteria of local residents as actors in approving the use of natural and cultural resources that Belongs to them.

Table 4. Development of the enterprises in the 4 CTC

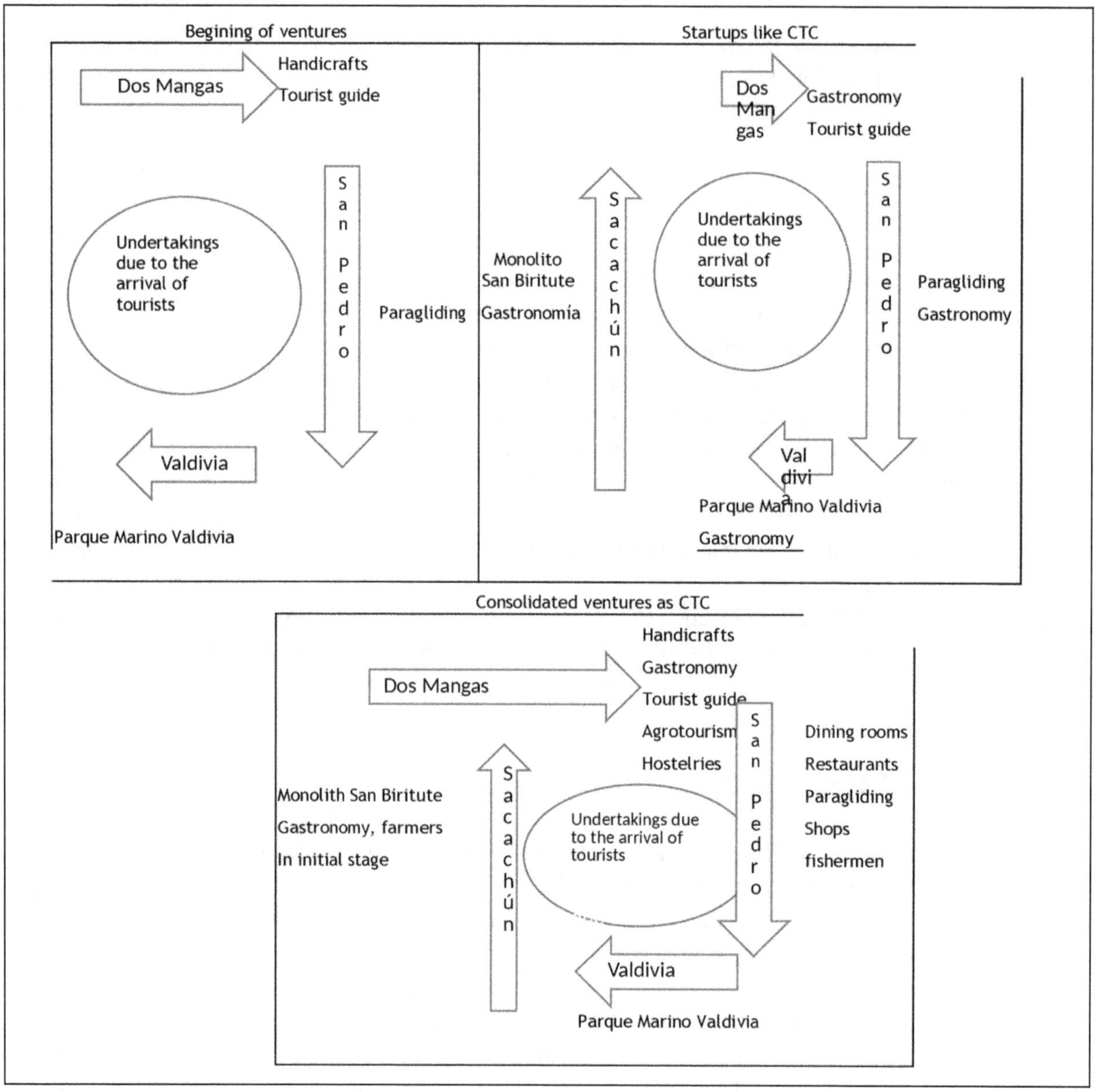

Own elaboration

REFERENCES

Almeida-García, F., Peláez-Fernández, M. Á., Balbuena-Vázquez, A., & Cortés-Macias, R. (2016). Residents' perceptions of tourism development in Benalmádena (Spain). *Tourism Management*, *54*, 259–274. doi:10.1016/j.tourman.2015.11.007

Ateljevic, I., & Doorne, S. (2000). Staying within the fence: Lifestyle entrepreneurship in tourism. *Journal of Sustainable Tourism*, *8*(5), 378–392. doi:10.1080/09669580008667374

Beritelli, P. (2011). Cooperation among prominent actors in a tourist. *Annals of Tourism Research*, *38*(2), 607–629. doi:10.1016/j.annals.2010.11.015

Bosworth, G., & Farrell, H. (2011). Tourism entrepreneurs in Northumberland. *Annals of Tourism Research*, *38*(4), 1474–1494. doi:10.1016/j.annals.2011.03.015

Bramwell, B., & Lane, B. (2011). Critical research on the governance of tourism and sustainability. *Journal of Sustainable Tourism*, *19*(4/5), 411–421. doi:10.1080/09669582.2011.580586

Butler, R. W. (1980). The concept of a tourist area cycle of evolution: Implications for management of resources. *The Canadian Geographer. Geographe Canadien*, *24*(1), 5–12. doi:10.1111/j.1541-0064.1980.tb00970.x

Cabanilla, E. (2017). *Configuración socio-espacial del turismo comunitario: caso República del Ecuador* (Tesis Doctoral). Universidad Nacional del Sur, Argentina. Accesado el 1 de mazo 2019 de:http://repositoriodigital.uns.edu.ar/handle/123456789/3430

Carlsen, J., Morrison, A., & Weber, P. (2008). Life style oriented small tourism firms. *Tourism Recreation Research*, *33*(3), 255–263. doi:10.1080/02508281.2008.11081549

Choi, H. C., & Murray, I. (2010). Resident attitudes toward sustainable community tourism. *Journal of Sustainable Tourism*, *18*(4), 575–594. doi:10.1080/09669580903524852

Davis, D., Allen, J., & Cosenza, R. M. (1988). Segmenting local residents by their attitudes, interests, and opinions toward tourism. *Journal of Travel Research*, *27*(2), 2–8. doi:10.1177/004728758802700201

Diedrich, A., & García-Buades, E. (2009). Local perceptions of tourism as indicators of destination decline. *Tourism Management*, *30*(4), 512–521. doi:10.1016/j.tourman.2008.10.009

Dodds, R., Ali, A., & Galaski, K. (2018). Mobilizing knowledge: Determining key elements for success and pitfalls in developing community-based tourism. *Current Issues in Tourism*, *21*(13), 1547–1568. doi:10.1080/13683500.2016.1150257

Frenzel, F., & Koens, K. (2012). Slum tourism: Developments in a young field of interdisciplinary tourism research. *Tourism Geographies*, *14*(2), 195–212. doi:10.1080/14616688.2012.633222

García, L., Sarabia, M., & Mendoza, A. (2017). Turismo comunitario y responsabilidad Social en Santa Elena: Un modelo para armar. *Estudios Turísticos*, *211-212*, 171–187.

Giampiccoli, A., & Mtapuri, O. (2014). The role of the international cooperation in community-based tourism. *Mediterranean Journal of Social Sciences*, *5*(4), 638–644. doi:10.5901/mjss.2014.v5n4p638

Goodnee, W. (2016). Corporate Social Responsibility as a Mechanism of Community Development: A Study of the Nigeria Liquefied Natural Gas Limited's CSR for the Bonny Kingdom, Niger Delta, Nigeria. *Journal of Corporate Responsibility and Leadership*, *3*, 50–68.

Gross, S., & Grimm, B. (2018). Sustainable mode of transport choices at the destination–public transport at German destinations. *Tourism Review*, *73*(3), 401–420. doi:10.1108/TR-11-2017-0177

Gursoy, D., Jurowski, C., & Uysal, M. (2002). Resident attitudes: A structural modeling approach. *Annals of Tourism Research*, *29*(1), 79–105. doi:10.1016/S0160-7383(01)00028-7

Hall, C. M. (2019). Constructing sustainable tourism development: The 2030 agenda and the managerial ecology of sustainable tourism. *Journal of Sustainable Tourism*, *27*(7), 1–17. doi:10.1080/09669582.2018.1560456

Hall, M. (2009b). The public policy context of tourism entrepreneurship. In I. Ateljevic & S. J. Page (Eds.), *Tourism and entrepreneurship* (pp. 243–263). Butterworth-Heinemann.

Hashemkhani Zolfani, S., Sedaghat, M., Maknoon, R., & Zavadskas, E. K. (2015). "Sustainable tourism: A comprehensive literature review on frameworks and applications", *Economic research-Ekonomska. Istraz̆ivanja*, *28*(1), 1–30.

Hernández, R. (2015). *La complejidad del concepto de desarrollo. Curso Planificación, Gobierno y Desarrollo*. ILPES/CEPAL.

Hjalager, A. M. (2002). Repairing innovation defectiveness in tourism. *Tourism Management*, *23*(5), 465–474. doi:10.1016/S0261-5177(02)00013-4

Hjalager, A. M. (2010). A review of innovation research in tourism. *Tourism Management*, *31*(1), 1–12. doi:10.1016/j.tourman.2009.08.012

Huh, C., & Vogt, C. A. (2008). Changes in residents' attitudes toward tourism over time: A cohort analytical approach. *Journal of Travel Research*, *46*(4), 446–455. doi:10.1177/0047287507308327

Hunt, C., & Stronza, A. (2014). Stage-based tourism models and resident attitudes towards tourism in an emerging destination in the developing world. *Journal of Sustainable Tourism*, *22*(2), 279–298. doi:10.1080/09669582.2013.815761

Jaafar, M., Rasoolimanesh, S. M., & Lonik, K. A. T. (2015). Tourism growth and entrepreneurship: Empirical analysis of development of rural highlands. *Tourism Management Perspectives*, *14*, 17–24. doi:10.1016/j.tmp.2015.02.001

Johns, N., & Mattson, J. (2005). Destination development through entrepreneurship: A comparison of two cases. *Tourism Management*, *26*(4), 605–616. doi:10.1016/j.tourman.2004.02.017

Kim, K., Uysal, M., & Sirgy, M. J. (2013). How does tourism in a community impact the quality of life of community residents? *Tourism Management*, *36*, 527–540. doi:10.1016/j.tourman.2012.09.005

Klein, L., & Dodds, R. (2018). Blue flag beach certification: An environmental management tool or tourism promotional tool? *Tourism Recreation Research*, *43*(1), 39–51. doi:10.1080/02508281.2017.1356984

Kline, C., & Milburn, L. A. (2010). Ten categories of entrepreneurial climate to encourage rural tourism development. *Annals of Leisure Research*, *13*(1/2), 320–348. doi:10.1080/11745398.2010.9686850

Koens, K., Steinbrink, M., & Rogerson, C. M. (2015). Slum tourism: State of the art. *Tourism Review International*, *18*(4), 237–252. doi:10.3727/154427215X14230549904017

Koh, K., & Hatten, T. (2002). The tourism entrepreneur: The overlooked player in tourism development studies. *International Journal of Hospitality & Tourism Administration*, *3*(1), 21–48. doi:10.1300/J149v03n01_02

Komppula, R. (2014). Role of individual entrepreneurs in the development of competitiveness for a rural tourism destination – A case study. *Tourism Management*, *40*, 361–371. doi:10.1016/j.tourman.2013.07.007

Lane, B., & Kastenholz, E. (2015). Rural Tourism: The evolution of practice and research approaches-towars a new generation concept? *Journal of Sustainable Tourism*, *23*(8-9), 1133–1156. doi:10.1080/09669582.2015.1083997

Lashley, C., & Rowson, B. (2010). Lifestyle businesses: Insights into Blackpool's hotel sector. *International Journal of Hospitality Management*, *29*(3), 511–519. doi:10.1016/j.ijhm.2009.10.027

Lee, T. H. (2009b). A structural model for examining how destination image and interpretation service affect future behavior: A case study of taiwan's Taomi eco-village. *Journal of Sustainable Tourism*, *17*(6), 727–745. doi:10.1080/09669580902999204

Lee, T. H. (2013). Influence analysis of community resident support for sustainable tourism development. *Tourism Management*, *34*, 37–46. doi:10.1016/j.tourman.2012.03.007

Lee, T. H., & Hsieh, H. P. (2016). Indicators of sustainable tourism: A case study from a taiwan's wetland. *Ecological Indicators*, *67*, 779–787. doi:10.1016/j.ecolind.2016.03.023

Lepp, A. (2007). Residents' attitudes towards tourism in Bigodi village, Uganda. *Tourism Management*, *28*(3), 876–885. doi:10.1016/j.tourman.2006.03.004

Liu, A., & Wall, G. (2006). Planning tourism employment: A developing country perspective. *Tourism Management*, *27*, 159–170.

Long, P., Perdue, R., & Allen, L. (1990). Rural resident tourism perceptions and attitudes by community level of tourism. *Journal of Travel Research*, *28*(3), 3–9.

López-Guzmán, T., & Sánchez Cañizares, S. M. (2009a). Desarrollo socioeconómico de las zonas rurales con base en el turismo comunitario. Un estudio de caso de Nicaragua. *Cuadernos de Desarrollo Rural*, *6*(62), 81–97.

Lundberg, E. (2015). *The level of tourism development and resident attitudes: A comparative case study of coastal destinations*. Academic Press.

Martíınez-Pérez, Á., Elche, D., & García-Villaverde, P. M. (2019). From diversity of interorganizational relationships to radical innovation in tourism destination: The role of knowledge exploration. *Journal of Destination Marketing & Management*, *11*, 80–88.

Meschkank, J. (2011). Investigations into slum tourism in mumbai: Poverty tourism and the tensions between different constructions of reality. *GeoJournal*, *76*(1), 47–62. doi:10.100710708-010-9401-7

Ministerio de Turismo de Ecuador (MINTUR). (2017). *Boletín de Estadísticas Turísticas 2011-2015*. Quito Accsesada el 20 de febrero del 2020 https://servicios.turismo.gob.ec/descargas/Turismo-cifras/AnuarioEstadistico/Boletin-Estadisticas-Turisticas-2011-2015.pdf

MINTUR. (2010). *Reglamento para los centros turísticos comunitarios, Pub. L. No. Acuerdo Ministerial 16 Ecuador*. Recuperado el 8 de marzo 2020: http://www.quito-turismo.gob.ec/descargas/lotaip-2013diciembre/BASELEGAL/REGLAMENTO DE CENTROS DE TURISMO COMUNITARIO.pdf

Mottiar, Z. (2015). The importance of local area as a motivation for cooperation among rural tourism entrepreneur. *Tourism Planning & Development, 12*, 1–16.

Mottiara, Z., Bolukb, K., & Klinec, C. (2018). The roles of social entrepreneurs in rural destination development. *Annals of Tourism Research, 68*, 77–88. doi:10.1016/j.annals.2017.12.001

Muler Gonzalez, V., Coromina, L., & Galí, N. (2018). Overtourism: Residents' perceptions of tourismimpact as an indicator of resident social carrying capacity-case study of a Spanish heritage town. *Tourism Review, 73*(3), 277–296. doi:10.1108/TR-08-2017-0138

Murphy, P. (1985). *Tourism: A community approach*. Methuen.

Muthuri, J. N. (2007). Corporate Citizenship and Sustainable Community Development. *Journal of Corporate Citizenship, 2007*(28), 73–84. doi:10.9774/GLEAF.4700.2007.wi.00008

Ngo, T., Lohmann, G., & Hales, R. (2018). Collaborative marketing for the sustainable development of community-based tourism enterprises: Voices from the field. *Journal of Sustainable Tourism, 26*(8), 1325–1343. doi:10.1080/09669582.2018.1443114

Nicholas, L. N., Thapa, B., & Ko, Y. J. (2009). Residents' perspectives of a world heritage site: The pitons management area, st. Lucia. *Annals of Tourism Research, 36*(3), 390–412. doi:10.1016/j.annals.2009.03.005

Nunkoo, R. (2017). Governance and sustainable tourism: What is the role of trust, power and socialCapital? *Journal of Destination Marketing & Management, 6*(4), 277–285. doi:10.1016/j.jdmm.2017.10.003

Nunkoo, R., & Gursoy, D. (2012). Residents' support for tourism: An identity perspective. *Annals of Tourism Research, 39*(1), 243–268. doi:10.1016/j.annals.2011.05.006

Nunkoo, R., Ramkissoon, H., & Gursoy, D. (2012). Public Trust in Tourism Institutions. *Annals of Tourism Research, 39*(3), 1538–1564. doi:10.1016/j.annals.2012.04.004

Pavlovich, K. (2003). The evolution and transformation of a tourism destination network:The Waitomo Caves, New Zealand. *Tourism Management, 24*(2), 203–216. doi:10.1016/S0261-5177(02)00056-0

Power, S., Di Domenico, M., & Miller, G. (2017). The nature of ethical entrepreneurship in tourism. *Annals of Tourism Research, 65*, 36–48. doi:10.1016/j.annals.2017.05.001

Reggers, A., Grabowski, S., Wearing, S. L., Chatterton, P., & Schweinsberg, S. (2016). Exploring outcomes of community-based tourism on the kokoda track, Papua New Guinea: A longitudinal study of participatory rural appraisal techniques. *Journal of Sustainable Tourism, 24*(8–9), 1139–1155. doi:10.1080/09669582.2016.1145229

Richards, L., & Dalbey, M. (2006). Creating Great Places: The Role of CitizeParticipation. *Community Development (Columbus, Ohio), 37*(4), 18–32. doi:10.1080/15575330609490193

Rivera, J. (2002). Assessing a voluntary environmental initiative in the developing world: The Costa Rican certification for sustainable tourism. *Policy Sciences, 35*(4), 333–360. doi:10.1023/A:1021371011105

Roux Fanny. 2013. Turismo comunitario ecuatoriano, conservación ambiental y defensa de los territorios. Federación Plurinacional de Turismo Comunitario del Ecuador (FEPTCE).

Ruhanen, L., Weiler, B., Moyle, B. D., & McLennan, C. L. J. (2015). Trends and patterns in sustainable tourism research: A 25-year bibliometric analysis. *Journal of Sustainable Tourism, 23*(4), 517–535. do i:10.1080/09669582.2014.978790

Ruiz Ballesteros, E., & Cáceres-Feria, R. (2016). Community-building and amenity migration in community-based tourism development. An approach from southwest Spain. *Tourism Management, 54*, 513–523. doi:10.1016/j.tourman.2016.01.008

Ruiz Ballesteros, E., & Solís Carrión, D. (2007). *Turismo comunitario en Ecuador. Desarrollo y sostenibilidad social*. Ediciones Abya-Yala.

Russell, R., & Faulkner, B. (2004). Entrepreneurship, chaos and the tourism area lifecycle. *Annals of Tourism Research, 31*(3), 556–579. doi:10.1016/j.annals.2004.01.008

Ryan, T., Mottiar, Z., & Quinn, B. (2012). The dynamic role of entrepreneurs in destination development. *Tourism Planning & Development, 9*(2), 119–131. doi:10.1080/21568316.2011.630747

Sarabia, M., García, L., & Bravo, O. (2019). Responsabilidad Social y Turismo comunitario en la provincia de Santa Elena (Ecuador). *Revista Espacios., 40*(30), 2.

Shaw, G., & Williams, A. M. (1998). Entrepreneurship, small business culture and tourism development. In D. Ioannides & K. G. Debbage (Eds.), *The economic geography of the tourist industry* (pp. 235–255). Routledge. doi:10.4324/9780203398425.ch13

Summers, G. F., & Wilkinson, K. P. (1992). Review: The community in rural America. *Contemporary Sociology, 21*(2), 212–213. doi:10.2307/2075444

Teh, L., & Cabanban, S. (2007). Planning for sustainable tourism in southern pulau Banggi: An assessment of biophysical conditions and their implications for future tourism development. *Journal of Environmental Management, 85*(4), 999–1008. doi:10.1016/j.jenvman.2006.11.005 PMID:17204361

Theodori, G. (2008). Constraints to the Development of Community. *Community Development (Columbus, Ohio), 39*(1), 63–81. doi:10.1080/15575330809489742

Theodori, G. L. (2005). Community and Community Development in Resource-Based Areas: Operational Definitions Rooted in an Interactional Perspective. *Society & Natural Resources, 18*(7), 661–669. doi:10.1080/08941920590959640

Thomas, R., Shaw, G., & Page, S. J. (2011). Understanding small firms in tourism: A perspective on research trends and challenges. *Tourism Management, 3*(5), 963–976. doi:10.1016/j.tourman.2011.02.003

Thompson, C., Johnson, T., & Hanes, S. (2016). Vulnerability of fishing communities undergoing gentrification. *Journal of Rural Studies, 45*, 165–174. doi:10.1016/j.jrurstud.2016.03.008

Tinsley, R., & Lynch, P. A. (2007). Differentiation and tourism destination development: Small business success in a close-knit community. *Tourism and Hospitality Research, 8*(3), 161–177. doi:10.1057/thr.2008.26

Tolkach, D., & King, B. (2015). Strengthening community-based tourism in a new resource-based island nation: Why and how? *Tourism Management, 48*, 386–398. doi:10.1016/j.tourman.2014.12.013

Tolkach, D., & King, B. (2015). Strengthening community-based tourism in a new resource-based island nation: Why and how? *Tourism Management*, *48*, 386–398. doi:10.1016/j.tourman.2014.12.013

Tönnies, F. (1988). *Community and Society: Gemeinschaft and Gesellschaft; with a new introduction of John Samples*. Transaction Publishers.

Vargas-Sánchez, A., Plaza-Mejia, M. D. L. Á., & Porras-Bueno, N. (2009). Understanding residents' attitudes toward the development of industrial tourism in a former mining community. *Journal of Travel Research*, *47*(3), 373–387. doi:10.1177/0047287508322783

Von Schuckmann, J., Barros, L. S., Dias, R. S., & Andrade, E. B. (2018). from slum tourism to smiley selfies: The role of social identity strength in the consumption of morally ambiguous experiences. *Journal of Consumer Psychology*, *28*(2), 192–210. doi:10.1002/jcpy.1016

Wang, C. C., Cater, C., & Low, T. (2006). Political challenges in community-based ecotourism. *Journal of Sustainable Tourism*, *24*(11), 1555–1568. doi:10.1080/09669582.2015.1125908

Weiermair, K., Keller, P., Pechlaner, H., & Go, F. (2010). *Innovation and entrepreneurship: Strategies and processes for success in tourism*. Schmidt.

Wilkinson, K. P. (1991). *The community in rural America*. Greenwood Publishing Group.

Woo, E., Kim, H., & Uysal, M. (2015). Life satisfaction and support for tourism development. *Annals of Tourism Research*, *50*, 84–97. doi:10.1016/j.annals.2014.11.001

Woo, E., Uysal, M., & Sirgy, M. J. (2018). Tourism impact and stakeholders' quality of life. *Journal of Hospitality & Tourism Research (Washington, D.C.)*, *42*(2), 260–286. doi:10.1177/1096348016654971

Yumisaca, J., & Mendoza, E. (2016). La hostelería y el sector intermediario como fortaleza del turismo en Santa Elena: caso Salinas y Montañita. In UPSE CECATERE, Investigaciones Turísticas del Ecuador: fortaleciendo la innovación de los destinos turísticos (pp. 69 -84). La Libertad: Imprenta López.

Ziegler, R. (2017). Social innovation as a collaborative concept. *Innovation (Abingdon)*, *30*(4), 388–405. doi:10.1080/13511610.2017.1348935

ENDNOTES

1 Ministry of Tourism.

2 Ministry of Agriculture and Livestock.

3 Ministry of Economic and Social Inclusion.

4 National Institute of Popular and Solidarity Economy.

5 Municipal Tourism Company.

6 (Community Assembly), groups in charge of the administration of all activities are recognized and authorized and the destination of the funds generated for the Community is regulated.

Chapter 3
The Attractiveness of Extreme Wind Sports Linked to Social Responsibility:
Innovations in Coastline Tourism

Luciano Barbosa Almeida
Universidade Estadual do Ceará, Brazil

Jakson Renner Rodrigues Soares
Universidade da Coruña, Spain & Universidade Estadual do Ceará, Brazil

ABSTRACT

The State of Ceará, located in the Brazilian Northeast, has a high tourist potential due to the beauty of its beaches and the great geomorphological differences within the territory. In addition, there are warm winds that allow the practice of wind sports throughout the state seacoast. This chapter presents three categories of analysis here interrelated with planning: wind sports, the use of social projects to improve the tourist experience, and the participation of the local community in social projects. The main objective of this chapter is to present innovations that might be implemented with the creation of a tourist route based on existing structures in the state of Ceará. The focus is to use the attractiveness of extreme wind sports linked to social responsibility (NGOs) and the strengthening of the tourist image. This chapter presents the social projects that relate tourism, social projects, and planning. Thru this job, the authors hope to give light to other destinations, which might perform the practice of benchmarking to face similar problems.

DOI: 10.4018/978-1-7998-4855-4.ch003

INTRODUCTION

The tourist activity is dynamic, complex and full of details, which requires careful planning. This is because tourist-receiving territories need to efficiently plan, organize, manage and disseminate tourist businesses in order to remain present in an environment of rivalry. Likewise, the economic activities of tourism generate socioeconomic improvements, such as new job opportunities, improved quality of life and increased purchasing power for the tourist population. However, we can also testify that uncontrolled exploitation causes irreversible damage to the socioeconomic and environmental structures of the regions that receive large tourist flows. Hence, the need for planning and dissemination that result in a conscious exploration of tourist resources through effective regulatory public policies, capable of curbing harmful and harmful tourism practices to the place. In addition, to achieve sustainable tourism development, the participation of all agents involved in the relationship with the activity is required. Thus, as it is common knowledge, the host community is a stakeholder in the development.

This work presents the Ceará's existing tourist potential and its innovations, such innovative practices are focusing on extreme sports and on the creation of a tourist route to manage paths and flows to be traveled by tourists who value sports practices known as extreme wind sports, such as kitesurfing, windsurfing, paragliding and hang gliding.

In order to elucidate the real functional characteristic of tourist routes, it is stated that these routes are designed to facilitate access, movement, organization and management of resources, in order to increase the travel time of tourists, thus causing the best experiences possible periods of travel. The creation of scripts would therefore be a categorization strategy, widely used in marketing and that is able to attract the eye to a product or service, in the individual's mind, the product offered stands out due to this route is an organized and prepared for his or hers enjoyment.

From the development of sports tourism on the coast of Ceará, we identified the inclusive factor that gave developed tourism to local labor. In this sense, young people at risk of exclusion from coastal beaches and from the outback are being professionalized, thus, these areas are achieving quality tourism. Today we can already witness in the regions of Cumbuco Beach, Guagiru Island and Titanzinho Beach successful examples in which wind sports rescue young people who were at risk of marginalization, for a life full of new positive experiences. In addition, we also identified in the proposed script the presence of the 4 "R" of the circular economy (reduce, reuse, repair and recycle) in the design of products and services related to sports tourism in this region.

However, the main objective of this chapter is to present innovations that might be implemented with the creation of a Tourist Route based on existing structures in the state of Ceará. And with the focus on the extreme wind sports linked to social responsibility (NGOs) and the strengthening of the tourist image of the State of Ceará, carried out by encouraging visitors to spread the good image of the experiences lived on their social networks.

Thus, this chapter will present readers with the possibility of encouraging the use of their individual electro-technological devices, such as laptops, smartphones and tablets that, using electronic media, have the ability to positively disclose or destroy a tourist destination almost instantly. Innovations in the dissemination of the image of the tourist destination will be presented, as there is a need for a well-defined and well-worked virtual image in technological communication networks.

The image must truly attract the potential tourist, as it is understood that a dissatisfied tourist will not return to the visited site and consequently he or she will disclose their negative experiences to everyone in their life cycle. To avoid a negative image wave, it is necessary to create and infuse messages that can add value to the destination of curiosity and the future visit, Perinotto et al. (2017). In this sense, the participation of the local community in the development and propagation of the image of Sport Tourism in Ceará is an innovation, since that one of the aspects that decant the look of the potential tourist to one destination or another is the authenticity of the place.

Finally, to identify potential differentiators of Ceará's Coast and categorize their sporting experiences, we will use as examples 3 successful Routes; the Emotions Route, it links the coasts of 3 Brazilian States passing thru Ceará / Piauí / Maranhão, the Coffee Route in the Baturité Massif Mountain Chain and the Route of Mirantes (Watch Point) on top of the Ibiapaba Mountain Chain. They came about due to the structure and experience in planning and managing tourist itineraries found in Ceará. In other words, competition and collaboration between tourism stakeholders are key aspects in the development of this territory, where tourism companies multiply their offerings and launch various products based on local and regional offers, whether cultural or geographical, it is tourism focused on new experiences such as cultural, gastronomic, sports, among others.

This chapter is organized into 6 sections. In this first one, the theme is to be addressed and the structure of the chapter are presented. The second point discusses the interconnection between planning as a solution to maximize the well-being of the local population who hosts tourism. In the third section, tourist destinations and wind sports are introduced as tourism inputs. Likewise, here is found the projects which are developed from these sports, it is also found innovating and the addition of value to the touristic product from the experience with the host communities. In the fourth section is exposed a discussion on improving the attractiveness of the destination through the union of wind sports with social projects applied to tourism. The last section presents the chapter's conclusions and, finally, the bibliographic references used.

PLANNING AND TOURISM, LOOKING FOR SOLUTIONS TO MAXIMIZE WELL-BEING

The Department of Economic and Social Affairs of the Statistics Division of the WTO by the International Recommendations for Tourism Statistics of 2008 published by UNWTO (2010), conceptualizes tourism as the set of activities that people perform during their travels and their stays in different places. Theses stays are in other surroundings than their usual home place for the purpose of leisure, business or other reasons, unrelated to a paid activity in the place visited. According to Barretto (1991), tourism is essentially the movement of people with the purpose of meeting their needs during travel. Thus, it is a phenomenon of interaction between the tourist and the host community, which outcomes in activities resulting from this interaction. In this sense, it is the event derived from the exchange and interaction relationship, where such exchange results in new experiences for travelers that differ from their daily lives.

Kim and Kim (2013) point out that the tourism business is one of the foundations for regional development and clearly emphasize that not only is economic growth fostered, but also the social and environmental sustainability of the regions, thus it requires control and planning.

Tourism is one of the activities that most contributes to the economic development of the state of Ceará (Fernandes, 2014).

De Masi (2000), in the book "Creative Leisure", states that humankind started the planning activity as soon as there was the discovery of grains and seeds, approximately 6000 years before Jesus Christ.

Beni (1999) states that tourism planning is related to three essential and distinct points: setting objectives, defining courses of action and determining feedback. Even so, the result of the competence to plan is the possibility of establishing favorable conditions to achieve pre-established goals and objectives. In addition, as Ruschmann (2001, p. 66) points out, "it is from planning that it is possible to establish favorable conditions to achieve proposed objectives".

Whenever planning tourism, one must first draw an elementary or preliminary plan of the present situation taking into account the local characteristics of all tourist links: attractions, transport, accommodation, food and services. When preparation of the elementary plan is done, and the second phase, which is the diagnosis, will begin, which is characterized by the analysis and scrutiny of the collected data. Thus reaching the final step, which is the prognosis and generation of future forecasts from the present already known (Beni, 1998). That is, first all the characterization of the place is made, therefore, the analysis of the data collected in the previous step and in the third step an attempt to predict the future based on the knowledge of reality for later proposal of guidelines. Finally, as Clavé & Calabuig (2005) recall, tourism can help to provide solutions to urban and social problems. It may also be able to diversify the economy and favor the social equality of a territory.

The economic aspects of this activity bring with it many positive as well as negative issues. Economic activities related to tourism generate socioeconomic improvements, such as new job opportunities, improved quality of life and increased purchasing power for the tourist population. On the contrary, we can testify that uncontrolled exploitation causes irreversible consequences in the socioeconomic and environmental structures of the regions receiving tourist flows, as pointed out by Silveira (2007). Likewise, it is understood that the problems and negative impacts of tourism can be overcome if the benefits produced are used effectively, in this sense, improving local development. That is why it is necessary that management be planned to implement adequate policies that benefit both the tourist and the local population.

To involve the local community in the planning, management and execution of tourism could be a suitable solution related to the sustainability of this sector. Many studies relate the intention of community participation to the perception of the effects of tourism on their lives. That is, while part of the community sees positive impacts as economic opportunities (social integration, improving the quality of life, increasing the number of attractions, infrastructure and services), another part of this community behaves just the opposite, disinterested in development tourism in their region. This is because the effects of tourism are more considered as injuries to their well-being. That is, they suffer due to a series of disturbances (pollution of the environment, water pollution, loss of their culture ...). Therefore, the commitment of tourism planning is related to the reduction of negative impacts and the increase of positive impacts of tourism, since, as Eshliki and Kaboudi (2012) point out, the negative effects of tourism on the community cause a lesser inclination of society to participate in tourism development programs. Hence the need for planning and dissemination that results in a conscious exploitation of tourism resources through effective public regulatory policies capable of constraining harmful and damaging tourism practices.

The object, the context and the appreciation of the potential of the existing tourist branch in Ceará are presented, focusing on this study on extreme sports, sports education as an opportunity for socioeconomic improvement of local youth and the analysis of the feasibility of creating a tourist route. Such tourist itinerary will have the intention of managing paths and flows to be followed by tourists who value sports practices known as extreme wind sports, such as “kitesurf”, “windsurf”, “paragliding”, “parasail” and hang gliding.

As seen previously, tourist routes are designed to facilitate access, movement, organization, resource management and identification of new opportunities. In this way, they are able to enhance the travel time of tourists, thus causing the best possible experiences during the period of the trip. That said; it allows one to say that tourist routes are also positive tools to leverage the attractiveness of a tourist destination and that is when the routes are well commanded. A well-managed route can become powerful appliances for social inclusion. Furthermore, they can also communicate a well lapidated, truthful and real image of the place that the tourists visit.

CEARA’S TERRITORY AND ITS WINDS: INPUTS THAT DIFFERENTIATE THE DESTINATION

Ceará is one of the most diverse destinations in the country due to its nature (coast, backlands and mountains), however, the winds are considered natural elements that are part of the attractions of this place. It is worth mentioning that the Route of Emotions which covers the coast of 3 states of Ceará, Piauí and Maranhão. It is a path through where tourists are indulged with kilometers and kilometers of white dunes, dotted by freshwater lagoons, which sometimes are blue and other times are green. Here, one starts to explore the scenarios, which are even more exuberant between the months of June and September, when the lagoons are completely full after the regional rain seasons has ended.

On the way it is possible to appreciate the beach of Jericoacoara, considered one of the 10 most beautiful beaches in the world, capable of attracting sportsmen from all over the world, mainly for the practice of “windsurfing” and “kitesurfing”. The beach is ideal for sailing and practicing radical maneuvers (Redação, 2009). In Jericoacoara’s region, the activities take place at sea and in the lagoons. In addition it, there are other coastal areas in Ceará which are intensely sought in its more than 570 coastal kilometers. One can start from the west end of the state and head towards the east end while crossing the beaches of Camocim, Preá, Guajiru Island, Icaraí de Amontada, Lagoinha, Paracuru, Taipa, Cumbuco, the coastal beaches of Fortaleza, Beach Park, Caponga, Barra Nova and Canoa Quebrada. Also according to this Redação (2009), in the water park “Beach Park”, located in Aquiraz / CE, the Brazilian stage of the “Kitesurf” World Tour of the Professional “*Kiteboard Rides Association*” was hosted for four consecutive years.

In this perspective, Ceará is for the winds just as like Hawaii is for the sea waves and, therefore, the State is known by the wind sports practitioners as the “Hawaii of the winds” (Noronha, 2010). According to the Cearense Foundation of Meteorology and Water Resources (Funceme, 2019), with wind gusts that reach up to 60km / h, in the peak season, the months of August and September, it is difficult to go to a beach in Ceará and not see someone practicing “windsurf” and kitesurf.

The municipality of Quixadá (located in the interior of the State, in the sertão region), Serra da Meruoca and Serra de Grande da Ibiapaba (located in the northern sertão of the State), are considered today one of the best places in the world for the practice of free flight (Neto, 2016). In order to attest and enjoy the unique conditions found in Ceará from north to south, the World Free Flight Rally, brought to the region 30 of the best world athletes in September 2003, the event went by the municipalities of Juazeiro do Norte, Iguatu, Quixadá, Sobral and Camocim (WebVenture, 2003). The region with constant temperature and wind provides ideal conditions all year round for hang gliding and paragliding pilots to fly at high altitudes and over long distances using the pockets of hot wind (Duarte, 2002). In addition to flights, there are trails to observe, archaeological sites and rocks for abseiling and climbing.

The media and specialized tourism agencies in the State highlight as a marketing advisements that the scenario for a "paragliding" flight at Canoa Quebrada Beach is more than favorable, with beautiful cliffs in reddish tones, a sea with brilliant waters, a sun always alive and a heaven to be envied due to its beauty (Vitorino, 2019). They also mention to the more adventurous tourists that they can enjoy these attractions on the beach or check entire coast out entire from the skies above on board of a paraglide.

According to Férias Brasil (2019), the constant wind that invades the wide beaches and with few reefs makes the Ceará coast a perfect point for the practice of nautical and extreme sports. Surfing gathers fans on the east coast beaches, such as Farol Beach, Porto das Dunas Beach, Prainha Beach and Fontes Beach, while the on the west coast attracts fans of "windsurfing" and "kitesurfing", which take waves from Praia do Futuro, Cumbuco Beach, to Taíba Beach to Jericoacoara Beach.

As Walraven (2009) points out, Cumbuco Beach, on the west coast, has hosted Brazilian stages of important World Kitesurfing Championships annually. For those who want to venture out and start practicing, it is important to know that it is considered one of the most aerobic water sports and can bring many physical benefits. However, it is important to understand that "windsurfing" and "kitesurfing" are a radical sport and as such, they need to be practiced safely (Andrade, 2010). Beginners should look for qualified people to teach the classes.

Reflecting on the data above, it is possible to see that spread across Ceará's territory is a true source for the practice of wind sports; therefore, it is possible to divide and regionalize the wind potential for extreme sports tourism. This organized division gives support to create a plan management that may become to be known as the Winds Route.

Sports such as "kitesurf", "windsurf", "paragliding", "parasail" and hang gliding are practiced in all areas studied in Ceará. Beach regions such as Canoa Quebrada (Municipality of Aracati), Pontal do Maceió Beach (Municipality of Fortim), Caponga Beach (Municipality of Cascavel), Morro Branco Beach (Municipality of Beberibe), Beach Park Beach (Municipality of Aquiraz), Praia do Futuro (Municipality of Fortaleza), Praia de Cumbuco (Municipality of Caucaia), Praia da Taíba (Municipality of Taíba) and the entire west coast from the Municipality of Paracuru to the Municipality of Camocim are perfect areas for these sports practices. In addition, the inland Municipalities of Quixadá, Sobral, Meruoca and Tianguá are suitable places for the practice of "paragliding" and hang gliding.

SOCIALLY RESPONSABLE TOURISTIC ROUTES: PRESENTATION OF PROJECTS

The aspect of being a tourist route focused on social responsibility and social inclusion arises from the possibility of integrating the local community in tourist activities and in the development of the place. In the development of tourism in the traditional coastal space of this region, better known as Sun and Beach Tourism, many times these local communities were taken for granted as obvious and they were left with underemployment or less qualified jobs. However, with these new tourist routes, the local idle labor, like the young people on the beaches of the coast of Ceará, receives opportunities that were not previously given to them. These young people ended up becoming an easy target for the hardships of modern life, among them, the crime of drug trafficking and addiction. Currently, it is already possible to witness social inclusion actions using water sports from the wind on some of the beaches of the Litoral Cearense, from Praia do Ilha do Guagirú, in the municipality of Itarema in the west of the coast, in Praia de Titanzinho in Fortaleza and up to Caponga Beach, in the Municipality of Cascavel. This union of social projects and wind sports is an innovation already existing in the destination that benefits countless families.

Vem Voar pro-Kitesurf Project

The "*Vem Voar Pro-Kitesurf*" Project was born in 2016 to teach kitesurfing to children and adolescents in situations of social vulnerability and people with special needs. As Soulkite (2019) points out, it aims to offer an opportunity to learn kitesurfing techniques, contributing to the psycho-physical-social development of children and adolescents; it focuses on the inclusion of the sport as forms of social, professional and social inclusion thru leisure and the practice of this sport. Finally, the Vem Voar Project pro-Kitesurf was nominated for the Best Social Project Award of 2019. It is located at Praia da Barra on Ilha do Guagiru in the municipality of Itarema. It has 1 fixed teacher and an oscillating number of volunteer teachers that at times can vary from 5 to 10 instructors. In its 4 years of existence, the project has benefited around 80 young fellas.

In the early years, more specifically in the year of 2016, the project reached only 3 students. However, in 2017, during a low season of strong winds, the project went around the beaches of the west coast including the Guajiru Island to gain new students. This trip was done along the vast coastal line with the help of a typical transport from the area known as buggy, it can easily moves through the abundant local sand dunes of the "landscape". During this TUR along the west coast of Ceará, the project was able to reach 13 children of local anglers from the secluded communities, such as Caetanos in Amontada, Maceió Beach in Camocim and Bitupita in Barroquinha. They all benefited from kitesurf lessons.

In 2018, the project was expanded in scope and it started to serve everyone as potential benefiters, all ages were included as well. That year, the project reached 25 users, and its routine became very intense especially in the first half of that year. In 2019, tourists who practice kitesurfing and were staying at the inns on Guajiru Island were invited to participate as volunteer instructors. These volunteer tourists train future teachers from various local communities, this year an exact number of 33 youngsters were assisted at the project. This opportunity to bring tourists closer to the local community categorizes the kind of segment of tourism in which these travelers are looking for. The wind sports and supportive tourism.

Escola Beneficente de Surf Titanzinho Project

The "*Escola Beneficente de Surf Titanzinho*" *Project* (E.B.S.T.) arose from the union of the desire to make a difference in the world and the love for sports on the part of its founder João Carlos Fera. He says that since he was very young he had the dream of going to poor and distant regions of the African continent to help people improve their lives. However, in the early 90s he made the decision to go to one of the most deprived regions of Brazil, which is the Sertão in Ceará State, an arid land, depleted and devoid of any structure for those who live there. He imagined that with this movement from the coast, where he grew up, to the arid interior of the state, he would be preparing for the future dream

Figure 1. Project Vem Voar pro-Kitesurf - Guagiru Island (Municipality of Itarema / Ce)
Source: Tomaz André Pinheiro's personal archive - Founder of Vem Voar pro-Kitesurf (2019)

Figure 2. Project Vem Voar pro-Kitesurf - Guagiru Island (Municipality of Itarema / Ce)
Source: Tomaz André Pinheiro's personal archive - Founder of Vem Voar pro-Kitesurf (2019)

Figure 3. Project Vem Voar pro-Kitesurf - Guagiru Island (Municipality of Itarema / Ce)
Source: Tomaz André Pinheiro's personal archive - Founder of Vem Voar pro-Kitesurf (2019)

of living in Africa to work with education, sports and thus contribute to the social improvement of the local population. He imagined that his time in the hinterland of Ceará had served him as an experiment to understand and rationalize if that was what he really wanted.

When he arrived in the interior of Sertão, Ceará's Outback, he was faced with such a shortage of quality live that he soon understood that what he longed-for, the possibility of helping others, was right there in Brazil. In his words, he said "I had arrived in Africa[1]" and I did not have to board any transatlantic flight. In early 1995, after he took this step, he decided to return to Fortaleza, Capital of Ceará, to his childhood neighborhood, which is called Vicente Pinzon, where is located the Praia do Titanzinho and on February 2nd, 1995, the E.B.S.T. was inaugurated.

According to EBST (2016), it aims to reduce unemployment time and rescue the citizenship of young people at risk through the sport of surfing, using it as a tool for social inclusion. At the same time, it promotes the best use of leisure and leisure time with social education and environmental responsibility. Young residents of the community at Praia do Titanzinho are the project´s target audience, they must be over 7 years of age and must be attending school and they show the willingness to play sports during idle hours. The Central Única das Favelas (CUFA) awarded this project the 2011 Anu Dourado Award, chosen as the most efficient philanthropic entity in the state of Ceará.

Currently, E.B.S.T employs 4 employees and in its 25 years of activity has served more than 500 needy young people. Surf lessons, surf kata (imaginary surf), capoeira, physical education and languages are some of the activities offered to young people served by the project. In order to broaden the scope of the actions implemented, it also started to provide services for young people's families, such as adult literacy, Yoga or psychological support for parents and children. Family services are provided by 8 volunteer professionals who donate their services free of charge to the entity's visitors (EBST, 2016).

Figure 4. E.B.S.T. Project - Praia do Titanzinho (Municipality of Fortaleza / Ce)
Source: Personal archive of João Carlos Fera - Founder of E.B.S.T. (2019)

Figure 5. Kata surf lessons (imaginary surfing)
Source: Personal archive of João Carlos Fera - Founder of E.B.S.T. (2019)

Figure 6. Surf lessons (Balance training)
Source: Personal archive of João Carlos Fera - Founder of E.B.S.T. (2019)

Surf Na Escola Project

The "*Surf na Escola*" Project at Paria da Caponga in the Municipality of Cascavel in Ceará, appeared when a childhood dream of three surfers, the Farney's brothers, materialized on August 17, 2009. The project left the three brothers imagination and it became a reality to promote social inclusion and the education of values, through sport.

The Surf at School Project is a non-profit organization that seeks to improve the sense of citizenship and social inclusion; it is aimed at children and young people from the public education network of Praia da Caponga (Waves 2009). Only students who maintain the required minimum performance average score enjoy the benefit of entering the sea. Through innovative actions, such as the monitoring of school performance and attendance by the institution, it keeps young people interested in school and at the same time interested in the practice sports. These criteria result in a greater school focus on the part of young people so that they can stay on the project.

According to Marcio Farney, in an interview for this chapter, the diversity of activities offered in the project aims to instigate the creativity of young people. Among them, we can check the theoretical and practical lessons of surfing, recreational activities such as music and theater classes, recreational tours.

All of them are aligned with pedagogical support made by volunteer teachers and with beach sports practices; these activities are available for children and young teenagers who are interested in participating. They just have to be enrolled and attending school regularly to be able to enjoy the project.

Cascavel and its beautiful beaches, including Praia da Caponga, are located in the RMF (Metropolitan Region of Fortaleza) which, due to their location belong to an area of high drug traffic, crime and other dangers. *Surf na Escola* sees its well-defined mission there, which is to keep the youth served and away from these afflictions that are typical of non-productive leisure. Occupying young people's free time, primarily with the sport of surfing, the Surf at School Project has a strong positive penetration in the local community. Since it serves more than 108 children and young people, in addition to 42 women and people with mobility restrictions.

Due to its valuable contributions, today the project has the support of the Municipality of Cascavel, the Praia da Caponga Surf Association (ASPC) and the surfwear brand Smolder.

Figure 7. Surf na Escola Project - Praia da Caponga (Municipality of Cascavel / Ce)
Source: Personal archive of Marcio Farney - Founder of Surf na Escola (2019).

Figure 8.
Source: Personal archive of Marcio Farney - Founder of Surf na Escola (2019).

All of these ideas started from the desire to contribute to the formation of young people and, today, they are successful projects. They link sports in the coastal area, social responsibility and the involvement of local communities as a tool to rescue youth who lived at risk for a life of fullness and full of new positive experiences. Connecting these social projects through the design of the Route of the Winds will generate benefits for both the tourist and the projects themselves. On the one hand, the tourist benefits from having authentic experiences, on the other hand, the local community finds itself inserted in a development route from the creation of the route.

IMPROVING THE ATTRACTIVENESS OF THE PLACE by CATEGORIZATION: ROTA DOS VENTOS PROJECT

In order to leverage the attractiveness of tourist destinations, pictures of positive experiences lived by tourists can change the negative perception of potential visitors. Currently, the widespread use of individual electronic devices, also known as "gadgets", for example, "laptops", "smartphones" and "tablets", such technological tools act in conjunction with electronic media have the ability to positively disclose or destroy the image of a receiving route for tourists almost instantly. Thus, there is a need for a well-defined and well-worked virtual image in technological communication networks.

The information from a tourist destination such as texts, images, photographs and videos speak vastly about potential of the place; thus, these contents can come to transmit favorable images or can even transmit harmful and destructive impressions in all cyber web and thus spreading quickly. Potential tourists, normally unaware of the reality of the tourist destination they wish to visit, may come to see these records and they quickly create superficial judgments about the place. In possession of their conclusions, when positive, they can be a key to igniting a future visits. But, if the impression arises from negative images, it can cause disillusionment and will permanently prevent them from going to the destination and consequently reduce the tourist flows. Therefore, according to Soares, Gabriel and Sánchez-Fernández, (2017) the construction and promotion of an attractive, coherent and sustainable image has a fundamental role in promoting the destination. With this in mind, it is true to say that encouraging visitors to the Rota dos Ventos to publish content on their social networks may strengthen the image of Ceará as a host pole for tourist flows.

The planning, as a whole, aims to supply the tourism fuel for its engines, the generators of jobs and income for the State as well as it aims to create elaborate plans that can preserve both tourists and host centers. The annual figures of the World Travel and Tourism Council (WTTC, 2019) show that the travel and tourism sector promotes 10.4% of the world gross domestic product (GDP), totaling US $ 8.8 trillion (eight trillion, eight billion American Dollars), which corresponds to almost 319 million direct jobs. It is concluded that 1 out of 10 jobs is from tourism and it is also concluded that 1 out of 5 new employees generated in 2018 came from the tourism sector.

The Winds' Route: a Value Proposition Based on Existing Inputs

Mimicking the already existing and successful Route of Emotions on the coast of Ceará / Piauí / Maranhão, the Route of Coffee on the Baturité Massif Mountain Chain and the Route of Mirantes (Watch Point) on the Ibiapaba Mountin Chain, can be affirmed the state of Ceará has structure and expertise in the planning and management of tourist routes. The Route of Emotions, the most well-known and the oldest of the three. It starts from Fortaleza going pass the Jericoacoara National Park, in Ceará's west coast, through the Delta of the Americas at Parnaíba River basin in Piauí and reaches as far as the Lençóis Maranhenses National Park. It was designed to meet the demands and yearnings for new experiences from new tourists. According to Bielinski et al. (2016), tourism businesses have multiplied in their offer and started to launch different products based on local and regional potential, be it cultural or geographic. It is important to point out that tourism focused on new experiences, cultures, gastronomy, sports, among others; thus, it has been presented as a highlight in the international tourist scene and this demand derives from the creation of tourist routes.

In addition to the innovation presented in the form of the welding of wind sports and social inclusion projects in coastal areas, this chapter presents the materialization of the Route of the Winds as a solution for the greater presence of the territory in the tourist market. The new Route of the Winds presents clearly and with easy logistics the main paths to be followed by the lovers of wind sports. Likewise, it aims to unite extreme sports tourism with the responsibility of social inclusion in the coastal area of Ceará. Among these paths the visitor will pass through several tourist regions: beach, backcountry, mountains and national parks, where for their leisure they can make the most of the warm winds that are abound in the State of Ceará, which is knwn as the world "Mecca" of radical sports wind. Likewise, and as an aggregator of value to the product, the potential tourist can participate in a positive way in social projects that imply improvement in the lives of many young people in the region visited.

The Route of the Winds has as its starting point the capital of Fortaleza, where you can visit the **Beneficial Surf School at Titanzinho Project**, immediately heading towards the east coast towards the renowned Beach Park water park, located in the Municipality of Aquiraz. Then the tourist stops at the

Surf at School Project. Hence the route passes through places such as Morro Branco Beach, Prainha do Canto Verde and consequently having as its first landing Praia do Pontal do Maceió in the Municipality of Fortim, a favorable place for "windsurfing" and "kitesurfing". The trip continues to the Canoa Quebrada Beach in the Municipality of Aracati, where it is possible to enjoy the contact with the winds through the sports "windsurf", "kitesurf" and "paragliding".

Towards the Sertão and the warm winds pockets, the "tourists of the winds" will head towards the Central Sertão, a little further southwest of the state to the city of Quixadá,where it is the place to find some of the best warm wind pockets on the planet, perfect for practice of hang gliding and "paragliding". Soon after, the route heads towards the Massif de Baturité, in the city of Baturité, which is a place of ramps for hang gliding and "paragliding". Once this is done, the traveler will move northwest, going deeper into the Sertão region, precisely to the a city called Sobral, where the Serra da Meruoca is the place go up for some hang gliding and "paragliding" flights. Finally, continuing, the route goes to the next stop in the Serra da Ibiapaba – to the Sertão da Ibiapaba Region, where it is possible to practice free hang gliding and "paragliding" flights and visit the Ubajara National Park with trails, abseiling and caves

Another section of the Route of the Winds that links sports tourism to winds, social responsibility and implies the local community is the stage on the West Coast, where the wind blows along one of the 10 most beautiful beaches in the world. The tourist takes as a destination the Northwest Coast region, the Municipality of Camocim and the National Park of Jericoacoara. In the last section, composed of the beaches Preá, Icaraizinho de Amontada and Praia da Barra on the Island of Guagiru, in Itarema, the tourist visits the Social Project **Vem Voar pro-Kitesurf**. In this stage, the tourist will spend a day to know the project, have fun, play sports and interact with local people. From there, head to Praia da Taíba and Praia do Cumbuco. Here it will be the last stop of the Route of the Winds round, where the tourist will always be on board air-conditioned 4 wheel drive vehicles.

In this innovative process of linking sports and social activities, what most stands out is the fact that everyone is interconnected to the coastal space. Among the countless attractive tourist products, the leisure that involves extreme sports which are mostly connected to the coastal space, they offers an excellent combination of sun, sea and wind, but lacks an action plan well designed to enable the creation of the Ceará Wind Route.

Annually several regional, national and international championships of extreme wind sports are held in the State. The events boost wind tourism as well as other economic activities that finance the social and environmental sustainability of the regions. We can mention the XCeara-Mundial de hang-gliding in Quixadá, the Superkite Brasil for "kitesurfing" at Praia do Futuro in Fortaleza. We can also point the in the waters of Camocim, on the west coast 365 km from the capital Fortaleza, the wet ways boil with the dispute of the International Wind Slalom and in Canoa Quebrada the International Wind Festival, with disputes of "kitesurfing", "windsurfing", paragliding and traditional *"Jangada"* local raft regatta

CONCLUSION

This work aims to bring attention to the debate that the participation of the local community in the process of local tourist progress is fundamental for a pole that induces tourist flows in combating the negative factors of the exploitation of this activity. It was found that the continued insertion and participation of the local community that receives tourist flows in a planned way is fundamental to provoke a cycle of continuous and spiral improvement that can develop aggregating results for the entire tourism production chain. The residents of the place, in turn, are valued and encouraged to continue with the tourist activities that support them. Thus, they also assume the responsibility of welcoming the tourist and of transmitting a positive image and experience to the visitors. Above all, for the natives of tourist locations, not only will they be left with the subjugation of underemployment or less qualified jobs of yore.

It is known that tourism can also cause problems, however, involving the native population in the benefits of this activity helps to protect the interests of the locals. The union of social projects with the development of tourism has shown results that benefit the progress of a healthy and positive environment, as is the case with the three projects presented. Thus, the young natives of the visited place are less and less exposed to crime, drugs and other evils. Which would be more of a benefit from responsible tourism.

It is noteworthy that the planned union of wind sports with social responsibility through the Route of the Winds brings multiple benefits to communities, as it will be receiving tourists who have bold interests in sports and positive contributions in their travel destinations. Therefore, on the other hand, the community will offer quality products / services to sports tourists, considering that they will be inserted in the environment accompanied by the best connoisseurs of the local reality. Thus, the practice of wind sports and satisfies both sides of tourism, hosts and tourists, as both are satisfied with having extremely valid experiences.

It urges to stress that the hosts will feel more important when they see themselves part of the local tourist activity in an active and properly valued way. On the other hand, tourists will feel fortunate to have more positive experiences because they are having fun and contributing well to the community that hosts them in a warm and unique way. The connection of the points, planning, duly represented community and scripted tourists focused on socially responsible sport contribute to the propagation of the positive images experienced. The word-of-mouth recommendation is a further consequence of this interconnection; tourists make use of instant tools, such as social networks and also the traditional recommendation, in conversations with friends upon their return. A positive experience lived in Ceará, on the Rota dos Ventos, has the potential to spread to different corners of the world given the diversity of origins of the tourists who visit this Brazilian state.

That said, the contact of tourists, whether of domestic or foreign origin, with local communities in a planned and structured way brings to the tourist experience a real authenticity of the local identity. This is the reason why the local people of a host site intensifies cultural exchange and customs, enriching the contact among tourists and residents.

ACKNOWLEDGMENT

Thanks to the informants Tomaz André Pinheiro (Project Ver Voar pro-Kitesurf Project), João Carlos Fera (Project Beneficial School of Surf Titanzinho) and Marcio Farney (Project Social Surf at School) who enriched the work with real and direct data from the source.

REFERENCES

Andrade, A. (2010). *O Nascimento do Windsurf no Ceará.* Retrieved November 21, 2019, from http://blogs.diariodonordeste.com.br/manobraradical/windsurf/o-nascimento-do-windsurf-no-ceara/

Barreto, M. (1991). Planejamento e Organização. In *Turismo*. Papirus Editora.

Beni, M. C. (1998). *Análise Estrutur.al do Turismo*. Editora Senac.

Beni, M. C. (1999). Política e Estratégia do Desenvolvimento Regional: Planejamento I. *Turismo em Análise*, *1*(1), 7–17. doi:10.11606/issn.1984-4867.v10i1p7-17

Bielinski, M., Jr., Disconzi, C. M. D. G., Knoll, K. R. H., & Fortunato, A. (2016). *Desenvolvimento de Roteiros Turísticos na Cidade de Santana Do Livramento/RS: Um Estudo sobre a Ferradura dos Vinhedo.* XVI Mostra de Iniciação Científica, Pós-graduação, Pesquisa e Extensão. UC.S.

Clavé, S. A. & Calabuig, J. (2005). *Planificación y gestión de destinos turísticos.* Master en Dirección y Gestión Turística. Madrid, IUP – Grupo Santillana (edición electrónica).

De Masi, D. (2000). *O Ócio criativo*. Sexante.

Duarte, C. (2002). *Céu de Quixadá está na rota do vôo livre.* Retrieved March 2, 2020, from https://www1.folha.uol.com.br/fsp/turismo/fx0306200222.htm

EBST. (2016). *A escola: identificação*. Retrieved March 20, 2019, from http://ebstitanzinho.blogspot.com/p/a-escola.html

Eshliki, S., & Kaboudi, M. (2012). Community Perception of Tourism Impacts and Their Participation in Tourism Planning: A Case Study of Ramsar, Iran. *Procedia: Social and Behavioral Sciences*, *36*, 333–341. doi:10.1016/j.sbspro.2012.03.037

Férias Brasil. (2019). *Esportes náuticos*. Retrieved January 20, 2020, from https://www.feriasbrasil.com.br/ce/fortaleza/esportesnauticos.cfm

Fernandes, L. M. M. (2014). O Ceará turístico: Política de regionalização e governança nos destinos indutores. *Revista GeoUECE*, *3*(5), 213–214.

Funceme. Fundação Cearense de Meteorologia e Recursos Hídricos. (2019). *Com temporada de ventos mais fortes, Guaramiranga registra rajada de quase 60 km/h.* Retrieved March 28, 2020, from: http://www.funceme.br/?p=5481

Kim, W., & Kim, H. (2013). Regional Development Strategy for Increasing - Cultural Tourism Business in South Korea. *Asia Pacific Journal of Tourism Research*, *18*(6), 534–548. doi:10.1080/10941665.2012.695285

Neto, J. A. (2016). *Voo livre colore o céu de Quixadá*. Retrieved March 28, 2020, from https://diariodonordeste.verdesmares.com.br/editorias/regiao/voo-livre-colore-o-ceu-de-quixada-1.1630237

Noronha, G. (2010). *Ceará: O Havaí dos ventos.* Retrieved March 18, 2020, from https://diariodonordeste.verdesmares.com.br/editorias/jogada/ceara-o-havai-dos-ventos-1.503959

Perinotto, A. R. C.; Muniz, R. A.; Muniz Brito, A. S., & Borges, D. M. (2017). Comunicação Turística (...). *Revista Hospitalidade, 1*(1), 1-28.

Redação. (2009). *Paraíso dos esportes náuticos.* Retrieved March 18, 2020, from https://diariodonordeste.verdesmares.com.br/editorias/verso/paraiso-dos-esportes-nauticos-1.52520

Redação Waves (Brasil). (n.d.). *Marcio Farney promove Projeto Social.* Retrieved March 10 2020 from: https://www.waves.com.br/arquivo/marcio-farney-promove-projeto-social/

Ruschmann, D. (2001). *Turismo e Planejamento Sustentável: a proteção do meio ambiente.* Editora Papirus.

Silveira, E. S. (2007). *Por uma Sociologia do Turismo.* Porto Alegre-RG: Zouk.

Soares, J. R. R., Gabriel, L. P. M. C., & Sánchez Fernández, M. D. (2017). A influência das fontes de informação na escolha dos turistas acadêmicos: O caso de Santiago de Compostela. *Investigaciones Turísticas, 14*, 130–153. doi:10.14461/INTURI2017.14.07

Soulkite. (2019). *Projeto Vem Voar! Capacidade Técnica-Operacional.* Portfólio Projeto Vem Voar pro-Kitesurf.

UNWTO. (2010)-*International Recommendations for Tourism Statistics 2008: Department of Economic and Social Affairs.* New York: Statistics Division. Retrieved Aug 25 2019 from https://unstats.un.org/unsd/publication/Seriesm/SeriesM_83rev1e.pdf#page=21

Vitorino. (2019). *O que fazer em Canoa Quebrada.* Retrieved March 19, 2020, from https://vitorinotur.com.br/o-que-fazer-em-canoa-quebrada/

Walraven, C. I. M. (2009). *Ceará é sede da 18ª BNTM no Nordeste.* Retrieved March 15, 2020, from https://www.skyscrapercity.com/threads/fortaleza-ce-estrutura-para-a-copa-de-2014.818858/page-4#post-37456328

WebVenture. (2003). *Rally Mundial de Voo Livre em Sobral.* Ceará. Retreived March 29, 2020 from https://www.webventure.com.br/rally-mundial-de-voo-livre-em-sobral

WTTC World Travel & Tourism Council. (2019). *Travel & Tourism Economic Impact 2019.* Retrieved March 29, 2020, from: https://www.wttc.org/-/media/files/reports/economic-impact-research/regions-2019/world2019.pdf

KEY TERMS AND DEFINITIONS

Coastal Tourism: Segment of tourist activity that uses activities in the coastal space, as the biggest attraction for the configuration of the tourist product.

Community Tourism: Tourist activity in which the community is involved in its community. The natives and heritage benefit from the "tourist" element and segmentations.

Extreme Sports: Action sports have the main characteristic of a high level of physical risk, dangerous activity with a high level of adrenaline for practitioners. Extreme sports conditions such as high speed, height and other factors are typical.

Social Project: A project or undertaking of solidarity actions aimed at improving one or more aspects of a society in which it exists, thus strengthening the sense of citizenship, or the social understanding of citizens.

Tourist Image: Images and imagery perceived by tourists about a specific tourist destination.

Tourist Planning: Action of planning, organization, direction, and control as a tourist activity in order to use the perennial in a sustainable way.

Tourist Routes: A route, itinerary or tourist itinerary is taken by someone who does tourism and in turn and a group of activities performed on a trip and overnight in another usual place, for leisure, work, and others.

Wind Sports: Extreme sports that have the use of the wind as their main resource, that is, they have the winds or the instrument of impulse of the sails.

ENDNOTE

1 Notes taken from the interview with the informant and founder of EBST on March 15, 2020.

Chapter 4
Responsibility and Social Innovation in Tourism Companies in Mexico:
The Case of Four Companies in Puebla

Maria del Carmen De la Luz Lanzagorta
Benemérita Universidad Autónoma de Puebla, Mexico

Edith Sarai Lozada Sánchez
Benemérita Universidad Autónoma de Puebla, Mexico

Jessica Abigail Cortés González
Benemérita Universidad Autónoma de Puebla, Mexico

Concepción Nancy De Cristobal González
Benemérita Universidad Autónoma de Puebla, Mexico

ABSTRACT

All over the world, institutions and organizations that recognize the social responsibility of companies are identified. They work on social innovation, which is companies' capacities to influence problems, generating synergies between the various sectors of society. In this research, a qualitative methodology was applied to a sample of companies from Puebla (Mexico) and the region to identify innovative practices of corporate social responsibility in the tourism sector and related companies. The central question that guides this research is, through which strategies or actions are companies in the tourism sector socially responsible and innovative? Therefore, the purpose is to show the good practices of different companies in the tourism sector in Puebla (Mexico) as well as their areas of opportunity and therefore strategies to strengthen responsibility and social innovation in the sector.

DOI: 10.4018/978-1-7998-4855-4.ch004

INTRODUCTION

In Mexico, some companies usually make sustainability reports, work with the ISO 26000 standard, align with initiatives such as the Global Compact or receive the Socially Responsible Company Badge granted by the Mexican Center for Philanthropy (CEMEFI) to those who comply with the established standards. In fact, the way of understanding social responsibility is usually different in each company. Some consensus has been achieved, but it has not been generalized. A long time ago, it was assumed that the greatest responsibility of companies was to generate profits. But, at present, that idea has been overcome, since it is known that, being a human group and sustained by people, the greatest responsibility is to care for interests of each one.

Some companies understand social responsibility as the "active and voluntary contribution to social, economic and environmental improvement by companies, with the aim of improving their competitive, value and added value" (Secretaría de Economía, 2016). Other companies do not assume that it is a voluntary activity, but precisely, as a responsibility, that is, something that must be fulfilled without waiting for it to be a requirement.

There are companies that understand this concept in the style of the Mexican Center for Philanthropy (CEMEFI): "a new way of managing and doing business, in which the company takes care that its operations are sustainable in the economic, social and environmental" (CEMEFI, s / f. page 2). With this vision, the importance of what has been called the triple results is highlighted, that is, results in the economic, social and environmental fields. With this approach, the interests of the different stakeholder groups are recognized. "It is a business vision that integrates respect for people, ethical values, the community and the environment with the management of the company" (CEMEFI, s / f. P. 2). Of course, this way of understanding social responsibility applies to companies of different sectors and sizes.

Some people claim that socially responsible companies have advantages over others, such as higher productivity, customer loyalty, lower turnover, access to markets, trust, better perception, among others. However, one aspect that has been highly questioned about corporate social responsibility is its aesthetic function: how far it can be considered that the company develops actions based on its internal and external stakeholders, as well as on the basis of the environment or its actions are aimed at fundamentally strengthening its image.

In this sense, it is worth talking about social innovation, which arises when there are unmet social needs and, consequently, people are forced to look for new ways to meet them. A fundamental aspect is the ability of companies to influence the problems generating synergies between the various sectors: government, business, civil society and even research centers or institutions of higher education, since in the latter knowledge is generated that must be useful to society "Here we show how the participation and social investment of the company promotes the development of social or community processes, and directly or indirectly benefits the company" (CEMEFI, s / f. P. 15). That is, social innovation strategies can be those that are implemented directly in the groups, either the internal or external community of the company, or those in which the company takes a leadership role and articulates efforts between different groups to promote initiatives that contribute to the common good.

Social innovation in companies is also understood in different ways; there are those who associate it with the possibility to offer jobs for the people; others understand it as the altruistic part of the company, by proposing innovative projects of social benefit, for the community surrounding the company; However, it is also important to consider that the social innovation of a company is an opportunity to transform the life of the company and its environment; an opportunity for personal growth for all the stakeholders.

Tourism in Mexico has diversified, as there are services aimed at various sectors of society. But it is important to recognize that the use of platforms for the use of hosting services has grown in an excessive way. Several times, it is cozy, comfortable and cheaper to rent a room in a house than a hotel room. Something similar has happened with transportation services, as the use of platforms allows to call a "safe taxi", which is monitored and offers various guarantees such as price, cleaning, accident insurance and others that some traditional services do not offer. It is also possible to share your own car with other travelers to reduce costs. To access food, platforms that offer a variety of home delivery, are also currently used. These business models can definitely be considered innovative. However, the question is how socially responsible are they or if their innovation has a social impact?

The "traditional" business models also have important contributions on issues of responsibility and social innovation, since they address relevant items. Among them, we can mention the care of the environment, the care of people in the organization (stakeholders), among others. But social innovation is also present, because the services they offer add value, not only for market purposes, but with the intention of really meeting the needs of the different groups of people with whom they interact. That is, triple results are achieved, having organizations with responsibility and social innovation.

SOCIAL INNOVATION

To speak of social innovation it is necessary to divide the concept. In 1903 the French sociologist Gabriel Tarde affirmed that "the social being, insofar as it is social, is essentially imitative and that imitation plays a role analogous to inheritance in organic life or to that of vibration between inorganic bodies". (p. 13) The sociologist stressed that the "invention" is what underlies these imitations; likewise, it emphasizes the free initiative of individuals in the process of social change, since they need to escape from their social environment. Due to this statement, he concludes that "innovation arises when there are fractures in the construction of sociability" (Domínguez, 2019 p. 81)

Years later, in 1912, the economist Joseph Schumptere elaborated an approach to the concept of social development based on the economic field, defining it as a "spontaneous and discontinuous change in the channels of the current, disturbances of the equilibrium, which always displace the state of the existing equilibrium previously "(Schumpeter, 1912, p. 75 in Domínguez, 2019, p. 81). Scientific research is essential to achieve technological leaps that break with tradition, in turn generating new paradigms applied to the economic, cultural and social spheres.

During the 20th century, innovation was understood only as the technological aspect, however, in recent decades, innovation has been understood in different fields such as social entrepreneurship, social movements and the social economy. For Cajaiba-Santana (2014 in Domínguez, 2019) what underlies the path of social innovation is not a social problem to be solved, but the change after that problem.

Rueede and Lurtz (2012) distinguish 7 categories in order to understand social innovation: to do something good in / for society; to change social practices and / or structures; to contribute to urban and community development; to reorganize the therapy processes; imbue technological innovations with cultural sense and relevance; to make changes in the social work area; and, to innovate through digital connectivity.

Social innovation has flexible and open conceptualizations, but it is difficult when establishing theoretical and methodological foundations. Many definitions coincide with: cohesion, well-being or social development. For this reason, on many occasions it is interpreted that civil society and social entities (especially those of the third sector) constitute themselves as spaces for processes of social innovation and that the public sector has a financing function. However, social innovation does not necessarily correspond to a sector of the population and neither is financing. There are projects in which various sectors of society participate in both functions.

Two types of companies are identified: those that are concerned with social justice and those concerned with social innovation practices. It is necessary to address a clear definition that incorporates the idea of novelty and change, and "that is oriented by a socially defined and progressive objective, and in which there is a rigorous treatment of social justice" (Cunha and Benneworth, 2013 in Hernández-Ascanio, & Rich-Ruiz, 2020 p. 75). Cajaiba-Santana (2014, in Hernández-Ascanio, & Rich-Ruiz, 2020) broadens the understanding of the concept in which it integrates theories with an institutional and structuring orientation. There are other current definitions that focus on "the understanding of agents (individualistic and behavioral approach where social innovation is developed from the actions of specific individuals) and those that focus on structure (social innovation is perceived as determined by the external structural context)" (p. 75).

Conejero-Paz and Redondo-Lebrero (2016, in Hernández-Ascanio, & Rich-Ruiz, 2020 p. 75) conclude that social innovation:

- It satisfies unmet human needs (due to scarcity of resources, political opportunity, they have not been identified as relevant by the various actors).
- It causes changes in social relations, especially in governance as it increases participation, generally in the most vulnerable groups.
- It fosters citizen empowerment based on increased socio-political capacity and access to resources.

When evaluating the different definitions we have that, in the scope dimension, there are two poles. The former would approach social innovation as a response when the market does not meet social needs. The latter would assimilate the concept to a classic innovation. In the field of application dimension, there is no consensus because they would relate the concept to social purposes or the cultural sphere or in the public sector.

People who promote social innovation are called managers of social innovation. These are social entrepreneurs, be they individuals, civil society (organizations), public administration (government) and social movements. The organizations or companies with the greatest possibility of making social innovation are those that show the capacity for social transformation. In other words, changes related to length, depth and significance that significantly affect society as a whole or different strongly interconnected social spheres. Social innovation capabilities are understood as "the distinctive competencies of the process of producing social innovations, the application of which is carried out in the context of social innovation ecosystems through the practice of one or more agents and that ensure the success of said processes" (Hernández-Ascanio & Rich-Ruiz, 2020 p. 76).

Social innovation is a widely used term today. Some large companies aspire that their management contributes to the success of social innovations that give them a certain image to guarantee their position vis-à-vis interest groups, others have a legitimate interest in designing spaces that allow the growth of the different groups that the By creating and creating relationships with value chains, they create spaces for personal, professional growth and social and family coexistence. Small companies are betting that social innovation is the springboard that will launch them into the spotlight of future partners that multiply their growth possibilities (Caro-Montero, 2018). Both companies must clearly establish how this approach connects with their history, with their current management model, with their future growth strategies, with their duty and desire to be and, mainly, with their organizational culture.

Innovation as a process is above all social, because it is only carried out by, with and for people who are grouped together in a family, in an organization, in a community, in a territory, in a country. This process directed ultimately to the socioeconomic development of society, is not only income and economic development, it is full and complete growth of human capacity; it is having access to the necessary resources to have a decent, long and healthy life; is to acquire knowledge (VV.AA., 2009 in Caro-Montero, 2018) and apply it according to that development. This is why innovation is intimately related to the ability to sustainably enhance economic, social, material and environmental factors, based on human well-being.

Given the breadth of interpretation of the concept of social innovation, another aspect is the versatility that it implies: it is an initiative, product, process or program that profoundly changes the basic routines, resources, authority or beliefs of any social system (Westley and Antadze, 2010 in Caro-Montero, 2018, p.4). In the innovative organization, the manager becomes a motivating entity, moderator, concentrator and integrator of efforts and visions. It has the responsibility of channeling people in the search and meeting of common goals. Among the premises raised by a group of Cuban authors (VV.AA., 2009 in Caro-Montero, 2018 p. 5) for a company to be innovative are: establish a clear sense of direction; reduce bureaucracy; establish open communication; instill a sense of belonging; make sure recognition and rewards are compatible; tolerance against risk and failure; and, eliminate projects and processes that don't work.

The European Commission (León, Baptista and Contreras, 2011) identifies levels regarding the social innovation of companies. The first level responds to the needs of the people most in need and not served by public institutions or the market. The second level must gradually eliminate the barrier between the economic and the social, since problems such as poverty, aging or others do not distinguish the difference between the social and the economic, in such a way that companies should contribute to these problems. A higher level of social innovation can occur when in companies, people have a level of empowerment and well-being. That is, it means profound changes in companies as traditionally understood and what is expected from social innovation, since it implies different ways of conceiving the purposes of the company, its way of working, the relationships between people, most likely from how a company is born.

Organizational values are guiding behaviors. Values underpin the organization's belief system, norms, and behaviors. For this reason, the development of values is considered as the basis for achieving true social innovation in companies, where the vision of the leader plays a fundamental role. Values form the basis for innovative behavior and allow the company's innovative initiative to be successful.

Díaz (2010, cited by Caro-Montero, 2018) defines organizational values as relatively stable mental structures over time that guide when a way of acting is better than its opposite, to achieve goals and welfare states desired. This definition is an attempt to relate two types of values with the results of the organization. Strategic values: they are the beacon, the guide, where you should and want to go to the organization. Therefore, they are related to the strategic purpose to be achieved and are identified with the mission and vision of the organization. In the proposed model, strategic values are conceived in the center of the triangle, where social innovation is the end to be achieved. Tactical values: they are the means that will allow achieving the organizational future outlined in the strategic values, in this case, a future of social innovation. These are located towards the vertices of the model triangle. They are called ethical values that indicate the way in which people relate to each other and themselves, practical values that indicate the way people act at work and development values that indicate how value is generated or created in the organization.

When ethical and practical values prevail, social innovation arises from conservative behavior. By dominating practical and developmental values, social innovation arises from entrepreneurial behavior; But if ethical and developmental values prevail, social innovation starts from dreamy behavior. Below are some characteristics of the behavior of the social innovator for each of the possible trends described (Caro-Montero, 2018).

1. Entrepreneurial-innovative behavior:
    a. Choose to resolve or act on something that may be better.
    b. It is operative, active and active.
    c. You have a well defined business model.
    d. It is rational, concise and structured in the generation of ideas.
    e. Use all the resources at your disposal
    f. Consider time as a key factor.
    g. He is interested in financial growth.
2. Dreamy-innovative behavior:
    a. He is visionary, challenging and sometimes utopian.
    b. Think of things that others do not assume to be objective, real, or feasible.
    c. High goals are set, not unattainable but difficult.
    d. He raises ideas knowing that he cannot execute them, so he asks others to materialize them.
    e. It is unreasonable in decision making.
    f. He lets himself be carried away by his desires and instincts, he is passionate, creative and unstructured in the generation of ideas.
    g. Does not pay attention to the resources required.
3. Conservative-innovative behavior:
    a. You are a cautious person.
    b. Think well before doing.
    c. Spend a lot of time on design and planning.
    d. He is afraid of taking risks, he worries a lot about failure and loss.
    e. It is difficult to give a final solution to a problem.
    f. It progresses slowly, although it may have partial results, these may be considered insufficient for others.
    g. Accurately evaluate and fear all options.

All social innovation implies an improvement by itself, an element with a superior quality to the one that gave rise to it. Because for an organizational result to be considered as social innovation, it has to be useful and promote the individual-collective, internal-external development of the organization that generates it.

Today's society presents challenges that allude to profound changes in social structures, behaviors, and forms of social development; and in other dimensions such as the economic and political. Social initiatives are examples of social innovation that generate new forms of social organization and offer new instruments to design a new society and solve problems in an alternative and participatory way. Social innovation has five dimensions (Correia, Oliveira and Gomez, 2016):

1. **Social Needs:** They are specific and must be met.
2. **Actors:** Acting in collaboration with institutions, organizations, etc.
3. **Process:** Managing social relations, above all, governance and participation.
4. **Innovation:** Innovative capacity of the implemented ideas and satisfaction of needs.
5. **Social Improvements and the Response:** Disseminate the improvements, and giving the possibility of replicating ideas in other contexts.

Due to the great variety of concepts, there is no clear epistemology or pragmatic consensus (Minks et al, 2012 in Correia, Oliveira and Gomez, 2016). By integrating different contributions from authors, it was reached that "social innovation has to do with generating social value for social actors and society. It is presented as a new response to an unsatisfactory situation, and aims at the well-being of people and communities through the satisfaction of needs, such as health, education, work, leisure, transportation and tourism" (Minks et al, 2012 in Correia, Oliveira and Gomez, 2016).

Two perspectives can be seen, one that focuses on processes as it analyzes inclusion, training, cooperation, as a way of obtaining transformations, with an emphasis on the participatory and collaborative nature of learning; the other, perspective of result, goes beyond specific needs, aims at social value and creation of meanings, social objective. For the integrated approach that is handled, the development process is not separated. For the authors, they understand that social innovations are collective processes that seek to satisfy social needs, disseminated by the actors to generate improvements or social responses, and from this definition an analysis of the dimensions of the concept is carried out.

1. Dimension of actors. The social actors can be members of any organization or association, it also includes companies and representatives of power. Social innovation is normally represented by organizations from the third sector. These initiatives should tend to emerge from the mobilization of civil society. A great measure of effectiveness is the motivation of the actors, strength to overcome obstacles, participation that depends on the relationship that one has, seeking to satisfy needs, creating public spaces for participation, and making decisions that facilitate the necessary changes.
2. Dimension of processes and collective activities. The objectives of social innovation are defined and constructed by society, and can be used to find answers to social, economic and environmental crises. Social needs can be classified as incremental, radical or systemic (Mulgan, 2006 in Correia, Oliveira and Gomez, 2016). In other cases, innovation occurs when civil society looks for ways to understand the context in which it is, keeps its members informed to overcome obstacles. The social actor can help integrate social claims that are not articulated with the needs during the data collection process, collaborative work with stakeholders is necessary to achieve this.
3. Dimension of social needs. During the creation and implementation process, a diversity of actors (network of multidisciplinary teams) and user participation are required. (Cloutier, 2003; Tardif & Harrisson, 2005; Benneworth & Cunha, 2015 in Correia, Oliveira and Gomez, 2016). Awareness of the problem, identification of causes and implementation of solutions are part of the process, in which the actors (in a network of multidisciplinary teams) and the users must necessarily participate. This approach requires introduction of new tasks, new division of labor based on skills and interests, new methods of coordination and learning to improve practices. Social innovation is an innovation of processes in which the dynamics of social relations, of power change; seek greater inclusion that results in social justice.

4. Dimension of social improvements or responses. The solutions that are reached must be "new", in the context sense, which has not been applied before, the solutions serve to improve existing practices and therefore improve the lives of the populations. Buckland and Murillo (2013 in Correia, Oliveira and Gomez, 2016 present two types of innovations, open innovation in which participants are free to copy an idea, reuse it and adapt it; and closed innovation, crushed by intellectual property.
5. Dimension of innovation. It focuses on the result of innovation, the latter being the systemic change it produces. It should generate improvements in matters of living conditions, satisfaction levels, life expectancy rates, reintegration into the labor market, the improvement of well-being and social cohesion (Westley, 2008; OECD, 2010; Caulier-Grice et al., 2012 in Correia, Oliveira and Gomez, 2016).
6. Key ideas: in social innovation there must always be the collaboration and participation of different parts of society interested in solving a social ailment using their knowledge, democracy and interdisciplinary collaboration to seek to improve the situation, with a view to contributing to the achievement of a systemic change.

Social innovation is derived from a social need, from which the improvement of the conditions in which people live is sought. In other words, it implies improving the democratic participation of society, reducing poverty, increasing attention to vulnerable people, among others. From this, two perspectives are created related to:

a. Knowledge management in Social innovation. Rodríguez-Gómez and Gairín (2015, in Rincón-Elías, Hermith-Patricia and Bautista-Molina, 2018) mentions that knowledge management is perceived as capital or assets, inciting innovation and learning to create organizational changes, allowing companies to take advantage of the knowledge available to act quickly in a variable market in the face of social needs or their prevention. This management (knowledge) seeks to cover the needs of the members of the organization, compares the external and internal environment to generate new knowledge. On the other hand, there is information management. This is in charge of focusing on structured data, in addition to considering the individual as an agent that generates capital. The innovation process may be linked to preserving the cultural and linguistic cultural spaces in development projects in their communities, because these spaces stimulate community participation and are used as a mechanism to solve their social problems. In addition, new information and communication technologies are changing the way in which people currently coexist or behave, thus providing opportunities to new contexts of citizen innovation. Atuesta-Venegas et al (2016 in Rincón-Elías, Hermith-Patricia and Bautista-Molina, 2018) They share that they should be associated with collaborative processes, and with co-creation processes in aspects of information and communication technology with themes of cultural tradition, artisanal methods, strategies applied for solving common problems. This will help to have a joint way of stimulating reflection and the use of learning spaces that serve as inputs to social innovation projects.
b. Citizen participation in Social Innovation. Herrera (2015 in Rincón-Elías, Hermith-Patricia and Bautista-Molina, 2018) says that social innovation means the establishment of new ties and new relationships that help to overcome the lines of fragmentation in places where social exclusion predominates, since it promotes spaces for reintegration into the labor market and in politics through active participation, encouraging changes in public policies. This allows us to find possible fields where we can provide solutions to various problems, and encourage citizen participa-

tion. The plurality of our society creates new entities and organizations with some independence from another power with an objective vision by creating scientific knowledge and analysis for the same diversity of people and situations that arise in society, and this serves to diminish, confront and solve the various conflicts between the characters and thus assess the knowledge generated to achieve it. The main idea of the article highlights that the participation of individuals is driven to make changes in the community, resulting in new proposals for the solution of their conflicts, their prevention and the communication channel that they must have and prevail. In the same way, this helps to improve the quality of life in a context where the community from its point of view reflects, participates and manifests its collective interest, achieving as an active community, solving its problems. Another way to promote this is through knowledge networks and social practices where research and creativity will help create a process of social innovation. Social innovation is a goal (solution of problems or objectives) that must be achieved together, in community, since citizen participation is essential to achieve it. As described, social innovation involves taking the needs of society, be it a community, an organization or a company, and generating response options. This process is not in charge of decision makers, authorities, owners, but must be a collaborative, participatory process that promotes democracy, in which each person takes responsibility for proposing and deciding.

For its part, social responsibility in companies refers to a way of life, a culture that is installed and lived from within and at all levels of it. The ISO26000 standard was developed with the intention that each organization uses it on a voluntary basis and not as a certifiable standard, with the intention of avoiding aesthetic use. Among the fundamental principles of the standard, the following are pointed out (ISO 26000, 2010): accountability, transparency, ethical behavior, respect for the interests of all stakeholders and respect for the principle of legality. And consider that the issues that every organization must address are (ISO 26000, 2010): organizational governance, human rights, labor practices, environment, fair operating practices, consumer affairs and active participation and community development.

It is important to consider that social responsibility consists of recognizing the impacts of the actions of the company as a whole and of each of the stakeholders. Consequently, the company must implement actions to reduce negative impacts. The tourism sector in Mexico has had a peculiar development, since it has benefited from aspects of its environment that have been extremely favorable, such as the climate, coastal resources and even its cultural and natural heritage in general. This has resulted in a well established tourism model. Over the past forty years, large investments have been made in coastal complexes, which have significantly changed the environments of these locations. On the one hand, they have had an important economic development, but they have not necessarily improved the quality of life of their inhabitants. In other words, the economic spillover has not in all cases met local needs or large groups of people. "It is unlikely that this model, in its current form, can sustain the objectives of a tourism policy that seeks to promote more inclusive and sustainable growth" (OCDE, 2017).

METHOD

To carry out this research, a qualitative methodology was applied through semi-structured interviews in an intentional sample of companies from Puebla (Mexico) and the region to identify innovative practices of corporate social responsibility in the tourism sector and related companies. For this, 3 companies were selected: Hotel La Descansería, Restaurant El Mural de los Poblanos, Fuente de Sodas Roketto, Tip Tours, which is a tourist transport company. The companies were selected, because in the sector they are identified as socially responsible, although none holds distinctions or awards. The interviewees were the owners of the companies, who are considered entrepreneurs, because they were the creators of the particular ideas that distinguish each of these companies from the rest of the branch to which they belong. The owners of these companies were consulted about the publication of the names of the companies and they agreed.

The interview had 9 questions organized in 6 categories: about the company (1. What is the main purpose of the company?), concepts (2. What does corporate social responsibility mean to you? 3. What does social innovation mean to you?), actions of the company (4. Do you consider that the company is socially responsible or do social innovation? 5. Who are the people involved in social responsibility actions and who in social innovation actions? 6. Who are the beneficiaries of both?), opportunity areas (7. In which areas do you think you should work to improve your actions of social responsibility or social innovation?), value added (8. What value does the company add to the development of its personnel?) and companies with and without social innovation (9. Are there differences between socially innovative and socially responsible companies and those that are not?).

RESULTS

The interviews with the owners of the companies allow to know, not only the activities that they carry out, but the true purposes with which they were created.

La Descansería and El Mural de los Poblanos

For La Descansería and El Mural de los Poblanos, Social Responsibility is the commitment you have towards others to help and return to society what you are having in life. He believes that if that vision of Social Responsibility were shared, there would not be so much inequality or so many needs. La Descansería y El Mural de los Poblanos belong to a group of companies that tries to look inward, that is, at the environment with the different actors, mainly with the workers. They must have good legal benefits, profit sharing, a better income level. The company is concerned about the environment of the collaborators, since they are not only conceived as a number in the list of workers, but as people, so it is about forming work teams where they feel part of a family. It seeks to work further with people. It is important to take care of the shareholders, since they are the people who invest in the company and must share this way of thinking. The company has a responsibility towards them and towards suppliers, fair purchases must be promoted and within the law. Purchases are made to be regional, suppliers are directly sought to support them, maintain better profitability and give value to their product. Other actors are external clients, to whom a service is provided. It seeks to fulfill the promise of offering excellent service. Finally, the neighbors, the community, the environment. Take care not to be a noisy or dirty neighbor.

They also recognize that, on many occasions, efforts are made that are not perceived by people. There are seasons a year when certain dishes are made, you go directly to the producers because you want to add value to each dish. The diners are presented with photographs of the producers, of their lands, so that they know that they are consuming dishes with products from the region and that the economy of the region is boosted. For example, the emblematic dish of Puebla, Mexico are the "Chiles en nogada";

There are producers that have three or four generations producing walnuts, apples, pears, poblano peppers. For the company it is important that the consumer not only sees the dish, but the whole context for its preparation, that people value the entire chain of effort behind a rich dish. Consumers feel related to a company that has these practices because they believe that the company's values are shared with them.

On the other hand, there is a commitment to contribute with different organizations. El Mural de los Poblanos is a supplier of cheeses and other products, which is a foundation that works with the street. It contributes by promoting these quality cheeses and selling them at the entrance of the restaurant. Other dairy products are consumed by producers in Tecamachalco, which is a place near Puebla; the cheeses are bought in Chipilo, also a place close to the city of Puebla. It is important to highlight that all these products are purchased at a regular price, no special prices are requested, since in this way, profitability is generated for the businesses.

In October two dishes are made: a main dish and a dessert. All those sold during that month are donated in full to the breast cancer campaign. Diners are informed of this campaign and many of them join in this campaign. In May 2019 a dinner was made with a specialized chef, 130 tickets were sold for a dinner with pairing. What was collected was donated in full to a foundation. It has been possible for some companies to donate wine for dinner and some other supplies. Work has also been done with the Teletón Foundation. A golf tournament is held, in which prizes such as nights in the Rest Room, food vouchers in the restaurant, food is served in the golf tournament. For this type of actions, other companies are invited to contribute. For this reason, diners are encouraged not only to consume due to the recognized quality of the food the restaurant has, but also because in various products and in various seasons there are actions for the benefit of society.

The owner of the Descansería and El Mural de los Poblanos considers that the dinners they have implemented are examples of social innovation, since they are a different way of seeking the well-being of people and organizations. On the one hand, quality food is offered to donors who bought a ticket, dinner is donated by El Mural de los Poblanos and other companies participate, for example, with wine. The total ticket sale is donated. The foundation collaborates by selling tickets. It is the result of the contribution of different people and organizations.

Another innovation is given by the incorporation of the new generations, the so-called millennials look for different forms of work, another reality is lived, if you do not work with the person there may be more turnover, people do not agree and leave. This makes the operation of the company expensive, but also the human factor is very interesting, that people stay and make a life project. It has gone from a time of slavery, in which the work was not paid for, to a time in which people not only seek to work, but also have a fullness in life as individuals and as families.

Every year there is an activity to improve the work environment, there is an opening to listen to the concerns and feedback given by the staff. They get together with the leaders and they work a lot with them so that they, in turn, work with their teams. There are two gatherings a year, one at the end of the year that takes place in January. Raffles are held and everyone takes a gift home, including gifts given by suppliers, but trips are also raffled because it is considered that, on many occasions, people are surviving and do not have the time or do not have the resources To travel to see a beach, there are people who do not know the beach. Sometimes they ask to change the trip for cash and you cannot make that change, but they are encouraged and those trips have been highly significant for families. The other party is held on a very important date: in the chili season in nogada. Travel raffles are held again. People have a good experience recorded and they commit themselves more, they value themselves and they realize that life is not survival, that it is worth the effort to grow as a person in the economic, spiritual and social areas.

There are other experiences for the staff. For example, when they invite the company to certain festivals, it is not only the chef who comes, as other people have come, such as the lady who works in production and narrated that it was the first time that she got on a plane. It may be insignificant for many, but for her it was an experience that helped her change her way of seeing life, to grow as a person. A pending issue is that all collaborators commit themselves, both in companies and organizations. It is necessary to include people who are benefited by donations, because when they receive, they can also commit to doing what is within their reach, such as donating money, hours, knowledge, etc. In the earthquake that Mexico suffered in 2017, the company stopped giving service and all the staff were afraid, they were paid their full salary and people made food for those who were in the "Zócalo" helping to collect food, but people told them They didn't need anything, so the staff went to the hospitals to see if food or water was needed. Additionally, they helped repair the company's damage. People are considered to be very aware of the importance of teaming up to achieve important things for everyone, it is a team, it is a family.

Rocketto

The company's main purpose is to create extraordinary and unforgettable moments for millions of people and for the communities that Rocketto serves. The purpose speaks precisely of three types of communities: external consumers, internal consumers and owners. External consumers, who access the services and products offered. The purpose is to make them live extraordinary and unforgettable moments in activities through the dynamics that the company has, through the product and the environment that is created from shopping experiences. It is a company that promotes the integration of the groups that attend, through board games and other activities. The internal customer is another community. Job opportunities are designed, the work environment is taken care of. It is intended that it is not just any job, for many collaborators working at Rocketto is their first work experience. The owner of this company considers that many people have bad experiences in their first job and that many times this is because significant experiences are not designed. It is important to teach young people to work and understand responsibility. The company encourages each person to question their life purpose.

A third community is the context where each Rocketto branch is located, 7 different cities. The company is expanding. Entrepreneurship talks are promoted, people are inspired by the history of the company. At Rocketto, a company's social responsibility refers to the fact that the company must be an organization that contributes and collaborates with and interacts with society. The company has a responsibility of remuneration towards society.

This company is considered innovative, because every day they have to generate new proposals on what to present to customers, although they recognize that it generally occurs as a result of intuition and making inferences about the needs of the market, but an innovation is necessarily organized with cause. They believe that over the years innovation is made not only for the external community, consumers, but also for the external public, but that it would be better to integrate the three types of communities to make the company evolve in a more sustainable way and with a common cause. Innovative ideas can be given by people from any of their communities; They consider that innovations are implemented almost every month, but the challenge now is to do it in a more comprehensive way for the benefit of the 3 communities. On the other hand, before implementing the innovations, they are filtered to analyze their feasibility, but once they are implemented, the external client also evaluates them, thus generating a comprehensive dynamic.

The most important actions that the company considers in terms of socially responsible innovation have to do with the surrounding community, that is, through talks with strong messages to young people that inspire them. Another action has to do with improving the relationship between the people who come to Rocketto, generating excellent experiences, there is a related culture. One of the brand differentiators is being a meeting point for a healthy coexistence for families and groups. For example, three generations gathered together living together. There are consumers who have appreciated the experience they have had: "I had never played a board game with my son, much less with my grandson"

Regarding employees, from the first day they enter the company, they are asked to write down their dreams, they are really motivated, because many people find it difficult to have a long-term dream and express it; The company believes that, if they can express it, they start working on it more than those who do not visualize themselves and do not express it. Many are scared and confined to small dreams, they are motivated to write their true dream, however ambitious it may seem. Sometimes they are invited to erase and rewrite until their true dream remains. Today in each branch there is a blackboard of dreams where all the people. The result is that there are incredible dreams, extremely interesting. The company lets employees know that Rocketto will do everything possible to help them fulfill their dream. Collaborators also have the possibility to help Rocketto achieve his dream; If at some point one does not contribute anything to the other, perhaps it is time to say goodbye and the relationship between the two is not damaged, because they have managed to learn from each other. This they consider very innovative and committed to collaborators.

Rocketto has found that an advantage of working in a socially responsible manner is that they can attract talent and millennial clientele, which is highly attractive to the company. Young people today seek meaningful experiences related to the human factor and social responsibility and innovation, with entrepreneurship. There are areas that could be improved on issues of responsibility and social innovation, such as human resources. There are always many projects, for now one is being evaluated and validated that consists of creating soft skills. This is a mutual benefit, as people with soft skills are more productive for the company, but it also means growth as a person. It is assumed that they are young people who in some cases will not stay long in the company and will have gained skills that will give them a competitive advantage in their professional environment, where they decide to develop. Rocketto's structure is prepared for that, so that there is a rotation, already assumed. And despite this, the company registers a favorable retention rate. People do manage to see in the company a place where they can develop and grow. Coaching is received in different areas and this is not just for the Management or the Coordinators, it can be for all the staff. Another area of opportunity is in terms of products, since these can be attractive, but innovation must lead the company to generate socially responsible products. That is, they do not oppose the health of the clients, they add value depending on the suppliers from whom the raw materials are purchased, among other aspects to take into account. What has worked in Rocketto is that the collaborators are given all the information and they are also taught to analyze the information, to interpret it, so that they achieve good results. Whoever takes advantage of this has managed to have a great school.

Tip Travel

This company is dedicated to creating experiences of happiness. Its purpose is to generate income, training and knowledge for both collaborators and shareholders. They are in charge of the trips in tourist buses to know the city of Puebla, as well as others that are nearby. They also organize tourist trips at the request of groups and companies. In Tip Travel, Corporate Social Responsibility is understood as the development of actions for the common good of society. While social innovation is creating different ways to improve the society in which you live. With the term society, in both cases, it refers to both its external customers, who are the ones who consume the company's services, and internal customers, who are the collaborators. The company considers itself socially responsible because it supports different causes: street children, low-income students who want to enter the business world, children with cancer, among other groups. In addition, there is internal support for employees in terms of wages, benefits and other specific supports. Equity between men and women is promoted. The people involved in the actions of social responsibility and social innovation are mainly the General Directorate and Operations. They are considered to be those who promote the company to participate or organize programs for the benefit of different groups of clients, both internal and external. As an area of opportunity in terms of responsibility and social innovation, Tip Travel identifies the need to have a program that covers more areas of knowledge for employees, which allows them to perform better in the company and make them grow as individuals. For this reason, the value that the company mainly adds to the development of its personnel is training and significantly promotes teamwork. Finally, Tip Travel considers that socially innovative and socially responsible companies are those that may be sensitive to the social and economic problems of its collaborators and society in general.

CONCLUSION

The companies with which this research was carried out show a high social commitment. In other words, their owners are interested in generating benefits for each of their stakeholders, groups that they identify very well. They are interested in benefiting from the company or group of companies that both the collaborators and customers, suppliers, partners, etc. have.

There is a concern to satisfy external customers, that is, consumers. This is logical, since every company seeks to meet the needs of its customers and approach potential customers. However, companies like those interviewed show that there is a genuine concern to give their customers the best quality in the products and services they offer, to listen to customers. Actions such as that of El Mural de los Poblanos should be highlighted, since they make known to customers that the dishes consumed, not only through photographs of the finished product, but also the effort involved in reaching such a dish and all the effort invested by various families. In this sense, culture is provided by knowing where the ingredients come from, but there is also a human element in highlighting the work of agricultural producers and the work that goes into making the dishes. Many of these foods have ancestral rituals of the communities behind them. For the restaurant it is very important that diners know this part of the culture, as it is important to appreciate the cultural load that each dish has.

All four companies have a particular interest in their collaborators. Maintaining a healthy work environment is one of his priorities. They are interested in listening to the needs they have as part of their team, but also as people. Rocketto stands out for his interest in promoting that his collaborators write a dream and work to achieve it. The company is concerned with contributing to the achievement of these dreams. Provide training, promote having coaches for staff at all levels. The company is prepared for staff turnover, especially for the youngest. Due to the type of work they carry out, in common that they are carried out by students, who work seasonally, but it is assumed that they will leave. Despite this, the company has a good retention rate. Something valuable is that they prepare young people for the future, involve them in various company activities and teach them to analyze and interpret information. So that when young people leave, they will have learned and developed a series of useful skills in working life. All four companies are strongly interested in staff development. Tip Travel invests in training its staff in different things. They must know about history, culture, etc. And all their guides are certified. But there is a particular concern to train them in such a way that they are aware of the problems they are experiencing and that, from their position, they contribute with useful solutions for the environment. Something that has characterized these companies is that they have shown ability to generate networks, either within their own branch or with other companies, in order to provide beneficial services for different communities. They have participated in philanthropic actions, such as donations, but they have also carried out actions that promote development. All are invited on several occasions to promote entrepreneurship with young people, who manage to inspire stories about how they started their businesses and how they have managed to keep them in a climate of strong competition. They also associate to contribute to the organizational development of foundations or other civil society organizations, whom they consider to be worth supporting, because they contribute directly to the solution of various social problems, which have a structure and that their own collaborators interested in contributing in different ways.

All four companies can be considered socially responsible. In fact, as stated at the beginning of this document, they are identified as such, although they do not hold awards or recognition. In other words, they do not make aesthetic use of corporate social responsibility. In the same way, the four companies have made innovation since their creation. The products and services that are offered make a differentiation with others. They consider that several of the strategies that have to do in their relationship with the different interest groups are innovative. Many of the practices that they carry out in a systematic way can be replicated by other companies, since they have given very good results. A recommendation may be to systematize these practices so that they can be replicated.

As mentioned by the European Commission, there are levels of achievement in terms of social innovation. It can be said that these companies are located in the first level, which makes a difference in relation to other companies and gives them the opportunity to grow in terms of social innovation. Although there are companies that are created from the base of society, with the intention of generating jobs and solutions for the most vulnerable, these companies do not arise from that perspective, but they are highly interested in developing high-level social innovation, in the that all stakeholders participate actively and democratically in the construction of scenarios that allow the economic development of the region, of the company, but in the same way, of each of the collaborators. In the same way, it is sought that the service they offer is truly consistent with the needs of all those involved, in such a way that an organizational culture based on social innovation is achieved.

It can be recognized that an area of opportunity is precisely to widely develop social innovation in the company, involving different stakeholders. It must be remembered that social innovation implies carrying out a participatory, democratic process, by means of which needs can be detected, not only for work or professionals, but also for personal needs, and with which the company develops plans. Such plans should not be separate, but rather aligned with the objectives of the company.

Making social innovation does not only imply creating the conditions to contribute to the development of others. Being innovative is also reinventing yourself in service to others, from the position that corresponds to each one. From the point of view of research in this area, the possibility of generating participation with the different groups involved in companies is glimpsed, allowing for a change in the way of working and conceiving companies. It is important to design participation scenarios, which can be achieved through participatory action research, promoting participation from the base to propose and test strategies and methodologies that allow raising the level of social innovation of companies

REFERENCES

Caro-Montero, E. (2018). *Desarrollo de valores organizacionales para la innovación social. Una mirada desde el emprendimiento en el contexto cubano actual.* Cuba: Universidad de la Habana.

CEMEFI. (n.d.). *El concepto de Responsabilidad Social Empresarial.* https://www.cemefi.org/esr/images/stories/pdf/esr/concepto_esr.pdf

Correia, É., Oliveira, V., & Gomez, C. (2016). Dimensions of social innovation and the roles of organizational actor: the proposition of a framework. RAM. *Revista de Administração Mackenzie, 17*(6), 102-133. doi:10.1590/1678-69712016/administracao.v17n6p102-133

Domínguez Orquendo, E. M. (2019), Innovación social: nociones de organismos multilaterales y concepciones universitarias. *Revista Virtual Universidad Católica del Norte,* (56), 72-88.

Hernández-Ascanio, J., & Rich-Ruiz, M. (2020). Caracterización de organizaciones del Tercer Sector desde el enfoque de la innovación social. Discusión a partir de un estudio de caso. *Innovar, 30(75)*, 71–86. doi:10.15446/innovar.v30n75.83258

Hernández-Ascanio, J., Tirado-Valencia, P., & Ariza-Montes, A. (2016). El concepto de innovación social: Ámbitos, definiciones y alcances teóricos. *C.I.R.I.E.C. España.*

ISO 26000:2010. Guidance on social responsibility. International Organization for Standardization. Geneva, Switzerland.

León, Baptista, & Contreras. (2012). La innovación social en el contexto de la responsabilidad social empresarial. *Forum Empresarial, 17*(1), 31-63. https://www.redalyc.org/articulo.oa?id=631/63124039002

OCDE. (2017). *Estudios en Turismo Estudio de la Política Turística de México.* Secretaría de Turismo de México.

Rincón-Elías, C., Hermith-Patricia, D., & Bautista-Molina, W. (2018). Innovación social y su importancia en la gestión del conocimiento y la participación ciudadana. trilogía Ciencia Tecnología Sociedad, Volumen 10. *Número, 18*, 51–61.

Rueede, D., & Lurtz, K. (2012). *Mapping the Various Meanings of Social Innovation: Towards a Differentiated Understanding of an Emerging Concept.* EBS Business School Research Paper No. 12-03.

Secretaría de Economía. (2016). *Responsabilidad Social Empresarial.* Gobierno de México. https://www.gob.mx/se/articulos/responsabilidad-social-empresarial-32705

Chapter 5
Tourism of Community Base as an Instrument for Local Development and Social Innovation:
A Case Study in the Extreme Coast West of Ceará, Northeast Brazil

Maria Sâmia de Oliveira
Ciência eTecnologia do Ceará, Instituto Federal de Educação, Brazil

Gil Célio de Castro Cardoso
Universidade Federal do Ceará, Brazil

ABSTRACT

Dealing with tourism, especially community-based tourism, requires an understanding of its concept and its relationship to local development. Additionally, community-based tourism presents itself as an example of social innovation, in view of its important function of aggregating values, social responsibility, and sustainable means of economic development. In this sense, this work aims to understand the relationship of residents with community-based tourism practiced in the coastal area of the extreme west of the state of Ceará, located in Northeast Brazil, and the perspective of residents on the importance of local tourism and its relationship with elements of social innovation. The research is classified as qualitative and the collection instrument was consolidated through the application of semi-structured interviews. The collected information was submitted to the treatment of the data through the technique of content analysis. The results are about the perception of these subjects about the tourist activities practiced in the city of Acaraú, Ceará.

DOI: 10.4018/978-1-7998-4855-4.ch005

INTRODUCTION

To report on the development of tourism, especially community-based tourism, which occurs in the interior of the state of Ceará, it is necessary to understand how the reality of Community-based Tourism in Brazil and its genesis was constituted through historical facts, experiences and its heterogeneous character. To conceptualize it, it would not be a simple task to talk about it, without combining all the modalities in which it develops in Brazilian territory. The analysis of culture in traditional communities also depends on their social agents, whose identity is constituted in the history of their places and, therefore, strengthen the way in which tourism has become an activity full of signs, full of representations and social values. Maldonado (2009) conceptualizes community tourism as "[...] any form of business organization based on property and sustainable self-management of community heritage resources. The author also states that the activity must be "[...] in accordance with the practices of cooperation and equity in work and in the distribution of the benefits generated by the provision of tourist services".

Thus, according to Maldonado (2009, p.29), "the distinctive feature of community tourism is its human and cultural dimension, that is to say anthropological, with the objective of encouraging dialogue between equals and intercultural encounters of quality with our visitors, from the perspective to know and learn with their respective ways of life".

Coriolano (2009, p. 280) reinforces the idea of a strong human characteristic when presenting community tourism as a new axis where communities are part of a large network through the prominence among these "social subjects, some social movements, communities, organizers of travel, fair trade operators, solidarity economies, environmental organizations and NGOs, all committed to this task of finding other paths for development".

The authors strengthen the idea that this modality is built collectively, thinking about the human dimension, a thought that was also shared when the definition brought up in Ricco (2012) about cultural tourism, in which tourism expresses as main attractive aspects of human culture. This, in turn, does not separate from the community, as culture is one of its main characteristics and there is no community tourism without the cultural representations and signs of the communities where it takes place.

Furtado (2012) still relates the role of the State and the importance of considering the collective in the implementation of policies when he points out that "[...] when implementing cultural policies, it is necessary that government officials have an acute perception of the historical moment, beyond what is essential: the collective will".

Furtado expands the reference of culture as one of the bases of development, and the notion for cultural policies, joining the idea that such policies can only be developed with the protagonism of the populations and that a "[...] development policy must be put at the service of the cultural enrichment process" (1984, p.32).

The author, who presents the belief that growth and prosperity should be extended to all, understands development as a process of transformation of the world carried out by human beings with the intention of meeting their needs, leading to reflect on how the policies created in the Northeast, like PRODETUR-NE[1], it sees Community Based Tourism, in which both the community and visitors who come to enjoy the local wealth are related to the objective of growth and what actions to preserve the areas where Community Tourism takes place. Such policies confront this view, since, by promoting hotel mega-enterprises, they significantly alter the social dynamics of the places where they have been installed, producing irreparable social and environmental inflows. Within this perception, the State, agent of the policies that should lead to development, must also be the intervener in the conflicts generated with the production and transformation of spaces, between the community and large owners of the means of production of the tourist activity.

State action is registered to reduce regional inequalities, in a constant dynamism, aiming at increasing growth and deconcentrating the income from Tourism, through regionalization, interiorization and segmentation of the tourist activity, when at the federal level, public policies were implemented in the first decade of the 21st century, through the National Tourism Plan (NTP)[2], in the context of which goals were established that gave rise to the Tourism Regionalization Program (TRP)[3].

Given the understanding of the authors above, we can see the intrinsic relationship between community tourism and the dimensions of social innovation that present new interdisciplinary perspectives. His first definitions were presented by Taylor (1970) when he expressed "social innovation as the search for answers to social needs through the introduction of a social invention, that is, a 'new way of doing things' a new social organization".

Thus, this investigation addresses the relationship of residents with community-based tourism practiced in the coastal area of the extreme west of the state of Ceará, located in the northeast of Brazil, as it presents itself as an instrument of local development and social innovation and the perspective residents about the importance of local tourism.

For this, the methodology used for the development of the research has a qualitative character. From the point of view of the objectives and the nature of the problem, the research is characterized as exploratory and descriptive, aiming to know the study variable as shown.

For data collection, a bibliographic and documentary survey was carried out by deepening the literature review and field visit. After this survey, semi-structured interviews were conducted. The collected information was submitted to the treatment of the data through the technique of Content Analysis, where will be treated information about the perception of these subjects about the tourist activities practiced in the city of Acaraú - Ceará.

Content Analysis according to Minayo (2001, p. 74), is "[...] understood much more as a set of techniques". For the author, it consists in the analysis of information about human behavior, allowing a very varied application, and it has two functions: verification of hypotheses and / or questions and discovery of what is behind the manifest content. For the author, these functions can be complementary, with application in both qualitative and quantitative research.

TOURISM OF COMMUNITY BASE AS AN INSTRUMENT FOR LOCAL DEVELOPMENT AND SOCIAL INNOVATION

Community Tourism and the Idea of Local Development

Nogueira (2008), in a discussion on the relations between development, State and society, their possible coalitions and the required institutionality, leads to thinking about the concept of development, placing it as complex and having the idea of society inevitably, political community, social justice, state and economy.

The concept of development has a strong "qualitative" component, with which it acquires another status. It is a much more comprehensive and complex fact that involves and demands an idea of society, of the political community, of social justice, of the State and of the economy (Nogueira, 2008, p. 47).

For the author, development is a social product and it is not limited to state regulations although it is responsible for that regulation, and for this reason, there is no way to talk about development without addressing coalitions and institutionalities. According to him, it is difficult to think of the concept of development in such a way that political, public terms and a dimension focused on society are not covered. Therefore, it is only possible to think of a "development project". At the same time, the author emphasizes, that even if working on a development project in which there are social and sustainable gains in order to guarantee legitimacy to social support, it would still be innocence to believe that everyone would benefit at the same time with development.

For Mbakogu (2004) when discussing development and its relationship with culture, and African cultural policies, he states that for a development task to be successful it needs to be rooted in people's culture. The author emphasizes that societies that "[…] truly aspire freedom from the shackles of underdevelopment, there is also an urgent need to break away from the confines of western cultural systems evident in our dressing, language, music, artwork, attitude to innovations […]".

Thus, Mbakogu (2004) presents culture as a preponderant factor when properly managed, considering it "an important vehicle for intercultural exchanges and sustainable development."

For debates about the concept of development, it is necessary to understand the distinction between development and economic growth. For Furtado (1961), from the point of view of the economic field, "development is basically an increase in the flow of real income, that is, an increase in the quantity of goods and services per unit of time available to a certain community" (p. 115-116). As a critical economist, the author also expresses the concept of growth as a simple quantitative variation of a product. It is understood with this, that there is no development without economic growth, however, there can be economic growth without effectively reaching development, since growth is linked to a process of quantitative change, while that for development to happen, changes must happen qualitatively, changing the economic and social structure and transforming the way of life of people, institutions and productive bodies, as Furtado (1961) denotes.

Oliveira (2002) agrees with Furtado (1961) when he reinforces that "[...] thinking about development is, first of all, thinking about income distribution, health, education, environment, freedom, leisure, among other variables that can affect society's quality of life". Thus, it is understood that the concept of development is complex and much broader, with growth as an indispensable condition.

In the field of tourism, concerns about local development go beyond economic development - there is a concern with social and sustainable development, among other strategies. And considering the complexity of the word development, it is fairer to think from the perspective of tourist activity, about a project for this purpose in which local characteristics must be considered, as well as possible coalitions, since such space is permeated by conflicts between the subjects involved.

In this sense, in Brazil, these efforts that seek the good of the community, through the development of tourist activities, triggered actions considered innovative, given the struggles for survival in the face of economic, social, environmental and political adversities and also in relation to conventional tourism (Coriolano & Sampaio, 2012).

Such actions are different from the mass tourism modality that sells vast territories, displaces populations and transforms people's living and working spaces. Thus, community tourism represents an alternative for the economic and social development of rural areas where the territory, social actors and the environment converge as fundamental elements of the activity and active participation of host communities, explains Villavicencio et al. (p.3, 2016).

Irving (2009), delineates about the concept of community tourism and presents premises that emerge as the main components of this activity, namely: it is the endogenous basis of the initiative and local development; considers participation and social protagonism in the planning, implementation and evaluation of tourism projects; it has limited scale and controlled social and environmental impacts; it generates direct benefits for the local population; it presents strong cultural and intercultural affirmation; mutual sharing and learning takes place between tourists and local actors. However, for the author "the development of community-based tourism can only occur if the protagonists of the destinations are subjects and not objects of the process" (Irving, 2009).

In a complementary way, Coriolano (2009) also points out that community tourism is one in which communities, in an associative way, organize local[4] productive arrangements, having effective control of the lands and economic activities associated with the exploration of tourism.

Such assertions lead to the understanding of the importance of dynamized tourism in an innovative way, causing significant social changes, given the development of activities that contribute to the organization of spaces, territories and productive arrangements aimed at local development.

When understanding about development, it is directed to the elucidation of points in order to understand the fulcrum of the impasses for local growth, with the focus of this research, the Municipality of Acaraú. Since this is considered a tourist corridor for those heading to Jericoacoara, the points to be clarified, considering the development due to the tourist activity, are the social impacts and the permanence or not of the cultural characteristics, due to the contact of the receiving communities with tourists.

In the understanding of Rios (2015), the development of tourism affects the demographic structure, creating jobs that, in turn, generate, in addition to the rural exodus, interregional migration. Also, according to the author, the creation of jobs due to the development of tourist activity, in turn, "[...] causes changes in the social structure of the communities visited, as well as in their social institutions".

The author consigns the idea that the social changes resulting from Tourism are mainly due to the transition from fishing or agricultural activities to those in the service sector, however he points out that in rural areas there is a need to plan local productive arrangements in the primary sector, ensuring their survival and the maintenance of the productive activity of the population of coastal areas.

The economic effects arising from Tourism are subject to quantification, such as the Gross Domestic Product (GDP)[5] and its influence on the expansion of the labor market, among other aspects. In Acaraú, the service sector is responsible for 30.35% of GDP, this being the sector of the economy where Tourism is inserted.

As for socio-cultural aspects and impacts, according to Rios (2012) "[...] Tourism can play a very important role in the preservation of cultural heritage and, in some cases, even in the rescue of this heritage". For the author, there is an interest on the part of the tourist in the history and local cultural wealth that is invested in economic value. According to Beni (2006), "[...] any progress has a social cost; with due precautions, its implementation or growth can minimize costs and increase benefits".

In this sense, it is noted that there are positive impacts on the communities that receive tourism activity, but that a planning policy that brings together potentials and values cultural aspects is necessary, as well as the care with social impacts that are the most complex, when implementing tourist developments and structures.

Another extremely important concept for understanding the dynamics of tourism that takes place in communities is the understanding of what development is. For Furtado (1984), "The idea of development is at the center of the worldview that prevails at the present time. From it, man is seen as a factor of transformation, both in the social and ecological context that is inserted as well as of himself". Thus, it is understood that local development, even if it is a dynamic and continuous transformation, cannot assume the character of traditional communities for the accumulation of capital.

It is not possible, in such locations, to make use of this activity as a means of appropriating surpluses for the surpassing enrichment of capital-holding entrepreneurs, just as it happened in the times of Brazil - Cologne.

[...] thus, despised by the elites, the people continue their formative process with considerable autonomy, which will allow the non-European roots of their culture to be consolidated and their creative strength to expand less inhibited, in the face of class culture dominant (Furtado, 1984, p. 23).

Indeed, Furtado (1974) defends the argument that "[...] the behavior of the groups that appropriate the surplus, conditioned by the situation of cultural dependency in which they find themselves, tends to aggravate social inequalities, due to accumulation advance". It is important to note that Furtado's criticism of the system carries on to reflect on what community-based tourism proclaims, with the affirmations of forms of cultural expression, development of the economy, communities, use of traditional activities, as attractions and the sustainable use of resources with environmental justice focused on collective concerns.

In agreement with the words of Furtado (1974) on the importance and relationship of the sustainable use of natural resources and how this happens in the communities that develop the tourist activity is Swarbrooke's (2000) statement that one of the cornerstones of sustainable tourism is the idea that the local community should actively participate in tourism planning, and perhaps control the local tourism industry and its activities.

Thus, it is necessary to analyze the sensitivity of these traditional communities with their multiple characteristics and, if they have a tourist vocation, the activity can, and should, be explored in favor of improving the quality of life of the community, in order to meet a sustainable development, with greater social inclusions and minimal impacts on the environment. It is essential that public policies do not have a mercantilist view, but that they are based on ethical principles that converge with the precepts of the National Tourism Plan - NTP (2013-2016), which denotes "[...] as a perspective the effectiveness of the activity's potential for sustainable, environmentally balanced and socially inclusive economic development". Such perspective, when showing itself as inclusive, dispels the idea of the extreme inequalities in which, sometimes, the tourist activity is inserted being this, a privilege for few.

Furtado elucidates the consequences regarding the domination of capital in companies and the interests of those who own the capital:

The lifestyle created by industrial capitalism will always be the privilege of a minority. The cost in terms of depredation of the physical world, of this lifestyle is so high that any attempt to generalize it would inexorably lead to the collapse of an entire civilization, putting at risk the chances of survival of the human species" (...) The idea of development has only been useful in mobilizing the peoples of the periphery and leading them to accept enormous sacrifices, to legitimize the destruction of 'archaic' cultural forms, to 'explain' and make 'understand the need' to destroy the physical environment, to justify forms of dependence that reinforce the predatory character of the productive system (Furtado, 1974, p. 75).

The NTP also states that tourism activity must first collaborate with the receiving community and Furtado reinforces the idea that this community must be the protagonist of the development process in an endogenous way: "[...] endogeneity is nothing else but the faculty that has a human community to order the cumulative process according to priorities defined by itself", (1984, p. 108).

With the vision that Ceará could become, with all its potential, a place full of opportunities and able to receive development through tourism activities, a vision beyond tourism focused on fostering the accumulation of capitalist production was needed, having a concern with the dialectical relationship between tourist, receiving community and the state, so that they are considered transforming agents and also regulators of social, environmental and cultural actions. In Furtado's ideas, culture and society are intrinsically related and this relationship can be the starting point to overcome the problems of underdevelopment, the neglect of the communities of the Interior potentially rich in tourist attractions and move towards the transformations where the core of endogenous development lies in invigorating and valuing cultural identities while exploring the creative potential of these communities so that there is possible cultural independence and avoiding the imitation of exacerbated standards typical of capitalism.

According to Coriolano (2003, p. 134), "tourism for the benefit of communities, or of local development, means adopting policies that create job and income opportunities for the majority, while giving the required social protection, placing man at the center of power, promoting his achievement"

Thus, this investigation addresses the way in which tourism is developed in other rural communities in the "*Acarauense*" territory and the behavior of the population in the face of demands that could increase socioeconomic growth, given the importance of the involvement of society in participating in the implementation of these guidelines, in addition to preservation actions in areas where community tourism takes place.

Community Tourism and its Relationship with Social Innovation

According to what was explained in the introduction, community tourism has its bases anchored in collective construction and in the use of local goods that include environmental, cultural and heritage resources to establish means of subsistence and local growth. Such actions find a strong relationship with the concepts of Social Innovation.

In this sense, there is no way to talk about community tourism without relating it to Social Innovation, since it can be understood as the development and implementation of new ideas (products, services and models) to meet social needs and create social relationships. or collaborations (Bepa, 2011).

Cloutier (2003) substantiates this relationship when he explains that Social Innovation is "a new response to an unfavorable social situation, aimed at the well-being of individuals and / or communities through action and sustainable change". The author further states that for the creation and implementation of social innovation, the process must meet the grouping requirements of the main categories of diversity of actors and user participation. For him "the diversity of actors and active participation are considered essential conditions for the creation and implementation of new solutions that have in their process the learning and knowledge creation."

Additionally, to conceptualize Social Innovation, Tardif and Harrisson (2005) systematized their concept by presenting five important dimensions that referenced the main discussions on the theme, they are: transformation, innovative character, innovation, actors and processes.

Understanding these dimensions presented by the aforementioned authors in order to relate social innovation with the existing relationships in the context of community tourism and social intertwines for regional development is of paramount importance. Let's look at each one of them according to the view of the authors Tardif & Harrisson in 2005:

Transformation: Social innovation results from the identification of economic and / or social problems, such as crises, ruptures, discontinuities and structural changes.

Innovative Character: Social innovation takes place from the context and the respective conditions of emergencies and changes; the actors are encouraged to develop solutions to mitigate a given problem situation.

Innovation: For the authors, this dimension seeks to meet the general objectives of those involved, seeking to reconcile the individual and collective objectives (common good) and there must be cooperation between the actors.

Actors: constituent agents involved in social innovation actions.

Processes: The way in which actors interact and coordinate the development of social innovation, the means involved and the restrictions on its implementation.

The Actors dimension presents important key concepts also present in the relations of tourist activity and, in particular, in community tourism, which are the social actors - represented by movements, cooperatives, associations, civil society and unions; the organizational actors - companies, social economy organizations, collective organizations, recipients - and institutional - states, identity, values and norms.

The key concepts presented in the Actors dimension above are also present in the research carried out when the readings of the constituent subjects on the contribution of the tourist activity to the socioeconomic development of the municipality of Acaraú - Ceará were sought. In the survey, the social actors were: a group of public authorities, a group of local entrepreneurs and a third group of representatives of community tourism in the localities of Acaraú, as well as families that make up the communities.

In the following chapter, there are the results about the residents' relationship with local tourism and their perception of this important activity developed in the municipality under study.

RESULTS AND DISCUSSIONS

This study was carried out with communities in the city of Acaraú and included semi-structured interviews with managers, locals and representatives of community tourism in the localities of Acaraú, as well as families that make up the communities. To expose the results, we present a little bit about the local characteristics and how tourist activities develop.

Community Tourism in Acaraú - CE: the Case of Curral Velho

The community of Curral Velho - Acaraú - CE, is known for its resistance in defense of its territory in order to guarantee environmental preservation. From the episodes of struggle for the protection of its ecosystem, in 2007, it became part of the Ceará Solidarity and Community Tourism Network - Rede Tucum, an integration that was consolidated when, according to Borges (2011), there was a crisis in fishing artisanal and the community realized that tourism could be a way of generating income and guaranteeing the permanence of residents.

In 2010, there were ten coastal communities in Ceará, including indigenous people, fishermen and residents of rural settlements, with emphasis on the Tucum Network, as shown in map 1 of this research. One of Tucum's articulators, Mônica Bonadiman, defends the argument that the chain offers "[...] Tourism that is sustainable, does not harm the environment or the communities that live there. The activity is theirs, the benefit is all of them, and it's a way of staying in the land itself ", as presented in the news story. Coriolano (2009 p. 284) reinforces the notion that Community Tourism acts in association "[...] with other economic activities, with initiatives that strengthen agriculture, fisheries and handicrafts, making these pre-existing activities to Tourism more sustainable", as already mentioned.

The following shows map 1 where these communities act to protect their spaces from 'other' explorations, showing increasing social and environmental responsibility and redesigning their territories.

Figure 1. Tucum Network Map
Source: REDE TUCUM, ADAPTED

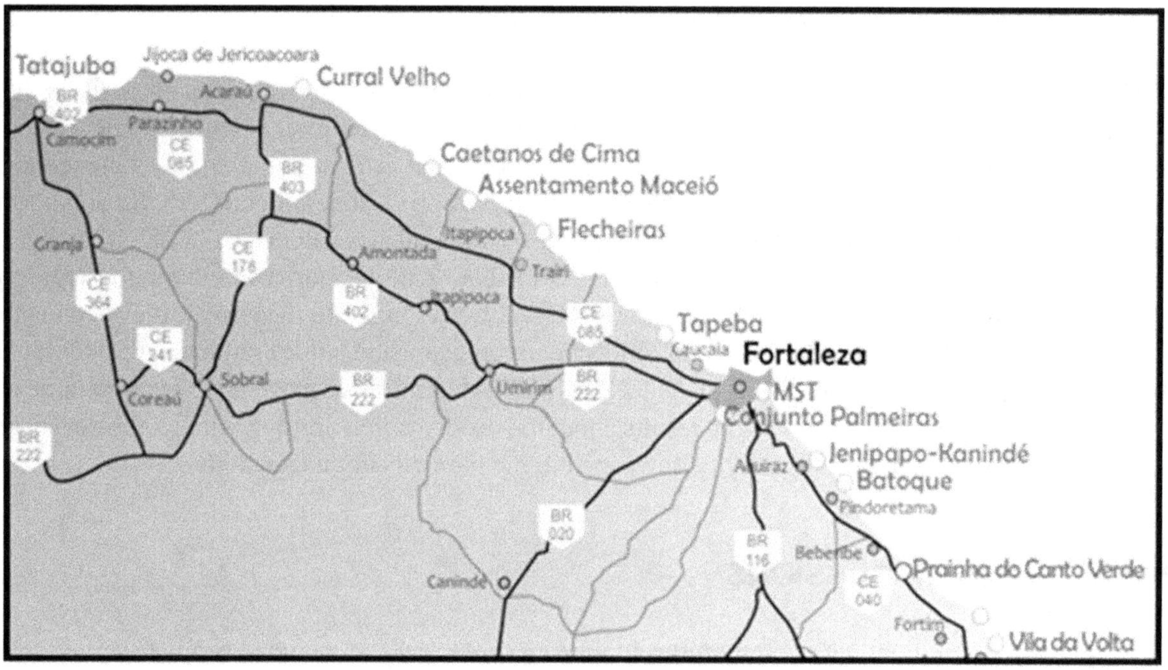

It is also possible to see on the map a significant number of these communities that struggle to strengthen the traditional models of survival of the local population in the State of Ceará.

According to Loureiro (2013, p. 12),

[...] The environmental preservation worked in the Community has consisted not only in the search for the balance of the existing ecosystems, but in the maintenance of the traditional models of survival and predominant culture that differ from the capitalist model, often exploratory.

Ahead, the broader study of these traditional models that make up Community-based Tourism and offer the provision of this service and leisure demands in a sustainable manner is entered. It is important to highlight another characteristic of these models reinforced by Moraes, Irving & Mendonça (2018) where they point out that the Brazilian case studies argue that community-based tourism management is the central point for the success of community-based tourism projects.

Curral Velho is 250 km from the capital Fortaleza and has approximately three thousand inhabitants. Develops fishing and shellfish as subsistence activities. Local residents have a close relationship with natural environments and, as a result, Curral Velho has been the scene of conflicts between local interests for the preservation of the environment and the territories of residents and owners of shrimp ponds, which define their relationship with the environment focused on the exploitation of natural resources.

Gomes (2010) expresses, in the book "*Historiando Curral Velho*", that in the community there is an environmental heritage composed of mangroves, dunes, beaches, "gamboas" and "apicum[6]. The same author mentions "*Reisado*" celebration and drama as the main cultural manifestations, in addition to other expressions that are currently more present, such as novenas, coronations and the practice of football, a sport that animates the community on weekends. On the traditional knowledge of the locality are "[...] the construction of fishing corrals and mud houses, the art of fishing, handicrafts in straw and lace, cuisine", in addition, "the craft of shellfish gatherers and of fisherman", "flour and the way of making "*mocororó*", a traditional drink from Curral Velho made from cashew filth - the latter two shown as inheritances of the way of life and culture of indigenous peoples (Gomes, 2010, p. 20).

This community was organized based on the intention to face the problems arising from the practice of shrimp farming. As a result, the "*Encante do Mangue*" Environmental Education and Community Tourism Center was created, which in 2007 became part of the Tucum Network. It was supported by this engagement with the Network that the locality realized that the development of tourist activities could give rise to income, improving the quality of life of resident families.

Indeed, the engagement of residents also foresaw the confrontation of these problems with the advancement of shrimp farming activities, which, according to Gomes (2010), are:

[...] land occupation; community division; violations of human rights, with cases of physical and psychological intimidation directed at people who resisted this activity; devastation of mangrove vegetation, carnauba tree areas and riparian forest, causing the landscape to become unstructured; disruption of the ecological functions of the mangrove; impairment of community activities, due to soil salinization; destruction of species that depend on this ecosystem to complete their life cycle and that are important for our food sovereignty; contamination of water; etc. (Gomes, 2010, p. 07)

With the development of the research, it was possible to verify that the community resists and that it is an example of struggle to meet the needs of families that survive from the activities of artisanal fishing, traditional agriculture and the production of handicrafts.

According to Borges (2010), the community's educational center, which symbolizes the strength of the local organization, offers training courses for young people and adults, free internet access and audiovisual workshops financed by the Ministry of Culture and *Banco do Nordeste*.

According to the local resident who also works with community tourism, the Environmental Education Center was an action that contributed to the development of local tourism. He mentions the main activities practiced in the community: "[...] the boat trip, the trails inside the mangrove, information about the knowledge of the local culture, such as the flour house and presentation of antique objects."

It confirms the fact that these activities are developed within the scope of community tourism and that their development is related to the struggles that occurred especially because of land tenure. Below, an excerpt from the interview.

[...] so all of this is developed with this practice of community tourism. our work is also related, our struggle! There was a great struggle because of the ownership of land because people wanted to take it and because of this struggle, because of this great struggle, the idea came to make it possible to concentrate more this center that today we call the environmental education center. what? in order to strengthen the community and make it possible for businessmen to see this region with good eyes, which they do not see. (Representative Of Community Tourism In Curral Velho, 2019).

It is also possible to perceive efforts arising from the government when local managers have taken other actions to take advantage of the tourist potential and achieve economic promotion. Among them, there is the bus terminal, which, according to the manager of the Tourism portfolio, is the "entrance gate" of the city. Upon being interviewed, the secretary expresses the credibility that the improvement in the living conditions of families in coastal communities has been perceived since the development of tourism in the region. As the secretary expresses,

[...] Tourism moves with many categories, it starts to provide living conditions for the greengrocer, it starts from there, right, there it goes ... there comes the question of taxi, moto-taxi, the hotel chain, inns, comes food ... messes with everything! [...] Those who knew Acaraú 10 years ago (sic,) today feel that there is already this return of tourism in our municipality, it is still not the way we want it, we have not even arrived in 40% that we want ... we just take let's say - the burr of the Tourism inductor that is Jeri. Even so, Tourism arrived late, so it has already influenced a lot here in the region and mainly in Acaraú because this is the region's pole city (sic).

The Acaraú Tourism Secretary also points out, in an interview given, that she tries to work with local tourism through the implementation of legal mechanisms that place "Tourism as the great development driver of Acaraú. Hence the Municipal Tourism Council was created, [...] the Municipal Tourism Plan and the Tourism Fund" (sic).

There is also a vision of care for managers with the social impacts arising from the development of tourism in the region. For the secretary of Social Assistance, these recent actions also awaken the awareness of regional growth in several aspects, but that the problems have also grown proportionally, and work must be done so that the region is visited by more tourists, but that it is necessary to do a job prevention, especially to avoid problems such as sexual exploitation and "real estate exploitation". For him, it is necessary to create, through actions by the local secretariats, "[...] a social protection wall, a wall that will reduce the major impacts".

With the above, it is possible to perceive the history of struggles and articulations as the engagement of families residing in Curral Velho to protect land and the environment and the adoption of sustainable practices that favored the development of tourist activities that predisposed a source of income, the which ended up revealing the regional tourist potential of the municipality of Acaraú. These local expressions, consist of the interpretation of the perspectives pointed out in the discourses disseminated by the community tourism networks pointed out by Moraes, Irving & Mendonça (2018) where under the endogenous perspective,

The aim is to enhance the natural and cultural heritage, respect and protection for ancestral territories, the development of organizational capacities, cooperation between social actors to facilitate network-

ing and technical and operational qualification for the benefit of the populations directly involved with community-based tourism.

On the other hand, it is also evident, the exogenous perspective to the process of formation of community tourism networks, pointed out by the authors. In this second perspective, "the aim is also to develop the promotion of alliances with external agents, in an attempt to create alternatives for the insertion of the proposals under development in the tourism market and, also, for the influence on the design of public policies" (Moraes, Irving & Mendonça, 2018, p. 14).

Thus, with the implementation of public policies in the municipality of Acaraú and the involvement of local managers, one can perceive the transformations established from the movements to defend territories, protect the environment and create alternatives for the survival and well-being of social actors.

CONCLUSION

With the conclusion of this study, the relevance of the research for the unveiling of how community tourism takes place in the coastal area of the far west of Ceará is confirmed, with emphasis on the history of struggles of the community of Curral Velho and sustainable action practices through the community tourism.

It is also noticed that such practices converge to the idea of Social Innovation presented by André and Abreu (2006) who describe it as actions idealized by a group of people who arrange to work together in the creation and implementation of new ideas, which become viable solutions in solving social problems, which mitigate people's ability to have better living conditions.

There is also a profound relationship between the case of Curral Velho and the generation of Social Innovation, in the light of these authors, who believe that the simultaneous fulfillment of three requirements is necessary: the satisfaction of human needs not met by conventional means, that is, the market; presenting as a result the promotion of social inclusion; and the training of subject social actors, potentially or effectively, at risk of social exclusion / marginalization, providing a more or less intense change in power relations.

At the same time, community tourism practiced in the Coastal Community of Acaraú is associated with the dimensions expressed by Tardif and Harrisson (2005). Let's look at the following dimensions / actions:

Transformations - preventing the increase in shrimp farms and local exploitation and facing the problems resulting from the practice of shrimp farming.

Innovative Character - development of tourist activities to generate income out of interest for the common good, society and sustainable local development.

Actors - involvement of social groups and families for the construction of the Environmental Education and Community Tourism Center "*Encante do Mangue.*

Still according to the dimensions presented by the authors, the dimension '***processes***' has to do with the way of coordination and mobilization and participation, thus, in the case of Curral Velho, the creation of the Environmental Education Center in 2007, which offers training courses for young people and adults, free internet access and audiovisual workshops financed by the Ministry of Culture and *Banco* *do Nordeste*, is a strong example of how groups are able to propose coordinated actions for the common good and development of the local community.

It is concluded that residents perceive the importance of tourism for local development and that there is a close relationship between history for building spaces to strengthen the community and the dimensions presented by Tardif and Harisson (2005) from the perspective of social innovation when there was a transformation of spaces and innovation with the purpose of guaranteeing the common good and general interest and cooperation.

It is also understood that, with this essay, there is a need to deepen the research through the analysis of public policies implemented in the municipality that complement these actions that generate local development and social transformations.

REFERENCES

André, I., & Abreu, A. (2006). Dimensions and spaces of social innovation. *Finisterra*, *41*(81), 121–141. Retrieved November 2, 2019, from https://revistas.rcaap.pt/finisterra/article/view/1465

Beni, M. C. (2006). *Tourism policy and planning in Brazil*. Aleph.

BEPA (Bureau of European Policy Advisers). (2011). Empoweringpeople, drivingchange - social innovation in the European Union. Publications Office of the European Union.

Borges, C. I. O. S. (2011). Community tourism in traditional communities of the coastal zone of Ceará: focus on the experience of the Red Tucum (Master in Geography). Universidad Estatal de Ceará, Fortaleza.

Cloutier, J. (2003). Qu'est-ce que l'innovationsociale? Cahier de recherchedu CRISES (n. ET0314). Montreal: UQAM.

Coriolano, L. N. (2003). *Tourism of Inclusion and Local Development*. Fortaleza: Funece.

Coriolano, L. N. (2012). Community tourism in Brazil. In L. N. Coriolano & C. A. C. Sampaio (Eds.), *Territories of Latin American Solidarity and Community Tourism in the Reduction of Transnational Mega Enterprises* (pp. 27–41). Ed. UECE.

Furtado, C. (1961). *Development and underdevelopment*. Culture Fund.

Furtado, C. (1964). *Dialectic of Development*. Culture Fund.

Furtado, C. (1974). *Essential*. Sao Paulo: Company of letters.

Furtado, C. (1984). *Culture and development in times of crisis*. Peace and Earth.

Furtado, C. (2002). *In search of a new model: reflections on the contemporary crisis*. Peace and Earth.

Gomes, A. O., & Vieira Neto, J. P. (2010). *Historian of Curral Velho (Acaraú / CE). Terramar*. Institute and the Cearense Community Tourism Network.

Irving, M. A. (2009). Reinventing reflection on community tourism, is innovation possible? In Community tourism: diversity of perspectives and Brazilian experiences. Rio de Janeiro: Letter and Image.

Loureiro, C. V. (2013). Community tourism as an alternative for the preservation of coastal ecosystems: the case of the Curral Velho Community, Acaraú-CE-Brazil. *Revista de Turismo Contemporâneo, 1*(1), 1 - 17.

Maldonado, C. (2009). Community rural tourism in Latin America: genesis, characteristics and policies. In R. Bartholo (Ed.), Community-based tourism: diversity of Brazilian perspectives and experiences. Academic Press.

Mbakogu, I. A. (2004). Is There Really a Relationship Between Culture and Development? *Anthropologist, 6*(1), 37–43. doi:10.1080/09720073.2004.11890827

Minayo, M. C. de S. (2001). Social research: theory, method and creativity. Rio de Janeiro: Vozes.

Minayo, M. C. de S. (2004). The knowledge challenge: qualitative health research (8th ed.). São Paulo: HUCITEC-BRASCO.

Ministério do Turismo. (2013). *National Tourism Plan 2013-2016: Tourism giving much more in Brazil.* Retrieved October 28, 2017, from: http://www.Turismo.gov.br/images/pdf/plano_nacional_2013.pdf

Ministério do Turismo. (2018). *PRODETUR*. Retrieved January 04, 2018, from: http://www.Turismo.gov.br/programas/5066-prodetur.html

Miranda, J. M. G. de. (2003). *Improvement of Clusters: Experiences from Asia and Latin America - "Evaluation and monitoring of APL Projects in SEBRAE: method and indicators"*. Term of Reference Sebrae.

Moraes, E. M. de., Irving, M. A., & Mendonça, T. C. M. (2018). Community-based tourism in Latin America: a network strategy. *Revista Tur., Vision and Action, 20*(2), 249-265.

Nogueira, M. A. (2009). Development, State and Society: The necessary relations, the possible coalitions to the required institutionality. In *Dialogues for development*. Brasília: Ipea.

Oliveira, G. B. (2002). A discussion on the concept of development. *FAE Magazine, 5*(2), 37–48.

Ricco, A. S. (2012). *Tourism as a social and anthropological phenomenon. Tourism*. Space and Local Development Strategies.

Tardif, C., & Harrisson, D. (2005). Complémentarité, convergence et transversalité: la conceptualisation de l'innovation socialeau CRISES. In CRISES. Centre de Recherche Sur Les Innovations Sociales. Cahiersdu CRISES.

Taylor, J. B. (1970). Introducing Social Innovation. *The Journal of Applied Behavioral Science, 6*(1), 69–77. doi:10.1177/002188637000600104

Villavicencio, B. P., Zamora, J. G., & Pardo, G. L. (2016). El turismo comunitario en la Sierra Norte de Oaxaca: Perspectiva desde las instituciones y la gobernanza en territorios indígenas. *El Periplo Sustentable, 30*, 6–37.

ADDITIONAL READING

Barddal, R., Alberton, A., Campos, L. M., & De, S. (2010). *The dimensions and methods of measuring sustainability and Tourism:* A theoretical discussion. *Social and Environmental Management Magazine.*, *4*(2), 138–155.

Bardin, L. (1995). *Content analysis* (70th ed.).

Bartholo, R., Sansolo, D. G., & Bursztyn, I. (2009). *Community Based Tourism: diversity of Brazilian views and experiences.*

Benevides, I. P. (2004). *Tourist practices and territorialities and governmental tourism planning in Ceará.* Doctoral Thesis, University of São Paulo, São Paulo.

da Veiga, J. E. (2008). *Sustainable development, what animal is this?* Associated Authors.

de Oliveira, A. A. R. (2015). *In search of Ceará: the convenience of popular culture in the figuration of Ceará culture (1948- 1983).* Thesis (doctorate) - Federal University of Ceará, Humanities Center, Department of History, Graduate Program in Social History, Fortaleza. Concentration Area: Social history.

de Oliveira, M. M. (2008). *How to do qualitative research* (2nd ed.). Editora Vozes.

Sessa, A. (1983). *Tourism and development policy.* Translation Lourdes Fellini Sartor. Uniontur.

KEY TERMS AND DEFINITIONS

Community-Based Tourism: The practice of this activity contributes to the conservation and development of a given region, with economic, cultural, and social benefits. It presents itself as a more democratic model of tourism that places local constituent subjects at the center of planning, implementing, and monitoring tourist activities.

Inland Communities: A group of people who live in the same place, usually with common social characteristics, with the same government, culture, and history and away from large urban centers.

Jericoacoara: Village considered as a tourism inducing nucleus, due to the territorial dynamics that promote development, of the modern activity that mobilizes visitor flows, having received equipment in accordance with State policies for Tourism, such as the construction of the Jericoacoara Tourist Pole Airport that serves to Tourism in the Municipality of Cruz e Jijoca de Jericoacoara, 62 km away from the headquarters of Acaraú.

Public Policies: Set of activities developed by the State with the help of public or private entities with a view to serving and guaranteeing the rights of society in general.

Social Innovation: According to the Stanford Social Innovation Review, it is a new solution to a social problem. A more effective, efficient, sustainable, or fair solution than the existing solutions and whose value generated benefits, as a priority, society as a whole.

ENDNOTES

1 The Tourism Development Program in the Northeast - PRODETUR / NE, is a policy that seeks to organize public interventions for the development of tourism activity, through previous planning processes in tourist regions (Mtur, 2015).

2 The National Tourism Plan - NTP, already in the first decade of the 21st century, was implemented with the challenge of contributing to the reduction of regional inequalities, in a dynamic and continuous process, aiming at expanding growth and deconcentrating the income derived from Tourism, for through regionalization, interiorization and segmentation of tourist activity.

3 With its legal framework in 2004, the Regionalization Program for Tourism - PRT defined tourist regions and categorized Municipalities for Tourism.

4 According to Miranda (2003), a Local Productive Arrangement - LPA is characterized by having clusters of companies, established in the same territory, presenting a productive specialization and maintaining links of interaction, cooperation and learning among themselves and with other local actors, these governments being business associations, credit, teaching and research institutions.

5 Municipal per capita percentage - which corresponds to all wealth generated by the Municipality's economy in a given year, relativized by its resident population.

6 It is an area around the mangrove also known as "Salgado". It is of great importance for the balance and maintenance of nature's biodiversity. To destroy them is to announce the death of the mangrove ecosystem, weakening the living conditions of more than 65% of fish species and other organisms that need estuaries to be born, grow and reproduce. It shelters lives like Chié, which has as one of its functions 'plowing' the land of the "Salgado" for the mangroves to develop (Gomes, 2010, p. 8).

APPENDIX

Table 1. Semi-structured interview - public power group **Interviewee: Date:**

Question	Answer
What is your perception of local tourism?	
Tourism in Acaraú started to have greater importance as of what fact, action or intervention?	
Is there Ecological Tourism and Rural Tourism in the Municipality? What activities inherent to ecological and rural tourism are developed?	
By whom are these activities organized?	
Are there disputes / conflicts between actors directly linked to tourism? Which groups of actors? What are the conflicting interests?	
Who are responsible for the protected areas and how do they interact with the changes due to the expansion of tourist activity in the rural area of Acaraú?	
Are there interinstitutional partnerships, with communities focused on the monitoring and planning of activities related to Tourism or for the implementation of initiatives related to Tourism?	
Are there participation forums focused on tourism planning and management? Who participates?	
Are there any support initiatives for the conservation, enhancement and rescue of local culture?	
Are there or did effective PRODETUR actions exist in the Municipality?	
How is the monitoring of PRODETUR actions in your Municipality?	
What were PRODETUR's main investments in the Municipality?	
What are the positive and negative impacts of economic activity?	
Does the Participative Municipal Master Plan exist?	
Is rural and cultural tourism encouraged? Through what actions?	

Table 2. Semi-structured interview - group of entrepreneurs and community actors related to tourism.
Interviewee: Date:

Question	Answer
What is your perception of local tourism?	
Tourism in Acaraú started to have greater importance as of what fact, action or intervention?	
Is there Ecological Tourism and Rural Tourism in the Municipality? What activities inherent to ecological and rural tourism are developed?	
By whom are these activities organized?	
Is the local community of Acaraú organized to receive tourists? How? Who are the leaders?	
Are tourism workers and entrepreneurs organized? Is there an incentive on the part of the local authorities to organize themselves?	
Are there disputes / conflicts between actors directly linked to tourism? Which groups of actors? What are the conflicting interests?	
Who are responsible for the protected areas and how do they interact with the changes due to the expansion of tourist activity in the rural area of Acaraú?	
Are there interinstitutional partnerships, with communities focused on the monitoring and planning of activities related to Tourism or for the implementation of initiatives related to Tourism?	
Are there participation forums focused on tourism planning and management? Do you participate?	
Do you know the implemented actions that are part of PRODETUR? Which are?	
Do you know tourism development programs at the municipal level?	
What are the positive and negative impacts of economic activity?	
Are natives employed by tourism enterprises? what types of activities?	
What has been done to increase the participation of natives in the workforce?	
Is rural and cultural tourism encouraged? Through what actions?	
Have you and your family experienced changes in the quality of life due to the development of tourism in the city?	
Is there a generation of jobs in your community due to the development of tourism?	

Chapter 6
Creativity and Entrepreneurship:
Gastronomic Tourism in Mexico

Nick Naumov
University of Northampton, UK

Gheorghia Costandachi
University of Northampton, UK

ABSTRACT

The relationship between food and tourism is well documented and a subject of increased academic interest. The role of gastronomy for the development of tourism is prominent from economic and socio-cultural perspectives, and there is sufficient evidence to demonstrate the critical role of food festivals, cultural traditions, and local customs from an intangible heritage perspective. Although food tourism has been explored and researched from a wide number of perspectives, its role in the context of creativity, innovation, and entrepreneurship has largely been neglected. This chapter seeks to address the gap in academic literature focusing on how food tourism can drive community development, stimulate local businesses, and inspire creativity and innovation. The chapter focuses on the development of food tourism and the role of gastronomy within the context of tourism development in Mexico. The results of this study highlight a number of creative and innovative approaches to food tourism and demonstrate the variety and diversity of food-related events in various parts of Mexico.

INTRODUCTION

Gastronomy is traditionally related to the practices of consumption and enjoyment of food. As Mulcahy (2019) explains, contrary to the popular perception, gastronomy is not a twentieth-century phenomenon and has a long history. In his book *La physiologie du goût* (The physiology of taste), Brillat-Savarin (1994) refers to gastronomy as 'everything connected with the nourishment of man' and explains how gastronomic traditions emerged in France and how they were accessible only to a certain group of privileged individuals. Published in 1825, his book is one of the seminar writings about gastronomy and among the first titles to introduce gastronomy as a culinary art and 'true science' (Courtine, 1996,

DOI: 10.4018/978-1-7998-4855-4.ch006

p.167). Gastronomy as a 'true science' but also as a discipline, has evolved substantially since Brillat-Savarin's contribution and has become a far-reaching discipline that includes anything related to what we eat and drink and it is no longer a privilege of nobility but also includes regional and local cuisines. For Santich (1996, p.2), gastronomy refers to the production of food, the preparation, cooking, the art of serving and enjoying fine food.

Gastronomy has always played an important part in the context of tourism management and development from both consumption and experience perspectives. According to the Global Report on Food Tourism (UNWTO, 2012), food constitutes an essential part of every tourist's budget and often accounts for 30-40% of the overall tourist expenditure (Boyne, Williams & Hall, 2002; Hall, Sharples, Mitchell, Macionis, & Cambourne, 2003; Horng, Liu, Chou & Tsai, 2012). According to Richards (2002), gastronomy has become a global industry and tourism plays a huge part in its development. The shift from a service-based economy to experience-based economy (Pine & Gilmore, 1999) has facilitated the emergence of 'memorable experiences' putting the role of food in the arena of experience production. Gastronomic activities which seek to provide more memorable, participatory and authentic experiences, have been developed at many destinations all over the world focusing for example, on cuisine, traditional food productions and dining experiences (see for example, Alonso & Northcote, 2010; Cohen & Avieli, 2004; Mak, Lumbers & Eves, 2012; Murayama, 2012; Pulido-Fernandez, Casado-Montilla & Carrillo-Hidalgo, 2019). Gastronomy has become a crucially important pillar of local and national economies (Kim, Yuan, Goh & Antun, 2009), a significant instrument for improving destination competitiveness (McKercher, Okumus & Okumus, 2008; Okumus, Okumus & McKercher, 2007) and a valuable tool used for marketing and advertising purposes (see for example, Lin, Pearson & Cai, 2011; Horng & Tsai, 2012; du Rand & Heath, 2006). Moreover, gastronomic tourism has emerged as a form of special interest tourism that represents the relationship between gastronomy and tourism and primarily referring to how food is marketed and promoted as a tourism product (Getz, Robinson, Andersson & Vujicic, 2014; Hjalager, 2002; Hall & Mitchell, 2001).

Gastronomic tourism, however, should not solely be considered in economic terms but also recognised for its wider socio-cultural implications. As Everett (2016, p.7) argues, food tourism is a 'multi-faceted', 'ever-changing' and 'complex concept' and it is difficult to frame it within rigid conceptual boundaries. Our understanding of the role food plays within our society has changed much since the first considerations of food tourism as an economic activity in the 1980s (see for example, Belisle, 1983). According to Richards (2002), gastronomy as a system of production, distribution and presentation of food can now be defined as 'cultural industry' as it involves a range of cultural practices such as presentation of food, the way food is consumed, etc. Scarpato (2002) argues that food consumption is a socio-cultural practice and even talks about the emergence of 'gastronomic culture'. In a similar vein, Hegarty and O'Mahony, (2001) refer to gastronomy as a form of 'cultural expressionism' and argue that the selection, preparation, presentation and participation with culinary and gastronomic aspects of food should be considered as 'fine art activity' and relate gastronomy not only as a consumption of food but also as a symbolic representation of status, prestige and aesthetics.

In line with this conceptualisation, the academic scholarship has explored the practices of food consumption from various sociological, cultural and anthropological perspectives where food has been defined not only as a sustenance but also as an intangible practice that relate to people in social, cultural and political dimensions (see for example, Arnott, 1975; Medina & Tresserras, 2018). As tourism has shifted from Fordist to post-Fordist approach (Ioannides & Debbage, 1997; Torres, 2003), gastronomic tourism has also shifted from consuming food and beverages to experiencing different cultures through participation and involvement at food-related and gastronomic experiences (Everett, 2009; Everett & Aitchison; 2008; Naumov, 2019; Santich, 2004). As previous studies have demonstrated (Everett & Slocum, 2013; Lee & Arcodia, 2011) gastronomic tourism is now conceptualised as a participatory and holistic experience that provides us with the opportunity to immerse ourselves into the culture and heritage of a region, making the experience memorable in a completely unique and personal way. In this context, gastronomic tourism is an experience in which we learn more about the local gastronomic traditions, appreciate the value of food from cultural/anthropological perspective (Metro-Roland, 2013; Montanari, 2009) and appreciate its importance in the context of cultural identity (Lugosi, 2012; Timothy & Ron, 2013).

Gastronomic tourism is often considered as a creative industry and as such, has the potential to stimulate local small-scale businesses, revive local traditions and customs and stimulate creative and innovative ideas to emerge (Richards, 2002). However, although gastronomy has been explored and researched from a wide number of perspectives such as place-making and consumption (Everett, 2012), local development and product diversification (Kim & Ellis, 2015;Kiralova & Hamarneh, 2017), and tourist experience (Kivela & Crotts, 2006), its role in the context of creativity, innovation and entrepreneurship has largely been neglected (Boesen, Sundbo and Sundbo, 2017). For instance, more research is needed to explore local food suppliers' network relationships and more specifically, in the context of hotel restaurants and micro-scale local restaurants (Rachão, Breda, Fernandes & Joukes, 2019).

This chapter seeks to address the gap in academic literature about the role of gastronomic tourism within the framework of innovation, entrepreneurship and creative industries focusing on how food tourism can drive socio -cultural community development, stimulate local businesses and inspire creativity and innovation. Our paper focuses on the development of food tourism and the role of gastronomy within the context of tourism development in Mexico. The increased popularity of Mexican food worldwide combined with a great geographical abundance of various tastes and flavours, have put Mexico at the forefront of the most popular destinations in the world. Historically, Mexico has been one of the first countries to focus on food and the first country where wine industry started in Latin America in 1597, long before the wine industries developed in Argentina, Chile and USA (see for example, Tejo-Prench, 2011). However, despite its prominent importance, gastronomic tourism is yet to be considered as an experience and there is much room for creativity, community initiatives and small-scale entrepreneurship. This chapter focuses on some of those initiatives exploring the potential of gastronomic tourism as a tourism experience while also highlighting some good practices of community co-operation, innovation and creativity.

The chapter begins with a critical overview on tourism and gastronomy focusing on the conceptual architecture of food tourism and its emergence and development as a field of academic enquiry. We then explore the creative and innovative approaches to the local gastronomic traditions in the context of Latin America. A brief overview of Latin America from a food production perspective is provided followed by examples of Ecuador, Peru and Argentina which demonstrate how local food, cuisine and cultural traditions have been creatively transformed to tourism events, products and experiences. The main body of the chapter explores the development of food tourism experiences in Mexico and focuses on few notable examples from the cities of Ensenada, Mérida, Puebla, Oaxaca and Guanajuato.

BACKGROUND

Recognising the multi-faceted and multi-disciplinary nature of gastronomy, Gillespie (2002, p.5), argues that gastronomy is 'about the recognition of a variety of factors relevant to the foods and beverages ate and consumed by a group, in a locality, region or even a nation'. Those factors could include marketing and management approaches (Henderson, 2009), cultural differences and more intangible culinary traditions (Hegarty & O' Mahony, 2001; Richards, 2002), and the existence of basic infrastructure (Williams, Williams & Omar, 2014). All those factors have shaped our understanding of the relationship between food and tourism and have also influenced the emergence of gastronomy as a tourism product.

The relationship between gastronomy, food and tourism has received a notable academic interest for the past few decades, but it has gained a momentum in the more recent years (Ellis, Park, Kim & Yeoman, 2018). For example, a number of special issues of tourism academic journals have emerged exploring food heritage (*Journal of Heritage Tourism* in 2013), food and tourism synergies (*Scandinavian Journal of Hospitality & Tourism* in 2017), sustainable food and wine tourism (*Sustainability* in 2019) and gastronomic practices in tourism (*Advances in Hospitality & Tourism Research* in 2019). This increased interest has enhanced the scope of gastronomic studies and has widened its conceptual boundaries. The wider academic literature on tourism and gastronomy considers 'food tourism' as an umbrella term and defines it as a 'food and drink motivated travel' (Everett, 2016, p.8). In this context, there has been a number of sub-sectors (micro-niches) that focus on the differences in terms of tourist motivation, tourist activities and tourist behaviour. Ellis et al. (2018) conceptualise food tourism from three different perspectives: activity-based, motivation-based and mixed. The *activity-based* perspective focuses on the involvement and engagement of the tourist in specific gastronomic experiences and includes any activities such as visitation to sites of food production, culinary workshops and cooking classes (see for example, Che, 2006; Hashimoto & Telfer, 2006; Presenza & Chiappa, 2013). The *motivation-based* perspective defines food tourism in the basis of tourist motivation and focuses more on the influence gastronomy plays on the choice of a destination to visit. For example, recent research undertaken by the Japan External Trade Organisation demonstrates the food is the number one motivation for international tourists to visit Japan. 58.5% of all international visitors to the country are strongly motivated to learn more about food production and taste the Japanese sushi, sashimi and other unique gastronomy products of Japan (Murayama, 2012). In this context, food can either be a 'principal resource' and the main purpose of the trip is for the tourists to enjoy the specific food products at a given destination (Smith & Costello 2009, p.49); or a 'secondary resource' for which the tourists can be partly motivated by the food experience (Bertella, 2011, p.355).

From a similar perspective, Hall and Sharples (2003) provide us with a classification of those activities into four different levels of interests. A the first level, *high interest*, gourmet tourism, gastronomic tourism and cuisine tourism, are considered as sub-niches of food tourism in which food is the primary motivation for travelling. At the second level, *moderate interest*, food-related activities provide a window the destination's lifestyle and traditions. For instance, this is exemplified with the sub-niche of culinary tourism, broadly defined as 'tourism trips during which the purchase or consumption of regional foods (including beverages), or the observation and study of food production (from agriculture to cooking schools) represent a significant motivation or activity' (Ignatov & Smith, 2006, p.238). At the third level, *low interest*, food-related activities are not the main motivation for travelling but they are considered as an opportunity for memorable experiences and still have a role to play in the decision-making process (Ottenbacher & Harrington, 2011). They are widely considered as a part of rural/urban tourism activities and often focus on visits to specific locations such as food festivals, restaurants and farms (Hall & Mitchell, 2001, p.38) or visits to primary/secondary food producers (Hall & Mitchell, 2005, p.20). At the fourth level, *low interest/no interest*, food is considered as a physiological necessity and thus only refers to eating and drinking.

The *activity-based* and *motivation-based* perspectives of Ellis et al. (2018) as well as the 'low interest' category of Hall and Sharples (2003) both put a strong emphasis on the need of having specific experiences which to engage the tourists in a more meaningful and personal way. This implies that local/national governments and destination managers in particular, should put a focus on creativity and innovation, and stimulate local entrepreneurship in order to facilitate the emergence of meaningful and memorable experiences. Creativity and innovation primarily refer to the production and implementation of new ideas, approaches and inventions (Schumpeter, 2008; Hall & Williams, 2008), while tourism entrepreneurship is broadly defined as a process that involves the creation and establishment of new, small-scale and local business opportunities that create value for both individuals involved and the society they live in (Page & Ateljevic, 2009; Shaw, 2014). As argued by Aeboleya et al. (2014), the gastronomic industry needs a strong emphasis on innovation in order to attract and retain customers, meet the ever-changing needs and demands of tourists and ensure long-term competitiveness. In the context of food tourism, this underlines the development of creative approaches to food production, 'experiential turns' to local food entrepreneurs, and the provision of authentic experiences (Sims, 2009). This is documented in previous studies which have reported positive impacts of food tourism development on destination branding, image and wider economic development. For instance, Karagiannis and Metaxas (2017) argue that culinary experiences can bring a number of potential benefits to a destination if promoted appropriately and if there is a good relationship between local entrepreneurs and public authorities. Their example is the Ontario Culinary Tourism Alliance (OCTA), a non-profit industry driven organization suggests that connects and coordinates all food tourism stakeholders in Ontario, Canada. In the same vein, Lane, Jones, Jones and Mitchelmore (2016) demonstrate how food and drink-based activities stimulate rural development and provide opportunities for small-scale entrepreneurship in the context of Wales while Ab Karim & Chi (2010) put an emphasis on the positive impact of food tourism on destination image and branding.

In this chapter, we consider 'food tourism' as an umbrella term and focus more explicitly on the development of food-related activities in Mexico from a holistic perspective. Our aim is to critically assess the current development of food-related activities and exemplify the important of food tourism from both economic and socio-cultural perspectives.

FOOD TOURISM IN LATIN AMERICA: FROM FOOD CONSUMPTION TO GASTRONOMIC EXPERIENCE

Latin America is a great example of a region where gastronomy plays a prominent economic and socio-cultural role. For centuries, there has an exchange between Latin America and other continents and this has enriched the variety of products and flavours.Latin American cuisine has been shaped by Spanish and Portuguese influences which have diversified the already rich gastronomic traditions of the local indigenous population (Schlüter, 2012). As Pilcher (2017) notes, the colonial rule across the continent has brought some unique flavours and ingredients from all over the world and this has greatly enriched the variety and availability of food products at present days. From Europe, ingredients such as wheat, olive oil and grapes have reached Latin America and these elements relate to the religious side of Iberian culture where bread (made of wheat) and wine (made of grapes) are one of the main ingredients (Earle, 2010). Other products such as barley, buckwheat, citrus fruits and meats (beef, pork, lamb) were imported and used by the European settlers to protect their bodies from potential disorders that indigenous products were thought to cause. Sugar cane from Africa and rice brought from Asia also became part of the everyday's main meals and by so innovating the gastronomy and social values exchanged among the different cultures. These new combinations led to the creation of traditional dishes such as 'burrito' in Mexico (using rice), 'carne asada' in Argentina (using beef) and 'Mote con Huesillos' in Chile (drink made from wheat) (Earle, 2010).

Similarly, the food that originated within the continent, such as potatoes, corn and tomatoes, have enriched the habits and tastes of all parts of the world. Corn (maize) has influenced the cuisine of Europe creating dishes such as polenta in Italy (Pilcher, 2017). Portuguese conquests have brought maize to West Africa, creating traditional dishes such as 'ugali' in Kenya. In Middle East, China and India, maize has become an ingredient with a huge impact on population growth bringing also many agricultural opportunities. Tomatoes and chilli have been quickly adopted by Europeans becoming a significant item within many European cuisines (Pilcher, 2017). Sunflower seeds are now a basic ingredient around the world, and the sunflower oil is now used around the world for many fried dishes (Echeverria, 2006). Chocolate is another great example considered to have originated from Latin American empires such as Maya and Azteca as a God beverage; coming to Europe this product has been adopted and consumed by the higher-end class, firstly in Spain and Portugal (Echeverria, 2006).

The importance of local food within Latin America has been declared in 2004 with the organisation of the *Forum on Hunger in the Andean Countries*, where local authorities supported by United Nations decided to 'improve nutrition with local produce' (Schuler, 2012). This initiative has reintroduced local products, which have since become major tourists' attractions. Henderson (2009) argues that one of the most effective ways to attract tourists is through organising events due to the expected positive socio-economic impacts.

In Ecuador, the government puts a particular emphasis on gastronomy and supports different regions to popularize food tourism. Ecuadorian gastronomy is characterized by its great variety of flavours and aromas. Ecuador is famous for its food markets called 'huecas', where one can find traditional Ecuadorian cuisine, community recipes and locally grown ingredients (López-Guzmán, López, Gálvez & Vélez, 2019). Food festivals are increasingly common and the best example is *The Raíces International Gastronomic Festival*, the most important one from a tourism perspective. Held since 2014 in Guayaquil, a famous tourism destination, the festival is organized but the city's tourism department and aimed at attracting domestic and international visitors. The festival helped reinforcing the local 'huecas' through public aid funds, also establishing culinary quality standards for community and tourists. Main dishes sold are 'seco de pollo' (chicken stew), 'guatita' (beef tripe stew) and 'caldo de bola' (dumpling soup). In 2015, the festival involved 128 huecas, attracted some 90,000 tourists and improved country's research work, while also applying innovation for social development through Ecuadorian gastronomy (López-Guzmán, Lotero, Gálvez, & Rivera 2017).

In Peru, *Novoandina* is another new gastronomic initiative in, where ancestral food is prepared using new techniques, in this way combining traditional cooking techniques with modern approaches (Mendoza, 2012). As Georgiadou, Gonzalez-Perez & Olivas-Lujan (2019) highlight 'this culinary explosion' in Peru comes from its colonial past, indigenous heritage and cultural variety the country has, in order to celebrate their traditional and diversified culinary treasures. This new gastronomic concept, particularly popular in Lima and Cuzco, combines well the aim to conquer the world culinary industry and highlights their national identity. The *Novoandina* initiative helps promote the traditional recipes such as 'Causa' (yellow mushed potatoes), 'Ocopa a la Arequipeña' (boiled potatoes with onion, cheese, peanut and Peruvian traditional herb 'huatacay') and 'Ceviche' (raw marinated fish with onion, chilli, salt and lemon) (Sainsbury, 2012). The new idea has helped to create new fusion foods as 'Quinotto' (where risotto takes a Peruvian swing being prepared with quinoa) and coca (bread that uses the controversial coca juice) (Eat Peru, 2016).

Another prominent example is Mistura, Peru's largest food fair. Organized by Apega, the Peruvian Gastronomy Association, Mistura started as a small local event in 2008 but grew up to become the largest gastronomic event in Latin America. The fair is a space where culinary experts, including famous chefs but also street vendors, of all backgrounds come together to showcase their Peruvian pride through what they cook (Bak-Geller Corona & Matta, R., 2020).

In Argentina, the small village of Tomás Jofre (after a Belgian engineer that introduced the railway to the village), has become a provincial gastronomic hub, due to the rich menu the area offers and also due to the reinforced *National Rural Tourism Plan* in Argentina in 2000 (Ellul & Schlüter, 2008). Tourists are attracted by the tranquil location of the village and the well-designed meals, where for example, breakfast consist merely of cooked pork or beef, homemade pastas and artisanal desserts (Ellul & Schlüter, 2008). A traditional dish of the village is 'Salame Quintero' (sausage made of beef and pork), for which a festival is being celebrated every year in January, known as *Fiesta del Salame*.

As the previous sections have emphasized, Latin America's cities have become culinary hubs due to their vast admiration for quality of food and importance kept towards the ancestral traditions. This is exemplified with the inclusion of many destinations in Latin America in the UNESCO Creative Cities Network which recognises different types of intangible heritage such as culture, literature and gastronomy. For instance, Cochabamba (Bolivia) is recognised for its food heritage and the city has hosted international food festivals such *as Expoalimenta* (2019). The city has also developed a House of Gastronomy Urban Development project which aims to transform vulnerable areas to local food production hubs (UNESCO, 2020a).

Panama City is another example which hosts one of the oldest food festivals, *Panama Gastronomica.* The event celebrates local food, crafts and traditions, stimulates culinary entrepreneurship and boosts local economic development (UNESCO, 2020b). In addition, Florianopolis in Brazil is widely known as 'Brazil's Capital of Entrepreneurship' due to its sustainable development projects. The city increases the importance of traditional cuisine by expanding exchange programmes between the students in culinary schools and industry's professionals (UNESCO, 2020c).

Despite the undisputed importance of food from a social, economic and cultural perspectives, there is a lack of academic research that focuses on how food stimulate economic development, revitalizes local traditions and customs and drives the development of urban and rural areas. As this section has illustrated, food takes a special places in Latin America and the variety of food festivals demonstrate the importance of food production as an economic driver but also a tool for social cohesion. The next section focuses on gastronomic traditions in Mexico and explores the significance of gastronomy and food traditions from a socio-economic perspective.

GASTRONOMIC TRADITIONS AND TOURISM IN MEXICO

A special case of its gastronomic heritage and gastronomic tourism is Mexico. In 2010, UNESCO inscribed the Traditional Mexican cuisine on its *Representative List of the Intangible Cultural Heritage of Humanity* highlighting the distinctiveness of Mexican cuisine as a 'comprehensive cultural model comprising farming, ritual practices, age-old skills, culinary techniques and ancestral community customs and manners' (UNESCO, 2019). The traditional Mexican cuisine is based on traditional ingredients such as corn, beans, chilli, tomatoes, peppers and avocados and includes some simplistic meals (e.g. tortillas) but also some special dishes such as 'tamales', traditionally eaten on special events such as 'El Día de Los Muertos (The Day of the Dead). The Mexican gastronomic traditions are deeply embedded within the local culture, history and heritage with many traditional recipes been passed from one generation to another. Food tourism in the country is in line with the understanding of food sharing as a social practice in which the 'guest' (the tourist) is treated with food by the 'host' (the local community), showing gratitude and respect to the local traditions (Kearney, 1970).

Gastronomic industry is very important in Mexico from a socio-economic perspective. It is estimated that there are around 221,000 restaurants in Mexico, of which 96% are small and micro enterprises which demonstrates the importance of small-scale businesses and local entrepreneurs, and the development of 'gastronomic entrepreneurship' (Sanchez-Gutierrez, Gonzalez-Uribe, Gutierrez-Govea & Garcia-Jimenez, 2012). For example, in various parts of the country, local entrepreneurs have established small food chains (e.g *Taqueria México*) promoting the variety and diversity through different events, festivals and other special occasions. Gastronomy is regarded as an opportunity for local people to establish and sustain a business as it provides opportunities for long-term development (Gonzalez, Husted & Aigner, 2017). Our analysis focuses on the cities of Ensenada, Mérida, Puebla, Oaxaca and Guanajuato. Each of these places has unique food traditions and food culture reflecting the customs, values and traditions of the local residents.

Ensenada, like many others coastal cities, is famous for its seafood, fish dishes and also locally grown grapes and olives. In additional to the famous 'fish taco', Ensenada also provides various opportunities to explore its ethnically diverse food heritage. For example, traditional and authentic bread making traditions of the local community of Russian immigrants or the French restaurants that provide Mexican-influenced French dishes. Ensenada is also the most important wine region in Mexico which produces 90% of the country's wine representing more than 500 labels and 50 wineries. Ensenada celebrates its gastronomic heritage by organising special events dedicated to its gastronomy. For example, the 'Ensenada for everyone' festival celebrates traditional food with indigenous dances bringing ethnic minorities, immigrant groups and locals together. La Vendimia, a grape harvest festival, is a Greek inspired event that involves ten days of various activities such as tasting, cultural events, educational seminars, etc.

The establishment of Ensenada's Gastronomic District has also supported the development of food tourism and has facilitated multi-stakeholders' collaboration and exchange of practices, including residence of celebrity chefs and gastronomy experts from other UNESCO Creative Cities of Gastronomy which is supported through the International Culinary Exchange programme. Moreover, food tourism has been enhanced by the introduction of the *Gastronomic Landscape of Baja California* programme which provides research funding to conduct research on regional cuisine, local traditions and ingredients and exchange of good practices (UNESCO, 2020d)

The Yucatan city of Mérida is another destination famous for its food culture and heritage. It was declared as a gastronomic hub by the Yucatan Government in 2013 and the importance of food as intangible heritage was declared. The local food production of the region has a cultural value to promote the indigenous and traditional flavours of the region in a more sophisticated way. Mérida's food scene is influenced by Mayan, Caribbean and Spanish cuisines and include a number of locally grown products such sour orange, sweet chillies, achiote, habanero pepper and oregano. Epazote (also known as ipazote, pazote, pazoli and paico), a very traditional for the region, is a fine aromatic herb with a distinctive strong flavour. It is widely used in quesadillas and some dishes such as 'Frijoles de Olla' (pot beans). 'Mulbil Pollo' is another traditional dish prepared especially for the Day of Dead festival (Ayora-Diaz, 2019). Mérida's added value to gastronomic industry is through delivery of food festivals. For instance, Sobremesa comes to promote Yucatan's food at an international level, and in 2018 there were 40 international chefs involved, including the Mexican chef Martha Ortiz (National Geographic, 2018).

Puebla is another city recognised as a culinary point, due to its varied foods and influences (Hinojosa, 2009). Traditional products grown in the region are green tomatoes, Chile Poblano (Puebla pepper), corn and cactus. Varied ingredients have led to the creation of dishes as 'cemitas' (bread in form of brioche with cheese, avocado and meets), molote (deep fried corn pastry), 'pelonas' (a bun shaped bread filled with meat, beans and lettuce), 'chalupas' (small 'taco' shaped corn tortillas with pork, onion and red salsa) and 'enfrijoladas' (tortillas filled in with black beans in black bean sauce). Influences from other states and countries have led to the creation of distinctive dishes as 'Taco Arabe (where meat is cooked following the Middle Eastern techniques) and 'Chipotle relleno' (where peppers are being filled with Manchego cheese that originates in Spain) (Ojeda, Camacho, Hernández & López, 2017). The rich ingredients within the area have led to the creation of some signature dishes – 'mole poblano' and 'chile en nogada' (chile in walnut mix). 'Mole poblano' contains more than 20 ingredients, main ones being red chilli and chocolate. In Puebla, 'Mole poblano' has a particular significance, being a special dish for the celebration of '5 de Mayo', where French troops lost a battle in Puebla in 1862 (Jiménez Medina & Escalante Conde, 2009). 'Chile en nogada' represents 'chille poblano' filled with shredded meat, walnut sauce and pomegranate. This dish is eaten usually in August and September when the pomegranate season is on. This corelates with the Mexico's Independence Day on 15th of September where this dish got special significance (Valencia, 2019). These traditional dishes of Puebla are found usually in the city centre as part of the 'street food' offering, and they provide an opportunity for visitors to taste local food. (Barrios & Pantoja, 2016). Barrios & Pantoja (2016) show that 80% of the tourists are likely to promote and recommend the food of the street vendors in Puebla.

The state of Puebla also celebrates its rich culinary traditions through different festivals such as *Festival de Nopales* and *Feria del Pan*. The former takes place in the small town of Tlaxcalancingo, celebrating foods made out of Nopal – salads, stews and even ice-cream. The latter is celebrated in Cholula, near the biggest pyramid base in Americas - Quetzalcoatl, celebrating the variety of 150 types of bread, which is followed by the celebration of Day of Dead (Mexconnect, 2020).

Oaxaca is another big culinary hub in Mexico and that has unique fruits and vegetables such as 'Chipilín', 'laurel', 'guapaque' and 'sapote', all of which are unique to the place and define state's heritage (Hinojosa, 2009). Oaxaca's signature dishes include 'moles', 'tlayuda' and 'tamales', together with the unique drink – mezcal (Bustos & Galván, 2015). Bautista & Smit (2012) describe Oaxaca as 'capital of mezcal', which is a traditional alcoholic drink made from smoked agave. This is a representative drink that bring Oaxaca's fame worldwide. Mezcal's importance is also proved through the creation of Mezcal National Board, an organization that brings together the State Committees of Durango, Zacatecas, Guanajuato, Guerrro, Oaxaca, San Luis Potosi and Tamaulipas. Founded in 2014, the board is responsible for recognising authentic tastes and protect authentic recipes of the 'probably the purest, most traditional spirit available on Earth' (Mezcal National Board, 2020).

The importance of ancestral food in Oaxaca is being celebrated in different festivals such as Guelaguetza celebration (Cocking, 2018). Held annually in July, this indigenous cultural festival originates from the pre-Hispanic times, including the elements of Zapoteco and aims to celebrate prosperity represented through traditional dances. In the present day, it is a tourist attraction, due to the food offered during the festival. Another event is the Mezcal Festival that takes place in July each year in Oaxaca and celebrates the rich tastes of mezcal from Oaxaca and other places such as Guerrero and Guanajuato.

In Guanajuato, the main dishes represented are 'tamales' and 'gorditas'. Tamales is a pre-Hispanic dish, made out of corn flour and filled with vegetables, meat or jam; usually eaten for celebrations as Christmas and Independence Day. 'Gorditas' are corn tortillas made in tortilla press and then filled with chilli, meat and 'frijoles' (beans) and then fried in a pan. These dishes are mainly served in street-food stalls. Guanajuato has implemented local policy to ensure secure vending, and food safety. The new policy promotes investment and small gastronomic businesses and requires hygiene documents from all food firms (UNWTO, 2017). The policy promotes the importance to preserve the traditional drinks and it also protects jobs as Jimador, Molcajetero (in Comonfort), Artisan baker (in Pénjamo) and Traditional cooks (UNWTO, 2017). New strategy and policy have helped Guanajuato to receive prizes for its gastronomic success as:

- 2014 Tourism Excellence Award: Mestizaje, cultura y sabor, Guanajuato Sí Sabe, FITUR.
- 2014 Iberoamerican Capital of Gastronomical Culture 2015.
- 2015 The best active tourist product by Tequila Circuit. FITUR
- 2015 Cooks Traditional Program. IPN National Polytechnic Institute School of Tourism

As this section demonstrates, food and gastronomic traditions in a wider context, are very nuanced, diverse and celebrated across various parts of Mexico. The examples of various Mexican cities have demonstrated the role food plays and more specifically, its cultural and even socio-economic importance. Food tourism undoubtedly has a great potential and it can stimulate the development of local/regional economies, revitalise traditions and customs. However, as it stands at the moment, there is a room for improvement and further development. First, in many areas of Mexico, including some of the destinations explained above, gastronomy is considered separable to regional and rural development. As illustrated by Cavicchi and Ciampi Stancova (2016), gastronomy has a great potential to develop rural areas and serve as a tool for regional development. In her work on Hungary, Csurgo, Hindley and Smith (2019) demonstrate the potential of gastronomic tourism for rural development and argue that gastronomy alone is insufficient to establish a rural tourism profile but can assist rural development. In the context of Mexico, it is essential to consider gastronomy not only as a separate economic activity but as a tool to be combined with other types of tourism such as rural tourism.

Second, the various examples of cultural festivals demonstrate the importance of food as intangible cultural heritage. This suggests the development of cultural and heritage tourism, particularly given the UNESCO recognition of the Mexican cuisine. However, we argue that gastronomy should be more explicitly put in the context of intangible heritage, identity and sense of place instead of commercial development of heritage and cultural activities. As argued by Timothy (2015), Lugosi (2012) and Richards (2002), gastronomy in an integral part of our intangible culture, it reflects certain cultural norms and values of people and places, and often represents our cultural identity. Given the popularity of the above-mentioned food-related activities and the efforts by public authorities to popularize these events, there are concerns of commercialisation and misrepresentation of cultural values and norms and in particular, concerns to the extent to which the events are authentic. As Gilmore & Pine (2007) argue authenticity is a major source of competitive advantage so it is in the interest of public authorities to control the extent of commercialisation of those events. From another perspective, as Sims (2009) argues, authenticity is integral for the service experience and thus particularly important for food-related activities such as ethnic dining and indigenous food products. It is important from a managerial perspective, to recognise the psychological factors that often play a key role in the decision-making process (see also, Baker & Kim, 2019).

CONCLUSION

This chapter provided a critical and contextual overview of the development of food tourism in Mexico. The paper examined the interrelationship between gastronomic traditions, food and tourism with a particular emphasis on how local and regional food traditions have been developed as intangible tourism products. The various examples from a number of cities in Mexico such as Puebla, Ensenada, Mérida, Oaxaca and Guanajuato, are illustrative of the important role food serves within the context of tourism in social, economic and cultural dimensions. Moreover, our work demonstrated how the development of food tourism and related festivals and special events, have contributed to the development of local communities, stimulated the preservation of their culture and intangible traditions while at the same time, stimulating the emergence and development of local tourism businesses and encouraging creativity, local entrepreneurship and innovation.

Despite the notable progress in the academic scholarship of food tourism, there are areas of research which are yet to be developed. This chapter has emphasized the critical role of tourism for the revitalization of local culture, but further research is needed to explore the extent to how local festivals preserve the authenticity and diversity, particularly in the context of Mexico. More empirical studies are also needed to examine the economic impact of food festivals and in particular, tourist expenditure and tourist spending from a quantitative perspective. Furthermore, more qualitative studies are needed to explore the perceptions of local businesses and residents, in particular, in the large tourist areas in Mexico.

ACKNOWLEDGMENT

The authors would like to express their gratitude to all faculty members at our Events, Tourism & Hospitality Subject Group at University of Northampton for their continuous support and encouragement. We would also want to express our gratitude to the World Food Travel Association, and in particular, Eric Wolf and Maria Athanasopoulou, for their support and insights about gastronomy and tourism, and Daniel Rodrigues Perales for his insights to Mexico an Mexican cuisine. A big thank you goes to our families and friends for their invaluable support and encouragement offered. Last but not least, we would like to recognise the very constructive feedback given by the two reviewers and the editor of this book.

REFERENCES

Ab Karim, S., & Chi, C. (2010). Culinary tourism as a destination attraction: An empirical examination of destinations' food image. *Journal of Hospitality Marketing & Management, 19*(6), 531–555. doi:10.1080/19368623.2010.493064

Alonso, A., & Nortchote, J. (2010). The Development of olive tourism in Western Australia: A case study of an emerging tourism industry. *International Journal of Tourism Research, 12*(6), 696–708. doi:10.1002/jtr.786

Arboleya, J., Garcia-Quiroga, M., Lasa, D., Oliva, O., & Luis-Aduriz, A. (2014). Effect of highly aerated food on expected satiety. *International Journal of Gastronomy and Food Science, 2*(1), 14–21. doi:10.1016/j.ijgfs.2013.12.002

Arnott, M. (1975). *The Anthropology of Food and Food Habits*. Mouton.

Ayora-Diaz, S. (2019). Making and Changing Yucatecan Taste in Yucatán: Innovation and Persistence in Yucatecan Gastronomy. In S. Ayora-Diaz (Ed.), *Taste, Politics, and Identities in Mexican Food* (pp. 101–115). Bloomsbury. doi:10.5040/9781350066700.ch-007

Bak-Geller Corona, S., & Matta, R. (2020). Las cocinas mestizas en México y Perú. Claves para interpretar el multiculturalismo en América Latina. *Antipoda: Revista de Antropologia y Arqueologia, 39*(39), 69–93. doi:10.7440/antipoda39.2020.04

Baker, M., & Kim, K. (2019). Heritage and Authenticity in Gastronomic Tourism. In S. Dixit (Ed.), *The Routledge Handbook of Gastronomic Tourism* (pp. 252–260). Routledge. doi:10.4324/9781315147628-29

Barrios, A., & Pantoja, C. (2016). Comida de la calle en el Centro Histórico. *Tlamati Sabiduria, 7*(2), 1–11.

Bautista, J., & Smit, M. (2012). Sustentabilidad y agricultura en la" región del mezcal" de Oaxaca. *Revista Mexicana de Ciencias Agrícolas, 3*(1), 5–20. doi:10.29312/remexca.v3i1.1477

Belisle, F. (1983). Tourism and food production in the Caribbean. *Annals of Tourism Research, 10*(4), 497–513. doi:10.1016/0160-7383(83)90005-1

Bertella, G. (2011). Knowledge in food tourism: The case of Lofoten and Maremma Toscana. *Current Issues in Tourism, 14*(4), 355–371. doi:10.1080/13683500.2010.489638

Boesen, S., Sundbo, D., & Sundbo, J. (2017). Local food and tourism: An entrepreneurial network approach. *Scandinavian Journal of Hospitality and Tourism, 17*(1), 76–91. doi:10.1080/15022250.2016.1218629

Boyne, S., Williams, F., & Hall, D. (2002). The Isle of Arran taste trail. In A. Hjalager & G. Richards (Eds.), *Tourism and Gastronomy* (pp. 91–114). Routledge.

Brillat-Savarin, J. (1994). The Physiology of Taste (A. Drayton, Trans.). Harmondsworth: Penguin. (Original publication 1825)

Bustos, J., & Galván, A. (2015). A Taste of Oaxaca: A Chef, a Recipe, and a Cookbook. *Diálogo, 18*(1), 177–178. doi:10.1353/dlg.2015.0057

Cavicchi, A., & Ciampi Stancova, K. (2016). *Food and gastronomy as elements of regional innovation strategies*. European Commission, Joint Research Centre, Institute for Prospective Technological Studies.

Che, D. (2006). Select Michigan: Local food production, food safety, culinary heritage and branding in Michigan agritourism. *Tourism Review International*, *9*(4), 349–363. doi:10.3727/154427206776330616

Cocking, L. (2016). *Oaxaca's Guelaguetza Festival Should Be on Everyone's Bucket List*. Culture trip. Available on: https://theculturetrip.com/north-america/mexico/articles/oaxacas-guelaguetza-festival-should-be-on-everyones-bucket-list/

Cohen, E., & Avieli, N. (2004). Food in Tourism, Attraction and Impediment. *Annals of Tourism Research*, *31*(4), 755–778. doi:10.1016/j.annals.2004.02.003

Courtine, J. B. (Ed.). (1996). *Larousse Gastronomique*. Lewis Esson.

Csurgo, B., Hindley, C., & Smith, M. (2019). The role of gastronomic tourism in rural development. In S. Dixit (Ed.), *The Routledge Handbook of Gastronomic Tourism* (pp. 62–70). Routledge. doi:10.4324/9781315147628-8

du Rand, G., & Heath, E. (2006). Towards a framework for food tourism as an element of destination marketing. *Current Issues in Tourism*, *9*(3), 206–234. doi:10.2167/cit/226.0

Earle, R. (2010). "If You Eat Their Food …": Diets and Bodies in Early Colonial Spanish America. *The American Historical Review*, *115*(3), 668–713. doi:10.1086/ahr.115.3.688

Echeverria, O. (2006). Europa Y Su Percepción Del Nuevo Mundo A Través De Las Especies Comestibles Y Los Espacios Americanos En El Siglo XVI. *Historia (Wiesbaden, Germany)*, *39*(2), 531–556.

Ellis, A., Park, E., Kim, S., & Yeoman, I. (2018). What is food tourism? *Tourism Management*, *68*, 250–263. doi:10.1016/j.tourman.2018.03.025

Ellul. D & Schlüter R. (2008). Gastronomía y turismo en Argentina polo gastronómico Tomás Jofré. *Pasos: Revista de turismo y Patrimonio Cultural*, *6*(2), 249-268.

Everett, S. (2009). Beyond the visual gaze? The pursuit of an embodied experience through food tourism. *Tourist Studies*, *8*(3), 337–358. doi:10.1177/1468797608100594

Everett, S. (2012). Production places or consumption spaces? The place-making agency of food tourism in Ireland and Scotland. *Tourism Geographies*, *14*(4), 535–554. doi:10.1080/14616688.2012.647321

Everett, S. (2016). *Food and Drink Tourism*. Routledge.

Everett, S., & Aitchison, C. (2008). The role of food tourism in sustaining regional identity: A case study of Cornwall, South West England. *Journal of Sustainable Tourism*, *16*(2), 150–167. doi:10.2167/jost696.0

Everett, S., & Slocum, S. (2013). Food and tourism: An effective partnership? A UK-Based review. *Journal of Sustainable Tourism*, *21*(6), 789–809. doi:10.1080/09669582.2012.741601

Georgiadou, A., Gonzalez-Perez, M., & Olivas-Lujan, M. (2019). *Diversity within diversity management-Community Based Perspective*. Emerald Publishing Limited.

Getz, D., Robinson, R., Andersson, T., & Vujicic, S. (2014). *Foodies & food tourism*. Goodfellow Publishers.

Gillespie, C. (2002). *European Gastronomy into the 21st Century*. Butterworth-Heinemann.

Gilmore, J., & Pine, B. J. (2007). *Authenticity: What Consumers Really Want?* Harvard Business School Press.

Gonzalez, M., Husted, B., & Aigner, D. (2017). Opportunity discovery and creation in social entrepreneurship: An exploratory study in Mexico. *Journal of Business Research*, *81*, 212–220. doi:10.1016/j.jbusres.2016.10.032

Hall, C., & Mitchell, R. (2001). Wine and food tourism. In N. Douglas & R. Derrett (Eds.), *Special Interest Tourism* (pp. 307–329). Wiley.

Hall, C., & Mitchell, R. (2005). Gastronomic tourism: comparing food and wine tourism experiences. In Niche Tourism: Contemporary Issues, Trends and Cases (pp. 89-100). Oxford: Butterworth-Heinemann.

Hall, C. M., Sharples, L., Mitchell, E., Macionis, N., & Cambourne, B. (Eds.). (2003). *Food Tourism Around The World*. Buttertworth-Heinemann.

Hall, C. M., & Williams, A. (2008). *Tourism and Innovation*. Routledge. doi:10.4324/9780203938430

Hashimoto, A., & Telfer, D. (2006). Selling Canadian culinary tourism: Branding the global and regional product. *Tourism Geographies*, *8*(1), 31–55. doi:10.1080/14616680500392465

Hegarty, J., & O'Mahony, B. (2001). Gastronomy: A phenomenon of cultural expressionism and aesthetic for living. *International Journal of Hospitality Management*, *20*(1), 3–13. doi:10.1016/S0278-4319(00)00028-1

Henderson, J. (2009). Food Tourism reviewed. *British Food Journal*, *111*(4), 317–326. doi:10.1108/00070700910951470

Hinojosa, M. del C. (2009). México y sus regiones gastronómicas. *Hospitalidad ESDAI*, (15), 109-128.

Hjalager, A. (2002). A typology of gastronomy tourism. In A. Hjalager & G. Richards (Eds.), *Tourism and Gastronomy* (pp. 21–36). Routledge.

Horng, J., Liu, C., Chou, H., & Tsai, C. (2012). Understanding the impact of culinary brand equity and destination familiarity on travel intentions. *Tourism Management*, *33*(4), 815–824. doi:10.1016/j.tourman.2011.09.004

Horng, J., & Tsai, C. (2012). Culinary tourism strategic development: An Asia-Pacific perspective. *International Journal of Tourism Research*, *14*(1), 40–55. doi:10.1002/jtr.834

Ignatov, E., & Smith, S. (2006). Segmenting Canadian Culinary Tourists. *Current Issues in Tourism*, *9*(3), 235–255. doi:10.2167/cit/229.0

Ioannides, D., & Debbage, K. (1997). Post-Fordism and flexibility: The travel industry polyglot. *Tourism Management*, *18*(4), 229–241. doi:10.1016/S0261-5177(97)00019-8

Jiménez Medina, L., & Escalante Conde, N. (2009). Miradas y reflexiones antropológicas sobre el desfile del 5 de mayo en la ciudad de Puebla. *Cuicuilco*, *16*(45), 227–245.

Karagiannis, D., & Metaxas, T. (2017). *Local Authorities' contribution on Gastronomy Tourism Development: the case of Ontario, Canada, MPRA Paper 79325*. University Library of Munich.

Kearney, M. (1970). The Social Meaning of Food Sharing in Mexico. *Kroeber Anthropological Society Papers*, *43*, 32–41.

Kim, S., & Ellis, A. (2015). Noodle production and consumption: From agriculture to food tourism in Japan. *Tourism Geographies*, *17*(1), 151–167. doi:10.1080/14616688.2014.978812

Kim, Y., Yuan, J., Goh, B., & Antun, J. (2009). Web Marketing in Food Tourism: A Content Analysis of Web Sites in West Texas. *Journal of Culinary Science & Technology*, *7*(1), 52–64. doi:10.1080/15428050902788352

Kiralova, A., & Hamarneh, I. (2017). Local gastronomy as a prerequisite of food tourism development in the Czech Republic. *Marketing and Management of Innovations*, (2), 15–25.

Kivela, J., & Crotts, J. (2006). Tourism and gastronomy: Gastronomy's influence on how tourists experience a destination. *Journal of Hospitality & Tourism Research (Washington, D.C.)*, *30*(3), 354–377. doi:10.1177/1096348006286797

Lane, E., Jones, R., Jones, A., & Mitchelmore, S. (2016). Exploring the potential of local food and drink entrepreneurship in rural Wales. *Local Economy: The Journal of the Local Economy Policy Unit*, *31*(5), 602–618. doi:10.1177/0269094216654669

Lee, I., & Arcodia, C. (2011). The role of regional food and festivals for destination branding. *International Journal of Tourism Research*, *13*(4), 355–367. doi:10.1002/jtr.852

Lin, Y., Pearson, T., & Cai, L. (2011). Food as a form of destination identity: A tourism destination perspective. *Tourism and Hospitality Research*, *11*(1), 30–48. doi:10.1057/thr.2010.22

López-Guzmán, López, Gálvez, & Vélez. (2019). Gastronomy, Culture and Tourism in Ecuador. In The Routledge Handbook of Gastronomic Tourism (pp. 39-47). London: Routledge.

López-Guzmán, T., Lotero, C., Gálvez, J., & Rivera, I. (2017). Gastronomic festivals: Attitude, motivation and satisfaction of the tourist. *British Food Journal*, *119*(2), 267–283. doi:10.1108/BFJ-06-2016-0246

Lugosi, P. (2012). Food, drink and identity. In D. Sloan (Ed.), *Food and Drink: The Cultural Context* (pp. 20–50). Goodfellow Publishers Limited.

Mak, A., Lumbers, M., & Eves, A. (2012). Globalisation and Food Consumption in Tourism. *Annals of Tourism Research*, *39*(1), 171–196. doi:10.1016/j.annals.2011.05.010

McKercher, B., Okumus, F., & Okumus, B. (2008). Food tourism as a viable market segment: It's all how you cook the numbers. *Journal of Travel & Tourism Marketing*, *25*(2), 137–148. doi:10.1080/10548400802402404

Medina, F., & Tresserras, J. (2018). Introduction. Gastronomy and Tourism: Social and Cultural perspectives. In F. Medina & J. Tresseras (Eds.), *Food, Gastronomy and Tourism. Social and Cultural Perspectives*. Universidad de Guadalajara.

Mendoza, Q. (2012). *Gastronomia novoandina del distrito de Puno, una alternativa para el turismo*. Universidad Nacional del Altiplano.

Metaxas, T., & Karagiannis, D. (2016). Culinary tourism in Greece: Can the past define the future? Dimensions of Innovation, Entrepreneurship and Regional Development. *Journal of Developmental Entrepreneurship*, *21*(3), 1–20. doi:10.1142/S1084946716500187

Metro-Roland, M. (2013). Goulash nationalism: The culinary identity of a nation. *Journal of Heritage Tourism*, *82*(2-3), 172–181. doi:10.1080/1743873X.2013.767814

Mexconnect. (2020). *Cuisine*. Mexconnect. Available on: https://www.mexconnect.com/cuisine

Mezcal National Board. (2020). *Unique Spirits*. Mezcal National Board. Available on: http://www.sientemezcal.com/2009/eng/espiritu_unico.php

Montanari, A. (2009). Geography of taste and local development in Abruzzo (Italy): Project to establish a training and research centre for the promotion of enogastronomic culture and tourism. *Journal of Heritage Tourism*, *4*(2), 91–103. doi:10.1080/17438730802366482

Mulcahy, J. (2019). Historical evolution of gastronomic tourism. In S. Dixit (Ed.), *The Routledge Handbook of Gastronomic Tourism* (pp. 24–32). Routledge. doi:10.4324/9781315147628-3

Murayama, M. (2012). Promoting Japanese food culture and products. In *Food and The Tourism Experience: The OECD-Korea Workshop* (pp. 147–153). OECD Publishing.

National Geographic. (2018). *Asiste al festival gastronómico más importante de Mérida*. National Geographic. Available on: https://www.ngenespanol.com/traveler/asiste-al-festival-gastronomico-mas-importante-de-merida/

Naumov, N. (2019). *Food Tourism as a Cultural Experience*. Paper presented at FoodTrex Thessaloniki, Thessaloniki, Greece.

Ojeda, A., Camacho, C., Hernández, C., & López, R. (2017). Caracterización de las principales tendencias de la gastronomía mexicana en el marco de nuevos escenarios sociales. *Teoría y Praxis*, *21*(21), 91–113. doi:10.22403/UQROOMX/TYP21/05

Okumus, B., Okumus, F., & Mckercher, B. (2007). Incorporating Local and İnternational Cuisines in the Marketing of Tourism Destinations: The Cases of Hong Kong and Turkey. *Tourism Management*, *28*(1), 253–261. doi:10.1016/j.tourman.2005.12.020

Ottenbacher, M., & Harrington, R. (2011). A Case Study of a Culinary Tourism Campaign in Germany: Implications for Strategy Making and Successful İmplementation. *Journal of Hospitality & Tourism Research (Washington, D.C.)*, *37*(1), 3–28. doi:10.1177/1096348011413593

Page, S., & Ateljevic, I. (Eds.). (2009). *Tourism and Entrepreneurship: International Perspectives*. Butterworth - Heinemann.

Peru, E. (2016). *Novoandina – Andean Cuisine's Rich Cultural History*. Eat Peru. Available on: https://www.eatperu.com/novoandina-novo-andean-cuisine/a

Pilcher, J. (2017). *Food in World history* (2nd ed.). Routledge Taylor & Francis Group. doi:10.4324/9781315718941

Pine, B., & Gilmore, J. (1999). *The Experience Economy*. Harvard Business School Press.

Prezenza, A., & Chiappa, G. (2013). Entrepreneurial strategies in leveraging food as a tourist resource: A cross-regional analysis in Italy? *Journal of Heritage Tourism*, *8*(2), 182–192. doi:10.1080/1743873X.2013.767810

Pulido-Fernandez, J., Casado-Montilla, J., & Carrillo-Hidalgo, I. (2019). Olive-oil tourism as a special interest tourism. *Heliyon*, *5*(12), e02975. doi:10.1016/j.heliyon.2019.e02975 PMID:31872132

Rachão, S., Breda, Z., Fernandes, C., & Joukes, V. (2019). Food tourism and regional development: A systematic literature review. *European Journal of Tourism Research*, *21*, 33–49.

Richards, G. (2002). Gastronomy: An essential ingredient in tourism production and consumption? In A. Hjalager & G. Richards (Eds.), *Tourism and Gastronomy* (pp. 3–20). Routledge.

Sainsbury, B. (2012). *Peru's Gastronomic renaissance*. BBC. Available on: http://www.bbc.com/travel/story/20120803-perus-gastronomic-renaissance

Sanchez-Gutierrez, J., Gonzalez-Uribe, E., Gutierrez-Govea, A., & Garcia-Jimenez, E. (2012). The effects of intellectual capital and innovation on competitiveness: An analysis of the restaurant industry in Guadalajara, Mexico. *Journal of Competitiveness Studies*, *20*(3/4), 32–46.

Santich, B. (1996). Looking for Flavour. Kent Town: Wakefield Press.

Santich, B. (2004). The study of gastronomy and its relevance to hospitality education and training. *International Journal of Hospitality Management*, *23*(1), 15–24. doi:10.1016/S0278-4319(03)00069-0

Scarpato, R. (2002). Gastronomy as a tourist product: the perspective of gastronomy studies. In A. Hjalager & G. Richards (Eds.), *Tourism and Gastronomy* (pp. 51–71). Routledge.

Schlüter, R. (2012). Promoting regional cuisine as intangible cultural heritage in Latin America. In *Food and the Tourism Experience: The OECD-Korea Workshop*. OECD Publishing.

Schumpeter, J. (2008). The Theory of Economic Development: An Inquiry into Profits, Capital, Credit, Interest and the Business Cycle. London: Transaction Publishers. (Original publication 1934)

Shaw, G. (2014). Entrepreneurial Cultures and Small Business Enterprises in Tourism. In A. Lew, C. M. Hall, & A. Williams (Eds.), *The Wiley Blackwell Companion to Tourism* (pp. 120–131). Wiley. doi:10.1002/9781118474648.ch9

Sims, R. (2009). Food, place and authenticity: Local food and the sustainable tourism experience. *Journal of Sustainable Tourism*, *17*(3), 321–336. doi:10.1080/09669580802359293

Smith, S., & Costello, C. (2009). Culinary tourism: Satisfaction with a culinary event utilizing importance-performance grid analysis. *Journal of Vacation Marketing*, *15*(2), 99–110. doi:10.1177/1356766708100818

Timothy, D. (2015). *Heritage Cuisines: Traditions, Identities and Tourism*. Routledge. doi:10.4324/9781315752525

Timothy, D., & Ron, A. (2013). Understanding heritage cuisine and tourism: Identity, image, authenticity, and change. *Journal of Heritage Tourism*, *8*(2-3), 99–104. doi:10.1080/1743873X.2013.767818

Torres, R. (2003). Cancun's tourism development from a Fordist spectrum of analysis. *Tourist Studies*, *2*(1), 87–116. doi:10.1177/1468797602002001098

Trejo-Pech, C., Arellano-Sada, R., Coelho, A., & Weldon, R. (2011). Is the Baja California, Mexico, wine industry a cluster? *American Journal of Agricultural Economics*, *94*(2), 569–575. doi:10.1093/ajae/aar080

UNESCO. (2019). *Traditional Mexican cuisine - ancestral, ongoing community culture, the Michoacán paradigm*. Available at: https://ich.unesco.org/en/RL/traditional-mexican-cuisine-ancestral-ongoing-community-culture-the-michoacan-paradigm-00400

UNESCO. (2020a). *Cochabamba*. UNESCO. Available on: https://en.unesco.org/creative-cities/cochabamba

UNESCO. (2020b). *Panama City*. UNESCO. Available on: https://en.unesco.org/creative-cities/panama-city

UNESCO. (2020c). *Florianopolis*. UNESCO. Available on: https://en.unesco.org/creative-cities/florian%C3%B3polis

UNESCO. (2020d). *Ensenada*. UNESCO. Available on: https://en.unesco.org/creative-cities/ensenada

UNWTO. (2012). *Global Report on Food Tourism*. UNWTO.

UNWTO. (2017). Second Global Report on Gastronomy Tourism. Madrid: World Tourism Organization (UNWTO).

Valencia, C. (2019). *Memoria culinaria de las mayoras en fondas del centro histórico de Puebla: Entrevista etnográfica y cartografía social*. Available at: https://repositorioinstitucional.buap.mx/handle/20.500.12371/4782

Williams, H., Williams, R., & Omar, M. (2014). Gastro-tourism as destination branding in emerging markets. *International Journal of Leisure and Tourism Marketing*, *4*(1), 1–18. doi:10.1504/IJLTM.2014.059257

ADDITIONAL READING

Fennell, D., & Bowyer, E. (2020). Tourism and sustainable transformation: A discussion and application to tourism food consumption. *Tourism Recreation Research*, *45*(1), 119–131. doi:10.1080/02508281.2019.1694757

Fuste-Forne, F. (2020). What do New Zealand Newspapers say about food tourism? *Tourism and Hospitality Research*, *20*(1), 82–92. doi:10.1177/1467358418810916

Jørstad, M., Roaldsen, I., & Ljunggren, E. (2020). Local Food in Tourism: A Systematic Literature Review. *Journal of Gastronomy and Tourism*, *4*(2), 97–111. doi:10.3727/216929719X15736343324887

Liberato, P., Mendes, T., & Liberato, D. (2019). Culinary Tourism and Food Trends. In A. Rocha, A. Abreu, J. de Carvalho, D. Liberato, E. Gonzalez, & P. Liberato (Eds.) *Proceedings of ICOTTS 2019 Advances in Tourism, Technology and Smart Systems* (pp.517-526). Frankfurt: Springer.

Rousta, A., & Jamshidi, D. (2020). Food tourism value: Investigating the factors that influence tourists to revisit. *Journal of Vacation Marketing*, *26*(1), 73–95. doi:10.1177/1356766719858649

KEY TERMS AND DEFINITIONS

Creativity: It describes the ability and skill to form something new. It involves development of tangible (equipment) and intangible (ideas) objectives which aim to have as a result a useful product. Creativity is a new approach that comes to fill in gaps in the knowledge and find solutions for problems.

Cultural Identity: The feeling that represents belonging to a group or society. It involves self-evaluation that makes each person categorise himself or herself in a group, based on religion, nationality, class, or any other social classification. Cultural identity involves individual's perception of him/herself whereby as a result the person becomes part of the group that share the same practices and values.

Food Tourism: It represents the type of tourism which has the main aim to experiment food from different countries and regions. It is also known as 'culinary tourism' which in the present days became a big and important part of overall tourist experience when visiting a destination.

Gastronomy: It is the science that studies the connection between food, tradition and culture of a certain region or area. It looks at different ways on how food is prepared and how cooking styles differ from region to region. Gastronomy examines the value of foods and the meaning it has towards the cultural practices.

Social Innovation: Idea that aims to improve the needs of a society. It implies studying and researching new approaches that give sophisticated solutions to a community's existing problems. The main purpose represents strengthening social unity through activities such as volunteering, learning and entrepreneurship.

Tourism Entrepreneurship: It defines any type of businesses that offer tourist products and services. This type of enterprise involves developing business activities such as accommodation, transport, shops, entertainment, and many others.

Chapter 7
Entrepreneurial Eco-Systems for Sustainable Community-Based Tourism Development in Albania:
Case Studies of Community Tourism Development

Aida Ciro
Polis University, Albania & Co-PLAN, Institute for Habitat Development, Albania

Merita Toska
https://orcid.org/0000-0003-3797-1057
Polis University, Albania & Co-PLAN, Institute for Habitat Development, Albania

ABSTRACT

The tourism sector in Albania has been upheld as a government development priority and a promising window for overall local economic development. Although these ambitions have started to shape the governance of the sector, the inherent challenges on a governance level have meant that most of the tourism development on a local level is being led by local initiatives and the private sector service providers. These initiatives are often fueled by entrepreneurial drive and are sustained by a network of local community actors, giving rise to applied models of sustainable tourism development, as shown by numerous agritourist enterprises emerging across Albania. Through a case study approach, this chapter will focus on the role entrepreneurial eco-systems rooted in local communities can play in the development of sustainable tourism models in Albania.

DOI: 10.4018/978-1-7998-4855-4.ch007

INTRODUCTION

Progressively increasing numbers, have shown tourism, has become one the world's most profitable industries, constituting a lucrative sector in terms of its job creation potential, with 1 in 10 jobs related to tourism. In 2019, as reported by the World Travel and Tourism Council (2020), tourism contributed a record $8.9 trillion to the world economy. More specifically, that means that the travel and tourism sector has been growing at a rate of 3.5%, has generated 10.3% of all global economic activity and is responsible for one in ten jobs globally. This direct correlation between tourism development and local economic growth has meant that, many places, regardless of how the concept of a 'place' is geographically defined, be it cities, regions, local communities or otherwise (O'Connor, Stam, Sussan, & Audretsch, 2018) have explored and developed tourism ambitions.

Correspondingly with the global picture, the importance of tourism in Albania, a small Mediterranean country in South-East Europe has also grown markedly fast in recent years. Once the most isolated country in the region until the early '90s, today Albania marks the fastest growth in the tourism sector in the region of the Western Balkans, by 15.8% (Ciro, 2020) with over 5.3 million reported arrivals of foreign citizens and about 2,193 million USD in international tourism receipts (United Nations World Travel Organization, 2019) accounting for an average of 2.8% to the Gross Domestic Product (GDP). Nominally, tourism receipts amount to a total of 45 billion ALL generated through tourism related activities including: accommodation, food and drink services, travel and tour operators, car rentals and other leisure activities (Institute of Statistics, 2019). This has increased pressure on places and local communities to innovate and embrace new opportunities for competitive development and at the same time ensure longevity of resources. Despite the renowned ambition of central and local government, to make of tourism the driving force of economic development, the contribution of the sector lags behind aspiration.

From 1992 – to date, Albania's vision and development strategy for the sector has been outlined in six strategies, of which two remained in a permanent draft form. With the exception of the current strategy in force, spanning from 2019 to 2023, none of the prior strategies were implemented in full. Although starting from 2013 the legislative and policy framework has been expanded and complemented with new laws, bylaws and fiscal incentives, the governance of the sector is beset by challenges. As a result, growth of the tourism sector has been led primarily by local initiatives and the private sector service providers in the form of entrepreneurial activity, often innovative, in various parts of the country. A plethora of economic activities have emerged as a result, echoing in numerous supportive activities and the creation of sustainable local microcosms. This has given rise to increased research interest into the topics of community tourism, entrepreneurship and innovation, as an alternative way to lead to growth of innovation-and-community-based tourism development with environmental considerations.

This chapter, aims to build on theoretical frameworks of entrepreneurial, innovative community tourism and its contribution to the economy and provide research findings in the form of case studies. Sourced from rural areas in northern and southern Albania, the case studies of 'Mrizi i Zanave' (north-west of Albania, Municipality of Lezhë), 'Caravan horse-riding Albania' (south of Albania, Municipality of Gjirokastër), and "Sotira Farm" (south-east of Albania, Municipality of Kolonjë) although different in approach and scale, present three clear cases of community tourism with strong entrepreneurial and innovation elements. Reliance on three single case studies may raise questions concerning limitations, yet as Kokkranikal and Morrison (2011) put it, such limitation is offset by the rich qualitative insight generated. This proves of particular value in the case of Albania, where tourism in all its aspects remains under-researched.

BACKGROUND

It is now clear that, for many, the sun alone is not enough (Welford & Ytterhus, 1998). In their tourist identity and experience, people want to shift away from their existing state of hectic, deadline driven days, overwhelmed by social pressure and affected by pollution of highly urbanized areas, into an almost antidotal experience of slow paced natural environment. The problem starts when there is too many of us longing for such an intimate and calming touristic experience, in places whose defining trait goes against mass volumes. This means that, pressure from tour operators to design such experiences and bring pristine, unexplored areas to the market, in turn lead to *reduced degree of diversity and the creation of a mono-culture, often remarkably similar to that of the home of the tourist* (Welford & Ytterhus, 1998, p. 194). Thus, the views on whether there is such a prospect as sustainable tourism, have evolved overtime, yet, remain divided. In the 1950s the increasing revenue generated by mass tourism was associated to increased prosperity; with the advent of an environment-sensitive culture, mass tourism was viewed more as a "this largely unregulated growth", although predominantly confined to a limited few, such were government authorities, academics and civil society actors. Concepts of alternative forms of tourism were coined, as more appropriate options for destinations with 'good' characteristics of (for example) community control, authentic attractions and small scales of operation championed against 'bad' mass tourism's corporate control, generic attractions and large scales of operation. This lead for mass tourism and alternative tourism to be considered as two opposite ends of the tourism development spectrum. By the early 1990s, for example, there was already growing support for the idea that mass tourism could be sustainable and alternative tourism unsustainable (Buttler, 1990).

Sustainable tourism has been defined in many ways, including elements of planning, environment, man-made heritage, ecology, social equity and participation, economic and social sustainability and longer-term future. The World Commission on Environment and Development (1987) in its report "Our Common Future", defines sustainable development as *development that meets the needs of the present without compromising the ability of future generations to meet their own needs* (United Nations, 1987). The definition of sustainability introduced by the otherwise known as the Brundtland Report (1987), captures two core ideas, as Creaco and Querini (2003) highlight: (i) *development has an economic, a social, and an environmental dimension and that* (ii) *the current generation has a moral obligation towards future generations to leave sufficient social, environmental, and economic resources for them* (Creaco & Querini, 2003, p.3). Lohmann & Panosso Netto (2017, p.132) assert that *sustainable tourism should not be considered a market segment but rather a new way of thinking and practicing tourism* and in proposing strategies and techniques for achieving sustainable tourism, they place community-based tourism at the meeting point of the three key dimensions: territorial, social and cultural. Welford & Ytterhus (1998), in their proposed framework list dynamic conditions, crucial to sustainable development: parallel development, futurity, consumer education, fiscally neutral local taxation, promoting conservation, the management of change, building on strong cohesion, participation, and destination management policy and strategy.

A key challenge in sustainable tourism is to develop economically viable enterprises that provide livelihood benefits to local communities while protecting indigenous cultures and environments (Simpson, 2007). A livelihood, as Chambers & Conway (1992) note, *comprises the capabilities, assets (stores, resources, claims and access) and activities required for a means of living*, (Chambers & Conway, 1992, p. 6). What makes it sustainable is the ability to *cope with and recover from stress and shocks, maintain or enhance its capabilities and assets, and provide sustainable livelihood opportunities for the next generation; and which contributes net benefits to other livelihoods at the local and global levels and in the short and long term (ibid.).* This would mean an equitable division of profits and benefits between external actors (i.e. *investors, business owners, corporates*, etc.) and the host community, generated in a setting that is considerate of the impact on the environment and cultural tessiture of the local community. In the long term, this is of particular importance to the future of the sector as the *communities' human, natural and cultural resources form the foundation of tourism destinations and touristic experiences* (Kokkranikal & Morrison, 2011, p. 138). In addition to the involvement of the local community, researchers believe that an alternative solution to the negative implications of rapid tourism development and catering for increasing demands for varied touristic products and experiences may very well be in encouraging and tapping into entrepreneurial energies and ecosystems. *Such energy when shaped into entrepreneurial activities, ranges from individual enterprises inspired by motives of wealth creation to altruistic endeavors with a community focus, as is the case in social entrepreneurship.* (ibid) Such view is far from the long held *perceptions of entrepreneurship as a rather lonely, or at least individualistic journey* (O'Connor et al, 2018, p.4). In fact, as Kline, Hao, Alderman, Kleckley, & Gray (2004, p. 306) points out, *local entrepreneurs that have a connection with the community and will reinvest ðnancially and emotionally in the local area.*

Entrepreneurship ecosystems are considered as a panacea for destinations wanting to maintain a competitive advantage, but also for peripheral territories with scarce resources and not necessarily endowed with a touristic profile, as it is possible to align real capacities within the tourism supply chain of neighboring municipalities by serving as suppliers of a particular good or service. By definition, *an entrepreneurial ecosystem refers to the interdependent set of physical, legal, cultural, ðnancial, human, and organizational elements within a community that has the potential to support or thwart an entrepreneur's activity* (Kline et al, 2014, p.306). O'Connor et al, (2018, p. 4) differentiates between *the agency (entrepreneurial) and the (human made) context (i.e. the ecosystem). Entrepreneurial activity, is the ðrst observable output of the entrepreneurial ecosystem and is considered to be the process by which individuals identify and pursue opportunities for innovation.* According to Chang (2011, p.467) usually, businesses in tourism sector are initiated by those entrepreneurs who significantly change the supply of leisure and recreational opportunities. Yet, the extent to which they get involved and establish further links are conditioned by a number of factors, such *typology of suppliers and producers, managerial capabilities, availability of ðscal resources and the societal structure of the destination* (Wall & Mathieson, 2006, p. 144), a tradition of tourism development in the area etc.

Schumpeter (1947) was focused on studying economic change, in defining the *entrepreneur as the economic agent,* included the connection between the entrepreneur and innovations, as did Grebel, Pyka, & Hanusch (2003), opening a discussion on invention and innovation. Hjalager (1997, p. 35) defines innovations as *further developments of the invention for industrial handling, together with the institutionalizing of the methods of production or the bringing of the new products to the market.* She classifies them into *product innovations, classical process innovations, process innovations in information handling, management innovations, and institutional innovations* (ibid.). Innovations introduced in the tourism sector in Albania are comparatively modest by international standards and gain meaning only within that specific context.[1] However, given the limited experience and tradition in the tourism sector, most innovations are observed in products and classical processes, driven by emerging entrepreneurial activity. As Power, Di Domenico, & Miller (2017, p. 39) note, entrepreneurship in the tourism sector is *shaped by smallness, informality and ñexibility* and Albania makes no exception.

Tourism potential for developing countries like Albania has been highlighted in several studies (Kristo, 2013; Ciro, Toska, & Nientied, 2018), since the country is endowed with an abundance of natural resources, history and cultural heritage, geography and rich territorial morphology, and proximity to other touristic destinations in the Western Balkans. While positive spillover effects of tourism industry in the economy are multifaceted including income and production, employment and entrepreneurship, and foreign exchange revenues (Creaco & Querini, 2003; Goeldner & Ritchie, 2012; Bunghez, 2016), there can be negative effects too. Tourism industry has a very strong relation with the territory (the destination) and if it is left to developed on laissez-faire model, uncontrolled and unplanned, tourism flows might increase pressure on resources contributing negatively on the environment (Creaco & Querini, 2003). Thus, its crucial in the long term perspective framework that benefits from the development of tourism must embed sustainability principles, socially, economically and environmentally (Bâc, 2012).

SUSTAINBALE TOURISM AND IMPACT: THE CASE OF ALBANIA

Albania is a small, Mediterranean country, in South-East Europe. Situated in the Balkan peninsula, Albania shares land borders with Greece, North Macedonia, Kosovo and Montenegro. Hall (2000) depicts it as *European yet Oriental*, aspiring to *EU membership yet racked by domestic and external instability* (p.34).

After being isolated for over 45 years and therefore little known to the outside world, in 1992 Albania stepped into democracy carrying a pristine, unshaped touristic offer of natural, cultural and historical heritage. In absence of sectorial governance and prior experience with tourism development, *for several years, tourism in Albania was driven primarily by the enthusiasm of the locals, nascent entrepreneurs, and first foreign visitors* (Ciro, 2020, p.69). In less than 30 years, amidst challenges, Albania has managed to develop a touristic image and shape its offer, marking the fastest sector growth in the Western Balkan region.

The fast-growing tourism industry represents one of the sectors upon which economic growth and sustainable development is expected to rely in the long term, based on national development priorities as summarized in the National Strategy for Development and Integration 2014-2020 (NSDI 2014-2020). The NSDI 2014-2020 (p.23) among others, aims at *to actively encourage development of middle to high-end sustainable tourism in Albania* through the establishment of an international standard certification system; incentivizing high-end tourism and extension of the touristic season, providing public infrastructure in areas where high touristic demand is present; and by promoting Albania's attributes internationally. Quantitatively, the government of *Albania aims at the increase of direct contribution of tourism sector in GDP to 8%* (NSDI 2014-2020, p.23).

In general, the tourism sector in Albania still suffers from structural deficiencies inherited from its communist past, and developed over the last three decades. Nevertheless, since an infancy stage at the beginning of the 90s, tourism sector has developed progressively and there is an increasing trend in the number of foreign visitors visiting Albania. Using as a reference the product life cycle model adapted by Butler (2006) and World Bank (2009), an assessment of Albania tourism life cycle can be constructed as shown in figure 1. Functional tourism area life cycle construction, Albania is considered as one tourist destinations, thus we do not account for the heterogeneity of tourism potential of the country (driven from the lack of detailed data).

Figure 1. Albania tourism area life cycle
Source: Instat & Authors calculations

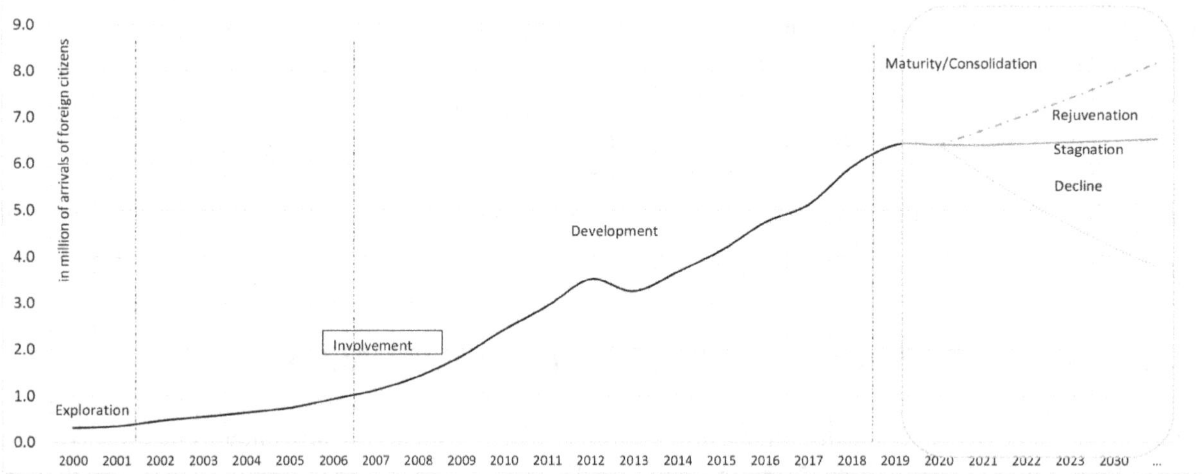

Figure 1 was built using as proxy for tourist number, the number of foreign arrivals in Albania over the period 2000-2019. During the first exploration stage, the country experienced the first visitors driven mostly by a curiosity factor and the urge to explore the unknown; some local enterprises provide services with no or minimal advertising conducted. While the number of tourists increased and the destination became known in international markets, the number of tourists increased and more enterprises entered the market providing for goods and services. Some advertising became noticeable in this stage. In the development stage, the destination became known internationally, growth in number of tourists accelerated and tourism receipts increased. Advertising and promotion became more intensive especially in international markets. In the maturity stage, growth rate in the number of tourists slows down, tourism receipts remain high, marketing intensifies and interventions are needed in order to keep on track and reap the benefits of higher financial inflows. To extend as much as possible the maturity stage and maximize benefits from the tourism sector, services and product innovation are important in order to avoid entering the decline/stagnation stage. In addition, government policy and commitment to develop tourism sector should materialize in higher investment levels and promotion of the sector along the tourism area life cycle.

Economic Impact of Tourism

Tourism sector is expected to be one of the drivers of growth in the Albanian economy, generating value added, employment, new businesses and foreign exchange revenues. The favorable geographic position and abounding natural, cultural and historical assets, augment the possibilities of this sector to provide for sustainable economic development. While expectations still remain high, the data on the gross value added (GVA) show for a marginal contribution of this sector in the Albanian economy. Until the year 2000, the weight of "accommodation and food services" sector to total gross value added was about 3.2%, shrinking to about 2.3% over 2001-2010. A further contraction with about 0.1 percentage points was marked during the period 2011-2017, registering a contribution of about 2.2% to total gross value added generated in the country.

Figure 2. Tourism relevance in GVA (share in % versus total GVA)
Source: Instat, Albania & Authors calculations

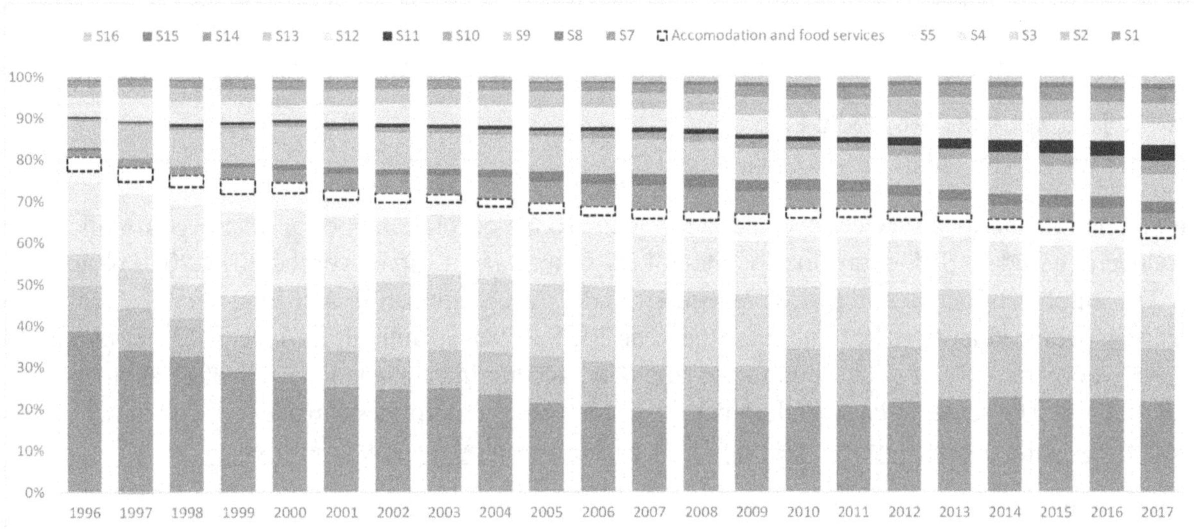

Other sectors, such as industry, construction and trade together generated about 35.6 percent of the total gross value added during 2017, whereasile, the "accommodation and food services" sector accounted for about 2.5% percent. The low contribution of the sector in value generation is assessed to be the result of several factors such as the declining general productivity and/or no efficient use of human resources in the economy (Çeliku & Metani, 2011; Çela, Metani, & Çeliku, 2019), and high informality rates in the economy, on average about 36.2% of GDP for the period 1996-2012 (Boka & Torluccio, 2013) and in average 32.7% of GDP for the period 1991-2015 (Medina & Schneider, 2018).

According to McKenzie & Brackfield (2008) labor productivity can be measured as labor productivity per employed person. Using data from the Structural Business Survey, labor productivity has been calculated as a ratio of the gross value added to the number of employees for the period 2014-2018. Results on labor productivity summarized in table 1, show that all market producers labor productivity remained unchanged during 2018 as compared to the previous year. While figures for all market producers do not show for high variation from year to year, within sectors variation in productivity levels is high. The highest labor productivity is marked in the mining and quarrying sector of the economy, followed by the electricity, gas, water supply and waste management. The lowest labor productivity levels are registered in the accommodation and food services, and manufacturing sectors of the economy.

The number of people employed in the sector has increased over the 2014-2018 period, but growth rate has moderated especially during 2017-2018. As a ratio of total labor force, employed in the ac-

Table 1. Single factor – labor productivity (in ALL million per employed person)[2]

	2014	2015	2016	2017	2018
All market producers	1.2	1.1	1.0	1.1	1.1
Mining & quarrying	5.8	3.7	2.4	2.8	3.0
Manufacturing	0.9	0.8	0.7	0.8	0.8
Electricity, gas, water supply & waste management	1.7	2.4	2.7	1.6	2.3
Construction	1.6	1.4	1.6	1.7	1.7
Trade	0.9	1.0	1.0	1.0	1.0
Accommodation & food services	0.4	0.3	0.3	0.4	0.4
Transport & information & communication	1.4	1.6	1.7	1.8	1.7
Services	1.0	0.8	0.8	0.9	1.0

Source: Instat, Albania & authors calculations

commodation and food services accounted for about 10.6% of total number of employed for all market producers (based on SBS data), and for about 4.2% of total labor force over the 2014-2018 period. s

The number of active enterprises registered under "accommodation and food services" shown in figure 4, followed an upward trend until the year 2015 (+28.4% annually). Since 2015 the number of active enterprises in this sector contracted annually, accounting for about 12.4% of total stock at the end of 2018. The sector is broadly dominated by small scale enterprises (86.5% of total stock of active enterprises with 1-4 employees) and only 0.3% of them employing +50 employees.

Figure 3. Employment by economy sectors (sector to total)
Source: Instat, Albania & Authors calculations.

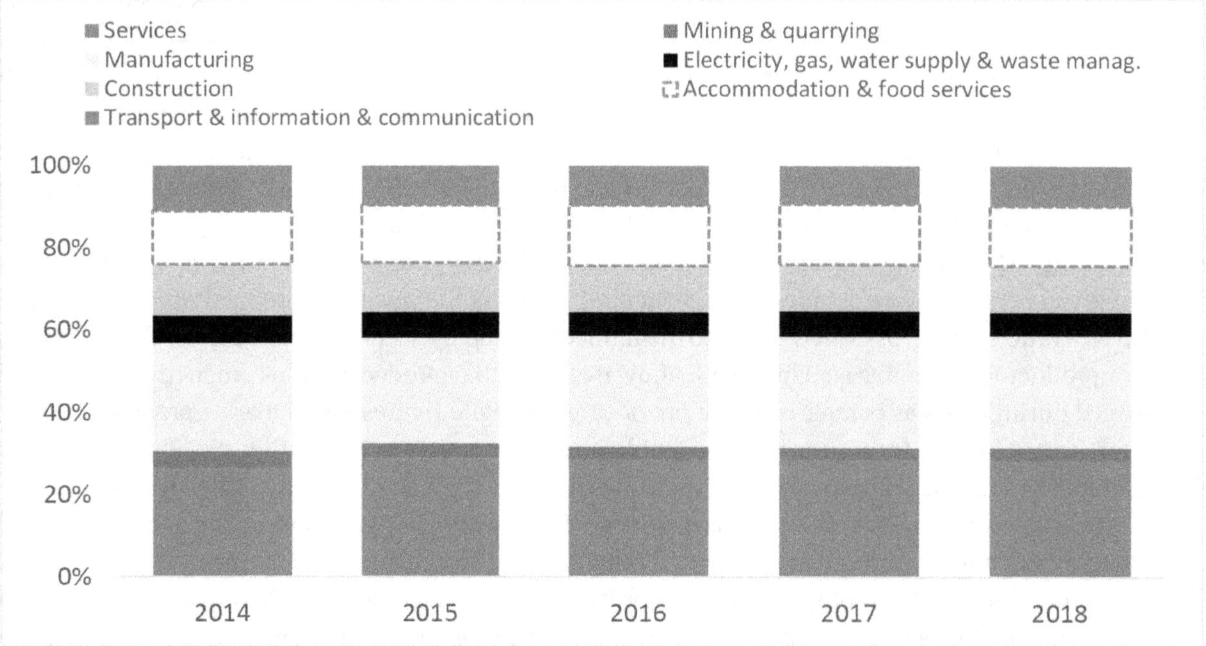

Figure 4. Stock of active enterprises
Source: Instat, Albania

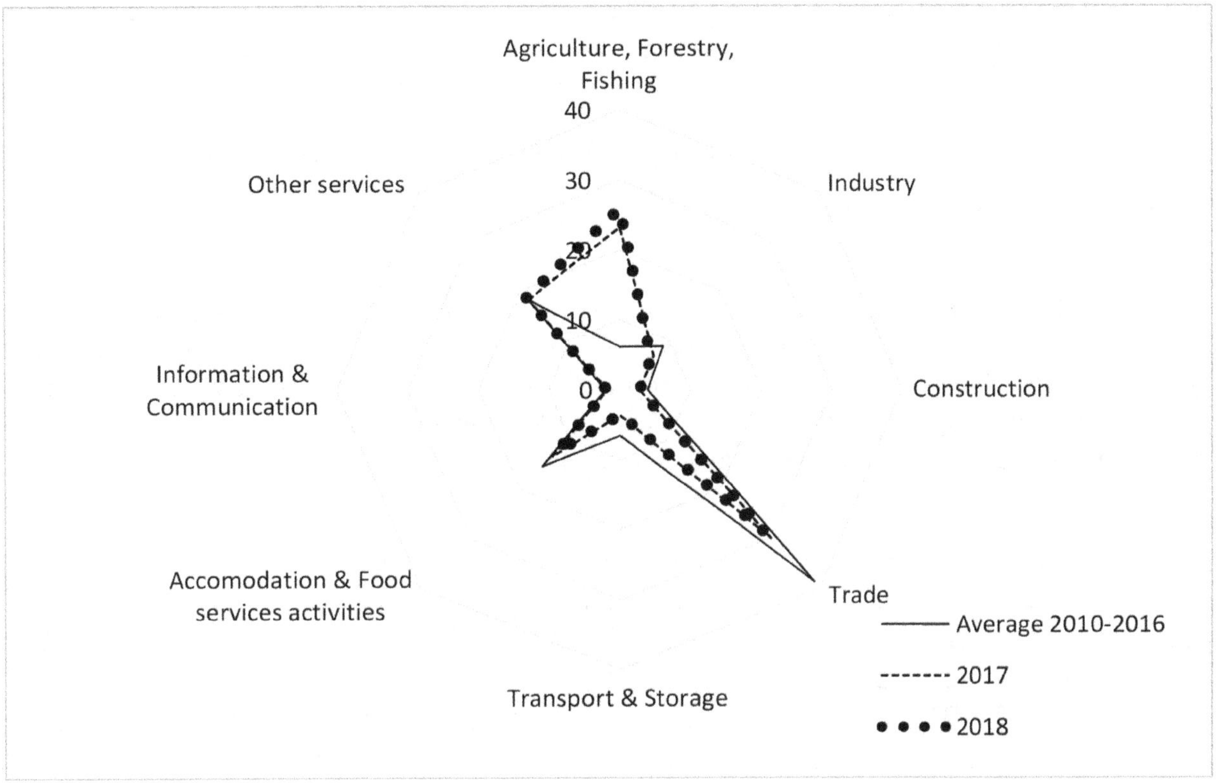

While the share to total stock shrank over 2018, there were about 3,120 new businesses registered in this sector as shown in table 2, out of which 86.3% employing 1-4 employees and only 4 businesses having more than 50 employees. Data on the number of closed businesses in the sector are not available, but a close look to the change in stock of active enterprises suggests for a relatively high death rate of businesses in this sector. The contraction of the number of active enterprises in this sector is surprising and arises despite the extended incentives provided by the government of Albania to promote tourism activities in accommodation and food services including VAT reduction from 20% to 6%[3], exemption from the payment of the infrastructure impact tax for new buildings for accommodation structures classified under "Hotel/Resort with 5 stars, special status" or holding international brand names[4].

The small size of enterprises is assessed to affect and determine the managing system of businesses in this sector, which are broadly family based and act more as property managers rather than hospitality managers (Kristo, 2013; Albanian Development Fund, 2018). Started as family enterprises, usually businesses in this sector are prone to employ extended family members especially in key positions, despite them non having adequate education, training and/or skills for the job position. Often, these emplyees are not registered in competent authorites and no social and health contributions are paid for them and in the case are registered, minimun salaries are declared.

The accommodation and food services sector provides for the lowest wages in Albania. The data included in table 3 show that gross wages in this sector are the lowest among economic sectors and well below the total sectors average despite the growing annual trend (especially due to the increase of

Table 2. Indicators on active enterprises in the accommodation and food services sector

	2013	2014	2015	2016	2017	2018
Stock of active enterprises *(no. at the end of the year)*	18,061	18,061	23,186	22,785	21,909	20,185
annual difference		0	5,125	-401	-876	-1,724
Newly created enterprises *(number)*	2,366	3,504	9,635	4,547	4,268	3,120

Source: Instat, Albania & authors calculations

the approved minimum wage in the country). Findings of the Albanian Development Fund (2018, p.9) suggest that staff engaged in this sector is usually *unskilled, under-productive, insufficient in number of specialists that lack the basic standard operating procedures*. The vicious cycle of low wages – unskilled workers hinder development of the sector and leaves little room for potential innovation in this sector, being it product, process, organizational and/or technological.

Table 3. Gross average monthly wage per employee by economic activities in ALL

	2014	2015	2016	2017	2018
Total	45,539	47,900	47,522	48,967	50,589
Agriculture, forestry and fishing	34,529	34,322	34,062	36,264	36,907
Industry	41,759	41,256	40,334	40,390	42,190
Construction	38,881	41,104	41,866	43,104	43,688
Trade & accommodation and food services	31,505	33,486	33,984	34,627	36,799
Information and communication	73,401	77,314	75,328	73,471	75,784
Financial and insurance activities	94,703	101,439	101,348	103,641	106,475
Real estate activities; professional, scientific and technical activities; administrative and support service activities	47,289	53,473	54,056	55,940	59,124
Public administration and defense; compulsory social security; education; human health and social work activities	54,828	55,566	56,139	59,921	61,082
Arts, entertainment and recreation, repair of household goods and other services	45,775	46,501	46,227	49,427	51,492

Source: Instat, Albania

Tourism activity contributes to the balance of payment through travel services account as shown in figure 5. It constitutes an important source of foreign currency contributing to the narrowing of the current account deficit (if net travel services mark a surplus). Over the considered period, travel services registered surpluses with exclusion of year 2013 which registered a deficit of about Euro 6 million due to added uncertainties in Albania (general elections were held in a highly debated context). After this year, net travel services account surplus witnessed an accelerated growth reaching the highest level in 2017 with about Euro 446.1 million. In the upcoming two years, net travel services account surplus shrank in average by 3.9% annually (in view of local elections held in 2019).

Tourism Supply

Figure 5. Travel services inflows and outflows, in Euro million
Source: Bank of Albania & authors' calculations

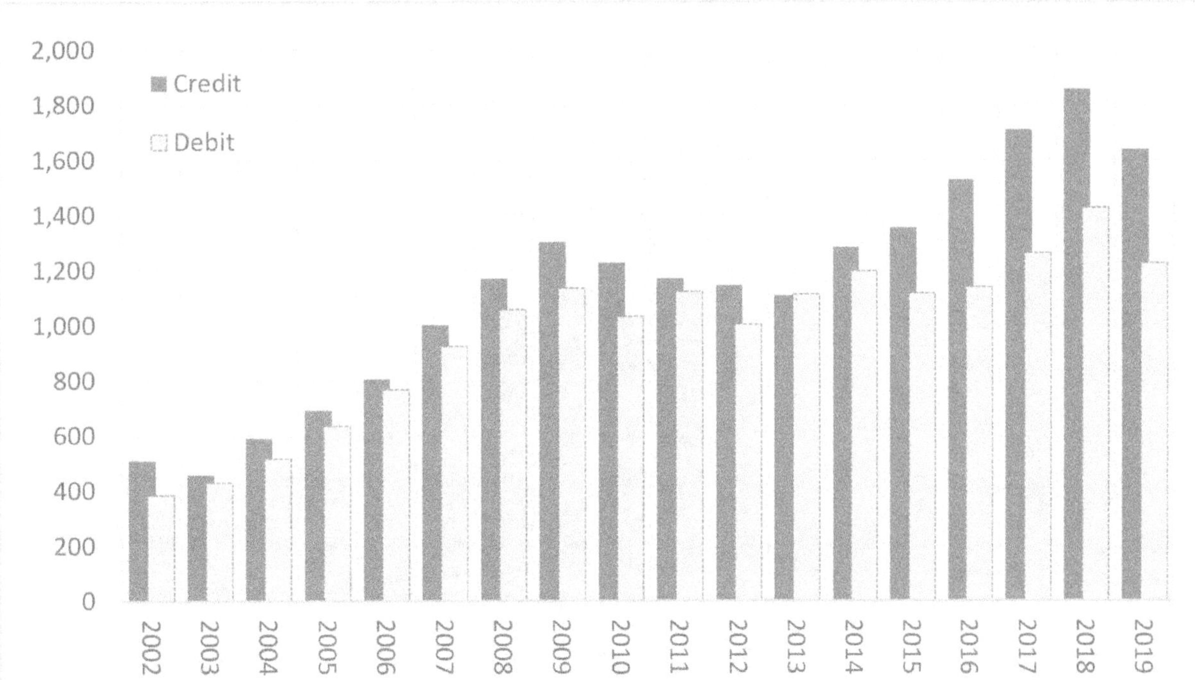

Tourism supply has been broadly defined as the supply of all assets, goods and services to be enjoyed and/or bought by visitors, contributing this way to the shaping of tourism demand in country. It includes the total of natural, cultural and environment, the infrastructure and superstructure and the operating sectors (transportation, accommodation, food services and attractions providers) as summarized in map 1 for the Albanian case.

The supply of accommodation capacities in Albania is broadly assessed as low and understated due to high levels of informality characterizing this sector (Kristo, 2013). Informality is a widespread phenomenon in the Albanian economy, assessed to be higher particularly in services sectors. In the case of accommodation units, informality could mean total lack of registration of accommodation structures and/or partial declaration of capacities (especially home-based renting) triggering also some form of unfair competition among actors in the market. According to Kristo (2013), informal accommodation capacities account for about 92.6% of total accommodation capacities. A study from the Albanian Development Fund (2018), suggests for a large gap in the number of the registered accommodation capacities and real figures, entailing for a considerable number of accommodation structures operating informally.

The most recent data summarized in table 4 show the number of accommodation structures in Albania (included in this category hotels, motels, camps, lodges, mountain shelters and other structures for short term stay). In year 2016, there were about 1,242 accommodation units exercising their activity in the

Figure 6. Touristic attractions distribution by 61 municipalities in Albania
Source: Ministry of Tourism and Environment, Albania

county totaling 17,692 rooms. The average ratio of total rooms to the number of accommodation units decreased from 23.2 in 2012 to 14.2 rooms per accommodation structure in 2016. The same is confirmed also from the ratio of the number of beds to total accommodation units, showing for a contraction from 50.6 in 2012 to about 23.3 beds per accommodating structure in 2016. Both aforementioned indicators and the average size of active enterprises in accommodation and food services sector suggest for an orientation of tourism services providers towards low scale structures. These findings confirm those of

Table 4. Accommodation capacities in Albania

	2012	2013	2014	2015	2016
Accommodation units (no)	632	718	836	905	1,242
Total rooms (no)	14,652	14,146	15,081	12,986	17,692
Ratio: total rooms to accommodation units	*23.2*	*19.7*	*18.0*	*14.3*	*14.2*
Beds (no)	32,004	29,737	32,879	29,903	41,376
Ratio: beds to accommodation units	*50.6*	*41.4*	*39.3*	*33.0*	*33.3*
Capacity (beds per 1,000 inhabitants)	11.0	10.3	11.4	10.4	14.4
Beds density (number of beds to total surface area 28,748 km^2)	1.1	1.0	1.1	1.0	1.4

Source: Instat, Albania

Kristo (2013) and Albanian Development Fund (2018) for the prevalence of small size accommodation structures.

The capacity of accommodating structures measured in number of beds per 1,000 inhabitants averaged at 10.7 during 2012-2015, and increased during 2016 at a level of 14.4 beds per 1,000 inhabitants in 2016. The density of beds for square km shows for a low touristic orientation of the country, registering only 1.4 beds for 1 square km of territory. While there are no available data for the location of these accommodation structures in the territory, Kristo (2013) suggests that 80% of accommodation structures are placed in coastal areas, while the rural areas are dominated by guest houses. The accommodation capacity utilization rate is assessed to be low, since tourism presents strong seasonal patterns – the touristic season in centered in the months of July – August as shown in figures 9 and 10.

Figure 7. Building permits issued for hotels and similar structures
Source: Instat, Albania & authors' calculations

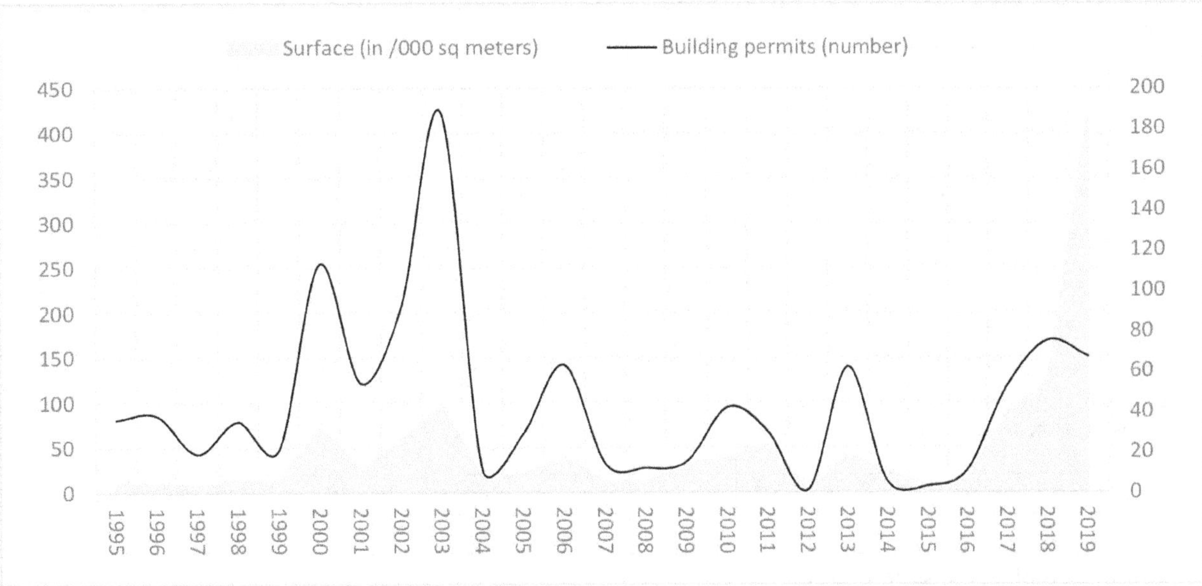

Adding to the existing stock of accommodating structures operating in Albania, figure 6 shows that especially over the last three years, accommodation capacities increased significantly, based on information on new building permits for hotels and similar being issued by authorities in Albania.

During the last three years 2017-2019 alone, authorities in Albania have approved about 195 building permits for hotels and other similar accommodation structures. That corresponds to about 630,000 square meters of hotel surface, over the three-year period. The stimulus provided by the government seems to have augmented the demand for building permits for the construction of hotels these last three years. Investment in new accommodation structures requires large amounts of financial resources, whether from internal or external sources. Bank lending for "accommodation and food services" sector slightly increased over the last three years representing in average about 4.1% of total stock of loans to the economy in Albania.[5] Bank lending to the economy is mostly directed to other sectors of the economy (like construction, industry, information and communication, financial activities and insurance etc.), rather tourism related ones.

Neither is this sector being supported or revealing attractive to foreign investors (see table 5) which disinvested about Euro 16 million over 2016-2017 and invested only Euro 1.2 million during 2018-2019Q3 (or 0.06% per cent of total inflow of FDIs in the country over 2018-2019Q3). Low attractiveness of the sector to foreign investor materialized in the low and decreasing weight of FDIs in total stock of FDIs, from 3.3% in year 2014 in about 1.5% in 2019Q3. Structural deficiencies such as the weak infrastructure, environmental regulatory framework, property rights disputes, absent coordination between central – local government, incomplete regulatory framework and frequency of changes in economic policies etc., are assessed to be among factors that have contributed negatively in FDIs levels. For the tourism sector to be able to compete in international markets the attraction foreign direct investments is more than necessary.

Table 5. Foreign direct investments and lending in tourism

	2014	2015	2016	2017	2018	2019Q3
FDI's stock (in Euro million)						
Transport and accommodation	152	158	164	127	104	123
Stock of FDI's	4,564	4,982	5,677	6,456	7,603	8,294
% of sector to total stock of FDI's	3.3%	3.2%	2.9%	2.0%	1.4%	1.5%
FDI's inflow (in Euro million)						
Accommodation and food services	-	-	(1.0)	(15.0)	0.3	0.9
Total inflow of FDIs	869.2	890.4	942.6	899.9	1,019.6	810.2
% of sector to total inflow of FDI's	-	-	-0.1%	-1.7%	0.0%	0.1%
Lending to the private sector (stock at the end of the period, in ALL million)						
			2016	2017	2018	2019
Bank loan to accommodation and food services sector			13,315	13,551	13,713	16,149
Total credit to the economy			369,934	364,278	337,848	360,067
% of sector to total credit			3.6%	3.7%	4.1%	4.5%

Source: Bank of Albania & authors calculations

The use of ICT has improved somehow as shown in table 6, with about 97.7% of enterprises in "accommodation and food services" use computers, 29.9% of employees use computers in their work, 96.7% of enterprises have internet access, and 9% do e-commerce sales over the year 2018. Other studies assess that the use of information technology in the accommodation and food services sector show for a weak use of ICT in general.

Tourism Demand

Table 6. Integration of ICT in accommodation and food services

	2015		2016		2017		2018	
	Sector	**Total**	**Sector**	**Total**	**Sector**	**Total**	**Sector**	**Total**
Enterprises using computers	93.0	95.0	100.0	95.6	91.0	96.0	97.7	97.3
Employees using computers at their work	31.4	24.4	31.0	28.0	29.9	28.1	30.8	260
Enterprises with internet access	97.6	96.2	92.7	96.8	90.3	96.9	96.7	97.5
Enterprises that did e-commerce sales	10.9	8.8	11.4	7.1	12.6	7.7	9.0	5.3

Source: Instat, Albania

In literature, tourism demand has been broadly defining as the total number of persons who travel or wish to travel, consuming touristic facilities outside their work and residence. In this chapter, tourism demand refers to the number of arrivals of foreign citizens as released by the Institute of Statistics, Albania and summarized in figure 8. Tourism industry early seeds can be found at the beginning of the 90s, when first visitors started to come. In 1992, there were about 130,000 tourist arrivals in Albania (World Bank, 2009) and in the year 2000 there were about 317 thousand arrivals of foreign citizens. Figure 8 shows the number of arrivals of foreign citizens for the period 2001 – 2019, clearly indicating for an increasing trajectory over time (excluding year 2013 where a decrease of 7.3% was registered in view of general elections held).

Starting from 2014, growth rate of foreign citizens' arrivals picked up at moderate levels. During 2019, there were about 6.4 million foreign citizens visiting Albania, up by 8.1% compared to the previous year representing about 2,244 foreigner arrivals per 1,000 inhabitants. Net tourism receipts have increased over the same period and present a positive correlation with the number of arrivals of foreign citizens. During 2018, net tourism receipts registered about Euro 412.3 million or about Euro 144.5 per capita.

The primary market for the existing capacities and tourism demand are neighboring countries (Kosovo, Macedonia and Montenegro and Albanian citizens residing abroad (especially in Italy and Greece). Quantitatively, over the last 5 years' arrivals of foreign citizens from South Europe account for about 75.5% of total arrivals.[7] This kind of demand is persistent and broadly driven by ethnic reasons and low prices. The later does not provide a breeding ground and incentives for businesses in the sector to innovate, improve the quality of their services, equip their working force with adequate trainings.

The tourism demand exhibits significant seasonality as shown in figure 9, in the months of July and August. During these months the number of arrivals of foreign citizen's triplicates compared to the rest of the year. Addressing such a high seasonality is another challenge for tourism market players in Albania.

Figure 8. Arrival of foreign citizens and net travel receipts, 2001-2019Q3[6]
Source: Instat, Albania; Bank of Albania; Authors' calculations

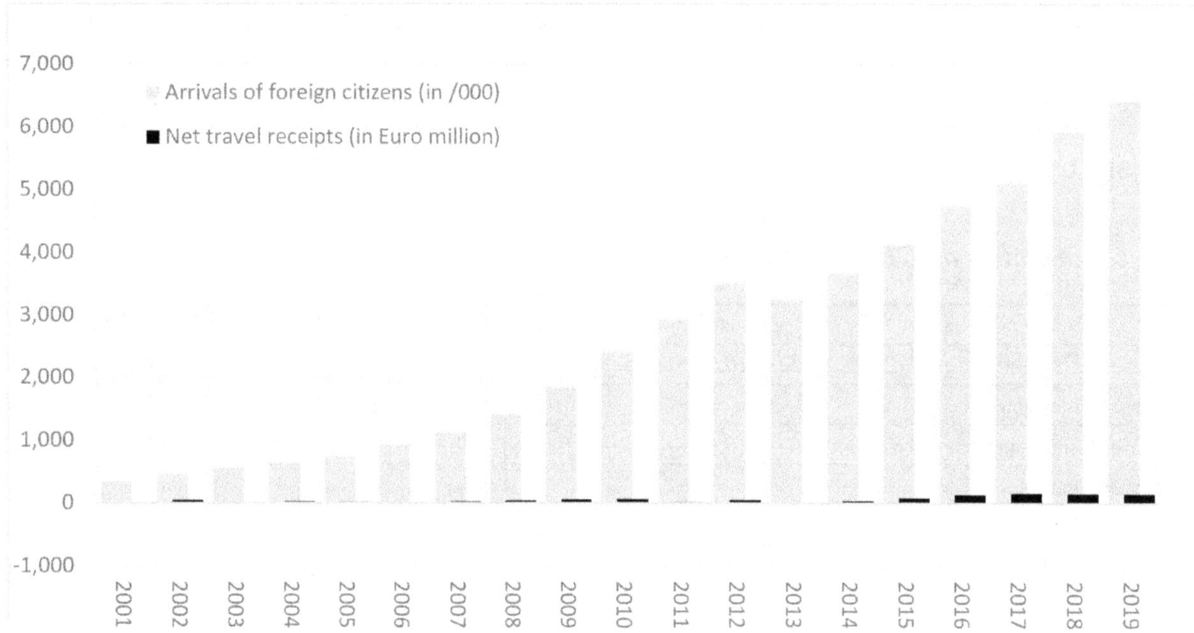

The breakdown of the data on arrivals of foreign citizens per purpose of travel as shown in figure 10 indicate that seasonality patterns are pronounced in the case of arrivals for personal reasons, including holidays, visits to relatives, health treatments, religious and transit. The number of foreign arrivals for holidays and visits to relatives represent about 94% of total arrivals, suggesting that these arrivals are in

Figure 9. Seasonal variation in total arrivals of foreign citizens
Source: Instat, Albania & Authors' calculations

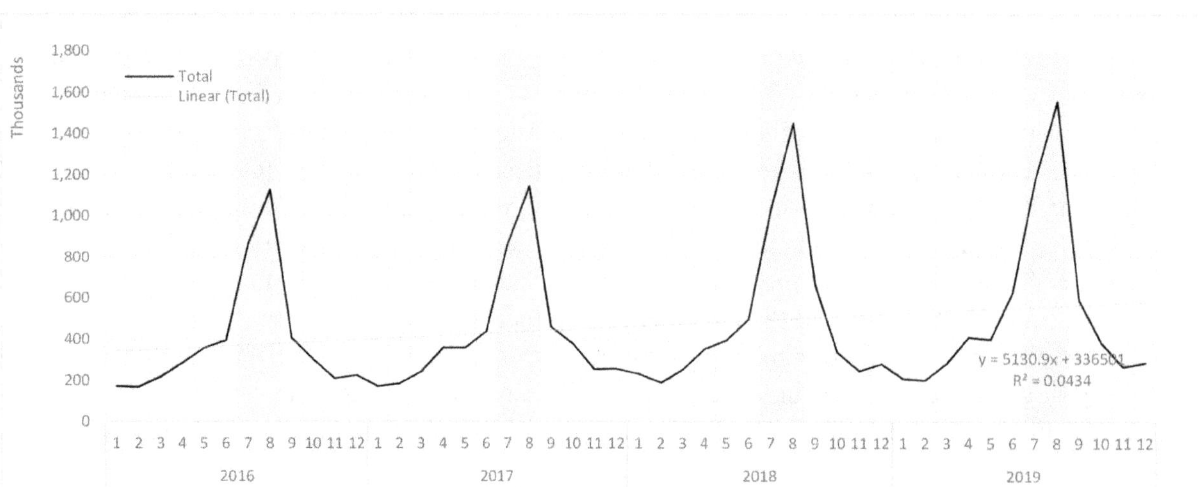

general Albanians living abroad. While, arrivals of foreign citizens for business purposes show for an increasing trend over time and high variation.

Place Based Tourism Experience in Albania

Figure 10. Seasonal variation in arrivals of foreign citizens by category: personal and business
Source: Instat, Albania & Authors' calculations

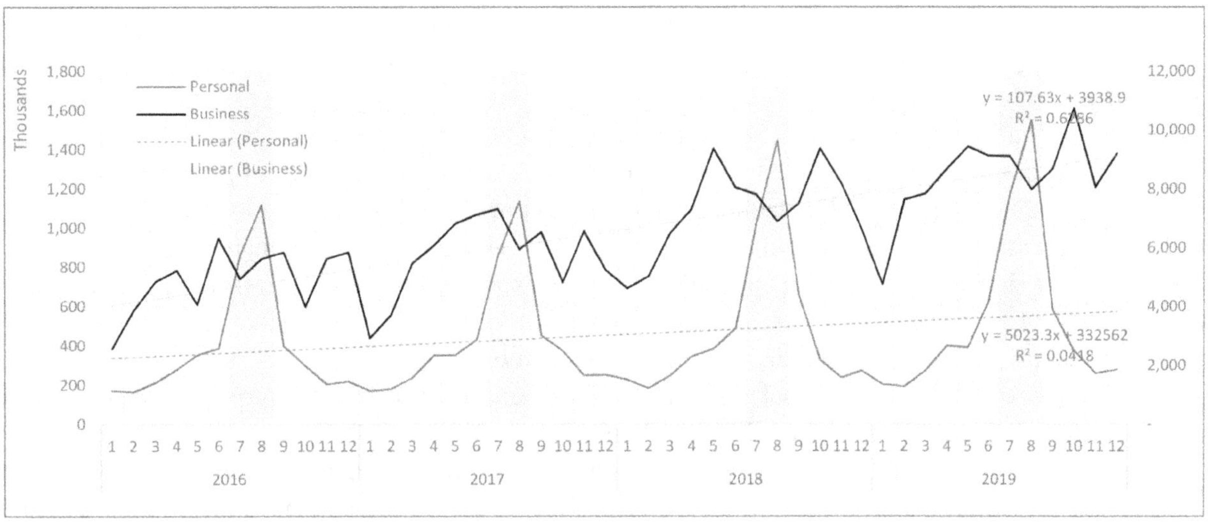

The data show that tourism in Albania is developing, but necessitates for streaming of efforts towards a sustainable scheme and formalization. While tourism potential has yet to be fully tapped into, to ultimately become competitive in international markets, community based tourism can be an important source for the survival and growth of local economies. It can work through the engagement of local population in economic activities ensuring for better living conditions and/or poverty alleviation; preservation of traditions and natural landscapes; promote better access to labor market through the creation of new job opportunities for females and vulnerable groups in their areas (Heesup, Taeyeon, Al-Ansi, Hyungseo, & Wansoo, 2019).While the dream of becoming a great tourism destination persists, some other innovative tourism models are being implemented in Albania, such as the cases of: (i) Caravan Horse Riding Albania; (ii) Farma Sotira; and (iii) Mrizi i Zanave. All the case studies were built based on the information publicly available in National Business Center and semi-structured interviews with managers through a questionnaire with open ended questions.

Case Study 1: Caravan Horse Riding Albania

Caravan Horse Riding Albania is an eco-tourism enterprise/farm focusing on equestrian activities and experiences, officially founded in 2017, having initially started in the form of private tours organized for close family and friends since 2011. The concept of Caravan Horse Riding Albania stems from an early childhood passion for horses and nature, for adventure and sightseeing of its founder, and is fueled by an almost idyllic natural environment of intertwined valleys, mountain ranges, rivers, and strong contrasts. Caravan Horse Riding Albania is based in the region of Gjirokastër, southern Albania bordering Greece, an inscribed UNESCO world heritage site for its cultural and natural heritage since July 2005, and a popular destination among tourists. CHRA proposes an innovative community-based tourism initiative for the local context. Its philosophy is simple and entirely related to nature and local people: authentic moments of horse-riding where the horse, horse-rider, and nature are in full harmony. Visitors can travel on horses through uncharted lands of beautiful unperturbed yet welcoming villages, and in the process, they can experience the vernacular architecture of the local houses, pastoral nature and the traditions of this part of the world from a horseback. Based on a comprehensive assessment of the territory in terms of distinct natural elements, resources, cultural heritage and willingness of the local community to host visitors, Caravan Horse Riding Albania designed 4 long itineraries as shown in figure 11. The first year of the businesses registered a deficit[8], due to a large extent to high current expenditures (staff cost, maintenance, equine food, tickets etc.), which account for about 80% of total revenues. The company has a website to displaying features and supply of services provided and can be found in different on-line booking platforms in the section of "Things to do".

The involvement of the local community in the Caravan Horse Riding Albania business value chain is essential (as showed in figure 12, appendix 1), as the experience promised to the visitors is shaped to a great extent by numerous services offered by different local hosts (about 18 families being part of the value chain), such as accommodation, food, insight into the local traditions culture, and history[9]; local vendors providing for food stock for horses, equipment, vet services; employment opportunities for locals; other services from city of Gjirokastër and Përmet.[10]

The business model that the Caravan Horse Riding Albania are applying is innovative for Albania, but well known in many countries world wise. The whole form of organizing and managing tours is a model of sustainable tourism centered on the experience of riding in an unknown place. The tours are designed to evoke emotions either from the adrenaline that the trip itself brings, combining all the levels of speed and terrain, or from the contacts the riders make with the host residents, or from the fact that the whole trip is located inside a historical fable that the visitors follow and unveil step by step. New elements are frequently introduced to the way the business operates and what it offers aimed at the technical and content improvement. Close observation of the visitor behavior and feedback and interaction with the local community (which is supportive to the initiative), are the key sources for feeding the process of designing and introducing new elements, which focus primarily on the following:

- Changes that bring greater comfort to horses and riders, including specific investments for certain horses in equipment for back protection, better weight depreciation, equipment that reduces the overall load on the horse, new rules that ensure better equine care, visitor participation in the treatment of horses, food, cleaning, equipment, etc.
- Enriching tours with simple episodes from the everyday life of residents including cattle shearing, cheese making, pressing coffee, milking cattle, baking in saç (a spherical baking sheet covered in burning coal), or attending any religious ceremonies.
- Designation of new itineraries, based on community and visitor suggestions, such is the Vjosa River trail, inspired by returning visitors to Albania, and christened as the "romantic" tour.
- Diversification of target visitor age groups by launching a new business direction inspired by children and those who want to become riders. The 'Adopt a Horse' program, which aims to take shape throughout 2020, will create Caravan's first club of members.

Figure 11. Touristic itineraries in Caravan Horse-riding Albania
Source: Author processing

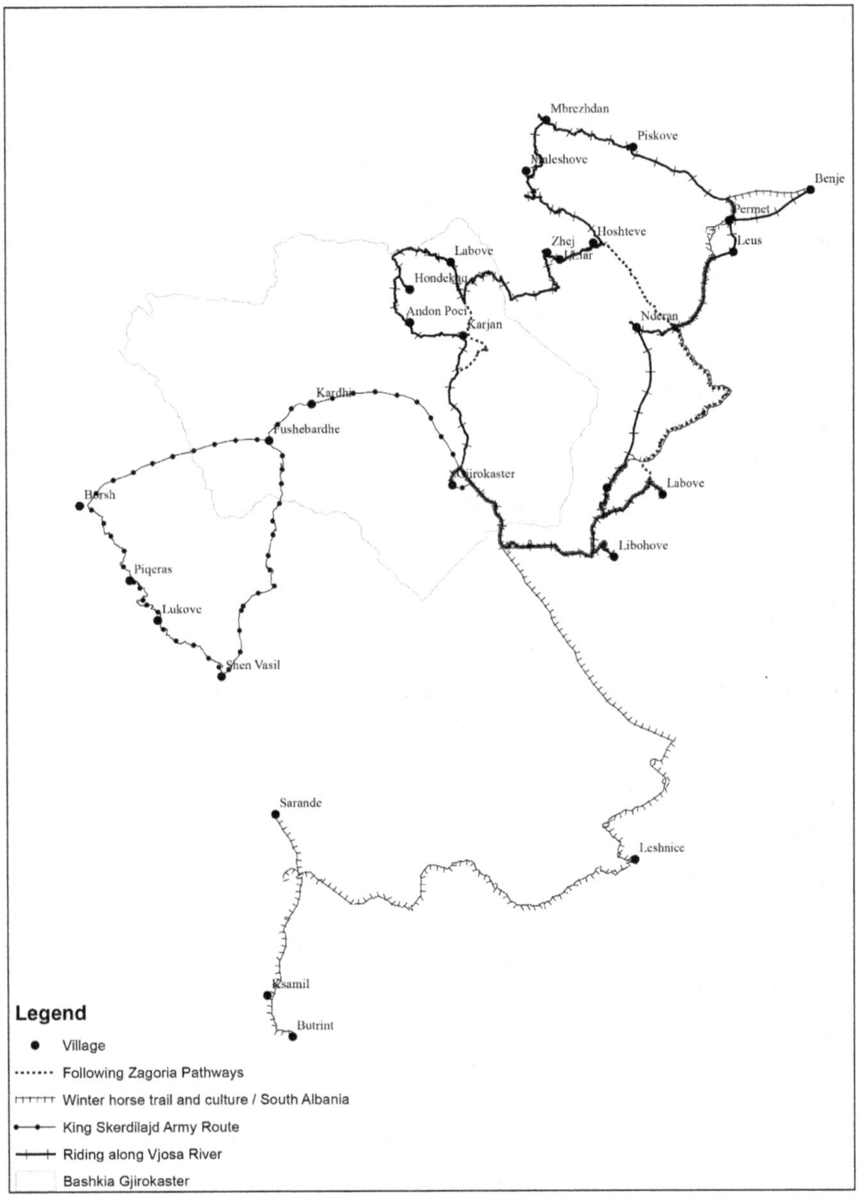

- Increased interaction between visitors and locals through different activities, including basic lessons for amateurs and horse lovers; short outdoor tours to bring horses closer to local people are initiatives that involve them.[11]

Visitors are inclined to pay only for authentic products and are sensitive and want to know where the money goes. Paying to motivate the growth of small business, small farms and supporting local life in the village is completely understandable to them. Visitors are sensitive to the critical state of monuments or nature and wish to contribute even if modestly, for improved preservation of the many elements of culture and tradition, and improved life of the local community. Most of them travel on the principles of sustainable tourism that generates income for families with limited opportunities. This aspect enhances the value of experiencing their journey.

Case Study 2: Farma Sotira

Farma Sotira was established back in 1998, making it arguably the first agritourism enterprise in Albania. The farm is located in the municipality of Kolonjë, an area in the south-eastern part of the country, renowned for its hospitality, culinary tradition, and rich cultural and natural heritage. Close to a number of tourist destinations, such as Korça, Pogradec, Voskopojë, etc., the location of the farm is an inland alternative to the standard, yet popular sun, sand and sea tourism model developed along the western part of Albania. Prevailing geographic elements such as the mountains of Kamenik, Omit, Cuka, and Pogonica that surround the farm, the river Vjosa – the last remaining free flowing river in Europe, at just 30 km distance, and the presence of several hot springs, such as Sarandapor, have all shaped the touristic offer of the farm. The first core commercial activity of the farm was fish-farming. Over the years, the farm has oriented its services towards nature based tourism activities, including: accommodation in bungalows and tents, restaurant, tour guides, hiking, mountain biking, local cuisine cooking classes, farm learning activities, etc.[12]

The driving philosophy of Farma Sotira is content over form, offered in a simple, homely way, offering the visitors a chance to feel as guests and have a glimpse into the everyday life particularities and local customs. Such an approach and optimistic assumption that visitors would travel all the way to the farm for the sheer pleasure of enjoying nature and culinary traditions, was met with skepticism by the local community. With it being one of the first such enterprises in Albania, at a time when tourism statistics were at the minimum levels[13], challenges were numerous, including poor infrastructure, lack of qualified human resources, limited access to capital or support from government funds, conflicting development priorities and investments such as marble quarries permitted in the vicinity of the farm (at a distance of 100m) and the resilience of the society in the aftermath of the 1997 crisis. About 22 years on, the farm measures 20 hectares of land, most of which preserved in its natural state. In terms of hosting capacities, the farm offers 9 two-bedroom wooden bungalows, space for camping tents, parking space for caravans, a restaurant, stables, and barns for its flock 150 cows, goats, sheep, etc. Its maximum serving capacity varies depending on the season, ranging from 40 guests during the winter, to 150 during the summer because of increased use of outdoor spaces such as tents, al-fresco eating, etc.

Among the main reasons to visit, the farm lists organic food, picturesque landscapes, and a number of decompressing activities that can be pursued within the farm premises, such as: walking, horse-riding, and photography. Although the farm is open to visitors all year, the farm's busiest tourist season May – September is slowly shifting to April – November amounting to an average of app. 14,000 visitors per season, out of which 70% are foreign visitors. The latter are offered a choice of complementing their stay at the farm with one of the 8 tours which range from 1-6 days, often including horse-riding. Because of the vast amounts of land at its disposal, compared to other agritourism enterprises, and their consolidated agricultural profile, the farm can generate most of the required produce in-house, quantified at about 70% of total need. The remaining needs, particularly local produce (such as vegetables, grapes, some types of meats, etc.) are addressed by the local community. Similarly, the farm taps into the local community heritage and engages with it on several levels:

- It recruits the farm staff which varies from 5 – 15, depending on the season;
- It acquires supplementary services, such as accommodation, and food, for its tour packages, including visits to traditional houses in the area;
- It "refers" local businesses to visitors, who are travelling in the area.

Close interaction with the visitors has yielded valuable feedback for most of the innovative elements the farm has introduced. The decisions to introduce camping and parking for caravans, cooking classes and visits to local houses, horse-riding tours and mountain biking have been all nurtured and shaped by the feedback obtained by the visitors.

Case Study 3: Mrizi i Zanave

'Mrizi i Zanave' (in English: fairy land*)*, named after one of the masterpieces in Albanian literature by the author Father Gjergj Fishta, is an internationally renowned sanctuary of the *slow food movement* in Albania. It was born from the initiative of a young romantic chef attached to his land, Altin Prenga, who having acquired ample culinary experience in Italy as a migrant from a very young age, returned to his village in Albania after 11 years, and did what most of the local residents considered absurd. In 2011 he opened a traditional restaurant with a modern twist in the village of Fishtë[14], which until then was known as the birth place of the Albanian poet – Fishta, and less well-known for its correctional facility for young offenders during the communist regime.

The business model that "Mrizi i Zanave" introduced was that of "from farm to table", meaning cooperation with local producers was essential, yet challenging. The fellow villagers of Fishtë, were not supportive of such an innovative initiative for the context, and were highly skeptical to the idea of Albanians driving for app. 90 min from nearby urban centers, and even more so from all other urban centers in the country. Initially they protested by refusing to sell their local produce to the restaurant. Instead, they would take their local products to the market in the city of Lezhë, and on many occasions Chef Prenga would follow them to the market and get the products from them there. Gradually, such persistence led to trust and the local community starting to get involved. The cooperation with the local community consists on two levels: (i) (i) acquiring all the produce locally through a pool households and (ii) recruiting local staff from Fishtë and the surrounding villages within a 10 km range, amounting to 65 in 2019.

In less than 10 years since its opening, Mrizi i Zanave is credited with taking Albanian food back to basics: fresh, organic farm-to-table produce and meat that celebrates the country's fertile terrain.[15] It also proved to be a successful case of community based tourism. It started with a small-sized restaurant which grew year by year into a broad value chain, with a snowball effect. The restaurant, and the hotel count an average of 100,000 visitors per year, or close to one million since first opening in 2010, of which 30-50% foreign visitors.[16] Passed initial doubts and skepticism, local community of producers engaged and integrated in the value chain determining the success of the initiative. A very positive contribution to the activity gave the conversion of the former prison building, into a factory – a poly-laboratory for the processing of local produce like milk (35 suppliers)[17], fruits, poultry, livestock and vegetables of the area. Food processing based on local produce contributed to the restoring of tradition and become an important source of income for the owners and local community. Figures show that the turnover registered by the company jumped impressively from ALL 1.4 million in 2011 to about All 84.8 million in 2018 (average growth rate 91.1%), with an average net profit rate of about 6.9%[18]. The company has a fashionable website (no on-line booking in the website is available), is registered in different on-line booking platforms and rating from guests is exceptionally high. The way the business evolved, from solely a restaurant trying to provide for inputs from local community to an agritourism broadly relying on a mix of own produced supplies with local inputs from different farmers in the area, has yielded somehow a mixed model of community-based tourism.

FUTURE RESEARCH DIRECTIONS

Entrepreneurial activity in the tourism sector in Albania is increasing, often oriented by the shifting trends towards nature based tourism experiences, in rural settings, "orchestrated" and fueled by local communities, which vouch for the authenticity of the experience and product, and feed local economy. Although most enterprises are in a nascent stage, as Albania's free economy history starts in 1992, based on the development of the tourism sector so far, the limitations on a governance level, and the panorama of the entrepreneur-introduced tourism development initiatives, the following future research directions can be explored further:

- Exploring the diversification of tourism supply: new touristic products to be defined especially building on cultural tourism. Such kind of touristic supply overcomes seasonality that generally characterizes the touristic classic offer and in peak season, soothes pressure on most visited sites (Vasiliadis, et al., 2016) – from south-west to inner/inland Albania. Also, frequent new proposals of touristic areas somehow streamline flows of tourists and help local economies to grow in a more balanced way.
- Integration of new technologies in tourism: applications to be developed including touristic attractions general descriptive information, touristic routes, food and accommodation facilities, safety measures and emergency numbers, etc. This can be done by local authorities (the municipalities with the support of donor community engaged in tourism field).
- Enabling conditions for start-ups: incentives and promotion schemes to be extended also to micro – small business start-ups in local economies in order to develop local entrepreneurial skills and enable integration in larger value chains;
- The hidden costs of tourism, if not accounted for, can outpace the benefits experienced in terms of income, production, employment and foreign exchange. Negative impact on the environment, on territory, socio – cultural identity and traditions can contribute in making them less attractive to future visitors. Thus, from the regulator being central or local, tourism should be seen as a long term industry, not an instrument to provide for short term liquidity in the economy.
- Skills and qualifications: the tourism industry is labour intensive (mostly bar and restaurant serving etc.) and poorly paid as shown by the data where wages in tourism sector are the lowest among economic sectors. That makes workers easily substitutable, thus they have no bargaining contractual power for higher payments and no prospects for carrier progression. Better education and training, aligned to industry sector need might contribute to smoothing inequality in compensation of employees in this sector.

CONCLUSION

The aim of this chapter was to build on the theoretical frameworks of entrepreneurial, innovative community tourism and its contribution in the economy, providing research findings in the form of case studies. Three case studies were explored from rural areas in northern and southern Albania, located in different stages of their business maturity cycle. "Mrizi i Zanave", "Caravan Horse Riding Albania", and "Sotira Farm" represent three models of community-based tourism, driven by entrepreneurial and innovation elements. Based on an overview of the core theoretical concepts of entrepreneurship and innovation in community-based tourism, and the analysis of the cases studies, several conclusions can be drawn.

- In absence of consolidated governance of the tourism sector, most development in this field is happening through a bottom-up approach, primarily driven by the private sector. The three cases under study, are but a few of numerous examples where visionary entrepreneurs have managed to spark the birth of local entrepreneurial microcosms or ecosystems, which rely on community involvement and sustainable practices.
- All three analysed cases are in different stages of community involvement and the nature of involvement of the community changes depending on the maturity state of the business. Sotira Farm, established since 1998, has gradually grown to a highly self-sufficient entity, meaning the scope and nature of services acquired by the local community are more specialized and more limited. Similarly, Mrizi i Zanave, with its rapid growth and investments, has diversified the range of products and services it offers in house meaning that, instead of purchasing ready produce, it now acquires ingredients or raw produce, and services. At what point in its growth trajectory a community based initiative becomes self-sufficient, is still unclear and deserves additional research.
- From the analyses, it emerged that all three cases had successfully managed to tap into the international market, ranging from 70% in the case of Sotira Farm, to almost 100% in the case of Caravan Horse Riding Albania. Domestic tourism numbers are particularly high in the case of 'Mrizi i Zanave', helping it extend the tourism season to an all-year-round one.
- The tourism industry is characterised by strong seasonal patterns, with great inflow during two months in the year and deserted all over the other 10 months. Also, the industry is largely exposed to internal and external shocks like wars, terrorism, natural disasters or like the Covid-19 pandemic (which effects and managing strategies in tourism sector are yet being developed).
- Albania's tradition and experience in tourism is very limited, hence most development qualifies as innovation when placed in such context. The use of the '*slow-food*' concept combined with the cultural heritage of northern Albania have produced a successful formula for 'Mrizi i Zanave' earning it a reputable place in the international culinary scene. Similarly, 'Caravan Horse Riding Albania' reintroduced the element of horse-riding, enriched with local natural and cultural heritage fragments to offer visitors a very local taste. That said, it should be noted that innovation intensity in the sector is low, and there is a predominance of incremental non-technological innovations.
- Public consultations and participatory planning are regulated by law. Yet, there are few examples of effective applications of genuine community-based tourism planning and development. Most of the tourism planning happens *ad hoc*, and it mainly initiated and driven by the vision and means of individual entrepreneurs. For the community-based tourism practices to become sustainable in the long-term, increased, genuine applications of community involvement need to be encouraged.

- Similarly, to other sector in the economy, informality in the sector is high. Although dimensions, causes and measures to curb informality are not subject of this research, it should be noted that this aspect is deserving of more research attention.
- The supply of touristic services shapes the typology of demand of tourists visiting the country and helps promote sustainable planning and use of resources. For such practices to be sustainable in the long-term, local community participation should be factored in the planning and development phase. Good value for money practices should also be explored.
- The topic of tourism, and aspects such as community-based tourism, the role of entrepreneurs in the development process in Albania, are understudied and additional research is needed.

REFERENCES

Albanian Development Fund. (2018). *Market research and value chain assessment*. Horwath HTL.

Bâc, D. (2012). *The impacts of tourism on society*. University of Oradea. Retrieved from http://steconomiceuoradea.ro/anale/volume/2012/n1/072.pdf

Boka, M., & Torluccio, G. (2013). Informal Economy in Albania. *Academic Journal of Interdisciplinary Studies*, *2*(8), 2012.

Bunghez, C. L. (2016). The importance of tourism to a destination's economy. *Journal of Eastern Europe Research in Business & Economics*, *2016*, 1–9. Advance online publication. doi:10.5171/2016.143495

Butler. (2006). *The tourism area life cycle*. Channel View Publications.

Buttler, R. (1990). Alternative tourism: Pious hope or Trojan horse? *Journal of Travel Research*, *31*(4), 40–45. Advance online publication. doi:10.1177/004728759002800310

Çela, E., Metani, I., & Çeliku, E. (2019). *Analiza dhe zhvillimet strukturore të ekonomisë Shqiptare: përqasje sipas matricave të kontabilitetit social*. Bank of Albania.

Çeliku, E., & Metani, I. (2011). *Pozicioni konkurrues i ekonomisë shqiptare në terma të produktivitetit e të kostos së punës*. Bank of Albania.

Chambers, R., & Conway, R. G. (1992). *Sustainable rural livelihoods: Practical concepts for the 21st century*. IDS.

Ciro, A. (2020). *Entrepreneurial Eco-systems for sustainable tourism development in Liminal Spaces*. IDAUP Pubblication.

Ciro, A., Toska, M., & Nientied, P. (2018). Social innovation and sustainable economic development: participatory torusim destination management. In M. Jaššo, M. Husar, & M. Finka (Eds.), *The role of public sector in innovative local economic and territorial development*. Springer.

Creaco, S., & Querini, G. (2003). The role of tourism in sustainbale development. In *43rd Congress of the European Regional Science Association: "Peripheries, Centres, and Spatial Development in the New Europe"*. Jyväskylä, Finland: European Regional Science Association (ERSA). Retrieved from http://hdl.handle.net/10419/115956

Goeldner, C., & Ritchie, B. (2012). *Tourism, principles, practices, philosophies*. John Wiley & Sons.

Grebel, T., Pyka, A., & Hanusch, H. (2003). An evolutionary approach to the theory of entrepreneurship. *Industry and Innovation, 10*(4), 493–514. doi:10.1080/1366271032000163702

Heesup, H., Taeyeon, E., Al-Ansi, A., Hyungseo, R. B., & Wansoo, K. (2019). Community-based tourism as a sustainable direction in destination development: An empirical examination of visitor behaviors. *Sustainability, 11*. Advance online publication. doi:10.3390u11102864

Hjalager, A.-M. (1997). Innovation patterns in sustainbale tourism. An analytical typology. *Tourism Management, 18*(1), 35–41. doi:10.1016/S0261-5177(96)00096-9

Institute of Statistics. (2019). *Tourism in figures*. Tiranë, Albania: Institute of Statistics.

Kasapi, I., & Koc, M. (2013). Changing tourism consumer behaviour: the impacts on tourism demand in Albania. *Creative and Knowledge Society, 2*(2), 16-34. doi:10.2478/v10212-011-0023-3

Kline, C., Hao, H., Alderman, D., Kleckley, J. w., & Gray, S. (2004). A spatial analysis of tourism, entrepreneurship and the entrepreneurial ecosystem in North Carolina USA. *Tourism Planning & Development*, 305–316. doi:10.1080/21568316.214.890127

Kokkranikal, J., & Morrison, A. (2011). Community networks and sustainable livelihoods in tourism: The role of entrepreneurial innovation. *Tourism Planning & Development, 8*(2), 137–156. doi:10.1080/21568316.2011.573914

Kristo, I. (2013). *Tourism, culture, hospitality, and agrotourism*. Academic Press.

Lohmann, G., & Netto, A. P. (2017). *Tourism theory: concepts, models and systems*. CABI. doi:10.1079/9781780647159.0000

McKenzie, R., & Brackfield, D. (2008). *The OECD system of unit labor cost and related indicators*. OECD Statistics Directorate. doi:10.1787/18152031

Medina, L., & Schneider, F. (2018). *Shadow economies around the world: what did we learn over the last 20 years?* IMF Working Paper.

O'Connor, A., Stam, E., Sussan, F., & Audretsch, D. (2018). *Entrepreneurial ecosystems. Place-based transformations and transitions*. Springer International Publishing AG.

Power, S., Di Domenico, M., & Miller, G. (2017). The nature of ethical entrepreneurship in tourism. *Annals of Tourism Research, 65*, 35–48.

Schumpeter, J. (1947). The creative response in economic history. *The Journal of Economic History, 7*(2), 149–159. doi:10.1017/S0022050700054279

Simpson, M. C. (2007). *An integrated approach to assess the impacts of tourism community development and sustainable livelihoods. Community Development Journal.*

United Nations. (1987). *Report of the World Commission on Environment and Development.* United Nations.

United Nations World Travel Organization. (2019). *International Tourism Highlights.* World Tourism Organisation.

Vasiliadis, L., Trivellas, P., Belias, D., Meleas, J., Kyriakou, D., & Koustelios, A. (2016). Cultural tourism revisited: the case of Thessaly. In V. Katsoni, & A. Stratigea (Eds.), Tourism and Culture in the Age of Innovation. Springer International Publishing.

Wall, G., & Mathieson, A. (2006). *Tourism: change, impacts and opportunities.* Pearson Education Limited.

Welford, R., & Ytterhus, B. (1998). Conditions for the transformation of eco-tourism into sustainable tourism. *European Environment, 8*(6), 193–201. doi:10.1002/(SICI)1099-0976(199811/12)8:63.3.CO;2-W

World Bank. (2009). Albania: building competitiveness in Albania. The World Bank.

KEY TERMS AND DEFINITIONS

Community-Based Tourism: Tourism activities involving the integration of local produce and local labor force into supply and value chain.

Destination: A place displaying enough characteristic to a determined number of visitors so that to be considered as attractive to a traveler.

Sustainable Tourism: Tourism activities that succeed in providing a balanced equilibrium between social-economic and environmental dimension.

Tourism Demand: The total number of persons travelling/visiting to a destination outside their work/residence, consuming touristic products and/or services.

Tourism Industry: Tourism industry will be approximated with accommodation and food services activities in the national statistics available in Albania.

Tourism Supply: The entirety of touristic attractions of all categories and goods and services from any form of organisation in hosting destinations.

Tourist: A visitor spending at least one night in an accommodating structure in the country visited.

ENDNOTES

1. The contribution of professional, scientific and technical activities account for about 2.8% of GDP over the last five years 2014-2018, based on National Institute of Statistics data.
2. Labor productivity was calculated as ratio of gross value added (in ALL million) to number of persons employed based on the data from Structural Business Survey, INSTAT.
3. Law no. 40/2018 for some amendments to law no. 92/2014 "On Value Added Tax in the Republic of Albania, changed.

4 Law no. 106/2017 for some amendments to law no. 9632/2006 "On local taxes system", changed.

5 Source: Bank of Albania.

6 Net travel receipts for year 2019 include only the first three quarters of the year.

7 Source: INSTAT and authors calculations.

8 Based on official Balance Sheets released at the National Business Center.

9 The enterprise/the farm currently has an optimal maximum serving capacity of 500 horse-riders per year, and a fleet of 42 horses, of whom 30 can be ridden. The remaining 12 are being trained and will constitute the future generation of horses. Visitors, who are almost entirely of international origin, learn about the farm primarily through the farm's online presence in various platforms, and through the numerous partnerships the farm has with at least 18 established partnerships with international agencies in Europe, USA, and Canada, such as: Pferd & Reiter (Germany), Unicorn, Far & Ride, Saddle Travel dhe Equus Journeys (United Kingdom), Hidden Trails në Kanada, Cheval d'Aventure, Cavaliers du Monde, Randocheval, Agence du Vouage a Cheval (France), Horsexplore (Sweden), Trailfinders, Centaur Trails (Belgium), Hippo Tours (Denmark), Paardenpass (Netherlands), Dirty Hooves (Poland), High Pointe (USA).

10 The relationship with the local community has been a mutually learning curve and has evolved in line with the business growth trajectory. At the beginning, in absence of a model and reference service provision standards, the cooperation with the local community, including both local inhabitants and businesses, was challenging and affected by numerous problems, such as: lack of capacities and quality service, deficient ethics, limited capacities to invest, poor access and digestion of information related various aspects of the touristic experience. Over the years, as the number of visitors grew, and itineraries/trails were consolidated and gained recognition in international markets, the relationship with the local community strengthened. In addition to the positive changes to the diversification of the livelihood sources, the establishment of the Caravan Horse Riding Albania, and the close collaboration with the local community, has led to significant changes both in the local mentality about tourism and in the international recognition of this part of Albania.

11 Local people are becoming growingly conscious of the various opportunities and role they can assume within the "tourism value chain", from cleaning activities to selling simple products/produce. A surprising, yet welcome "side-effect" has been that of curbing deforestation as people are reluctant to cut trees illegally because they feel monitored/watched, resulting in paths/trails becoming greener over the years. Village centers are cleaner because "off-the-grid" *forgotten areas* for many years, are turning into tourist attractions. Similarly, the protection and awareness of cultural monuments along the trail has changed significantly over the years. The conscious decision to include in the trails churches, medieval bridges, roads and ancient ruins has in turn promoted the protection of nature, monuments and spiritual and material heritage, but it has also drawn attention to the critical state of conservation, resulting in positive institutional response and restauration of the church of Hoshteva, that of Labova the Cross and the Nivan Bridge.

12 Farma Sotira has a website and is registered in different on-line booking platforms with an average high scoring by its visitors.

13 Based on INSTAT data, in year 2000 there were about 317.1 thousands of arrivals of foreign visitor.

14 In the Zadrimë Region, Municipality of Lezhë.

15 https://www.lonelyplanet.com/articles/lezhe-to-lunder-northern-albania-for-foodies

16 Prenga, personal communication in May & October 2019.

17 https://ata.gov.al/2019/02/13/modeli-i-agroturizmit-mrizi-i-zanave-ne-fishte-te-lezhes-merr-certifikaten/

18 Based on balance sheet data available in the National Business Center, Albania.

APPENDIX 1

Figure 12. The supply chain in Caravan Horse-riding Albania
Source: Authors processing based on interview

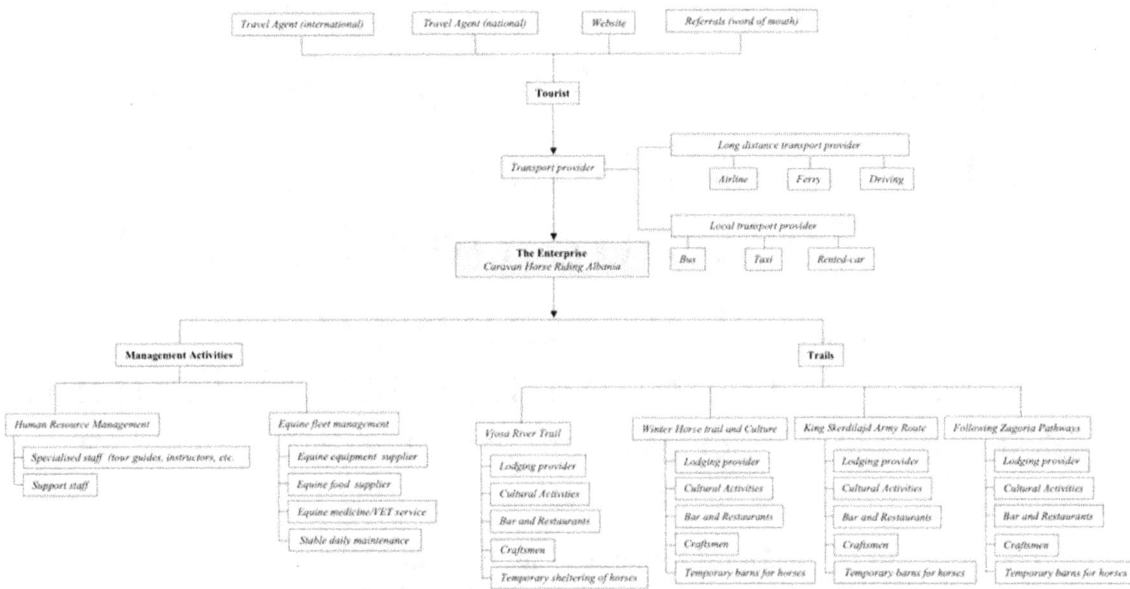

Chapter 8
Fostering Resilience by Empowering Entrepreneurs and Small Businesses in Local Communities in Post-Disaster Scenarios:
The Case for Community-Based Tourism in Puerto Rico After Hurricane Maria

Maria Francisca Casado-Claro
https://orcid.org/0000-0002-4395-9620
Universidad Europea de Madrid, Spain

ABSTRACT

In September 2017, Hurricane Maria caused widespread devastation in Puerto Rico. In its aftermath, Puerto Ricans envisioned destruction as an opportunity to renovate and innovate. In this context, community-based tourism has become a key factor to foster economic recovery and build resilience in local communities. This chapter aims at analyzing some of the recovery efforts in the field of community-based tourism in the island after Hurricane Maria. Whereas on the one hand, tourism and the visitor economy occupy a prominent position in the governmental "Disaster Recovery Plan," on the other hand various non-profit organizations are working hand in hand with local communities to tackle untapped economic opportunities. In this chapter, the authors will analyze the efforts of the Foundation for Puerto to help communities around the island transition from relief to recovery. The lessons extracted from this case could be useful for destinations that are exposed to natural disasters, or those who wish to retain population by creating tourism-related business and employment opportunities, among others.

DOI: 10.4018/978-1-7998-4855-4.ch008

THE IMPACT OF HURRICANE MARIA IN PUERTO RICO

In September 2017, in the term of a couple of weeks, the island of Puerto Rico was hit by two major hurricanes (Irma and Maria)[1], which wreaked havoc and caused important damage to its natural environment and infrastructures, with the ensuing economic and social impact. The archipelago, that had already been facing a recession for over a decade, was thrown in the doldrums in the aftermath of the hurricanes. The government estimates that the damages produced by the hurricanes to the tourism industry amount to $ 547 million in direct income (Gobierno de Puerto Rico, 2018: 43), while a survey of municipalities showed a decrease in economic activity, particularly in key industries such as agriculture, tourism and retail.

Hurricane Maria, which was the worst hurricane to hit the island in 80 years, caused more damage and had a deeper impact in the island's economy, as tourist arrivals decreased sharply, and more than 10.000 small businesses were forced to close, the situation being even more serious in communities outside the San Juan metropolitan area. By 2018, according to the UNWTO, Puerto Rico's visitor arrivals had decreased by 18% and tourist revenues by 15%.

In the aftermath of a disaster, communities need to fix the damage and rebuild. The discourse of change is a common feature in such scenarios, as disasters present both challenges and opportunities. They are an opportunity to start again, to replace the things that did not work, to renovate technologies that were obsolete, to create and innovate. Puerto Rico was already suffering the effects of an economic and financial crisis before the hurricane, economic activity is unevenly distributed, and some areas are still in need. As a result, according to the US Census Bureau, the population had been decreasing since 2004 due to migration, and the hurricane has accelerated the population loss trend.

In 2017, before the hurricanes, it was estimated that 3.4 million Puerto Ricans lived in the island and about 5.6 million lived in the U.S. Since then, in the following two years, about 130,000 Puerto Ricans migrated as a direct result of Maria's devastation, and the government reported that it expects an additional 8% population loss by 2024. To make things worse, the ones leaving are the young generation outside urban areas, the ones staying the older one, which result in a dramatic human capital loss.

Even though Community-based Tourism (CBT, hereafter) is not the answer to all the island's problems, it can be part of the solution. In July 2018, the government published its recovery plan "Transformation and Innovation in the Wake of Devastation: An Economic and Disaster Recovery Plan for Puerto Rico", in which sustainable tourism plays an important part as a key to boost local economies while preserving unique natural, cultural and historical assets. In the same month, the non-profit DMO *Discover Puerto Rico* started operations with the mission of promoting Puerto Rico abroad as a premier destination for business and leisure travelers. Whereas, amongst grassroots initiatives, Foundation for Puerto Rico's *Bottom Up Destination Recovery Program* identifies and invests in four impact areas in its effort to support communities to become more resilient and active in the creation of economic opportunities, one of those lines focuses in "developing branding and marketing strategies to bring more tourists to the community in a sustainable manner".

Although recovery has been slow and taken longer than expected, Puerto Rico is an example of resilience in the face of adversity. CBT was already shaping public and private efforts in this Caribbean island before hurricane Maria, and it is expected to be playing an even more important role in the mid and long-term post-disaster recovery efforts. Whereas in the 1950s, governmental policies aimed at attracting big investment to build infrastructures and large leisure complexes, some of them with serious environmental impact, the present development of CBT is more sustainable since it takes place at a smaller scale. Furthermore, the involvement of civil society in CBT can help "turning local communities from resource users to resource protectors" (Zacarias & Loyola, 2017: 142).

As a small Caribbean island rich in culture and nature, community-based ecotourism is a win-win combination for all stakeholders. On the one hand, it offers visitors the opportunity to get to know the real Puerto Rico and one of its most valuable assets: the people and their heartfelt hospitality. On the other hand, communities can benefit from direct expenditures and their multiplier effect within the local economy. In this chapter, a literature review of CBT and the visitor economy sets out the stage to explore how the two have been applied in Puerto Rico to build resilience in local communities around the island, after the devastation caused during the 2017 hurricane season, using a descriptive research methodology. The focus of the chapter is set on third-sector initiatives implemented in order to strengthen local communities and foster entrepreneurship. The socio-economic effects and the virtuous networks generated by these dynamics, geared towards building CBT, results in more resilient communities. In the conclusion, some lessons will be extracted on how to transition from relief to recovery and how CBT can contribute to build more resilient local communities.

WHAT IS "COMMUNITY-BASED TOURISM" AND HOW DOES IT RELATE TO THE "VISITOR ECONOMY"?

Since its inception in 1985, in Murphy's seminal book *Tourism: A Community Approach,* the conceptualization of CBT has evolved driven by both theory (academic contributions and empirical studies) and practice (numerous projects carried out in all continents). Ruiz, Henández et al. (2008: 400) argue that under the CBT label very different realities are hidden and that, even though CBT cannot be associated to any tourist product in particular, it is undoubtedly linked to ecotourism. Whereas Mullo, Vera and Guillén (2019: 179) consider that CBT has emerged as an "international brand", and Guzmán-Lopez, Burgos and Castillo (2011: 434) contend that as a novel concept CBT presents new research lines on tourist development in developing countries, adding up to other existing approaches such as Pro-Poor Tourism (PPT), Community Benefit Tourism Initiatives (CBTIs) or donor-assisted community-based tourism (DACBT).

In a socio-economic context quite different from the present one, Murphy saw in CBT an opportunity to leverage the prospects and challenges that the rapid growth of mass tourism presented. In his view, "a community emphasis would temper the economic concerns with the environmental and social considerations" (Murphy, 1985: 1). Therefore, the author advocated for the adoption of an ecological approach to tourism planning on the basis that since a community's natural resources are essential to attract tourism, their stewardship was crucial for its long-term success and survival. Furthermore, an ecological approach to tourism planning would allow tourism to be integrated into general community goals.

In the same line, Guzmán-López et al (2011: 435) remark that in developing countries CBT has emerged as a possible solution to the negative effects of mass tourism, that allows for the social organization of the local community by integrating within the community itself various hospitality services: accommodation, food and complementary offer. In this regard, Thériault's (2015: 1) suggests that CBT initiatives are authentic laboratories where communities take charge of tourism, besides being a relevant and innovative tool for the development of certain regions.

Governments consider CBT as a viable alternative to exploit in a sustainable way local resources, which allows communities with limited economic resources to generate jobs and new forms of income. By transforming social, environmental and cultural assets into competitive opportunities for the communities involved, CBT finds practical solutions to complex problems, such as rural development (Burgos-Doria, 2016: 210) or population loss, amongst other. Although it must be said that, despite its good intentions, not all CBT-related projects present positive results. CBT initiatives sponsored by local and national governments, tourist institutes and academia, and non-profit organizations have spread throughout the world: there are experiences in Asia, Africa, Europe, Australia and New Zealand, Canada, the United States and Latin America.

Guzmán-López et al (2011: 433) emphasize that the management of natural, heritage and cultural resources must be in the hands of the local community itself, who should also set limitations to the tourist development of the area. Furthermore, they consider that in order to be sustainable, tourism can should be a complementary economic activity, an alternative source of income for the local community, never the main and only one.

By contrast, Mullo et al (2019: 180) consider that the most outstanding component of this type of tourism is its human dimension, since it encourages a true encounter and dialogue among people from different cultures, facilitating mutual knowledge and learning. Whereas, Burgos-Doria (2016: 211) concludes that CBT is an activity that brings development to and empowers local communities, that get actively involved in caring for natural and cultural resources.

The role played by individuals and local communities is critical to the success of CBT, as Kibicho (2008) points out amongst the key factors for the development of CBT: the recognition of individual and collective benefits, the formulation of tourism-related objectives and that the local community takes accountability for the implementation of the plan (in Guzmán-Lopez et al, 2011: 435). Nevertheless, to get community support, the project must first and foremost respond to a local need, that it is a project by and for the community (Thériault, 2015: 2).

Mullo et al (2019: 179) also see in CBT a strategic field for social, economic and cultural development that is successfully facing the present demands of today's world for authentic tourist experiences. Tourist experts observe a change in the behavior of tourists. Globalization has homogenized cultures, therefore in their travels tourists seek to explore destinations where local culture and customs, ethnic attributes and historical heritage enable them to experience cultural diversity as a form of personal enrichment.

The profile of this kind of alternative tourists is plural and the type of activities demanded by visitors depends on their motivations to visit the local community: enjoying the nature, gastronomy, traditional architecture, folklore and popular traditions, (Mullo et al, 2019: 182), but also getting to know the archaeological and historical heritage, dance, arts & crafts, trekking or practicing sports (rafting, canoeing, rappelling, tubing, snorkeling, scuba-diving, etc.), amongst other.

In Table 1, the main characteristics of CBT covered in this literature review are summarized for refence, as further discussion on their application will be carried out in this chapter.

As for the concept of *Visitor Economy* (VE, hereafter), Kyriakidis et al. (in Hristov, 2015: 222) consider that it revolves around two key components: the direct contribution of tourism, linked to the Tourism Satellite Accounts (TSA), and the indirect contribution from other sectors of the economy – reliant on, or supporting tourism activity. A 2009 study carried out by Deloitte and Oxford Economics revealed that the UK tourism industry's direct contribution to its economy amounted to £52bn or 3.7% of the GDP, but when the direct and indirect contributions were considered, the total came up to £114bn or 8.2% of the GDP. After two *Visitor Economy Forums* were held in London, in 2012 and 2013, this new approach was adopted as a fundamental framework in England, the most successful tourism destination in the United Kingdom, (Hristov, 2015: 220) by the DMO VisitEngland.

Table 1. Summary of the main characteristics of CBT covered in the literature review

Some characteristics of CBT	Authors
It leverages the benefits of mass tourism and ecotourism/local tourism, it is a complement of other economic activities	Murphy (1985); Guzmán-López et al (2011: 433)
It adds to other approaches in their search for practical solutions	Guzmán-Lopez et al (2011: 434); Burgos-Doria (2016: 210)
Community management of local resources	Murphy (1985); Guzmán-López et al (2011: 433); (Thériault, 2015: 1-2); Burgos-Doria (2016: 211)
Social and human dimension A true encounter between locals and visitors	Guzmán-López et al (2011: 435); Burgos-Doria (2016: 211); Mullo et al (2019: 180)
It offers authentic experiences and varied activities	Mullo et al (2019: 179 and 182)

The VE highlights the multiplier effect of tourism in the local economy, accounting for not only direct tourism expenditure in goods and services, but also indirect consumption in agriculture, business services and communications, amongst other components of the supply chain. In contrast to mass tourism, most of these businesses are small and medium sized enterprises managed by locals. Although there isn't an agreed definition for visitor economy, it is irrefutable that at the heart of it lays the economic activity of "visitors" (not only tourists) in the destination (Hristov, 2015: 222). Moreover, the visitor economy uses existing tourism infrastructure (natural and built attractions, culture and heritage, as well as the services that cater for the needs of visitors and residents) tapping into unexploited opportunities to bring economic benefits into the local community.

As can be inferred from this conceptualization, both the VE and CBT revolve around the economic interest of the local community and aim to putting to work the same tourist assets in order to exploit untapped business opportunities. However, even though the means are the same, their goal is slightly different. As an economic concept, the VE seeks about all short-term economic gains, whereas in CBT the long-term social well-being of the community plays in principle a more central role, notwithstanding the economic benefit. In the section that follows, we will set in perspective the Puerto Rican tourism sector, so that the reader will be in a better position to understand the post-hurricane Maria's advances presented in this chapter.

THE PUERTO RICAN TOURISM SECTOR BEFORE AND AFTER HURRICANE MARIA

From a political standpoint, Puerto Rico is a Free Associated State (FAS) to the United States of America. Geographically, it is the smallest of the major Antilles and it is located in the northeast Caribbean Sea. it boasts a mild climate throughout the year, and its white sand beaches, the historic city of Old San Juan, and El Yunque National Forest are amongst its most appealing and popular tourist attractions. But Puerto Rico has much more to offer.

One of the big challenges that Puerto Rico faces is changing the image of tropical paradise by the sea and getting visitors to explore the island beyond San Juan, experience its nature and culture, and spend their money in local communities to revitalize local economies. For years, the tourist strategy of the island has encouraged tourists to "explore beyond the shore" in order to diversify tourism as well as to bring economic and social development to areas and places off the beaten tourist track, without ever losing sight of mass tourism. As a matter of fact, San Juan is one of the main cruise ports in the Caribbean and, to a certain extent, day trippers represent a meaningful source of tourist income.

Diversifying tourism is one of the greatest challenges of all times for the Puerto Rican tourism industry. Puerto Rico is a small island, but it boasts a wide array of ecosystems, both marine and terrestrial. From dry forests to rainforests, from coastal plains to karst systems where caves and caverns proliferate, wetlands, mangrove forests, coral reefs, bioluminescent bays and much more (Miller & Lugo, 2009). It is an island rich in biodiversity, well suited for ecotourism.

But, despite all these natural resources, most tourists stay in the San Juan area and its surroundings and it is mostly local tourists the ones that venture off the beaten track and patronize the local businesses and visitor attractions. Another source of contrast is the dominating type of business: tourism in the north and east of the island is dominated by large resorts managed by US and multinational corporations, while the tourist industry in the south is locally based, predominantly characterized by small firms and local entrepreneurs. Even though there are attractive community-based tourism programs all around the island, they are not well known.

Tourism was a priority in the New Economic Development Model (NEDM) set out at the turn of the 20th century aiming for an increase in the number of hotel beds and tourist facilities available. Tourism was then identified as a substitute for the so-called "936 industries"[2] (companies that moved to the island attracted by fiscal advantages, which expired in 2006), as well as one of the solutions to the unemployment problem.

The Legal Framework for Tourism in Puerto Rico

Puerto Rico was already pursuing a tourism diversification strategy before hurricane María struck the island in 2017, but the need for diversification became more pressing in its aftermath as a means to alleviate the local population, help rebuild the local economies and recover a normal pace of life. Tourism development in the island was reinforced in the aftermath of the disaster by governmental and non-governmental recovery efforts promoted at several levels: federal, national, local and business level.

Prior to the hurricane, tourism was framed in two laws: the "Tourism Development Law" of 2010 (*Ley para el Desarrollo Turístico, Ley núm. 74 de 2010 según enmendada*) and the "Ecotourism Law" (*Ley de Política Pública para el Desarrollo Sostenible de Turismo en Puerto Rico, Ley núm. 254 de 30 de noviembre de 2006*).

The 2010 "Tourism Development Law" is a framework regulation that revises the prior 1993 tourism law, extending the definition of eligible businesses in order to recognize the importance of the development of new tourism products and setting the stage to reach a number of major long-term goals stated in its preamble, which include, amongst other: (a) Turning Puerto Rico into a world-class tourist destination; (b) Promoting the appropriate conditions to ensure the continued development and competitiveness of its hotel industry; (c) Fostering proper conditions for local and foreign capital investment in tourism projects; (d) Supporting companies in their efforts to diversify tourist products.

Puerto Rico passed its Law on the Sustainable Development of Tourism on 30 November 2006, which identifies tourism as a priority sector for the country and shapes a number of mechanisms for its sustainable development. In consonance with the principles of sustainable tourism development defined by the World Tourism Organization, the law commits to the "conservation of natural resources and the preservation of critical habitats, taking into account the Carrying Capacity, and the Limit of Acceptable Change of ecosystems, as well as the ecological footprint of development, before authorizing the development of new tourism projects".

In its Article 11, the Law addresses community involvement by entrusting the Executive Director of the Tourism Company with the mission of incorporating "cooperative movements, environmental organizations, owners of *Paradores*[3], inns, hotels, lodges in general, private tour organizers and other institutions interested in the industry", as well as bestowing the ability of "granting special incentives to owners of lands surrounding natural areas of high ecological value, so as to develop sustainable (or green) companies that provide support services to the [tourism] industry".

Under the umbrella of the UN Sustainable Development Goals, the Puerto Rico Tourism Company (PRTC) had created a set of guidelines and tools for the planning and operation of ecotourism and sustainable tourism businesses. Thus, the PRTC boasts to have developed the first government sponsored Green Certifications Program in the Caribbean region. The certificate encompasses four categories: accommodation, agrotourism, ecotourism, and community-based tourism.

The Disaster Recovery Plan for Puerto Rico

Pursuing a visitor economy strategy, the *Economic and Disaster Recovery Plan* considers fundamental for recovery "to develop a solid and sustainable visitor economy to help position Puerto Rico as a global destination for tourism, investment, production and wealth" (Gobierno de Puerto Rico, 2018: 145). Taking as a base the island's natural and cultural assets, the government proposes the creation of tourist centers around the island to decentralize tourism and generate economic growth in local communities, that are a vital part of the process as people should be empowered to be active participants, through training courses that enable them to become tourism entrepreneurs or work in tourism-related businesses.

The document starts sketching the vision, goals and objectives of the plan. The long-term vision framing all efforts is stated in the following terms: "building the new Puerto Rico to meet the current and future needs of the people". Four goals stem from this vision: to improve resilience, to strengthen critical infrastructure, to promote sustainable economic development and to promote a society based on the needs of the public. Objectives materialize in transversal strategic initiatives, eight groups of actions in various fields, such as the maritime economy and the tourism economy.

The preservation of natural, cultural and historical assets takes pride of place in the plan. This heritage is a first-class tourism asset that can be used to conform sustainable community-based tourist experiences, which contribute to its conservation by educating locals and visitors alike.

Since Puerto Rico is an island, special attention is devoted in the plan to the marine ecosystems and marine economy, the so-called "blue economy", which encompasses a range of economic activities ranging from fishing and tourism to emerging industries, such as marine biotechnology and sea-based renewable energy. To take advantage of these resources properly, initiatives should focus on ensuring health and sustainability of the oceans and coastal areas, while also contributing to economic growth (Gobierno de Puerto Rico, 2018: 144).

Among the actions prescribed in the plan, those aiming at improving the image of post-hurricane Puerto Rico in the media must be pointed out. The Recovery Plan proposed to launch an "Open for Business Campaign" with an estimated budget of 67 million dollars, the ECN 29 states it in the following terms: "Design and launch a marketing strategy to inform the world that Puerto Rico is ready to resume receiving tourists and inviting visitors to see that the natural resources have been preserved and that the tourism industry is ready to serve them" (Gobierno de Puerto Rico, 2018: 374).

In their study on post-disaster marketing efforts, Walters and Mair (2012; 2013) analyze nine major narratives that tourism destinations use in their communications in order to attract tourists back: open and ready for business, empathy or solidarity, community readiness, restoring confidence and change misperceptions, curiosity enhancement, visitor testimonials, celebrity endorsements, short-term discounts, and product diversification. The carefully orchestrated communication campaign, that tourism authorities in Puerto Rico launched in both mass and social media, employed most of these strategies.

Culminating its efforts, in January 2019, The New York Times placed the island at the top of its highly reputed and referenced list of "52 Places to Go in 2019", naming Puerto Rico the number one destination to visit in 2019 (Navarro, 2019). A movement both praised and loathed by the newspaper's readers.

Tourism Entrepreneurship and Innovation as a Byproduct of the Disaster

Entrepreneurs and small businesses are a key piece in tourism diversification. Small communities can neither host big resorts and tourist facilities, nor attract big amounts of tourists. Small communities are a natural habitat for small businesses and tourism-related ventures, that can develop a more sustainable model and offer authentic experiences to visitors that are looking for more than the Caribbean sun, sea, and sand. For years, the Puerto Rico Tourism Company has tried to attract European visitors, because they stay longer, display a higher expenditure and have diverse interests, but without the necessary capacity (for example, in terms of accommodation) and infrastructures, without an attractive offer, it would difficult to succeed.

In the literature review, the leveraging role of CBT as a complement to mass tourism. Although both of them create employment, mass tourism is the domain of big companies and it entails migration. CBT creates employment and entrepreneurial opportunities at a smaller scale and without the need to migrate. Rural areas are the ones that suffer the most from depopulation. Therefore, CBT could contribute to attenuate the population loos that the island has been suffering for over a decade, owing on the one hand to the 2007 global economic crisis which in Puerto Rico overlapped the tax breaks that the US Congress granted to companies that established business in the island in order to attract investments and, on the other hand, to the 2014 financial crisis that placed the island on the verge of bankruptcy. Since Puerto Ricans hold US passports, they can move freely to the US mainland, where many of them have family and friends.

Some people are leaving, but others remain, and still others return to their homeland. Various kinds of family and community-based microbusinesses have started operations after the hurricane. At the helm of those small tourist business, one finds the stories of effort and personal sacrifice of entrepreneurs that love what they do: small guest houses (*hospederías*), agritourism ventures (the focus on locally produced, sustainable, and ecologically responsible agriculture is a major component of the ecotourism movement) catering as well to local restaurants and cafés that incorporate proximity products in their business models, small catering businesses that empower women, arts & crafts shops sourced by local artists (the most traditional one are carved wooden figures of saints made by *santeros*), local guides and locally managed cultural and natural tourist attractions. Far from accommodating, some Puerto Ricans have seized the opportunity to let their entrepreneurial spirit flow and start new tourism-related ventures.

THE PATH FROM RECOVERY TO RELIEF: THE ROLE OF COMMUNITY-BASED TOURISM

Tourism is amongst the most vulnerable economic sectors to the impact of natural hazards. Disasters destroy tourism infrastructure at the destination, including various types of accommodation and sightseeing facilities, transit routes and supply chains. Media intensify the image creating a "ripple effect" that spreads the detrimental impact of disasters and, in the case of tourism destinations hinders their recovery due to negative consumer perceptions (Filimonau & De Coteau, 2019: 2). Tourism plays an important role in the in the efforts to dynamize the economy, both locally, regionally and at country level.

Seeking short and mid-term results, the non-governmental civil society organization Foundation for Puerto Rico (FPR, henceforth), is implementing various programs around the island, geared towards communities located outside the San Juan metropolitan area, that boast outstanding cultural and natural tourism attractions, involving local communities in both planning and execution.

The FPR was born in 2011 with the vision of integrating the island of Puerto Rico into the global economy, through sustainable and inclusive economic development. Its mission is to transform Puerto Rico into a world-class destination for people to visit, live, work and invest in. FPR is a pragmatic organization that moves from knowledge to action, involving communities in planning and managing their own development. In 2015, before hurricane Maria, the Foundation had already been working with the concept of Visitor Economy and later succeeded in including it in the political agenda, so that from an high-level institutional perspective it could work "top down", whereas the Foundation would work from the "bottom up". This way, the efforts of both sides would contribute to reach the goals in a cooperative way, meeting somewhere in the middle.

Table 2. Relief and recovery initiatives of the FPR described in this chapter

Relief initiatives	Description
230 relief missions	Bring electricity, drinking water and basic supplies to local communities
Small Business Cash Grant	Economic help to small businesses to prevent them from closing
Recovery initiatives	**Description**
Bottom Up Destination Recovery Initiative	Identifies and invests in four impact areas. One of them focuses in bringing more tourists to the community in a sustainable manner
Whole Community Resilience Planning Program	Community Cohesion
Destino Empresarial	Entrepreneurship program
Emprende 360°	Entrepreneurship program

When the hurricane hit the island, the FPR focused on the immediate relief efforts. The FPR coordinated more than 230 relief missions, that had an impact in 150,000 individuals (FPR, 2018A: 13). By working with communities, it became apparent that what communities outside the metropolitan area needed the most was electricity and drinking water, as well as mattresses, appliances and basic supplies, whose distribution they channeled. On the other hand, since small businesses needed economic support to be able to continue operations, they created the *Small Business Cash Grant*.

After completing the relief tasks, the recovery efforts started. And this is where the FPR deployed the Bottom Up Destination Recovery Initiative, the Whole Community Resilience Planning Program and the entrepreneurship programs Destino Empresarial and Emprende 360°, which will be discussed in the following sections. *Table 1* summarizes these relief and recovery initiatives, whereas some of these programs carried out throughout the island are located in the map in Figure 1.

Figure 1. FPR Relief and Recovery programs around the island of Puerto Rico
Source: https://www.foundationforpuertorico.org/hurricane-efforts

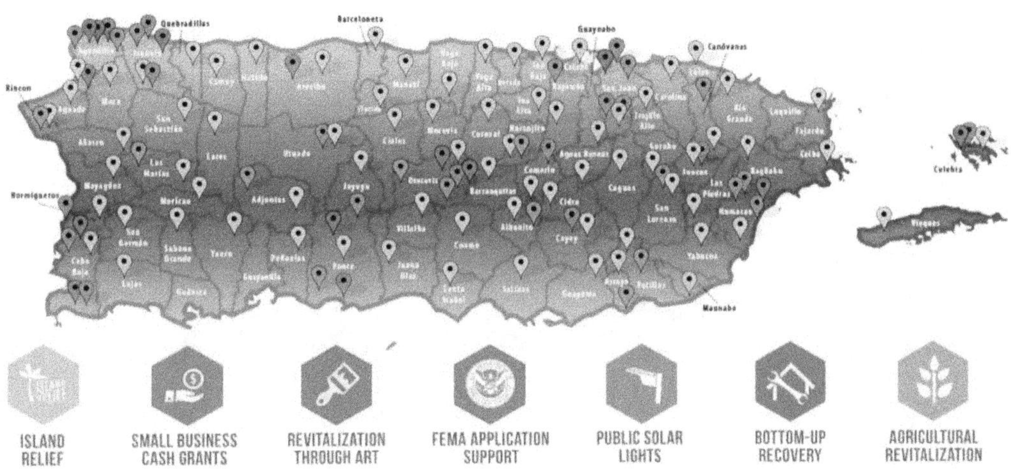

The Bottom Up Destination Recovery Initiative

In its effort to support communities to become more resilient and active in the creation of economic opportunities, the *Bottom Up Destination Recovery Program* identifies and invests in four impact areas. One of them focuses in "developing branding and marketing strategies to bring more tourists to the community in a sustainable manner". Launched in 2018, the Program is one the *Commitments* in the Clinton Global Initiative of the Clinton Foundation, it has an estimated duration of 3 years and an estimated total value of almost 7 million dollars. From July to December 2018, the planning phase of program was carried out in the municipality of Orocovis, where a *Local Plan for the Management of the Destination of Orocovis* (*Plan local para la gestión del destino de Orocovis*) was created through a co-design process facilitated by the FPR in which groups of local leaders took part. Until 2021, it is expected to expand to 25 communities in six regions, around the island.

Three candidates were identified, but Orocovis was chosen for the pilot program because it is located outside the metropolitan area, in the central mountain range, no other NGOs were working in the area and in its vicinity both natural and manmade attractions can be found. The Toro Negro State Forest stands out amongst the natural ones, whereas the biggest manmade attraction is a 2.5-kilometer-long zipline called "The Monster", the second longest in the world, located in Toro Verde Nature Adventure Park.

Getting inspiration from different sources, taking bits and pieces from here and there, and adapting valuable "concepts to the realities and particularities of the island" (FPR, 2018A: 26), the FPR created its own framework (Figure 2) to bring about economic development to the island through the visitor economy.

Figure 2. Bottom up operational framework
Source: Foundation for Puerto Rico, (2018A: 32).

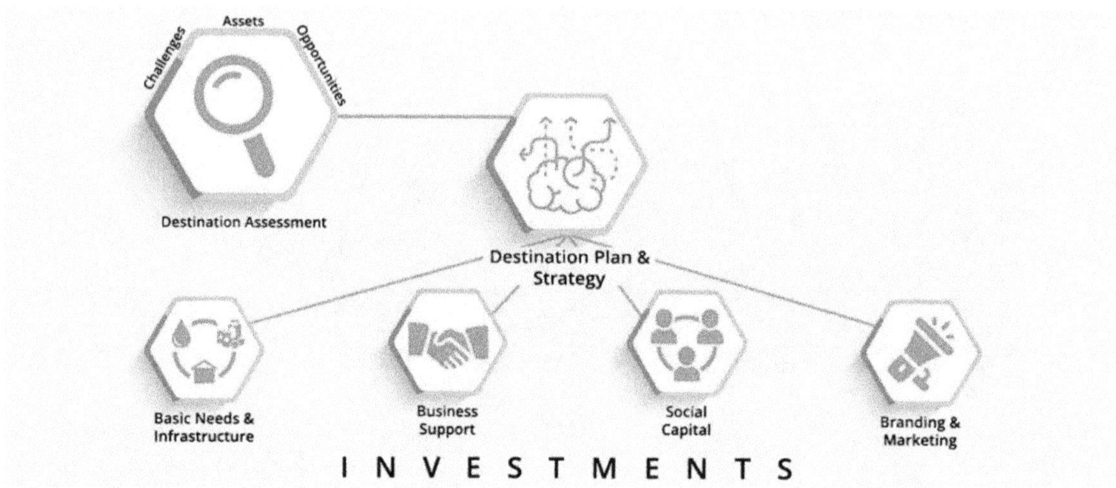

This framework, which later supported the *Bottom Up Destination Recovery Initiative* is based on five *Bottom Up* guiding principles (Figure 3), which revolve around the chore principle of Visitor Economy.

Following the project management methodology of the Project Management Institute (PMI), the program is deployed in five phases: conception and initiation; planning; execution; performance/monitoring; and project close.

Figure 3. Bottom up destination recovery principles
Source: Foundation for Puerto Rico, (2018A: 28).

In the initiation phase, a FPR team made up of 10 people moved to Orocovis for six months, in order to identify opportunity areas and needs. The team takes contact with the community, in order to identify business clusters and local stakeholders that will participate in the planning phase, where an Asset Map will be created with everyone's cooperation using visual tools. The program is partly funded with community donations, mostly in kind, and FPR's funds. The initation phase culminates with a document where the program objectives are articulated, the Project Chart.

In the execution phase, one of the challenges the Foundation encountered was the lack of capacity. As the Bottom Up Program Manager Michelle Torres herself put it in an interview with the author of this chapter, "you have to teach them how to fish, you cannot fish for them" (Torres, M., personal communication, March 4, 2020). Therefore, in order to ensure the sustainability of the project, it was necessary to train and coach participants, in various business aspects, from marketing and finance to human resources and legal issues.

Branding and marketing efforts are integrated into the project as well, as Torres acknowledged by asking: "what comes first? The visitor or the infrastructure?" (Torres, M., personal communication, March 4, 2020). Capacity building has been addressed by identifying, on the one hand, ecotourism attractions that might have gone unnoticed without the professional help of Local Guest, a sustainable tourism company; and, on the other hand, facilities that could be used as accommodations (this was the case of the Finca Oro Rojo, which offers an "agro-ecological experience").

Platforms like *AirBnB*, *HomeAway* or *Join and Join,* as well as social media like *Facebook's Marketplace,* are changing not only the way travelers want to stay in destinations, but also the possibilities of a small business owners to market their business to the world. And it is not only about accommodation, it is also about restaurants and experiences. Social networks, like Instagram, have also been essential to show Orocovis's hidden gems and make them visible to a new generation of travelers: the millennials, who seek authentic experiences and are the travelers of the future.

At the time of writing, the deployment of the *Local Plan for the Management of the Destination of Orocovis* is a work in progress, Key Performance Indicators (KPIs) in four areas (basic needs and infrastructure, business support, branding and marketing, and social capital) will have to be monitored in the following phase, before closing the project. Meanwhile, new projects will be deployed in 14 municipalities and 25 communities around the island. The reader can find more information in the web site of the FPR.

Small Business Cash Grant (SBCG)

Employing more than 80% of the workforce, small businesses are the backbone of the local economy in the island. Businesses are not only directly affected by disasters, but they also suffer indirect losses, which have economic ripple effects. According to FEMA (Federal Emergency Management Agency), after a disaster, between 40% and 60% of small businesses never reopen, whereas another 25% fail within one year (FPR, 2018B: 6).

While working on the relief operations, FPR noticed that many local businesses were either closed or experiencing difficulties to operate. In order to help "business owners to remain open, retain employees, and optimize their business through immediate assistance and relief" (FPR, 2018B), the Small Business Cash Grant Program (SBCGP) was created three weeks after the hurricane and, even though it was intended to run shortly as a means of relief, grants were distributed for about ten months, through July 2018.

In their report, the FPR states some lessons learned from the program which prove useful to policy makers in post-disaster scenarios:

1. Developing more short-term financing vehicles
2. Pre-selecting geographic areas of impact, instead of vying for open solicitation
3. Supplementing the cash assistance with technical support to ensure business continuity
4. Conditioning part of the grant to receiving technical support. The merger of cash and coaching was crucial in the program design
5. Integrating economic resilience planning within the technical support
6. Work with partner organizations instead of establishing new program structures

Fostering Entrepreneurship: Destino Empresarial and Emprende 360º

The FPR has been active in new business generation as a means to improve business fabric in the island and foster business ecosystems, that promote employment and attract investments. New businesses are also necessary in order to innovate in tourism and be able to respond to the increasingly sophisticated demands of present-day tourists. Rather than opening a new restaurant or souvenir shop, prospective entrepreneurs are encouraged to think out of the box from the moment they start generating business ideas and come up with innovative businesses.

Two initiatives have been put in place in this line: *Destino Empresarial* and *Emprende 360º*, program names that speak by themselves and could be roughly translated as "Business Destination" and "Start up 360º". Both programs promote entrepreneurship by creating the right atmosphere and training would-be entrepreneurs in practical business issues.

Instead of building its own capacity, the Foundation has partnered with INprende, an organization that facilitates innovation and entrepreneurship using its own business generation methodology, which is deployed through nine modules. Once they complete the training phase, participants are encouraged to present their project proposals in a contest in order to get seed capital to fund them. The selection is exacting and only and if the proposal is not ready, it will be sent back.

Destino Empresarial is a new business ideas incubator operating in 12 municipalities around the island, which aims at recruiting 144 local entrepreneurs and funding 18 business ideas related to the visitor economy with seed capital contributed by FPR. Whereas, *Emprende 360º* is funded by the Ewing Marion Kauffman Foundation, it is intended for the southern municipalities of Ponce and Coamo, and its goal is to recruit 200 prospective entrepreneurs and help 15 ideas become businesses.

CONCLUSION

It is widely recognized that disasters present not only challenges, but also opportunities for change and improvement. This chapter has explored how, in the post-hurricane Maria scenario, the cascading effect of government policies (Top Down), third sector support (Bottom Up) and entrepreneurial efforts have aligned in developing CBT in local communities around the island. CBT can contribute to diversify tourism (both from the offer and from the demand side) and become an alternative source or income, as it was argued in the literature review, however it would be unrealistic to believe that CBT could replace mass tourism.

Governments can regulate, non-profit organizations can offer support, but at the end of the day it is local leaders, small business owners and entrepreneurs the ones who bring about change in local communities. By reinterpreting resources in the community to foster its potential as a tourist destination and empowering local actors to tackle untapped business opportunities in the tourism sector, CBT fosters resilience in local communities. Yet, in the aftermath of a disaster, before tourists can arrive, local economies need to be rebuilt.

This chapter has explored some good practices in the transition from disaster relief to recovery, that can be applied elsewhere. First of all, in the immediate aftermath of a disaster, it is necessary to preserve the local business fabric and support local communities in their efforts to recover social and economic stability. On the one hand, relief missions can help alleviate basic needs such as drinking water, electricity and basic appliances, while on the other hand small local businesses need economic support to stay open and thus retain local hospitality jobs, since many of them do not have resources to face not only the damages produced by the disaster but also cope with several weeks or months of inactivity. As we have seen in the chapter, the Small Business Cash Grant served to this aim.

Next, in the path to recovery, you need to strengthen social capital and local economies in order to foster resilience. A second lesson that can be extracted is that you need to involve local communities in tourism planning, which in turn will create bridging capital through social and business networks. Furthermore, building consensus from the bottom around what needs to be done instead of imposing plans from the top, provides a greater assurance of achievement. Some practical ways to promote participation have been explored in the chapter through the discussion of the Bottom Up Destination Recovery Initiative. In parallel, programs for building social cohesion could be staged, as the mutual support amongst neighbors is essential for communities to bounce back after a disaster.

Additionally, new businesses need to be created to cater to CBT demand, therefore it is essential to foster entrepreneurship, involving in particular young generations, that can act as change agents in mid and long-term. This is the goal of the programs *Destino Empresarial* and *Emprende 360º*, discussed in the chapter. Equally important is on-the-job training for present day managers, as leadership training will have positive effects in overall improvement. Furthermore, investments in physical infrastructures need to be made, and not simply to rebuild damaged ones.

In summary, through the case study of CBT in Puerto Rico after hurricane Maria, this chapter has explored how building CBT destinations contributes to create more resilient communities and four master lines of action have been identified: (1) Retention of hospitality jobs and business, (2) Programs for building social cohesion, (3) Community involvement in CBT planning, and (4) Entrepreneurship and leadership training. Practical lessons that could be applied in communities that wish to retain population by creating tourism-related business and employment opportunities, which in turn contribute to create social cohesion and make communities thrive.

ACKNOWLEDGMENT

This article is the product of independent research, which has not been funded by any grants.

The author wishes to thank Michelle Torres and Annie Mayol, from the Foundation for Puerto Rico, for their opportune and enthusiastic help and support in the research phase, as well as the two anonymous reviewers of this chapter for their insightful comments and recommendations.

REFERENCES

Burgos-Doria, R. (2016). El turismo comunitario como iniciativa de desarrollo local. Caso localidades de Ciudad Bolívar y Usme zona rural de Bogotá. *Hallazgos*, *13*(26), 193–214. doi:10.153321794-3841.2016.0026.08

Filimonau, V., & De Coteau, D. (2019). Tourism resilience in the context of integrated destination and disaster management (DM 2). *International Journal of Tourism Research*. Advance online publication. doi:10.1002/jtr.2329

Foundation for Puerto Rico. (2018A). *Bottom up destination recovery: Orocovis case study*. Retrieved from https://www.foundationforpuertorico.org/research

Foundation for Puerto Rico. (2018B). *Small Business Cash Grant Program*. Retrieved from https://www.foundationforpuertorico.org/research

Gobierno de Puerto Rico. (2018). *Transformación e innovación luego de la devastación. Un plan de recuperación económica y de desastres para Puerto Rico*. Retrieved from https://recovery.pr/document-library

Guzmán-López, T., Borges, O., & Castillo, A. M. (2011). Desarrollo económico local y turismo comunitario en países en vías de desarrollo. Un estudio de caso. *Revista de Ciencias Sociales*, *17*(3), 432–446. doi:10.31876/rcs.v17i3.25574

Guzmán-López, T., Borges, O., & Castillo Canalejo, A. M. (2011). Desarrollo económico local y turismo comunitario en países en vías de desarrollo. Un estudio de caso. *Revista de Ciencias Sociales, 17*(3), 432–446. doi:10.31876/rcs.v17i3.25574

Hristov, D. (2015). Tourism versus the Visitor Economy and the shifting landscape of Destination Management. *Tourismos: An International Multidisciplinary Journal of Tourism, 10*(1), 219–234.

Miller, G. L., & Lugo, A. E. (2009). *Guide to the ecological systems of Puerto Rico*. Gen. Tech. Rep. IITF-GTR-35. San Juan, PR: U.S. Department of Agriculture, Forest Service, International Institute of Tropical Forestry.

Mullo, E., Vera, V. M., & Guillén, S. (2019). El desarrollo del turismo comunitario en ecuador: Reflexiones necesarias. *Universidad y Sociedad*, *11*(2), 178–183. http://scielo.sld.cu/scielo.php?script=sci_arttext&pid=S2218-36202019000200178

Murphy, P. E. (2012). *Tourism: A Community Approach* (4th ed.). Routledge.

Navarro, M. (2019, January 9). Why Puerto Rico Is No. 1 on our Places to Go list. *The New York Times*. Retrieved from https://www.nytimes.com/2019/01/09/travel/visit-puerto-rico.html

Ruiz, E., Hernández, M., Coca, A., Cantero, P. A., & del Campo, A. (2008). Turismo comunitario en Ecuador. Comprendiendo el community-based tourism desde la comunidad. *PASOS Revista de Turismo y Patrimonio Cultural*, *6*(3), 399–418. doi:10.25145/j.pasos.2008.06.031

Thériault, G. (2015). Le tourisme communautaire : quand le développement touristique est une affaire de communauté. *Observatoire Québécois du Loisir, 12*(8), 1–4. Retrieved from www.uqtr.ca/oql

Walters, G., & Mair, J. (2012). The effectiveness of post-disaster recovery marketing messages. The case of the 2009 Australian bushfires. *Journal of Travel & Tourism Marketing*, *29*(1), 87–103. doi:10.1080/10548408.2012.638565

Walters, G., & Mair, J. (2013). Post-Disaster Recovery Marketing Messages for tourism destinations. In Routledge Handbook of Tourism Marketing. Routledge.

Zacarias, D., & Loyola, R. (2017). How ecotourism affects human communities. In *D. T Blumstein, B. Geffroy, D. S. Samia. & E. Bessa (Eds.), Ecotourism's promise and peril. A biological evaluation* (pp. 133–151). Springer.

KEY TERMS AND DEFINITIONS

Business Incubator: An organization that helps entrepreneurs shape their business ideas into startups and early stage companies, and/or accelerating their launch to the market.

Carrying Capacity: The quantity of simultaneous visitors that a tourist destination can support without seeing negatively affected its physical, economic, and socio-cultural environment.

Civil Society: The organized non-state, non-market sector.

Community Benefit Tourism Initiatives (CBTIs): A tourism concept which raises the need to search for economic benefits for the community, regardless of other socio-political issues, defending that the community should have ownership, management, control over tourism projects.

Community-Based Tourism (CBT): A tourism concept that seeks a suitable balance between the environmental, Economic, and socio-cultural aspects of tourism development, by placing in the hands of local communities the development and management of visitor facilities and attractions, whose benefits revert in the community itself.

Destination Branding: Marketing practice which aims at making the destination easily identifiable in order to increase its value and attractiveness in the public eye.

Destination Management Organization (DMO): Tourist organization in charge of promoting the assets of destination and creating brand awareness in order to attract visitors, who in turn will dynamize the economy with their expenditures.

Donor-Assisted Community-Based Tourism (DACBT): Promoted by national and international cooperation agencies as well as NGOs, that assists local communities immersed in subsistence economy to generate their own micro-businesses by providing cash funding and technical support.

Limit of Acceptable Change in Ecosystems: A framework aiming at preserving an appropriate balance in ecosystems by establishing a threshold, that is considered acceptable, in the variation of a particular ecological component or process.

Paradores: Privately operated family-owned small inns, which conform a network that has succeeded in building a strong recognizable brand.

Pro-Poor Tourism (PPT): A tourism concept that advocates for tourism as a tool to fight against poverty. It aims at generating net benefits and improving livelihoods. Benefits are not necessarily economic, they might as well be social, cultural, or environmental.

Resilience: The ability of human or natural systems to cope with adverse events and be able to effect a quick recovery.

Tourism Diversification: A destination's efforts to create various tourist products that suit different visitor needs and desires, in order to improve tourism-generated income or balance seasonality.

Visitor Economy: A tourism concept that highlights the multiplier effect of tourism in the local economy, accounting not only for direct tourism expenditure in goods and services, but also for indirect consumption in agriculture, business services and communications, amongst other components of the supply chain.

ENDNOTES

1 On September 6, hurricane Irma passed about 97 km north of Puerto Rico, affecting mostly the northern part of the island and the central mountainous region, as well as the islands of Vieques and Culebra. Two weeks later, on September 20, when the country was still trying to recover, a second and more violent hurricane stormed through. Maria was the worst natural disaster in recorded history to hit the island.

2 Section 936 of the Internal Revenue Code gave US companies an exemption from Federal taxes on income earned in Puerto Rico.

3 Unlike Spanish Paradores, which are government owned accommodations many of which are located in historical buildings or emblematic localities, often defined as a "public hotel chain". In Puerto Rican *Paradores* are privately operated family-owned small inns, which conform a network that has built a strong brand.

Chapter 9
Innovation Opportunities and Gender Equity in Community-Based Tourism:
Raízes and Vivejar (Brazil)

Karla Juliana Pinheiro Melo
Universidade Estadual do Ceará, Brazil

Eliana Coelho da Silva
Universidade Federal do Ceará, Brazil

Raquel Santiago Romo
Universidade de Santiago de Compostela, Spain

Williany Loren Gaia Ferreira Austrégesilo
Universidade Estadual do Ceará, Brazil

ABSTRACT

This chapter aims to demonstrate that there is a fruitful relationship between innovation and business generation within community-based tourism. Continued innovation in the tourism sector is one of the pillars to establish its sustainability. With this approach, Brazilian companies Raízes and Vivejar will be presented, which are innovating as promoters of socioeconomic development in community-based tourism, based on a focus on gender equity. The example of these organizations strengthens the understanding that, over the years, women have positioned themselves to transform the reality of their labor occupations, leaving positions of low hierarchical levels, little valued to put themselves in business management, which, consequently, collaborates for the transformation of social structures.

DOI: 10.4018/978-1-7998-4855-4.ch009

INTRODUCTION

Tourism is an economic activity very sensitive to any change that may occur in the socioeconomic environment, as such, it needs to be in a continuous transformation adapting to new challenges as society and the economy advance. (Reyes, Ortega & Machado, 2017). This continuous transformation requires innovation to increase the sustainability and competitiveness of the sector (World Tourism Organization)

For this reason, observing which companies and how they face these challenges is at least indispensable.

The companies Raízes e Vivejar, object of this study and headquartered in Brazil, show themselves as innovative businesses that promote community tourism with an emphasis on experience. Both enterprises, developed and managed by women, are presented as innovative businesses that promote Community-Based Tourism. (CBT onwards).

The Raízes organization presents itself as a social enterprise committed to change towards a better world, working with projects that provide business planning and professional qualifications for communities in community-based tourism. (Raizes, 2019).

The company Vivejar has a similar business objective, committed to innovation and the development of community-based tourism, and seeks to encourage travellers to participate in cultural and social experiences. During the itineraries, this company promotes the search for meaning, reflection and opinion of the trips made to the traditional communities of Brazil (Vivejar, 2019).

As they are companies with innovation and gender perspective, some aspects should be clarified in this regard.

In the first place and in terms of innovation, given the complexity of the concept, we will refer to this term as the creativity and the implementation of processes that together generate new products, services, forms of business organization, etc. in tourism enterprises (Huete & Mantecón, 2017; Espinoza-López et al., 2019).

Secondly, with regard to the gender perspective, some general characteristics of the sector should be indicated.

Data from the Institute of Applied Economic Research (IPEA, 2014) indicate that proportionally more women are working in the tourism sector. Women represent 45 per cent of the total workforce, which is more than the representation of women in the economy (38 per cent of the total).

Tourism being a feminized sector in almost all its activities, the positions of power remain mostly occupied by men and and still present equality gaps in the promotion of women (Huete, Brotons & Sigüenza, 2016; Díaz-Carrión et al.,2014; Mejía-Vázquez et al., 2017; Melo & Soares, 2019).

Nevertheless, according to the Brazilian Micro and Small Business Support Service (Sebrae, 2016), in Brazil women undertake more than men, their rate is 15.4%, while that of men is 12.6% and they are more innovative.

Women's entrepreneurship in tourism could create not only a path towards overcoming gender inequality, but could also be a way to build creative and innovative processes through which other women are inspired. .According to Steuernagel Assis (2017), from the Impact Women program, "Women's empowerment is the collective conscience, expressed by actions to strengthen women and develop gender equity".

Regarding the CBT research production in Brazil, it does not have a wide bibliography despite being a country with a strong presence of this type of tourism (Graciano & Holanda, 2020). In fact, since 1997, the National Meeting on Local-Based Tourism (ENTBL) has been held to increase scientific interest in the study of the strategies and benefits of this type of tourism in Brazil. (Arruda & Gonçalves, 2020).

Given this lack of research, we set out to qualitatively characterize two Brazilian business models led by women who carry out innovative projects. The objective was to identify how these innovative processes could influence gender equality in the CBT. Also, to see if these innovations in CBT generate synergies between innovation, gender equality and business initiatives.

In Ceará (Brazil), the region where the Vivejar and Raízes companies are located, the weight of tourism in the GDP is 5% (IPECE, 2018).It is also relevant, as the Brazilian Minister of Tourism points out, that the tourism sector represents 8% of the Brazilian economy and that it is an important generator of employment (7 million workers). These two factors increase the interest in analyzing two companies located in this territory.

To achieve this purpose the chapter is structured in four parts: The first one in which an introduction is made to the most relevant concepts for its understanding. The second part of the book presents the companies and their actions. Finally, the conclusions obtained and possible future lines of research are presented, after an analytical observation of both institutions and through a careful bibliographic review using Scopus and Web of Science databases.

COMMUNITY-BASED TOURISM AND GENDER EQUITY: CONCEPTING

The conceptualization of community-based tourism is complex; there is no consensus or a concrete definition of it. However, in successive lines, inherent characteristics are developed to understand this tourist phenomenon.

In view of the classifications that involve the different activities belonging to the tourism industry, the term community tourism or community-based tourism is derived from tourism modalities, known by the terms cultural tourism or etotourism, ecotourism and agrotourism.

Several studies (Sampaio,2005;Ruiz-Ballesteros, 2011; Arruda &Gonçalves, 2020;Graciano & Holanda, 2020) indicate that community tourism is a strategy for traditional populations to be protagonists of their own ways of life, becoming a possible alternative to the materialist / consumerist way of life, even with the hegemony of urban societies. They also point out that community tourism gives local residents the opportunity to progress locally, preventing foreign exploitation by appropriating spaces, ensuring environmental sustainability and local characteristics, such as material and immaterial culture of the region.

Community-Based Tourism (CBT) allows visitors to get in touch with the material and immaterial culture of each community. With themes related to the preservation of nature, architectural works, and, at the same time, the conservation of traditional ways of life (cultural diversity and social relationship) (Soares &López-Vidal, 2015). This contact goes beyond living with foreign visitors, as it allows for closer ties between the community and the visitor, contributing to solidary consumption and the demystification of preconceived ideas.Similarly, increasing the affirmation of the territory at the cultural and anthropological level and strengthening the visibility of the identity of the local population, as a tourist destination (Emmendoerfer, Moraes & Fraga,2016).

Then,a Community-based experience tourism means deepening local customs during travel and interaction, in addition to experiencing traditions, memorable experiences and learning about the culture of the visited community. According to SEBRAE (2005), “The individual leaves his usual place to live in the space of others, different from his routine.”

However, according to Grünenewald (2003), the idea of tourists visiting predominantly traditional communities in search of experimentation, can generate acculturation in these communities, however, this new axis of tourism, community or alternative tourism, comes against the installation of large investments hospitality, preventing external entrepreneurs from dominating communities, strengthening the maintenance of local culture.

In Brazil, the term Community Tourism appears in 1997, at the I National Meeting of Local Tourism (ENTBL), held in the city of São Paulo, southeast region of the country, later, in 2003, it was added to a scientific debate in Fortaleza, city belonging to the northeast of Brazil.

With an inclined look at different contexts: economic-administrative, historical, philosophical-political-sociological and environmental, in addition to others, companies that practice community tourism, as is the case with the Raízes and Vivejar institutions, based in southeastern Brazil, foster sustainable development through community-based tourism, connecting communities, residents and tourists, providing development within these contexts.

What is fundamental to the development of this activity is the participation and awareness of the community and the training of professionals in the field of tourism. Experiential tourism can generate wealth, sustainability, preservation of cultural assets, generating profit in communities

In Brazil, this new way of doing tourism, where there is real interaction with the visited space, is recent; in other parts of the world, it is already quite explored. The beginning of this activity took place in Rio Grande do Sul, in the south of the country, starting with experience tourism in wineries, inspiring the same activity in other places in Brazil, with visits in communities with producers of coffee, cheese, cachaça and even beer. There is also interest in experience tourism for the historical part, the part of enograstronomic tourism, health tourism, ecological tourism, religious tourism, agrotourism, among other segments that may behave for this type of tourism, which allows the immersion of tourists and exchanges with the population of communities. (Gosling, 2019)

The great challenge for the development of experience activities is the question of structures, mobility, roads, communication, in addition to the greatest challenge, training tourism professionals or tourism professionals, to pass on to visitors the respect for diversity, preservation, identity local and community involvement; the people's awareness of the reality and the benefits of this practice.CBT also requires the generation of synergies through strata coordinates between different sectors and agents. (Transport, accomodation, etc.) (Burgos & Mertens, 2016).

Companies such as Raízes and Vivejar, in addition to being configured as Tourism Community, as stated before, they also present the differential to be women-led companies.

Although they start new businesses more than men, women are a minority in leadership positions in companies (Hryniewics, 2018) and this may be tied to "the sexual division of labor".Hirata & Kegoart (2007) explain that the term "sexual division of labor" has origin in France in the 1970s within the feminist movement. Initially, the concept explained that the work that women did in women's homes was free work.

Therefore, "there is a designation priority of men to the productive sphere and women to the reproductive sphere and, at the same time, the appropriation by men of the functions with greater social value added"(Hirata & Kegoart, 2007:599). Hirata (2009) clarify that contemporary feminism is a continuity of expectations of 19th-century feminism, but incorporates the "issue of autonomy female sexuality; motherhood is not the only horizon of women and, more still, the desire for "non-motherhood", after the advent of female contraception begins to express itself in a positive way and no longer as a lack."(Hirata, 2009:145).

Sex, culture and patriarchy were object of study of anthropology that investigated gender, term this that would be used only after the 1980s. (Louro, 2007).

Anthropologist Margaret Mead (1901-1978) researched female and male patterns in three New Guinea societies, resulting in the ethnographic research that grounded gender studies. The book "Sex and temperament in three primitive societies" (1935), reveals that the patterns of femininity and masculinity were neither universal nor much less biologically based, but that would be social constructs.

In philosophy, Simone de Beauvoir (1908-1986), publishes in 1949 the classic work of feminism "Le deuxième sexe"in which it infers that the femininity was a social construction based on the difference between the sexes.

Betty Friedan (1921-2006), a North American feminist activist, published in 1963 "The feminine mystique", another important book for theoretical construction about the feminist movement. The book "has become one of the main triggers of the so-called second feminist wave that swept the West" (Duarte, 2006). In this work, the author clarifies how the female existence would be based on frivolity, consumerism and motherhood, but that these would not be innate characteristics of women, but social expectations required of them and inculcated by socialization.

Hereupon, the contemporary feminist movement is a reflection of the transformations of the original feminism and is configured as a multiple discourse, of various theoretical trends, but with common foundations (Narvaz & Koller, 2006:649). Historically, the gender studies starts from the feminist perspective (Piscitelli, 2009), but some themes were expanded and new questions arose.

Questions like those posed by the historian Joan Scott (1995). She says that gender "is a constitutive element of relationships social based on perceived differences between the sexes, and gender is a form of first to mean power relations" (Scott, 1995:88). In his text "Gender: a useful category of historical analysis" (1995), she says is in agreement with Gayle Rubin about psychoanalysis having an important theory about gender reproduction, but claims that this is not enough to define a concept history on gender (Scott, 1995:87 - 88).

According to Rubin (1993), sex/gender is a product of historical human activity (Rubin, 1993). In his essay "Trafficking in Women: Notes on "Political Economy of Sex" (1993), it runs through authors from the Social Sciences who thought inequality of the sexes, such as: Marx, Engels and Lévi-Strauss until reaching the psychoanalysis.

In the 1990s, the post-structuralist debate on gender was made by the philosopher Judith Butler. According to Butler (2015), the emphasis of gender relations would not be on physical structure of the difference, but in the subjectivity and uniqueness of the experiences of gender. Guacira Lopes Louro, gender researcher in Brazil, studious of Butler, considers that "gender and sexual identities are therefore composed and defined by social relations; they are shaped by a society's power networks". (Louro, 2007:11). That said, gender relations are power relations established from sexual differences, these being also social constructions and cultural.

The bussines led by women break with these social expectations, of the historic sexual division of labour, establishing women in positions of power. Moreover, the companies listed in this chapter also innovate because promote actions that aiming at gender equity.

The consequences of women's empowerment actions can be felt in the following dimensions: economic; psychological; sociocultural; political and family (Fernandes, et al., 2016). Thus, women having autonomy financial, echoes in building emotional independence from social expectations created for them and, consequently, reaching gender equity. Putting these objectives in practice, having the perspective of establishing gender relations more is also to innovate in the tourist business.

Reviewing, thus, historically and culturally, the women are seen as confined to domestic spaces and they occupy predominantly positions of low hierarchical levels (Gómez-Urrutia & Gómez- Figueroa, 2015)however, when they head business, their eyes are less addicted to the traditional management way and they lead ventures more boldly, as is in the case of Raízes and Vivejar, both are community tourism businesses, created and managed by women. As previously said, this data is not random since women are more enterprising than man is and their rate is 15.4% while that of men is 12.6% (Sebrae, 2016)

A LOOK AT THE ENTERPRISES RAÍZES AND VIVEJAR

With the purpose of presenting enterprises that innovate and accommodate opportunities through the practice of community experience tourism, we expose the institutions Raízes and Vivejar. Both are organizations founded and managed by women living in Brazil. The following data exposure on the institutions was developed from the consultation of information made available by companies on their websites.

Raízes was founded in 2006 by 4 enterprising women, driven by the desire to innovate in tourism. Initially the institution explored ecotourism and adventure tourism. Over the years, it has diversified its activities developing multiple works in different sub-areas of tourism and strengthened its focus on the following specialties:

- Sustainable tourism and associated production, fostering the development of handicrafts, gastronomy, agroecology and culture.

"We believe that collective solutions are more resilient than individual ones. Managing challenges together allows you to access other angles and see more creative and richer alternatives. Working in groups or networks also has the benefits of task distribution for a lighter construction and the possibility to achieve greater results. " (Raizes, 2020)

- Entrepreneurship and income generation - work of autonomy and empowerment from the development of people to management processes of local enterprises, focusing on the creative economy, focused on areas related to tourism, such as crafts, gastronomy and culture.
- Gender equity and women's empowerment - based on the fifth United Nations (UN) Sustainable Development Goal (SDS) that should result in gender equality and empowerment of all women and girls by 2030, still considering that Brazil is the fifth country in the world in the ranking of domestic violence, to Raízes prioritizes work with the female public, in the quest to combat the structural machismo that continues to interfere in the lives of many women, preventing them from having access to basic rights.
- Network governance, involving associations, cooperatives, collective businesses and traditional communities - Raízes believes that collective solutions are more efficient than individual ones, therefore it invests in the management of joint challenges that allow access to other fields of development. "Working in groups or networks also has the benefits of distributing tasks for lighter construction and the possibility of achieving greater results. " (Raizes, 2020). The company continues with co-creations developed in a personalized way to the enterprising groups. With respect

to the identity of each organization, it collaborates with dozens of associations and business cooperatives of informal networks.

- Traditional communities - Brazil is one of the countries with the greatest cultural diversity, namely: indigenous people (of hundreds of ethnicities), quilombolas, riverine people, sertanejos and caiçaras, for example. Each one of these groups has its way of life marked by particularities in crafts, clothing, food, leisure choices, rites of passage, beliefs, buildings, expressions and language. With the principles of appreciation and respect for cultural diversity, to Raízes works with intervention models in its projects that strengthen and place at the center the autonomy and identity of the participants. In this way it is possible to adapt solutions from other contexts to the needs and specificities of traditional peoples who so desire.

In 2017, the company developed a partnership with the Vale Foundation, in the state of Minas Gerais, with lectures and mini-courses for a group of weavers from Senhora do Carmo, in Itabira. In the same year, it became a partner of the company Korui, a producer of menstrual collectors located in Florianópolis, State of Santa Catarina, starting the "Owner of my Flow" project, with the purpose of donating the collectors, initially applied to women in Minas Gerais, precisely in the Jequitinhonha Valley and then expanded to the Amazon region in 2018, also favoring women from the riverside communities of the State of Pará.

Still in 2018, the Raízes signed a partnership with Natura to support projects to strengthen women's empowerment, with a strong contribution to income generation and autonomy from collective businesses.

Currently, the company develops the following activities: elaboration of income generation projects; participative strategic planning; entrepreneurial qualification and business modeling; creation and governance of networks; strategies for strengthening social capital; integrated development of tourism and its associated production; creation of content, workshops, lectures and engagement campaigns; benchmarking strategies and inspiration visits; project monitoring and creation of qualitative indicators.

Raízes develops the work activities from an innovative methodology, based on flexibility and adaptable to diverse needs. For a working consensus, the company preaches the premises of focus on people and the needs of individuals in different cultures and circumstances, develops the projects in partnership with clients and beneficiaries, generating development based on interconnection, all without losing sight of the care with sustainability.

The following table (1), gathers the information of the work program developed by Raízes and explains how the steps for developing the projects are conducted.

The following illustration shows the places where Raízes has already operated and continues working with the development of various programs, in Brazil alone have been more than 50 projects, distributed throughout the country: Pará, Maranhão, Rio Grande do Norte, Alagoas, Sergipe, Minas Gerais, Rio de Janeiro, São Paulo, Santa Catarina, Rio Grande do Sul, Mato Grosso do Sul and Mato Grosso. The company also provided services in Cape Verde, a country belonging to the African continent.

In the table (2) The following is a summary of the company's website, in chronological order of some of the projects already developed by Raízes in the last 10 years.

As we can see in tables 1 and 2, in all projects developed by Raízes, innovation is present in actions to promote sustainable community tourism. The occupation of the space commonly carried out by foreigners in the processes of classic tourist activities, give space to the development of local human capital and strengthening of the enterprises of activities that contemplate tourism directly and indirectly. It is

KNOW	This is the first step and is based on the investigation of secondary data, through surveys from interviews, field visits, image surveys, records on maps, reports and any other sources that allow a closer understanding of the space and its inhabitants. The purpose of this stage is to develop projects that are in accordance with the reality of the region.
INSPIRE	In order to inspire, Raízes develops courses and lectures that contemplate, among others, themes of entrepreneurship, tourism, innovation and collaborative management. The company also bets on the benchmarking tool, with the purpose of always presenting the best practices.
ARTICLE	At this stage, the company works on the strategic planning as well as on the implementation of the plan, seeking to identify and allocate the right people in the ideal functions, through direct contacts, meetings and communications in general to define the forms of access to projects, as well as their development.
COCRIATE	Through creative meetings, the company develops participatory planning and engages people in promoting collective intelligence in order to strengthen the cooperation of all to obtain ever more positive results.
REALIZE	Raízes contributes with the realization of projects of community principle or private companies, offering action plan, training of teams, monitoring and evaluation of work.
ACCOMPANY	Through monitoring, Raízes assists organizations and regions in the creation of result monitoring indicators. In order to guarantee the permanence of positive results achieved.

Source: RAIZES website.

also possible to verify the growth of the women's empowerment policy through the data presented as results of the implementation of the projects that compose the history of Raízes.

It was in this context that the Brazilian Marianne Costa, one of the women members of the creation of the company Raízes, has decided to expand her field of activity, undertaking with the Vivejar business, a community experience tourism operator that promotes trips with itineraries where people can appreciate the culture of traditional Brazilian communities.

The Vivejar project is guided based on three points:

- Dreaming - thinking strategically about how to generate change;
- Monitor - monitoring data and results, the context and the assumptions made;
- Assess - Measuring whether the intended change has taken place or analyzing how much remains to achieve it.

Figure 1.
Source: RAIZES website.

Touristic Development in Mato Grosso - Project *Raízes* Araguaia (2019) The Araguaia League - a group of rural producers in the Araguaia Valley region in the state of Mato Grosso, is a movement for the adoption of sustainable livestock farming practices and aims at the promotion and economic and social development. In partnership with Raízes, the group developed the Raízes Araguaia project, which is focused on implementing good social and environmental practices on its members' farms. The project identified potential of the Serra do Roncador and Vale do Araguaia regions for actions of a touristic nature, such as natural areas that are poorly explored and rich in attractions, where mystic-religious tourism is already developing. Considering the tourism potential for local development, the project requires considerable structuring and organization to achieve the expected results, they are: inclusion, generation and diversification of income, fixation of people in the countryside, strengthening of links with the territory, promotion and enhancement of environmental assets within the properties. To this end, identification and mapping of the main elements of the region, visits to recognize the territory, interviews, study of secondary data and construction of a tourism development program were carried out.
Project Dona do Meu Fluxo 2019 (2019) For the second year running, in 2019 Raízes in partnership with the company Korui, producer of menstrual collectors, implemented in the Amazon the Dona do Meu Fluxo Project. Eight workshops were held that benefited about 400 women between the ages of 13 and 74, all of whom received donation collectors for themselves or to provide for friends and relatives and learned about menstruation, relationships with their own bodies, beauty patterns, judgments, sorority and support networks. All the work was composed by a team of 10 women.
PESC Embroiderers of Ipoema (2019) Social project that brings together 13 women embroiderers in the town of Ipoema (Minas Gerais). Based on the creative economy, the initiative aims to promote the strengthening of local culture, through the production and commercialization of handicraft products that carry the identity, knowledge and tradition of the region. In this process Raízes develops training and qualification for business management, design concepts, lectures and workshops for learning contemporary and traditional embroidery techniques. The Embroiderers plan to launch their own collection to be marketed.
Entrepreneurship and Tourism in Resplendor (2018) In partnership with the Vale Foundation, Raízes contributes to the social and economic development of the Resplendor region, a city with extensive tourism potential, located in the BARI Territory - a delimited area that includes four municipalities, three of them in the state of Minas Gerais (MG) and one in the state of Espírito Santo (ES), Brazil. They are them: Baixo Guandu/ES, Aimorés/MG, Itueta/MG and Resplendor/MG. Despite the tourist profile, development in this area is low due to factors such as the bankruptcy of a dairy cooperative, migration of the youngest to more developed regions and aging population. Thus, the work developed by Raízes and Fundação Vale collaborates with the generation of income for the population through the strengthening of tourism enterprises and intensification of the increase in the municipality's production chain. Positive results are obtained from the technical qualification of entrepreneurs, promotion of the destination in the tourism market and improvement of local tourism governance, for example. Those who enjoy the services offered, experience train travel, know the traditional gastronomic varieties of Minas Gerais and enjoy the view of the Rio Doce, which passes right in the middle of the city. •
Equity Weavers of Carmo (2016) Equity is a work developed between the Vale Foundation and Raízes, with the aim of speeding up the development of the social business of women living in the rural town of Senhora do Carmo, district of Itabira (Minas Gerais). In addition to the beautiful mountain landscapes and gastronomy, the crafts developed by women are an important attraction in the region. For the Equity project, a group of women artisans was selected through individual visits that resulted in the identification of leaderships, knowledge of production and understanding of the profiles and individual skills of the weavers, indispensable elements for the diagnosis of the group's profile. From the identification of their limitations and main conflicts, a clear image can be constructed to deliberate the starting point for the implementation of the adjustments of the community work activities. •
Gender equity and income generation in Periquito (2018) Raízes, together with the Vale Foundation, develops in Periquito (Minas Gerais) a work with women who weave and sell handmade fabric rugs. In promoting gender equality and income generation, these women receive professional qualification in entrepreneurship and design, access to the market and advice for business creation.
Sustainable Culture in Vale do Jequitinhonha (2015) In order to promote the environmental, cultural and economic aspects of the Jequitinhonha Valley region (Minas Gerais) in community tourism, Raízes has implemented the Sustainable Culture project for groups of artisans who make conscious use of local biodiversity. This has resulted in the generation of work and income compatible with the new demands influenced by the formatting of itineraries and promotion of the destination as a cultural tourism point.
Rural Villages Route of Santo Antão (2015) Through an initiative of the Ministry of Tourism of Cape Verde, in Africa, Raízes has implemented training actions in Rural Tourism Business, on the island of Santo Antão. These were based on individual technical assistance visits to entrepreneurs with a focus on helping them develop their businesses. The project provides, among other benefits, the strengthening of local entrepreneurship from producers in the region.
Income Generation in Jequitinhonha (2009) In 2009, Raízes started a project for artisans at Vale do Jequitinhonha (Minas Gerais). The work strengthened the dissemination of culture and the generation of income from local crafts and the practice of fair trade. In addition to the empowerment of women, the products that gained exclusive packaging and space in the market were also highlighted, as well as conquering a specific identity and cultural value.

Source: RAIZES website.

The activities developed by Vivejar are aligned with the 2030 sustainable development agenda of the United Nations (UN). The company is currently managed by a team of six (06) competent women.

In order to obtain positive results on the implemented projects, the company values sustainability, focusing on three pillars:

- Economic - focus on the inclusion and dimension of new businesses;
- Social - Protagonism in communities and women's empowerment;
- Environmental - Rational and sustainable use of natural resources in tourism.

Such premises evoke three questions: is it economically viable? Is it socially fair? Is it environmentally friendly? And when positive responses are obtained to such questions, the work is developed from the perspective of strengthening contact networks that are based on fair trade, with adjustments occurring whenever necessary.

Vivejar currently operates in the following Brazilian states: Pará (Cotijuba Island), Minas Gerais (Tourmaline Communities - Jequitinhonha Valley), Rio de Janeiro (Morro da Babilônia), São Paulo Grajaú (Bororé Island) Amazonas (Rivers and Bioregion). According to the sustainability report released by the company in 2018, the positive impact in financial terms for the locations where activities were carried out in 2017, its second year of operation, was approximately R$ 20 thousand in revenue. It is worth mentioning that in addition to the financial impact, the operations intensify the appreciation of local cultures and traditions, provide women's empowerment, foster self-esteem and above all, influence subjective personal transformations in travelers.

We live in such a large, diverse and contradictory country, where promoting change is not an easy task. But every small change is capable to generate positive impacts in a special way. Even though at the beginning, Vivejar already has beautiful stories of transformations to tell. And that does not let us give up (Marianne Costa, founder of Vivejar)

The company innovates by promoting tourism activity, in partnership with people belonging to the communities, who participate from the planning process to the reception and monitoring of the itineraries with tourists. According to Vivejar, this context promotes in people the need to recover traditional knowledge of the region in which they live, which promotes the self-management of the natives, together with the valorization and cooperativism between the local culture of the community and the company, resulting in further development local entrepreneurship, as well as the role of the individuals involved.

The organization works with a focus on sustainable development objectives (ODs), among which is gender equality; decent employment and economic development; sustainable cities and communities. Thus, the company-community partnership takes place based on the following commitments:

1. Provide alternative / complementary economic activity;
2. Promote and encourage social inclusion;
3. Offer opportunities to generate income;
4. Strengthen women's empowerment;

The company directs its efforts based on the values of local investment, respect for the environment, protection of wildlife and preservation of culture. Tourist activities are developed by the company together with the residents of the communities. In order to have a successful performance, the initial reality of the locations where the institution operates is monitored to obtain information that guides future steps and decisions.

The residents of the visited communities take advantage of what is offered by the space where they live to build local 5. Valuing material and immaterial, historical and natural heritage.

Development based on the offer of tourist experiences, projecting a positive change in the profile of these communities, when they are integrated with few alternatives for work and income.

With a constant focus on the simultaneity of community development and gender equity, female labor is prioritized. It should also be noted that according to the report, from 42 passengers who were on the Vivejar itineraries in 2017, 62% of them were female tourists. This data reinforces the importance of female recognition in the work and consumption of tourism in Brazil.

In Turmalina, located in the state of Minas Gerais, the artisanal production group "Do barro a arte" is constituted 100% by women, located in the Jequitinhonha Valley, where the poverty generated by drought and the lack of jobs contrasts with a rich culture and strong social capital. ''The state's inventive and artistic breadbasket, the Valley is the cradle of a strong tradition of craftsmanship made in ceramics. From dolls to kitchen utensils, tradition is passed on from generation to generation (Vivejar, 2018)."

In the state of Pará, through the itinerary called "Secrets and spices of the Amazon", Vivejar develops cultural activities in Belém and in the island of Cotijuba, exploring the natural beauty of the region, providing tasting of all its gastronomic variety, in addition to direct contact with people from the local community.

In compliance with its purposes, Vivejar highlighted in its 2017 sustainability report how women participated in the projects that were carried out. For demonstration purposes, below are graphs with the results of work and female participation carried out in 2017 by the states of Minas Gerais and Pará.

When analyzing Figure 2, it is possible to identify the important collaboration of women in the formation of family income.

Figure 2. Percentage of women's earnings share in enterprise income at Minas Gerais and Pará, 2017
Source: adaptation of sustainability report of the company Vivejar, 2018

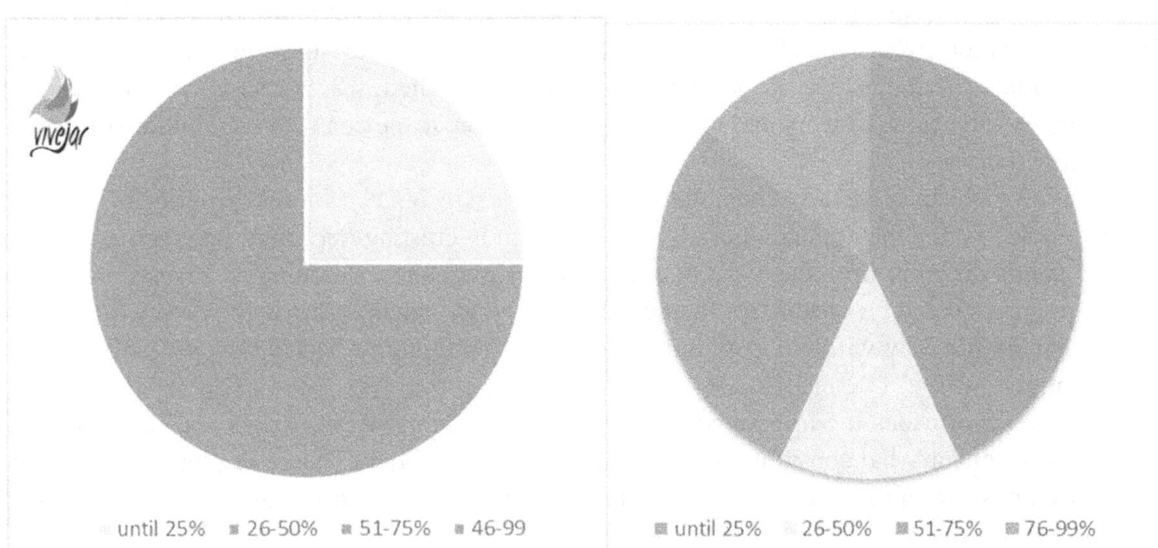

Figure 3. Level of studies in both activities
Source: Sustainability report of the company Vivejar, 2018

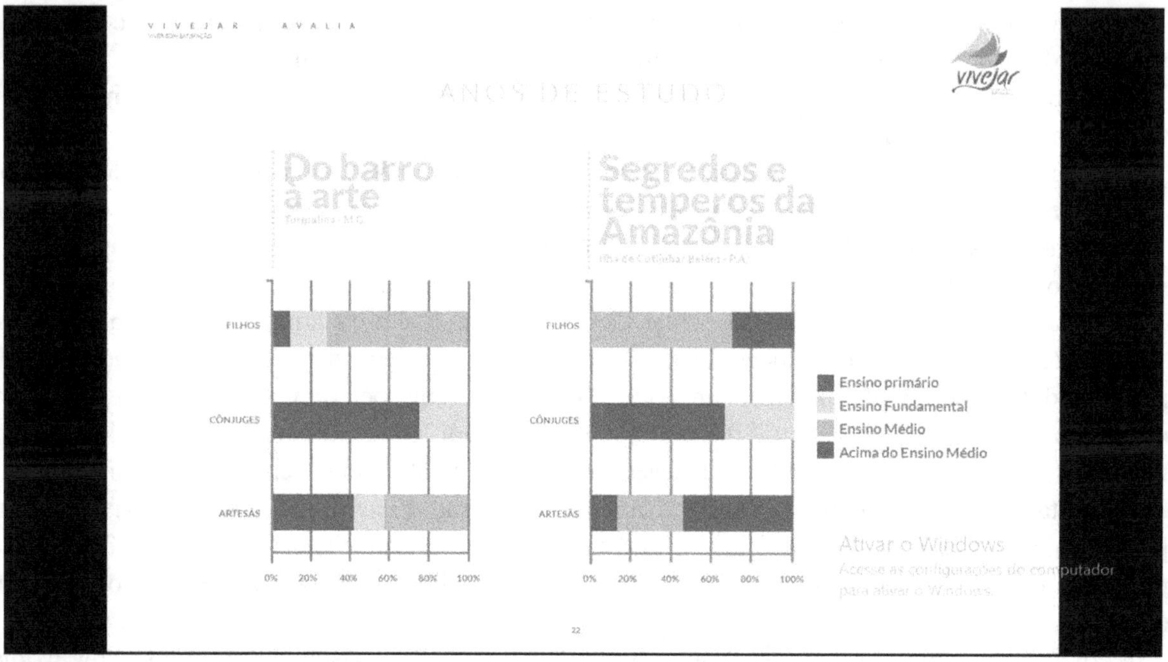

Figure 4 shows the level of education that makes up the profile of families in the communities. It should be noted that women in general have a higher level of education than their spouses and it is also possible to conclude that although the communities are located in remote areas, the children of the artisans had more opportunities to access school.

It is suggested from the observation of the graph (3), that wives and children had more opportunities to choose the type of work they do. Unlike spouses, who possibly, because they have a lower level of education, started to adhere to work activities that emerged regardless of choices.

Based on this collection of information, it can be seen that in the locations highlighted in the graphs of the Vivejar company sustainability report, the following situation exists: while men have restricted access to studies and the choice of their professions is also limited, there are a rise of women in access to studies and choice of work and this positive projection also extends to the children of families, regardless of their gender.

In addition to the positive facts about the locations chosen by Vivejar for tourism itineraries, it turns out that people visiting the communities also benefit from interesting opportunities, such as: knowing other life habits different from their reality and acquiring new ones. Learnings. "Vivejar conducts its travel experiences based on conductive threads. In other words, themes that help visitors to make tangible the experience they will live from immersion in the community. That's why we call it personal transformation tourism." (Vivejar, 2020).

Thus, Vivejar provides the involvement of the hosts (residents) with visitors and the community. This context results in what constitutes experience tourism in practice. Therefore, this classification of tourism becomes part of the routine of local activities and provides positive gains for all stakeholders.

Figure 4. Options for choosing professions
Source: Sustainability report of the company Vivejar, 2018

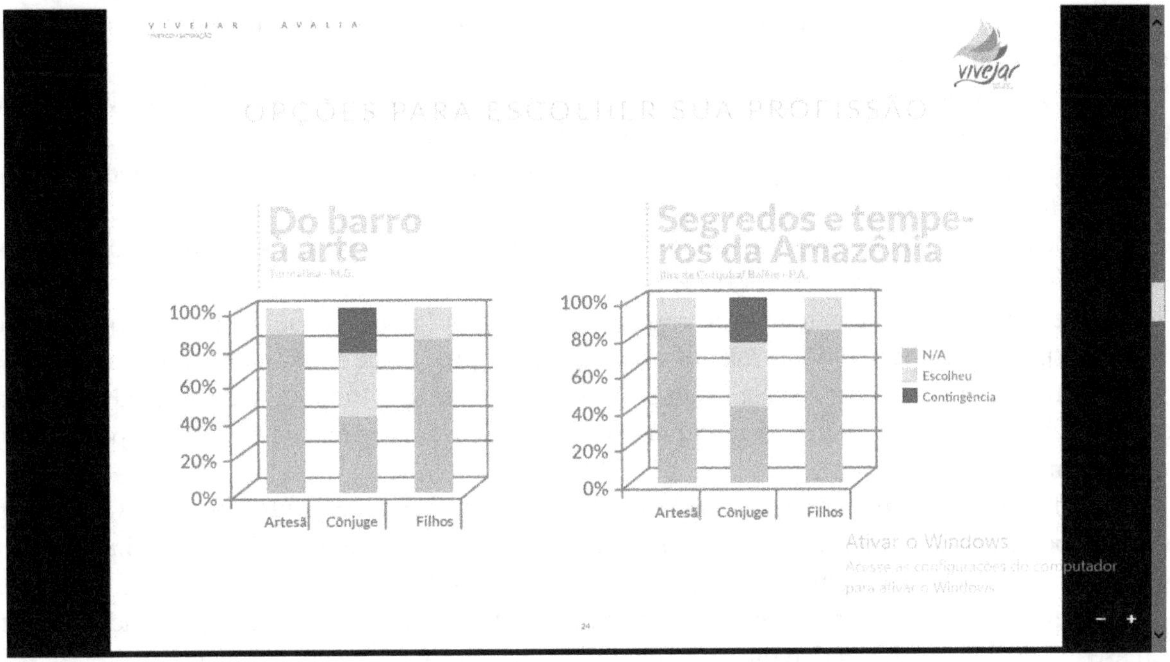

The hosts plan the interaction and sharing of their cultures, values, stories, constructions and arouse interest in people to participate in this construction in practice. This interactive exchange between those who visit and those who are visited, constitutes in practice what we know as community tourism. From this context, the valorization of the cultural and natural heritage is favored and, at the same time, favoring human development.

CONCLUSIONS AND FUTURE LINES OF RESEARCH

Once the companies Raízes and Vivejar have been characterized, and we know their lines of action through the projects carried out, several premises are concluded.

Firstly, leaving aside the fact that the creation of the projects is carried out by an entity external to the community, it highlights as a positive feature the management of women to take into account issues exclusive to women in the community.

For example, the Dona do Meu Fluxo 2019 Project might not have taken place if it had not been programmed by women.This type of community-based progress makes women's lives easier, in the case of communities that are isolated or do not have easy access to such progress. In this case, other women can more easily detect certain sensitivities and perspectives beneficial to women

It could also generate a new perspective in the imaginary of the women of the community, a very useful tool when it comes to empowerment. Observing how other women lead and direct creates a figure in which they can project themselves. It is in this fact that community women acquire protagonism, an element sought in the CBT (Fraga&Holland, 2019). In this line, the PESC projects Bordadeiras de Ipoema (2019), Equidade Tecelãs do Carmo (2016) and Equidade de gênero e geração de renda em Periquito (2018) also stand out. In these projects, besides providing women with training tools in management and leadership, their work is revalued, their self-esteem is strengthened and their autonomy is promoted through economic income.

Similarly, Vivejar, by encouraging tourists to reflect on the importance of the culture and customs of the communities, can have an impact on the visibility of women's work.

From a gender perspective, this type of innovative project is an impulse to the empowerment of women in the communities (Montaño Armendáriz et al, 2018) and at the same time allows the consolidation of the perpetuation of their customs through CBT. Feedback on the results obtained could be used to publicise the possible extrapolation of these projects to other communities.

However, there is some controversy in certain plans, in some cases it could be pointed out that they continue to perpetuate the traditional roles of women, as in the case of the embroiderers.Similarly, Vivejar notes that the women who participate in its projects achieve greater training and this leads to greater access to schools for their children. A possible indicator that care is carried out by women, perpetuating the division of labour by sex.

The difficulty in separating these roles lies in the fact that they are a source of empowerment and that the typology of developed tourism requires that traditions and the community are not infected with external elements that can destroy the purpose and cause of CBT for all its stakeholders.

However, there is a circular synergy: women entrepreneurs promote actions based on gender equality, increase innovation in CBT and in turn, community women can become entrepreneurs.

Based on the approaches that have emerged, it would be interesting to open a debate with a gender perspective to investigate the limits between tradition and the perpetuation of gender stereotypes in CBT and whether these activities do not increase women's workload as is the case in rural tourism (Sparrer, M., 2003).

REFERENCES

Arruda, D. O., & Gonçalves, J. P. (2020). Limits and possibilities in the development of Community-Based Tourism strategies in quilombola territory. *Interações (Campo Grande)*, *1*(21), 107–123. doi:10.20435/inter.v21i1.1968

Beauvoir, S. (2009). *O segundo sexo (2nd ed.)*. Nova Fronteira.

Burgos, A., & Mertens, F. (2016). Collaborative networks in community-based tourism: Implications for participatory management. *Tourism & Management Studies*, *12*(2), 18–27. doi:10.18089/tms.2016.12203

Butler, J. (2015). *Problemas de gênero: feminismo e subversão da identidade. Col.Sujeito & História* (8th ed.). Civilização Brasileira.

Carmo, H. (Ed.). (2014). Desenvolvimento comunitário (3rd ed.). Lisboa: Universidade Aberta.

Coelho, M. H. P., & Sakowski, P. A. M. (2014). *1938: Texto para discussão - Perfil da mão de obra do Turismo no Brasil nas atividades características do Turismo e em ocupações. Instituto de Pesquisa Econômica Aplicada*. Ipea.

Díaz-Carrión, I.A., Verján-Quiñones, R. & Castrejón-Alonso, B.E. (2014). Estereotipos de género y turismo: análisis cualitativo del "suelo pegajoso" en hoteles de 4 y 5 estrellas en Tijuana (México). *Turydes Revista Turismo y Desarrollo Revista Turydes: Turismo y Desarrollo, 17.*

Duarte, A. R. F. (2006). Betty Friedan: Morre a feminista que estremeceu a América. *Revista de Estudos Feministas, 14*(1), 287–293. doi:10.1590/S0104-026X2006000100015

Emmendoerfer, M. L., Moraes, W. V., & Fraga, B. O. (2016). Turismo Criativo e Turismo de Base Comunitária: congruências e peculiaridades. *El periplo sustentable*, (31), 1-18.

Espinoza-López, P.C., Moreno-Dena, J.M., Robles-Parra, J. M., Borbón-Morales, C. G., & Salazar-Solano, V. (2019). Innovation processes for micro, small and medium tourism enterprises development in the Sonora River region. *Revista de alimentación contemporánea y desarrollo regional, 29*(53), 1-23.

Fernandes, T. S., Lopes, G. S.C., Watanabe, M., Yamaguchi, C. K., & Godoi, C. K. (2016). Dimensões do empoderamento feminino: autonomia ou dependência? *Revista Alcance -Eletrônica, 23*(3), 391-413.

Foucault, M. (1998). A história da sexualidade: a vontade de saber (18th ed.). Graal.

Friedan, B. (1971). *A mística feminia*. Editora Vozes Limitada.

Gómez-Urrutia, V., & Jiménez Figueroa, A. (2015). Family co-responsability and the work-family balance: Ways to improve gender equity *Polis. Revista Latinoamericana, 14*(40), 377–396.

Graciano, P. F., & de Holanda, L. A. (2020). Bibliometric analysis of the scientific literature on community-based tourism from 2013 to 2018. *Revista Brasileira de Pesquisa em Turismo, 14*(1), 161–179. doi:10.7784/rbtur.v14i1.1736

Grünewald, R. A. (2003). Turismo e etnicidade. *Horizontes Antropológicos, Porto Alegre, 20*, 140–160.

Hirata, H., & Kergoat, D. (2007). Novas configurações da divisão sexual do trabalho (Tradução: Fátima Murad). *Cadernos de Pesquisa, 37*(132), 595–609. doi:10.1590/S0100-15742007000300005

Hirata, H., Laborie, F., Le Doaré, H., & Senotier, D. (2009). *(Orgs.). Dicionário Crítico do Feminismo*. Editora UNESP.

Huete, R., Brotons, M., & Sigüenza, M. C. (2016). La desigualdad entre mujeres y hombres en el sector hotelero español. *Estudios y Perspectivas en Turismo, 25*, 73–87.

Huete, R., & Mantecón, A. (2017). Landscape is the key. Exploring mass tourism alternatives. *Arbor, 193*(785), a397. doi:10.3989/arbor.2017.785n3003

Louro. (2007). *O corpo educado*. Autêntica.

Mead, M. (2015). Sexo e Temperamento em Três Sociedades Primitivas (5th ed.). Perspectiva.

Mejía Vázquez, R., Serrano-Barquín, R. C., López Carré, E., & Serrano-Barquín, H. P. (2017). Participación de las mujeres en empresas de hospedaje del municipio de Toluca, México. Análisis de su empoderamiento. *Investigaciones Turísticas, 14*(14), 21–44. doi:10.14198/INTURI2017.14.02

Melo, K. J. P., & Soares, J. R. R. (2019). La participación de las mujeres en los emprendimientos turísticos: Una mirada sobre la realidad brasileña. In Turismo y Género.Thomson Reuters.

Montaño Armendáriz, A., Trejo Briseño, N., Pérez Conche, J. C., & López Torres, V. G. (2018). Emprendiminetos de turismo de naturaleza. Una vía para propiciar el empoderamiento de de la mujer em la zona rural de Los Cabos, Baja California Sur. In *Turismo y género. Una mirada desde Iberoamérica* (pp. 75–100). Universidad Autónoma del Estado de México.

Narvaz, M. G., & Koller, S. H. (2006). Metodologias feministas e estudos de gênero: Articulando pesquisa, clínica e política. *Psicologia em Estudo*, *11*(3), 647–654. doi:10.1590/S1413-73722006000300021

Piscitelli, A. (2009). Gênero: a história de um conceito. In Igualdades. São Paulo: Berlendis & Vertechia.

Reyes Vargas, M.V., Ortega Ocaña, Á.F., & Machado Chaviano, E.L. (2017). Model for integrated community tourism in Ecuador, case study Pastaza. *REVESCO Revista de Estudios Cooperativos*, *123*, 250-275.

Rubin, G. (1993). Tráfico de mulheres: Notas sobre a "Economia Política do Sexo". In SOS Corpo. Recife.

Ruiz-Ballesteros, E. (2011). Social-ecological resilience and community-based tourism: An approach from Agua Blanca, *Ecuador. Tourism Management*, *32*(3), 655–666. doi:10.1016/j.tourman.2010.05.021

Sampaio, C. A. C. (2007). Turismo como fenômeno humano: Princípios para pensar a ecossocioeconomia do turismo e sua prática sob a denominação turimos comunitario. *Turismo em Analise*, *18*(2), 148–165. doi:10.11606/issn.1984-4867.v18i2p148-165

Scavone, L. (2008). Estudos de gênero: Uma sociologia feminista? *Revista de Estudos Feministas*, *16*(1), 288. doi:10.1590/S0104-026X2008000100018

Scott, J. (1995). Gênero: Uma categoria útil para análise histórica. *Educação &amp. Realidade*, *20*(2), 71–99.

Soares, J. R. R., & López, M. P. V. (2015). O uso do etnoconhecimento para o desenvolvimento do turismo sustentável. *Tourism and Hospitality International Journal*, (4), 19-28.

Souza, M. L. (2004). Desenvolvimento de comunidade e Participação (3rd ed.). Cortez.

Sparrer, M. (2003). Género y turismo rural. El ejemplo de la Costa coruñesa. *Cuadernos de Turismo*, *7*, 181–197.

ADDITIONAL READING

Gosling, R. (2019, 11 FEBR). A opinião de Gosling [video]. YouTube. https://www.youtube.com/watch?v=7OwU34I8tVs&t=1451s

https://www.ipece.ce.gov.br/2018/06/29/turismo-cearense-tem-participacao-de-5-no-pib-estadual-e-indice-deve-aumentar-nos-proximos-anos/

Impact Hut Curitiba. (2017) Mais ó final que é o empoderamento feminino? Retrieved from: https://impacthubcuritiba.com/empoderamento-feminino/

Martins, A. (2019). Cresce a participação do Turismo no PIB nacional. Ministério do Turismo, Retrieved from: http://www.turismo.gov.br/

Naçoes Unidas Brasil. ((2016) ONU: Taxa de feminicídios no Brasil é quinta maior do mundo; diretrizes nacionais buscam solução. Retrieved from: https://nacoesunidas.org/onu-feminicidio-brasil-quinto-maior-mundo-diretrizes-nacionais-buscam-solucao/

SEBRAE. *Turismo de experiência.* Recife, 2015. Retrieved from: https://m.sebrae.com.br/Sebrae/Portal%20Sebrae/UFs/PE/Anexos/turismo_de_experiencia.pdf

Steuernagel Assis, J. (2017) mas afinal o que é empoderamento feminino? Visible body: Justiça de Saia. Retrieved from: http://www.justicadesaia.com.br/mas-afinal-o-que-e-empoderamento-feminino/

Turismo e inovação (2017). Porque buscar inovação no turismo. Retrieved from: http://turismoeinovacao.com/conceito/porque-buscar-inovacao-no-turismo

Chapter 10
Female Entrepreneurship and Marine Tourism:
Innovative Practices on the Coastline

Iria Caamaño-Franco
https://orcid.org/0000-0001-7983-8620
Universidade da Coruña, Spain

Antonia Pérez-García
Universidade da Coruña, Spain

María Andrade-Suárez
Universidade da Coruña, Spain

ABSTRACT

Marine tourism activities represent an innovative proposal that, responding to the diversification needs of the professional fishing activity, provides a new alternative tourism product. The aim of this chapter is to analyse women's entrepreneurship in the frame of marine tourism as an innovative and territorial development dynamics that collaborate with other sectors (fishing-tourism, gastronomy, canning) promoting in turn a sustainable economy based on blue growth and development of endogenous resources. This exploratory and descriptive research intends to increase the tourist work carried out by women in coastal areas more and more public, highlighting and dignifying their role within the sector, as well as their contribution to tourism innovation and social and economic development. In order to meet the objectives, some secondary information with qualitative-quantitative data has been collected and primary information through a qualitative study by means of in-depth interviews to 10 entrepreneur women in the tourism field in Galicia (Spain).

DOI: 10.4018/978-1-7998-4855-4.ch010

INTRODUCTION

The evolution of socio-economic, political and environmental conditions is affecting the fishing activity in the European Union and, in Spain, they have generated a profound crisis situation in the sector (Molina, González and García, 2010). Diversification strategies are thus urgently needed in coastal areas whose economic driving force is fisheries. They are seeking to improve living conditions and strengthen local economies by providing them with complementary sources of income (Farnet, 2011). According to this, it is necessary to promote activities that contribute to the socio-economic dynamization of coastal areas in an integral way through the realization of projects that promote synergies between the fisheries sector, tourism and the use of local resources (coastal tourism activities, marine tourism, cruise tourism, culinary heritage, use of fishery by products, etc.). In this way, the search for alternatives that guarantee the survival of artisanal fishing with great roots in coastal populations, and with the necessary economic, social and environmental sustainability, is being encouraged.

Marine tourism activities represent an innovative proposal that, responding to the diversification needs of the professional fishing activity, provide a new alternative tourism product increasingly demanded by the market that can become a valuable example of revitalization and sustainable coastal development (Sagittal Project, 2007). In order to achieve this goal, this model of local development considers it essential and necessary that economic, social and institutional local agents intervene in the diversification process. Thus, they create a system of productive, commercial, cultural and institutional relations whose density and innovative character benefit the processes of growth and change (Vázquez and Rodríguez, 2015). Consequently, entrepreneurship has become a key option for economic incentive, rural development and even equal opportunities programmes.

In this scenario, the Fisheries Local Action Groups (FLAGs), responsible for managing a European programme in the area, aim at the implementation of a bottom-up strategy of axis 4 of the European Fisheries Fund for the promotion of the sustainable development of fishing areas and play a key role in supporting women's entrepreneurship initiative (Martínez García, 2017). Therefore, many of the initiatives promoted by fisheries professionals are developed, which demonstrate a great dynamic capacity as promoters of innovative projects through which they respond to the demands of the sector and to its own ones as a collective (Spanish Network of Women in the Fisheries Sector, REMSP, 2017).

Due to these considerations, nowadays social tourism undertaking, which is focused on recognizing opportunities to achieve the creation of a social mission that is sustainable over time (Moreira and Urriolagoitia, 2011), have become increasingly important. However, the study about its impact is still absent in literature (Power, et al., 2017; Mottiara et al., 2018), in particular, the nature of the personal and environmental factors that influence this process (Wang et al., 2019). In addition, the way in which gender impacts on entrepreneurship of women working in rural areas has received hardly attention (Forsberg and Lindgren, 2015) so researching in this field also becomes particularly important (Ortiz García, 2017).

Bearing the tourist female entrepreneurship in the countryside in mind, particularly in coastal fishing areas, the aim of this paper is to analyse women's entrepreneurship in the frame of marine tourism as an innovative and territorial development dynamics that collaborate with other sectors (fishing-tourism, gastronomy, canning...) promoting in turn a sustainable economy based on blue growth and development of endogenous resources, in this case, of Galicia's coastal regions.

This exploratory and descriptive research intends to increase the tourist work carried out by women in coastal areas more and more public, highlighting and dignifying their role within the sector, as well as their contribution to tourism innovation and social and economic development. For this reason, we are trying to understand and going in depth in the entrepreneur profile, such as the factors that trigger these business behaviours, as well as the motivations and main obstacles that they may be found. In order to meet the objectives, some secondary information with qualitative-quantitative data has been collected and primary information through a qualitative study by means of in-depth interviews to 10 entrepreneur's women in the tourism field in Galicia (Spain).

BACKGROUND: WORKFORCE AND THE ROLE OF WOMEN AS VECTORS OF TOURISM ENTREPRENEURSHIP: THE UNIQUENESS OF FISHING COMMUNITIES

In Galicia there is a historical relationship with the sea; coastal women have played the main role in the fishing activity, becoming very important for the sector from a socioeconomic point of view, since they traditionally participate in all the activities and processes related to the production chain, from the extractive subsector to the final marketing of the product (García and Zotes, 2006). However, a key aspect to be highlighted is the difference in representativeness depending on the sector, as some of the processes in the fish production chain are mainly carried out by women. For example, the female presence is negligible when it comes to fish unloading from a vessel, while, on the contrary, there is a highly female presence in all those support activities (net menders, packagers or "neskatillas", (this last name refers to those women responsible for the reception of vessels at dock to assist fishermen with the landing process and the transport of fish for the first sale), canning and marketing, retail as well as in the shellfishing, etc. (CETMAR, 2014; REMSP, 2017). At the state level, the importance of women working in the fisheries sector is 16.04%, according to 2016 REM (Special Regime of the Sea) data, mainly working as net menders, shellfish gatherers and packagers. In the case of the fisheries and aquaculture subsectors, this figure has increased to 32.2% and in the specific case of Galician women related to this sector. This number accounts for 55% of the total (La Opinión de A Coruña, 09/03/2018).

Taking this scenario into account, rural development policies have opted to promote and support women-led business initiatives with the dual objective of creating labour alternatives and increasing female labour participation and, therefore, achieve greater diversification of the rural economy (Alario and Morales, 2016).

In the case of Spain in 2010, the only national Europe's network of women working in the fisheries field (REMSP) has become a reality thanks to the Ministry of Agriculture, Environment and Food. It still keeps a pioneering networked platform to boost not only the role of women in the fisheries sector, but also associationism, communication and exchange of experiences among women in the fisheries field. Also, in 2016 another historic milestone for women's association took place with the creation of the National Association of Women in Fisheries (ANMUPESCA). Bearing this strategic plan in mind, the Directorate-General for Maritime Affairs and Fisheries has developed the first "Plan for Gender Equality in the Fisheries and Aquaculture Sector (2015-2020)" through REMSP. It should be noted, however, in order that these political initiatives can contribute to social change and gender equality, the results depend on the premises behind the approach of these policies (Pettersson et al., 2017).

On the other hand, literature highly appreciates the role of women within the social and economic development strategies of the rural environment, while considering that the female population plays key roles and is considered a local own resource and a possible comparative advantage that must be enhanced and monetized in terms of income and employment (Rico and Gómez, 2009; Osuna-Rodríguez et al.,

2015). Women's labour in rural areas is considered necessary for the start of new economic activities within the process of productive diversification, to maintain population, increase activity rates, generate new jobs and, in general, to effectively promote new development strategies in rural areas (García and Baylina, 2000).

Nowadays, it can be said that the economic infrastructure of the rural environment is changing and is no longer the primary sector, which was the main driving force of growth until a short time ago, but the tertiary, the one that plays the most important role and the one that employs most of the population, especially the female one (Rico and Gomez, 2009). In this sense, the changes in the functions assigned to rural areas and, consequently, in the strategies followed by rural policies have made entrepreneurship an agent of social change and a key element in its economic dynamism (Alario and Morales, 2016).

The diversification of the fisheries sector in search of complementary revenue pathways through new business ideas based on segments with growth potential, such as tourism, is seen as a labour and growth opportunity. It must be claimed that women's entrepreneurship is vital and that women in the fisheries sector have long demonstrated a natural talent for entrepreneurship, showing a clear and consistent business vision by being adjusted to the changes taking place over the past decade (REMSP, 2016).

The important role of women in rural development has often been highlighted in literature but the role of women's entrepreneurs in this field is more novel (Heon et al; 2014). Studying entrepreneurship from a gender perspective has not been systematically explored. Although in recent years this line of analysis has become pretty relevant, there are not many studies that are focused on identifying substantial gender differences in entrepreneurship (Ortiz, 2017).

Tourism entrepreneurship has also received hardly attention (Solvoll et al., 2015; Movono and Dahles, 2017) and while the role of women as social entrepreneurs is attracting more and more attention (Datta and Gailey, 2012), there is also a shortage of research on social women entrepreneurs in tourism.

In this sense, we should remember that the role of women as vectors of social entrepreneurship in tourism is increasing (Kimbu and Ngoasong, 2016) and that, in fishing communities, the female entrepreneurial initiative in this area stands out, as well as their high capacity to innovation and their commitment to diversification (REMSP, 2017) by simultaneously pursuing these economic and social activities, that is to say, by meeting market needs and contributing to local development. A central element of social entrepreneurship is that it allows individuals to find their own balance between the desire to make a social contribution and the personal need to capture an economic return from an activity (Seelos and Mairb, 2005). Social entrepreneurship thus encompasses the set of activities and processes carried out to discover, define and exploit opportunities with a motivation not only related to economic benefits but to improve social welfare (Tortosa et al; 2018; Mottiara et al, 2018).

Based on all these considerations, this type of female entrepreneurship could even be classified in the fishing areas as pro-sustainable entrepreneurship ("sustainopreneurship") which, as Gessa and Toledano (2011) point out, is still a novel line of research that is characterized by its priority orientation towards the preservation of nature and support to the local community through the search and exploitation of opportunities that result in new products, processes, services, etc. What is more, it is the type of entrepreneurship characterized by adopting a bottom-up approach towards the interaction and cooperation of the different agents (small businesses and public institutions), systems (business, social and natural) and resources to analyze and implement local solutions that embody the problems (Abrahamsson, 2008).

In short, tourism entrepreneurship, together with positive social, environmental and economic impacts on local development, has been identified in the literature as both a potential vehicle for women's empowerment (Datta and Gailey, 2012; Kimbu and Ngoasong, 2016) and an increase in gender equality (UNWTO, 2011).

APPROACH TO DIVERSIFICATION STRATEGIES IN RURAL COASTAL COMMUNITIES: MARINE TOURISM AS AN ALTERNATIVE

The current fishing system is suffering a profound crisis not only an economic one but also an ecological and social crisis (Miret-Pastor et al, 2015). The European Union itself has recognised the failure of its Common Fisheries Policy (European Commission, 2011) and the need to reformulate it. Taking this approach into account, the European Fisheries Fund (EMFF) considers it necessary to finance a restructure of the sector and its brand as well as one of its five priority axes, Axis 4, the "sustainable development of fishing areas" through the support of "measures and initiatives aimed at diversifying and enhancing economic development in areas affected by declining fishing activities" (REMSP, 2017). It has raised the opportunity to link the latent human potential in fishing communities to the economic and environmental opportunities that these areas have (CETMAR, 2014).

The European Union is seeking to promote an ecosystemic vision for its Integrated Maritime Policy (IMP). In this sense, the potential synergies between tourism and fisheries have been unnoticed by fishermen, the administration and by the tourism sector (Chen, 2010; Herrera-Racionero et al, 2018). It is in this crisis framework and with the financial support from the EMFF that different initiatives have appeared in Spain aimed at creating an incipient marine tourism (Molina et al., 2010).

As a means of implementing these policies, as noted by Martínez García (2017), the Participatory Local Development (PLD) stands out as one of the most important developments of the Common Fisheries Policy in the Spanish coastal areas. The PLD is led by the Fisheries Local Action Groups (FLAG), i.e. public-private partnerships formed by representatives of the fisheries and aquaculture sectors, as well as other members of the local community who act as dynamizers of the area to encourage entrepreneurship, blue economy, the fight against climate change, etc.

The associations recognized as FLAG are working to implement a Local Development Strategy and allow the integration and representation of the following sectors: (a) Fisheries sector: associations and entities with collective fishing interest, seafood and aquaculture with its own legal feature; (b) Social sector: socio-cultural, sports, environmental, recreational and educational associations, women's associations, social inclusion associations and similar; (c) Economic sector: employer and entrepreneurs associations, associations of economic agents and trade union organisations; and d) Public sector: composed of local public administrations and public sector entities linked to or dependent on them.

From that moment on, these groups were responsible for the planning and implementation of local development strategies and the dynamizers of the territory, taking direct responsibility for the management and implementation of the European Maritime and Fisheries Fund (EMFF). FLAG thus represents a new system of territorial governance more complex, plural and inclusive, with new formulas of management more decentralized and based on a bottom-up approach (Felicidades and Piñeiro, 2017) that can also be considered an important vehicle to increase the visibility of women's role in the fisheries sector (REMSP, 2017).

In Spain, marine/fishing tourism has thus been developed in coastal communities' thanks mainly to the boost of Fishermen's Guilds, Fisheries Local Action Groups (FLAG), private companies and local councils (Moreno, 2018). From this perspective, it is seen as an alternative to the sector crisis and source for sustainable development and stands as a novel and recently implemented initiative that can encourage sustainable coastal development (Molina et al., 2010; Herrera-Racionero et al, 2018; Moreno, 2018).

The growing demand for this type of tourism is given by the interest in rediscovering the environmental values and cultural traditions of our environment (Molina et al, 2010) and, taking this perspective into account, the uses and traditions linked to the culture of the sea offer an opportunity for the development of complementary tourist products (Besteiro, 2004). The potential of marine tourism as an innovative tourism model according to the integrated management of coastal areas is therefore confirmed as it respects the conservation of biological resources and fully integrates with its environment (Sagittal Project, 2007).

However, the development of fishing tourism in Spain is low and irregular. As recent studies show (Miret-Pastor et al., 2015; Herrera-Racionero et al., 2018), fishing-tourism in Spain, although it has advanced in recent years, is still in a fledgling phase. In fact, as the above-mentioned authors point out, many of the Spanish projects have encountered various cultural, financial and, above all, restrictive legislation. Proof of this are the different names with which the legislation refers to this activity (fishing tourism, fishing-tourism, marine tourism, etc.).

Only recently, the Law 33/2014 of 26 December amending Law 3/2001 of March 26, on Maritime Fisheries of the Spanish State explicitly has recognized this activity and has introduced the concepts of "fishing and aquaculture diversification" as well as those of "fishing or seafaring tourism", "aquaculture tourism" and "fishing-tourism" in national regulations. Although at the regional level, with the exception of Catalonia, Galicia, the Balearic Islands and the Canary Islands, many of the initiatives still lack of an appropriate legal framework (Herrera-Racionero et al, 2018). Law 11/2008 of 3 December on Fisheries of Galicia was the first autonomous regulation that defined and regulated marine tourism and that has led Galicia to be the region where it has been developed most (CETMAR, 2010; Moreno, 2018) and where there was a Guide to Marine Tourism of Galicia that gathered all the information about the activities and companies that carry out activities related to marine tourism as far as 2012.

OBJECTIVES AND METHODOLOGICAL DESIGN

This paper is based on identifying and describing the development of innovative entrepreneurial practices and initiatives linked to marine tourism implemented by women through the support of Galicia's Fisheries Local Action Groups (FLAGs). Galician coastline has been divided into 8 different sections: 1. A Mariña-Ortegal; 2. Golfo Ártabro Norte; 3. Costa da Morte; 4. Finisterre-Ría de Muros y Noia; 5. Ría de Arousa; 6. Ría de Pontevedra; 7. Ría de Vigo-A Guarda; 8. Golfo Ártabro Sur (See Figure 1). In each one of these areas, there is a management entity called FLAG, which has been established according to the European Maritime and Fisheries Funds (EMFF). These entities have established specific objectives to support not only diversification within or outside the commercial fisheries sector, but also permanent learning and job creation within different activities such as fishing or seafaring tourism, gastronomy, the environment or the processing and marketing of fishery products. Each one of them, independently and by taking the intrinsic characteristics of each area into account, develops a Participatory Local Development Strategy (PLDS). The latter ones are designed for the period 2014-2020 coinciding with the programming of the EMFF funds.

The methodology applied in the study has been exploratory-descriptive with a mixed-type data analysis, since in-depth semi structured interviews have been conducted as well as the review and collection

Figure 1. Distribution of FLAGs in Galicia
Source: https://galp.xunta.gal/

of some interesting quantitative and qualitative data. Two specific objectives are set out in turn to mark the two phases of information collection, taking the methodology described into account.

The first specific objective of the study is based on the methodological design that supports the project. It thus consists of a first phase of bibliographic review and collection of information from secondary sources of a qualitative-quantitative type, coming from the official website of FLAGs, which allows framing the field and the object of studying and carrying out a detailed inventory of business activities done by women linked to the fisheries sector. This is the reason why a descriptive sheet model (Table 1) has been designed to be completed with the purpose of identifying, categorizing and evaluating the mentioned entrepreneurial initiatives. In particular, 10 women's business projects financed between 1999 and 2019 for the 8 Galician FLAGs are analyzed in depth.

The selection criteria of the project sample has been carried out through a sampling for convenience, by bearing the following factors in mind: productive projects related to the marine tourism sector, projects representative of the 8 FLAGs, covering the entire Galician coast, and projects that were currently active.

The second objective is to obtain first-hand information from different woman profile involved in the subject of studying to go in depth the issues related to the reasons that lead them to start up, the

Table 1. Project sample fact sheet feature

CHARACTERISTICS SHEET OF ENTREPRENEURIAL INITIATIVES
. General sector of activity . Company dimension (number of workers)
Project description . Objectives . Phases and start-up (Origin of initiative and evolution over time) . Financing (Investments made and grants received) . Use of technology (website, social networks, etc.) . Innovative orientation . Environmental interest

Source: Own elaboration

benefits, the barriers to the challenges, achievements, future prospects, etc. This qualitative phase of research includes the conduct of 10 semi-structured in-depth interviews with women entrepreneurs who are developing business projects linked to marine tourism, who are the topic of the selected sample for this paper. The content analysis of the collected discourses has been carried out by taking the categorical variables of analysis into account, in addition to the anonymity of women's entrepreneurs, as well as the specific name of the company, which has been respected so they could use identification codes for them. A deeper knowledge has been obtained through this analysis from the vision and perception of women as central leading agents in this study.

From this perspective, a script has been prepared for interviews (Table 2) based on the three dimensions (demographic profile, psychological profile and sociological profile) from which the figure of the entrepreneur is usually addressed (Fuentes and Sánchez, 2010). It should also be noted that the thematic blocks of interviews have been established according to the various aspects that condition the entrepreneurial activity (Ortiz García, 2017).

Table 2. Thematic blocks on which the interview script is based

BLOCKS	TOPICS
SECTION A	Entrepreneur's sociodemographic and work profile
SECTION B	Motivation and perception about the entrepreneurial process
SECTION C	Innovative contribution to the sector
SECTION D	Future prospects

Source: Own elaboration

RESULTS AND DISCUSSION

On the coastline, economic dynamics have been transformed. One of these changes has been the move of the primary sector to tourism. In this scenario, women, as well as in fisheries, are playing a main role. The diversification of tourist products and their marked differentiating aspect is a key element for the development and competitiveness of tourism of any destination. Undoubtedly, the coast has traditionally been one of the favourite destinations for tourists and, therefore, the main consolidated destinations are located on the coast. With the aim of reconverting or boosting their local economy they have opted for taking advantage of the natural resources of the area and promoting marine tourism, a tourist typology considered a low impact one (Mohonk Agreement, 2001; Farnet, 2013) which also encourages the involvement of the locals in tourism development (Chek et al., 2018).

Table 3 shows the total number of tourism companies promoted with EMFF funds with the help of FLAGs in Galicia. In total, there have been 221 tourist companies that, in some way, are related to the value of the cultural fishing, tangible and intangible, used as a tourist resource. The data corroborate the presence of female entrepreneurship in coastal areas even though it accounts for only 20% of the total number. In Galicia, it is noted that the activities for which more women apply is the hotel sector (47.8% in comparison with 42.15% of the total) where, mainly, rural tourism houses, inns and / or tourist apartments stand out. The results of the interviews also show that this decision is made because of the ease of family reconciliation rather than hospitality or complementary activities.

Table 3. General sector of activity

Activity	Total Female Entrepreneurship	Female Entrepreneurship
Hotel sector	51 (42.15%)	**22(47.8%)**
Catering sector	27 (22.31%)	12 (26.1%)
Complementary Activities	43 (35.54%)	12 (26.1%)
TOTAL	221 (100%)	46 (100%)

Source: Own elaboration

The study shows that most tourism companies promoted by Galician FLAGs were created from 2016 (97%) (See Table 1) detecting two peaks in 2017 (50 companies) and 2019 (63 companies). Likewise, it is noted that in 2019 there was a greater number of companies with female initiative, a total of 17, although in global terms 2017 can be considered the best year of female entrepreneurship with 28%. Undoubtedly, in Galicia entrepreneurship in this sector has the possibility of becoming an agent of change and a driving force for social and economic development as well as a tool to enhance many resources of rural fishing communities (Patiño, 2016) .

In order to address the study, the area where the projects were developed has also been taken into account (see Table 4) finding that the areas where there has been a higher rate of female entrepreneurship are: Golfo Ártabro Norte (FLAG 2) and Costa da Morte (FLAG 3), corresponding both areas to the North of Galicia. Likewise, the organization of the Galician coastline according to the new distribution allows observing the effects of this help in a model of tourism development more focused on the value of sea culture.

The option of entrepreneurship among women as an alternative to precariousness and as a possibility of finding job niches concerning their training and needs has become an alternative in rural areas (Heldt Cassel and Pettersson, 2015; Alario and Morales, 2016) and, specifically, in coastal communities. In this context, it should be emphasized that women have been playing a key role in the development of the fisheries sector, either through professional activities or through a more complementary role that has no statutory recognition, and which has made them invisible. As an example of the mentioned comple-

Figure 2. Female entrepreneurship evolution 2016-2019
Source: Own elaboration

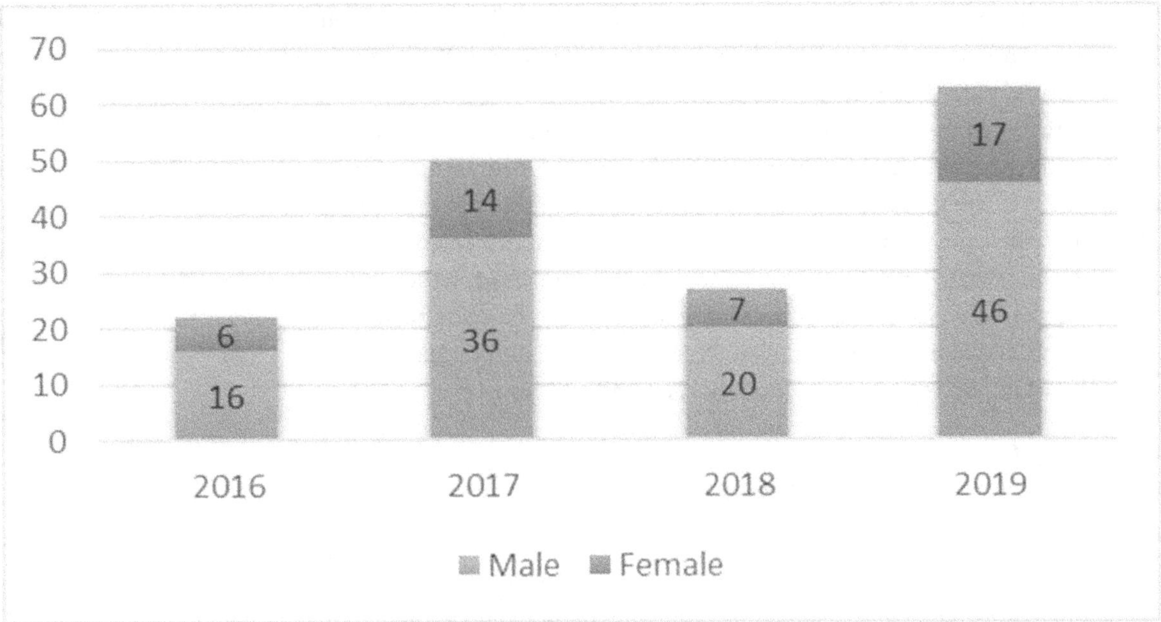

mentary role, the interviewee VM_CR points out that in the case of the shellfish gatherers they have an average salary of €600 to which they have to subtract €200 for their Social Security, which shows that they are mainly seen as a complement for the family economy.

According to the GEM Galicia REPORT (2018), women as an autonomous worker in Galicia are an emerging social reality, constantly growing and of obvious qualitative importance. The report indicates that the entrepreneur is a woman between 41 and 42 years old in 57.66% of the cases (in this paper the highest percentage is between 30 and 40 years old -see Table 5) and that the percentage of abandonment of the initiative is lower than the one among men. According to the TEA index corresponding to 2018,

Table 4. Percentage of entrepreneurship in THE FLAG areas

FLAG	TOTAL	WOMEN
FLAG 1. A Mariña-Ortegal	22 (18.18%)	5 (10.56%)
FLAG 2. Golfo Ártabro Norte	**17 (14.04%)**	**11 (23.92%)**
FLAG 3. Costa da Morte	**20 (16.53%)**	**11 (23.91%)**
FLAG 4. Fisterra-Ría de Muros-Noia	26 (21.49%)	7 (15.22%)
FLAG 5. Ría de Arousa	14 (11.57%)	6 (13.04%)
FLAG 6. Ría de Pontevedra	9 (7.44%)	3 (6.52%)
FLAG 7. Ría de Vigo-A Guarda	12 (9.92%)	3 (6.52%)
FLAG 8. Golfo Ártabro Sur	1 (0.83%)	0 (0%)
TOTAL	121 (100%)	46 (100%)

Source: Own elaboration from Xunta de Galicia (Galicia's government): https://galp.xunta.gal/

this rate of female entrepreneurship has been increasing since its very start. In fact, the general increase in entrepreneurial activity in Galicia has resulted from the increase associated with female entrepreneurial activity, which has increased from 4.52% to 6.43% in 2018. The women entrepreneurial activity continues its upward trend since 2016, reaching the peak in the period 2005-2018 and surpassing Galician male entrepreneurial activity (4.76%), and also the Spanish average. Around a third of entrepreneurial initiatives are promoted by people with university studies (32.97%), a large majority is developed in an urban environment (83.04%), and the main motivation for their implementation is the opportunity to business (64.86%). As it can be seen in Table 5, concerning the study sample, in terms of university studies, the results show the same number (40%) in VET (Vocational Education and Training) and university studies, representing a small 10% in studies related to tourism. As for the motivations for the implementation of the project, asked in the interviews, in general it can be summarized in the need to start (or in other cases expand) a different business idea, creative and related to marine tourism, coinciding with the data in the GEM Galicia report, and secondly for family reasons.

Table 5. Sociodemographic profile of the sample

Sociodemographic Characteristics	Percentages
Age	Under 30: 0% Between 30 and 40: 40% Between 40 and 50: 30% More than de 50: 30%
Training	Basic Training: 20% Vocational Education and Training: 40% Universities: 40%
Tourism studies	Yes: 10% No: 90%

Source: Own elaboration

As for the future prospects, an issue also addressed in the interviews, the following responses have been obtained: there is an obvious concern about unfair competition, the need to go beyond the local clientele and attract tourists, achieving a minimum profitability of the business and interest to give greater visibility and promotion to the area and the business itself, since they sometimes regret that there is no greater effort from authorities to promote the territory. With the exception of an interviewee, all of them have a positive perception of the future as a business, despite the difficulties, and as a dynamizing element of the economic development of the territory, since employment has been created.

As indicated in the methodology, the fact sheet (Table 6) designed to identify, categorize and analyze the 10 women's business projects sampled is completed.

With regard to the proposed objectives, correlated with the type of project, a repetition is observed in most of them (Table 6). The following established objectives could be summarized: creating jobs, launching new products, activities or initiatives and promoting the area economically through marine tourism.

As for the funding received, by taking FLAGs can subsidize up to 50% for this type of productive projects into account, the sample shows that they mean between 42% and 50% of the approved public funds, except for the case of AMDA whose data are unknown. It can therefore be said that in the cases analyzed the subsidy has been quite high or complete.

Table 6. Characteristics of the projects analyzed

COMPANY NAME	OBJECTIVES	PHASES AND COMMISSIONING	FINANCING FLAG	TECHNOLOGY USE	INNOVATIVE GUIDANCE	ENVIRONMENTAL INTEREST
ALA	Creating an innovative marine project by leveraging environmental and heritage elements for diversification Create jobs	Creating an innovative restaurant 2018-2019	€123,823.62 47%	YES	YES	YES
ACRM	Creating jobs through the diversification of local fishing areas Creating a new intersectoral tourism product	Creating a sustainable restaurant 2017-2018	€53,185.53 50%	YES	YES	YES
MNM	Creating jobs through diversification of the local economy of fishing areas Making Mariña-Ortegal a differential tourist destination highlighting the potential of marine tourism	Expansion of a themed hotel 2017-2018	€93,396.08 50%	YES	YES	YES
CGMS*	Creating jobs through the diversification of the local fisheries area economy Promoting marine tourism as a new economic driving force in the territory	Integral reform 2018	€22,475.00 50%	NO	NO	NO
AMDA*	Adding value to fishery products	Start-up of activities 2010	-- --	NO	NO	NO
VM_CR	Promoting potential initiatives and preserving natural resources	Launching an active tourism project 2019	€14,942.14 --	YES	YES	YES
MGSL	Supporting diversification of commercial fisheries Supporting permanent learning	Creation of accommodation 2017-2018	€186,323.22 45%	YES	YES	YES
AOMA	Improving the competitiveness of the fisheries sector Creating jobs Guaranteeing a sustainable activity	Launching new service lines 2017	€6,633.34 45%	YES	YES	YES
LNC	Supporting and creating a new accommodation network Promoting economic diversification and the creation of jobs	Hotel rehabilitation 2017-2018	€135,506,12 42%	YES	YES	NO
PPTAM	Creating jobs through diversification of fishing areas local economy Increase the diversification of the fishing sector towards tourism activities	Culinary offer of catering services 2018-2019	€75,675.35 50%	YES	YES	YES

*Note: After interviews, some of the companies were not really innovation-oriented, so their case is not presented as an innovative practice.

Source: Own elaboration based on the information obtained from FLAGs.

As for the variables, the use of technology and orientation to innovation, 80% of cases are positive in their compliance. With regard to the variable environmental interest, it has decreased to 70% because this variable was not taken into account in three cases in three cases and two of them lacked innovative orientation.

Eight of the 10 interviews conducted detail clearly innovative practices in their businesses, always related to marine tourism. In order for this work also serves as a manual of good practices, each of the cases is described individually:

Case 1: Restaurant (ALA)

Fresh and traditional seafood are offered, the restaurant furniture is already innovative, since it recreates a market.

The auction is a very innovative offer that is being launched: the auction is a recreation of the market. There are some push buttons on the tables and some screens where the dishes are offered. Those clients who want to auction on a dish or a product that is currently being offered, press the button from their table. "For example, if some clients want the same Tuna Tartar or the same Turbot they will participate in the same auction and the one who offers the highest price will get it." They are still trying this model but it is already announced it on the web. They claim, "it's like a kind of entertainment or game for our customers"

Case 2: Restaurant (ACRM)

Its innovative condition is that they are all women, which was not easy because when it came to searching a highly trained chef they found it "extremely complicated" (...) "In this sector men are better trained than women"

Its main innovation was to develop its entire offer based on the sustainable, ecological, traditional and local product. Not only the building itself, which has been rehabilitated by respecting the traditional structure and by taking sustainability into account, but also the offer of the menu with a profound relation to the sea (they go directly to the local market to buy). The dishes offered are based on traditional recipes with the intention of recovering traditional local recipes. They argue: "They are top quality local products and from here in the area, nothing is offered that is not purchased in this area".

In the backyard there are about 2,500 m2 where all vegetables offered in their dishes are grown without chemical agents of any kind. They are collected when the dish is going to be cooked, and to avoid insects they use natural remedies (such as nettling water). The fertilizer they use is the restaurant own organic waste that they compose them. They're also building in a greenhouse to extend their production all year round.

Another innovation related to their philosophy of sustainability is that they do not use plastic packaging, they must be returnable. They say, "no long-life plastics". They have also top quality water that they offer their customers in a jar on the table.

Case 3: Hotel (MNM)

It was the first hotel to open in Spain in 2004 where rooms are related to cinema films. At first, it was not well understood, but then the idea was considered very innovative. Each room has been decorated by an expert designer that has to do with the cinema and artistic world. There are 10 different atmospheres based on specific films. The last 3 suites (double rooms), opened with the help of the FLAG in 2018, are related to the seafaring cinema, specifically: "*20,000 leagues of underwater travel*", "*The Blue Lake*" and "*The Great Wednesday*". They say, "We make a great difference, for example, the best surfers in the world have stayed here."

They have also launched sea-related and nature-related activities that offer their clients in collaboration with other local companies. The objective is to enhance the development of marine tourism in Ribadeo (boat trips, visits and knowledge about sea traditions, etc.). They have enhanced the seafaring rooms and activities related to marine tourism through the social media. For example, those guests who have chosen these activities receive a discount when booking.

Case 4: Active Tourism Project (VM_CR)

This project wants to promote coastal cycling through the design of 'Bicycle Routes' that run through various towns of the Ria de Pontevedra (Pontevedra tidal inlet). The idea is to promote the heritage and natural and cultural resources of this area. They claim that the innovative contribution was the "use of tourist resources for a less massive and intrusive type of tourism in comparison with the usual in Rías Baixas (tidal inlet). Incorporation of the Galician market into tourism within the region itself".

Three types of services are offered: half-day walks lasting about four hours. The Picnic offered by the company is optional. The depart is adjusted to customer's requests as much as possible, bearing in mind that the group can be a minimum of five people and a maximum of eight. Full-day routes: With a length of approximately eight hours, there is also food and drink as well as a support van and tailor-made routes: the customer can choose the route and its features. In addition, walks and excursions are classified according to their difficulty, taking aspects such as the type of road, the gap or mileage into account.

Case 5: Accommodations (MGSL)

New tourist accommodation project that gives the client access to all cultural and traditional aspects linked to the sea, which usually runs out of visibility for conventional tourism. There are five accommodation options that have a private garden and small swimming pools with light that evoke a phenomenon of luminescence in the form of spheres of light caused by marine microorganisms in a certain phase of the moon. They are apartments which are innovative because of their construction, by respecting the typology of traditional seafaring villas. They are environmentally friendly buildings, with minimal impact and energy consumption close to zero.

These accommodation options include themes related to the sea and the marine world, furniture, painting exhibitions related to the sea, literature, poetry... Buildings are "smart." The innovative function SPA (Sailor Personal Assistant) in the accommodation allows customers to organize the day according to their tastes, the weather, activities, local festivals, parties and local events, always related to the sea as the origin of traditions. Another particular feature of SPA is its interactivity, since it provides plenty of visitors' experiences which allow improving and optimizing the service.

The activities of the SPA are: creation of exhibitions and libraries in which the sea is the protagonist; local festivals throughout the region related to the sea and its products; invitation to the discovery of the gastronomy of the area, based on the products of the sea; visits and enjoyment of natural and artificial heritage; practice of sea-related sports (paddle, canoeing, surfing, windsurfing, scuba diving, fishing...); artistic encounters of painting, sculpture, poetry, dance and others, all linked to the sea; routes on sailor boats and visits to the markets and small ports.

Case 6: New Visitor Services (AOMA)

Two new and improved service lines have been created so that visitors to the place and schoolchildren can fully enjoy the destination, contacting the environment, with its culture and spreading such a good thing like fishing. These lines are the marine world through the water bike, making known the seafaring environment of the Ría de Arousa (Arousa tidal inlet) by performing water activities with an innovative and differentiated system; conferences on marine tourism for school children. As they have said in the interview "this is a tour that leads them to visit the fishing port, do a water activity and an informative marine workshop", all of this in the same day.

They have also said, "We have a mobile application to have our users geolocated at all time since they are performing activities in the middle of the sea. They have also detailed that "this year we have created health tourism as innovation, by combining bike routes, kayaking, canoeing, etc. combined with meditation, stretching (...), on the other hand, we have innovative gift packages that change every year and that we send national and internationally. For example, canned food, a scroll to give as a present, the fact of opening the can wrapped in nets; bottles with a message; and even a treasure chest with chocolate coins."

Case 7: Hostel (LNC)

An old traditional house has been rehabilitated on Camino de Santiago (the pilgrims' way to Santiago), to be converted into a three-star inn with seven rooms and a center of social dynamization used by both tourists and the locals population. Guests can enjoy various exhibitions and cultural activities offered by companies in nearby towns such as photo safaris or English lessons by means of a guided tour around the Costa da Morte (the Death Coast), as well as seafood tastings.

Case 8: Catering (PPTAM)

It is an innovative gastronomic proposal with a takeaway offer, food delivery and catering services, always with local fish and seafood from the Ría de Arousa (Arousa tidal inlet). Packaged products are also made, mainly canned local discards.

It has a space for showing and selling products and a "gourmet" shop where canned seafood made there are on display and sold. Workshops, cookery courses, cooking shows, etc. are offered. The dissemination of activities and online sales is done through the website. They have also said, "We are in charge of a school catering to provide more fish in their diet. We teach them about the sea in (...) we take them to the fish market so they can meet and recognize it (...) to see the fish unloading, the auction and then they cook it."

CONCLUSION

Since 2007, the Common Fisheries Policy has encouraged the diversification of activities aimed at the economic development of fishing communities by funding fishery tourism and/or promoting local fishery products. Once the report has been reviewed, supported by pioneering reports and studies, such as the 2017 REMSP (Spanish Women Fisheries Network) report, it can be ensured that marine tourism is shown as a business alternative along the Galician coast, where women pop up as an entrepreneurial, creative and naturally talented person.

After the research carried out, it is empirically shown that the female entrepreneurial initiatives in tourism along the Galician coast are still minority (20%) although it is a feasible and interestable way of women's entrepreneurship. In this female venture and through the financing by FLAGs, we are more committed to the catering sector, with 47.8%, in comparison with other sectors such as complementary activities, which represent 26.1% respectively. It has also been found that Golfo Ártabro Norte (FLAG 2) and Costa da Morte (FLAG 3) stand out among the total 8 FLAGs because of a higher female entrepreneurship index, representing almost 24% of the total in both cases. It is also worth noting the high percentage of female entrepreneurship without studies in tourism which shows a lack of specific training and the lack of professionalization of the sector.

It is noteworthy to mention that the economic contribution of FLAG funding is considered a very important factor for the start or continuity of the project, since the fund received, in the cases analyzed, is around 50% of the total cost of the productive project. In order to achieve the maximum amount of funds, they must meet a number of premises based on innovation and sustainability. In the case study analyzed, 80% of the projects have an innovative, participatory and responsible nature whose added value is based on the local cultural and natural heritage paying particular attention to the intangible value and their linkage with maritime culture. In these projects, the variable use of technology and environmental interest accounts for 80% and 70% respectively, showing concern and adaptation for new trends and social needs.

In short, the consolidation of those business projects promoted by women will undoubtedly be a decisive factor in the progressive reduction of the gender gap in the professional field. Entrepreneurship is presented as an option chosen by women for their professional incorporation and, probably, as a good formula for the work-life balance, as well as work-life alignment. In addition, the launch of their businesses has allowed keeping and creating new female jobs, has led to revitalization and development in coastal areas contributing to the socio-economic and environmental sustainability of these areas, it has enhanced women fisheries has risen social awareness on the importance of their contribution to the sector and it has also reinvented coastal tourism, among other benefits.

REFERENCES

Abrahamsson, A. (2008). Sustainopreneurship –business with a cause: the promise of creative business organizing for sustainability. In Bridging the functional and territorial views on regional entrepreneurship and development (pp. 137-155). Swedish Foundation for Small Business Research.

Alario Trigueros, M., & Morales Prieto, E. (2016). Iniciativas de las mujeres: Emprendimiento y oportunidades en el espacio rural de Castilla y León. *Documents d'Analisi Geografica*, *62*(3), 613–637. doi:10.5565/rev/dag.369

Besteiro, B. (2004). El desarrollo del Turismo Náutico en Galicia. *Cuadernos de Turismo*, *13*, 145–163.

Centro Tecnológico del Mar (CETMAR). (2014). *Análisis de la participación de la mujer en la actividad pesquera y acuícola*. Ministerio de Agricultura, Alimentación y Medioambiente.

Chek Sulaiman, F., Jaini, N., Rouzee Jamaluddin, E., & Idzhainee Hashim, N. (2018). The impact of marine tourism towards the local community at Palau Pangkor, Perak. *Asian Journal of Quality of Life*, *3*(13), 168–176. doi:10.21834/ajqol.v3i13.174

Chen, C. (2010). Diversifying fisheries into tourism in Taiwan: Experiences and prospects. *Ocean and Coastal Management*, *53*(8), 487–492. doi:10.1016/j.ocecoaman.2010.06.003

Datta, P. B., & Gailey, R. (2012). Empowering women through social entrepreneurship: Case study of a women's cooperative in India. *Entrepreneurship Theory and Practice*, *36*(3), 569–587. doi:10.1111/j.1540-6520.2012.00505.x

FARNET. (2011). *Guía 5 Diversificación de las zonas de pesca.* Retrieved from: https://www.mapa.gob.es/es/pesca/temas/diversificacion/analisiseconomicodeladiversifiacionpesquerayacuicola_tcm30-290383.pdf

FARNET. (2013). *Uniendo la pesca con la economía del turismo.* Retrieved from: https://webgate.ec.europa.eu/fpfis/cms/farnet/files/documents/FARNET_Magazine_09_ES_0.pdf

Felicidades-García, J., & Piñeiro-Antelo, M. A. (2017). Competitividad, Sostenibilidad y Diversificación en la Fachada Atlántica de la Península Ibérica: La Gestión Local del Fondo Europeo de Pesca (2007-2013). *Revista de Estudios Andaluces*, *34*(1), 350–379. doi:10.12795/rea.2017.i34.12

Forsberg, G., & Lindgren, G. (2015). Regional policy, social networks and informal structures. *European Urban and Regional Studies*, *22*(4), 368–382. doi:10.1177/0969776412474663

Fuentes García, F. J., & Sánchez Cañizares, S. M. (2010). Análisis del perfil emprendedor: Una perspectiva de género. *Estudios de Economía Aplicada*, *28*(3), 1–28.

García Negro, M.C. & Zotes, Y.N. (2006). El trabajo de las mujeres en el sector pesquero gallego. Análisis de los problemas relacionados con su tratamiento estadístico. *Revista galega de economía*, *15*(1), 1-25.

García Ramón, M. D., & Baylina Ferré, M. (2000). *El nuevo papel de las mujeres en el desarrollo rural*. Oikos-tau.

Gessa Perera, A. & Toledano. (2011). Turismo, emprendimiento y sostenibilidad en los espacios naturales protegidos. El caso de Andalucía (España). *Estudios y perspectivas en turismo*, *20*(5), 1154-1174. Retrieved from: https://dialnet.unirioja.es/servlet/articulo?codigo=3739754

Global Entrepreneurship Monitor. (2018). *Informe Global 2018.* Retrieved from: https://www.gemconsortium.org/report

Heldt Cassel, S., & Pettersson, K. (2015). Performing gender and rurality in Swedish farm tourism. *Scandinavian Journal of Hospitality and Tourism*, *15*(1), 138–151. doi:10.1080/15022250.2015.1005333

Herrera-Racionero, P., Miret-Pastor, L., & Lizcano, E. (2018). Viajar con la tradición: Los pescadores artesanales ante la pesca-turismo en la Comunidad Valenciana (España). *Cuadernos de Turismo*, *41*(41), 279–293. doi:10.6018/turismo.41.327031

Kimbu, A. N., & Ngoasong, M. Z. (2016). Women as vectors of social entrepreneurship. *Annals of Tourism Research*, *60*(1), 63–79. doi:10.1016/j.annals.2016.06.002

Martínez García, P. (2017). Mirando al mar desde los feminismos. Comunidad y supervivencia en torno a las trabajadoras de la pesca. *Prima Social, 19*, 183-214. Retrieved from: http://hdl.handle.net/10251/61538

Miret-Pastor, L., Muñoz Zamora, C., Herrera-Racionero, C., & Martínez Novo, R. (2015). Análisis regional del turismo pesquero en España. *Revista de Análisis Turístico, 20*, 23–28.

Molina, A., González, J. F., & García, C. (2010). El turismo pesquero como instrumento de apoyo al desarrollo sostenible en zonas litorales: la experiencia del proyecto Sagital. *XIV International Congress on Project Engineering*.

Moreira, P., & Urriolagoitia, L. (2011). El emprendimiento social. *Revista española del tercer sector*, (17), 17-40. Retrieved from: http://www.injuve.es/sites/default/files/2018/29/publicaciones/5._emprendimiento_social.pdf

Moreno Muñoz, D. (2018). Aportación a los conceptos de turismo marinero/pesquero y pesca-turismo. *Cuadernos de Turismo, 42*(42), 385–396. doi:10.6018/turismo.42.17

Movono, A., & Dahles, H. (2017). Female empowerment and tourism: A focus on businesses in a Fijian village. *Asia Pacific Journal of Tourism Research, 22*(6), 681–692. doi:10.1080/10941665.2017.1308397

Ortiz García, P. (2017). The discourse on entrepreneurship of women from a gender perspective. *Vivat Academia, 140*(140), 115–129. doi:10.15178/va.2017.140.115-129

Osuna-Rodríguez, M., Rodríguez García, L. & Gómez Parra, M. (2015). *Mujeres, Turismo Rural y Brecha Salarial de Género*. Córdoba: UCOPress, Editorial Universidad de Córdoba.

Patiño Romarís, C. A. (2016). El turismo marinero: un producto diferenciador y emergente de la oferta turística del litoral gallego. In X.M. Santos, P. Taboada, & L. López (Eds.), *Actas del X Congreso Internacional de Turismo Rural y Desarrollo Sostenible* (pp. 401-410). Santiago de Compostela, España: Universidad de Santiago de Compostela.

Pettersson, K., Ahlb, H., Berglundc, K., & Tillmard, M. (2017). In the name of women? Feminist readings of policies for women's entrepreneurship in Scandinavia. *Scandinavian Journal of Management, 33*(1), 50–63. doi:10.1016/j.scaman.2017.01.002

Power, S., Di Domenico, M., & Miller, G. (2017). The nature of ethical entrepreneurship in tourism. *Annals of Tourism Research, 65*, 36–48. doi:10.1016/j.annals.2017.05.001

Proyecto Sagital. (2007). *El papel de la mujer en el sector pesquero: potencialidades en el ámbito del turismo pesquero*. Madrid: Fundación General de la Universidad Politécnica de Madrid. Retrieved from: http://www.cesmuamfar.com/pdf/El_Papel_de_la_Mujer_en_el_Sector_Pesquero._Potencialidades_en_el_%C3%A1mbito_tur%C3%ADstico-pesquero.pdf

Red Española de Mujeres en el Sector Pesquero (REMSP). (2017). *La igualdad en cifras: Diagnóstico sobre la situación de la mujer en el sector pesquero y acuícola*. Ministerio de Agricultura y Pesca, Alimentación y Medio Ambiente. Retrieved from: https://www.mapa.gob.es/es/pesca/temas/red-mujeres/diagnosticoencifrasjunio2017revision_tcm30-425320.pdf

Rico González, M., & Gómez García, M. J. (2009). La participación empresarial de la mujer en las iniciativas de turismo rural en Castilla y León. *Economía Agraria y Recursos Naturales*, *9*(2), 51–77. https://core.ac.uk/download/pdf/6449164.pdf

Seelos, C., & Mair, J. (2005). Social entrepreneurship: Creating new business models to serve the poor. *Business Horizons*, *48*(3), 241–246. doi:10.1016/j.bushor.2004.11.006

Solvoll, G., Alsos, A., & Bulanova, O. (2015). Tourism entrepreneurship: Review and future directions. *Scandinavian Journal of Hospitality and Tourism*, *15*(sup1), 120–137. doi:10.1080/15022250.2015.1065592

Tortosa Conchillo, J. M., Valenciano, J. P., & Uribe Toril, J. (2018). Innovación social como motor del desarrollo local y el emprendimiento social. *Equidad y Desarrollo*, *31*(31), 11–25. doi:10.19052/ed.4416

UNWTO. (2011). *Global report on women in tourism 2010*. World Tourism Organization.

Vázquez, A., & Rodríguez, J. (2015). La política de desarrollo local: los desafíos de los territorios de desarrollo tardío. *Ciudad y Territorio: Estudios Territoriales*, *186*, 625-638. Retrieved from: https://recyt.fecyt.es/index.php/CyTET/article/view/76436

Wang, S., Hung, K., & Huang, W. (2019). Motivations for entrepreneurship in the tourism and hospitality sector: A social cognitive theory perspective. *International Journal of Hospitality Management*, *78*, 78–88. doi:10.1016/j.ijhm.2018.11.018

Ziene Mottiara, K., Bolukb, K., & Kline, C. (2018). The roles of social entrepreneurs in rural destination development. *Annals of Tourism Research*, *68*, 77–88. doi:10.1016/j.annals.2017.12.001

KEY TERMS AND DEFINITIONS

ANMUPESCA: National Association of Women in Fisheries in Spain it is a non-profit entity established in 2016 that brings together more than 10,000 women linked to the world of the sea from all over the national territory in the association.

Blue Economy: In economics it is a term which encourages better stewardship, exploitation and preservation of our marine environment.

EMFF: European Maritime and Fisheries Funds is the fund for the EU's maritime and fisheries policies for 2014-2020. It is one of five European Structural and Investment (ESI) Funds which together, promote a growth and job-based recovery in Europe.

FLAGs: Fisheries Local Action Groups (FLAGs), responsible for managing a European program in the area.

Marine Tourism: Segment that uses activities that involve the enhancement of the maritime culture.

REMSP: Spanish Network of Women in the Fisheries Sector which promoting communication and the exchange of initiatives and best practices among professionals working in fishing and aquaculture activities, in any of its different areas, promoting their entrepreneurship and associationism, as well as their empowerment and leadership.

Sustainopreneurship: Is a business model which aims to solve problems related to social and sustainability agenda.

Chapter 11
Entrepreneurship and Innovation in Communities for Women Who Travel

Luana Moreira de Araújo
Universidade Federal do Delta do Parnaíba, Brazil

Cintia Martins
Universidade Federal do Delta do Parnaíba, Brazil

André Riani Costa Perinotto
Universidade Federal do Delta do Parnaíba, Brazil

ABSTRACT

This study addresses innovation in the community of women who travel alone by analyzing a platform where women share lodging facilities. This platform, in addition to providing accommodation services, allows users to share experiences through collaborative tourism and contact between female travelers. This chapter focused on the applicability of this innovation to the accommodation market. For that, it is essential to understand the many reasons that lead women to rely on a unique tool for accommodation, understanding the relationship between market, platform, and users in the modern world. This study used primary sources, interviews with the platform's founder, and questionnaires administered to users to understand innovation from the perspective of a sensitive community that wants to travel safely and the strategic positioning of the platform within the market, in addition to identifying concepts that describe the relationship of users with the platform.

DOI: 10.4018/978-1-7998-4855-4.ch011

INTRODUCTION

Technological development has revolutionized the way companies position themselves, steadily increasing access to information and allowing new communication technologies to modify habits and practices and stimulate new sociabilities (Kaufman, 2015). It is not just learning and information that represent technological revolution, but also the usefulness of this learning for creating new knowledge, in a cycle between innovation and its application. New information technologies are not just tools to be used, but also procedures to be developed from the moment users and creators can become one and the same, taking control of technology (Castells, 1999).

For Kaufman (2010), when it comes to users and creators, the key is to understand that information technologies are not simply tools. There is a relationship between users and creators, who ultimately merge during the collaborative process. Collaborative models reinvent traditional market forms, promoting trust among people and experience sharing through applications and online platforms. Spaces that can be shared, exchanged, rented, or bought are created, and we adopt new ways of understanding consumption, where what matters is not ownership, but access to these goods or services (De La Encarnación, 2016).

Electronic commerce refers to any business practice within the electronic environment. It is a business model where the product is seen, demonstrated, and sold through technology (Guimarães & Borges, 2008). At Fórum E-commerce Brasil 2016, Phocuswright data were presented showing that the tourist trade represented 11% of Brazil's GDP, which corresponded to 65 billion Brazilian reais (BRL) (USD 12.5 billion) in total sales, of which BRL 17 billion were online. This indicates that the level of internet penetration is 26% in this country. While tourism will grow 3% in the next few years, online travel will grow 12%, reaching 48% of internet penetration in 2020.

Digital economy transversality drives outcomes that change the tourism value chain, promoting media that trigger innovation and increasingly influential advances in the tourist trade (Moreno-Izquierdo, Ramón-Rodríguez, & Such Devesa, 2016). Thus, a platform focused on travel experiences for women travelers has recognized the importance of a support network among women as a way to satisfy their needs and reduce the obstacles they face such as through collaborative tourism. This new business model has changed the form of marketing and, as a result, there is a new interest in research aiming to understand this innovation in the online tourism lodging market in Brazil, focused on the obstacles faced by women when taking a trip and the aspects that induced the tourist accommodation market to create a travel platform exclusively for women.

With that in mind, our study was based on confirming the applicability of this platform (which will not have its name revealed for ethical reasons) by analyzing the relationship between market, platform, and female users. The platform functions as an intermediate and, at the same time, as a research tool. The survey was distributed with the contribution of the platform, which made it possible to contact its users, allowing this study to reach a significant number of people. Data were collected between December 2018 and October 2019. Data were analyzed using a word cloud created in the IRAMUTEQ software and a descriptive analysis of the answers given by platform users. Both the interview with the creator of the platform and the survey aimed to collect data to verify the importance of an exclusive lodging service for a female audience and to present the experiences of the users as women travelers within a network of collaborative lodging. Lastly, we analyzed the shared experiences and reflections reported during the study to understand the relationship between tourism and women in the modern world.

Market Segmentation, Motivation, and Gender

The growth of the tourism industry caused changes in the behavior of consumers, who are increasingly informed and independent. Given the advances in digital platforms, companies must compete with their own consumers, who benefit from the largest amount of information and feel capable of offering services to themselves (Perinotto & Siqueira, 2018). A tourist destination encompasses different consumer profiles, therefore, segmentation is the most precise way to reach a desired audience (Ansarah & Panosso Netto, 2010).

The goal of tourism segmentation is to classify consumer groups with different profiles and similar needs to offer tourism products that are suitable for the target audience, both to satisfy them and to gain their loyalty (Colantuono, 2016). Therefore, it combines a set of consumers who share the same needs to develop a market strategy capable of reaching the target market. To segment, it is essential to identify the factors that intervene in the purchase decisions made by the consumers (Ferreira, 2000). "The more we know the characteristics of the target market, the greater the effectiveness of the marketing techniques for publicity, promotion, sales, and public relations" (Lage, 1992, p. 62).

According to Kotler and Armstrong (2008), through market segmentation, the marketing professional must analyze the market as a whole and subdivide it into smaller segments. An example is the segmentation of consumer markets, although it should be noted that there is more than one way to segment the market. For Colantuono (2016), supply chain segmentation comprises fishing tourism, social tourism, ecotourism, adventure tourism, rural tourism, and cultural tourism, among others. Demand segmentation is characterized by geographic, demographic, psychographic, and behavioral variables. Geographic segmentation has geographic units, such as localities, countries, regions, states, and cities, as variables; companies may decide to tailor their product marketing to each one of these areas, with the goal of meeting demands and desires based on the geographic context of each consumer (Lage, 1992; Kotler & Armstrong, 2008). With demographic segmentation, the market is divided into groups according to factors such as age, sex, income, religion, social class, race, and nationality, among others. Demographic variables are one of the most common approaches for assigning groups, based on the concept that consumer preferences are most often associated with demographic factors (Kotler & Keller, 2006).

Psychographic factors represent the personality of the consumer and their perspective in decision making. They are a more modern approach compared to other criteria, stimulating and expanding numerous industries. Consumers are subdivided into groups based on lifestyle, personality, and values; individuals belonging to the same demographic group may have different psychographic characteristics (Swarbrooke & Horner, 2002; Kotler & Keller, 2006). Behavioral segmentation, on the other hand, defines consumer groups with regard to their behavior towards decision-making. It takes into account behavioral factors, considering, for instance, client loyalty, availability, interaction, and other factors that characterize the profiles of these consumers (Cunha, 2018). Accordingly, demographic and psychographic segmentation are the starting point to address the connection between motivational factors and gender issues, given that demographic and psychographic aspects raise questions regarding consumer sex and personality.

Berdychevsky, Poria, and Uriely (2013) believe that "gender relationships affect the way men and women create their tourism experiences". In research reviewed by Ryan, of fourteen motivations related to men and women, three were different between the two: women valued travel experiences more to avoid the daily rush, relax physically, and relax emotionally (as cited in Swarbrooke & Horner, 2002). The number of women who travel alone is increasing; as they become more and more independent financially and emotionally, women tend to prioritize personal satisfaction (Carvalho, Baptista, & Costa, 2015).

In the reports published by Maneze and Pacheco (2019) of women traveling alone, through the experience carried out in an "alternative way", such as asking for a ride even as a woman, "[...] people think that the automatic stop is prohibited, and it is not prohibited. It's dangerous! As women, we hear a lot of nonsense, don't we? " Early in the trip, she confesses to having been afraid and thought about the possibility of some risk. However, throughout the experience she believed that being a woman was a "myth" of being dangerous factor.

For women, traveling means positioning themselves in control of their choices, of the decision of what to do and where to go. This behavior reveals the importance of a women-only trip. It shows that trips that are planned by a woman or a group of women provide comfortable experiences, which is related to the fact that they are in control and can make their own decisions regarding their trips (Antonioli, 2015). Importantly, women are gaining visibility in the tourism sector, motivated by added concerns relative to other consumers. In addition to having issues in common with other tourist profiles, women who travel alone need initiatives that not only offer a product, but are also a collaborative network among women who wish to travel alone so that they feel safe and supported by other women who share the same interest.

Collaborative Tourism and the Digital Tool

The collaborative model is very important for the tourism sector, which has been including these practices more over time, continuously expanding the increase of online community platforms. Thus, the way people travel has changed considerably based on the services provided by collaborative platforms (Heo, 2016). Accordingly, the term collaborative consumption refers to people who perform the acquisition and distribution of a resource for compensation in the form of exchanges and transactions (Belk, 2014), since consumers share access to goods and services instead of transferring ownership. According to the National Confederation of Shopkeepers of Brazil (Confederação Nacional de Dirigentes Lojistas, 2019), 89% of Brazilian consumers who experienced some form of collaborative consumption approved this model, and only 2% of the surveyed said that they were dissatisfied.

Collaborative tourism startups have appeared as a result of the opportunity for a sharing economy, creating a new form of doing tourism through tools that are being adapted for these innovations. This allows the tourist to take autonomous and personalized trips, offering authentic and unexpected experiences that can be shared with other people and encompass several tourism segments (Albertini, 2017).

Collaborative or sharing economy produces new business models that combine economic aspects with environmental and social sustainability and take advantage of technological changes, especially the internet. There are many initiatives that could be framed within the context of shared economy and collaborative consumption, such as Zipcar, Uber, Freecycle, and other initiatives around the world (Silveira, Petrini, & Santos, 2016). As a model of collaborative tourism, the platform creates an opportunity for a female audience, through a tool that recognizes the needs of women travelers for a service that goes beyond being an accommodation booking platform to work as a support network among women who share the same interests. The increase in the number of women who travel alone has led to new ways of examining the confluence of gender and tourism issues, which directly influence the economy, culture, politics, and society (Carvalho, Baptista, & Costa, 2015).

Only women can join the network, although there may be men either accompanying the female tourist or at the host's house. However, they are always indicated by "I am traveling with a male companion(s)/ there are men living with me at home", and each woman can decide whether to accept it or not. In addition, the platform's website also features "pillars" that must prevail among the "sisters", and the advantages of being a host or a traveler.

This network creates a new option for women who wish to travel autonomously and safely, based on the concept that women travelers connect with female hosts through collaborative lodging at other women's homes, promoting comfortable and safe experiences and forming collaborative relationships through unique local experiences, which is advantageous for both parties. In addition, the women develop friendship bonds based on the choice of a profile that shares the same interests and with whom they may have an affinity.

The platform expands its support network through an exclusive blog to introduce some subjects to the sisters—female hosts and users who are interested in these topics—such as cultural diversity, traditional gastronomy, cultural itineraries, tips for traveling solo, sustainable tourism while traveling, how to receive a traveler well, traveling mums, among others.

The relationship of the platform with the market is exciting and growing. From the moment the consumers' need is identified, numerous alternatives emerge to create innovative services and maximize the market. The investigation of women who travel alone has increased in the last few years because of women who did not travel and live new experiences before because they did not have a companion. The difficulties faced by women in taking trips, and specifically the murder of two friends who were traveling together, raised some questions regarding the conditions faced by women when traveling. This resulted in movements and hashtags such as #Itravelalone, #viajosozinha, and #viajosolo as a way to encourage the largest number of women to travel.

An analysis of a Digital Platform for Women to Travel

A survey was conducted with the goal of learning about the general travel experience of users, in addition to determining the relationship of the users with the platform, to understand their use of the digital tool. The survey included the following questions:

1. How do you feel regarding your experience as a woman who travels?
2. What does it mean to be a woman traveler?
3. What motivated you to use the platform?
4. How do you feel as a user of this platform?
5. In your opinion, has traveling become safer for women with this platform?
6. What is it like to be a woman and use the platform, which is a support network to women when they are traveling?
7. After the survey was filled out by the users, an electronic interview was conducted via email. This interview was specifically directed to the platform founder. In it, she was asked about the platform's positioning within the market, through an analysis of the marketing mix of the company. Questions included:

With respect to product policy, the concept of the platform is based on bringing advantages to women who intend to travel using a platform that offers collaborative lodging that is compatible with their financial situation. Depending on the information obtained, the development of new procedures depends on approval processes, which guarantee the services provided. To create the service, the platform combines safe and friendly lodging with photographs that embody credibility by being closer to reality. These

Table 1. Topics for the interview with the platform owner

Product policy	Service concept and what attributes the product should have
Price policy	Setting a price, how the product compares to the competition
Consumer policy	Marketing strategies in communication channels, consumers targeted, and affiliated message
Distribution policy	Where to find the product, distribution channels, sales effort, reservation systems and competition

Source: Tocquer and Zins (1999), adapted by the authors.

photographs convey the actual interaction between the two parts; they are images that depict locations or landscapes and fuel the imagination, the desire, and the need of travelers to visit them (Perinotto, 2013).

According to our data, the largest age group among the platform users included women aged between 25 and 34 years, corresponding to 37.09% of the users. A research survey carried out by Voopter (2018), a site for searching and comparing airfare, revealed that women were the majority in terms of searches on the site (68.70%), and 36.22% of the site searches were by people in the 25–34 years age group. With these data, three key points explain the fact that more women aged between 25–34 years search for trips. This may be an indication of more digital readiness, due to the large number of current initiatives in the virtual environment that facilitate and optimize the daily lives of people, increasing access to travel information. The second reason is tied to increased financial independence, since it is assumed that this age group has already entered the job market. Lastly, this is generally the moment that precedes marriage and motherhood.

After characterizing the age profile of the platform users, we introduce the elements that comprise the survey presented to the users with the goal of obtaining information that conveys the sentiments of these women travelers towards traveling with regard to the platform. The experiences of women travelers are marked by a feeling of insecurity that refers to the socially constructed image of women as the fragile and dependent sex. According to Antonioli (2015), "From a gender perspective, for a long time there were constraints over women and their freedom of movement, in addition to factors related to class conditions and tastes, that were intrinsically tied to the fact that they were women" (p. 3).

The ratio of free time of men relative to women is uneven, based on the concept that men have more free time than women. Even when taking a vacation from work, women carry out domestic chores and care for children (Huete, 2013). Women have multiple activities that occupy their moments of leisure, leaving fewer opportunities and less time for them to use for activities outside their obligations.

A Tripadvisor market research carried out in 2015 found that when women were asked about the measures they usually took when alone, safety was the main factor. The majority said they avoided deserted places (61%) and tried not to behave as a typical tourist (45%).This lack of security permeates all steps of the women. When it comes to independent travel, in addition to physical fear, there is the fear of being misinterpreted as someone who is traveling to seek company, framing women's decisions as inseparable from men. Another term used by the women surveyed was "free", referring to the effect of the trips on their lives. For Brazilian women, traveling solo helps gain more independence (50%) and more confidence (52%), and stimulates them to learn about new cultures (51%). Their favorite activities are cultural attractions (88%), local gastronomy (70%), and shopping (48%) (Tripadvisor, 2015).

When contemplating what it means, in fact, to be a woman traveler, words such as "liberty" and "freedom" are contemplated. Through the word liberty, women express their pleasure in traveling freely, independently. Over the years, women are claiming their freedom, which results in them taking control of their own lives with the goal of achieving cultural and professional enrichment (Rodrigues, 2015). The "knowledge" cited by the women surveyed represents the consequences of traveling, based on the prospect of knowledge and experiences. During a trip, an individual comes into contact with new experiences and is introduced to different cultures, expanding their opportunity for self-knowledge and freedom.

For most women, the main reason that contributed to their use of the platform was that they trusted the network to provide experiences that were safe for these women who travel alone. The platform provides security to its users, since they stay with female hosts, offering collaborative lodging, meaning that both sides gain an advantage from using it. This way, women can be safer while traveling, and female hosts can meet people and have new experiences through this contact. In light of this, the surveyed women consider their experiences in the platform as welcoming and safe, given the features directed towards comfort, satisfaction, and new traveling opportunities. These are women who are interested in supporting the trips of other women, creating connections and encouraging each other.

Analyzing the capability of the platform to provide what the users seek and whether the platform achieves its main objective, which is safety, the women surveyed acknowledged that they were satisfied with the platform, for instance stating that "I feel like a pioneer facing a more intelligent and interesting possibility of travel". We observe that, when women understand the proposition of the platform, they feel represented by a platform that encourages them and creates the possibility of autonomy. When it comes to facing the condition of being a woman traveler and a platform user, the connection between "happy", "safe", and "support" is established, which creates statements such as: "I am happy that initiatives like this are available to me", "Much safer", "It is amazing to belong to this network", and "I feel that I belong to a group that promotes and implements sorority among women". These are feelings that women share in a network that interconnects bold women and encourages them to travel.

With regard to user responses about their experiences, we may consider that these words express the female empowerment present in their behavior, based on the idea that empowerment represents a process sustained by the collective construct, respect to female diversity, the cooperative albeit unique establishment of processes (Marinho, 2016). The tourism sector adapts daily to the needs and desires of women who travel, so that they can travel alone for longer (Huete, 2013).

The information collected from the survey was organized based on a concept map, or word cloud. According to Correia et al. (2016), "[...] a concept map may be understood as an interconnected set of propositions that contains intelligible messages with the goal of expressing conceptual relationships" (p. 42). Thus, the ideas in the survey are organized and classified in a way that helps us understand the analysis, displaying the most relevant elements among the themes reported by platform users. The emphasis, represented by the size of the term, is related to the frequency of its use.

When on their own, women feel vulnerable and exposed to dangerous situations. As a result of violence against women, the safety issue is noticeably one of the most important topics for women, and it is a concern when choosing a destination. As one responder said, "I love to travel, it makes me happy, but I feel unsafe". Violence against women represents a clear manifestation of the sexist standards established by society, generating fear in women towards everything and everyone, from their clothes to the way they act, and directly interfering in their choices (Reis, 2017).

The relationship between women and freedom has always been marked by prejudice. Despite all these separations, women achieve personal and professional independence and are present in universities and in the job market, attaining more egalitarian conditions (Rodrigues, 2015). When it comes to traveling, women are overcoming their own fears to travel and experience new things. As one responder said, "Free and motivated. Traveling is a life lesson at each stop", an expression that demonstrates the relationship between women and freedom. Thus, contemplating women who travel alone reveals how much this practice is growing among women. In a research survey conducted by Booking.com (2019), 72% of men said they had traveled abroad by themselves before, but women were not too far behind: 62% had traveled abroad with no companion.

The act of traveling gives people an opportunity to search for knowing and being. In this sense, social tradition may indirectly interfere on and affect people's identities. Even as everything is changing there is complexity, it is not a simple process. The essential characteristics that define independent travelers are a spirit of adventure and freedom, traveling autonomously (Maneze & Pacheco, 2019).

As technology becomes an increasing part of people's lives today, it is common to use information and apps that help improve the tourist experience as much as possible. A study shows that four in five (79%) Generation Z travelers in Spain now want virtual or enhanced reality experiences to help plan their trips. When choosing a destination, almost all (86%) of the Generation Z travelers in Spain reviewed videos and photos published by travel brands in social networks (Travelport, 2019). Given the simplicity that technology brings to planning and executing trips, it is easier to enhance experiences and get to know a destination in different ways.

For women who use the platform for support, be it as tourists or as hosts, there is still a possibility that they can travel safely within a support network. As one woman surveyed said, "I feel that I belong to a group that promotes and implements sorority among women, and that is very important and interesting". Such expressions demonstrate how women feel supported and the platform's influence on the achievement of trips by women. By understanding the outlook of user feelings and the factors that drive their decisions, the platform reinvents itself and creates ways to reach the largest number of users and stand out in the tourism market, based on a set of strategies that help identify and direct its services towards the target audience.

The focus of the platform is to create services that connect women and encourage them to travel more often through support, information, and inspiration, following the notion of a streamlined startup. The Brazilian Association of Startups (Associação Brasileira de Startups, 2019) defines startups as "early-stage companies developing innovative products or services that have the potential to grow quickly". The methodology for a streamlined startup is based on a set of techniques that contribute for their development; through software, they operate in management systems or online platforms (Sousa, 2016).

Based to the price policy analysis of the platform, it acts as an intermediary in the negotiation between users, offering collaborative lodging at a fair price. The entire process undergoes cost-benefit validation; when the negotiation is considered unfair by the platform's evaluation, platform intervention in the deal is needed as a way to balance both sides. Therefore, it is the platform's responsibility, according to article 25 of the Brazilian Consumer Protection Code (Código de Defesa do Consumidor) of 1990, to provide the service requested and ensure that it is performed, responding in case there are safety failures during the intermediation service.

In addition, we observed that users search for some variables that represent this essential service of lodging. Some criteria were identified at the moment the sisters made their decisions, such as location, cost, rapport, cleanliness, and a negative rating in the network. For the responders, the following issues would stop them from choosing a given host: "location and price", "lack of rapport", and "bad feedback". The presence of positive feedback is fundamental, since services and quality undoubtedly increase productivity and lead to better market performance (Martins, 2018).

The market offers similar platforms, which use the sharing economy to facilitate travel services for women that help both sides of the multilateral platform, obtaining a benefit from using it. Based on a study conducted by the Multilateral Investment Fund in April 2016, the sharing economy is most important in Brazil (32%), Mexico (13%), Argentina (13%), and Peru (11%), countries that lead the number of sharing economy initiatives in Latin American and the Caribbean (Fondo Multilateral de Inversiones, 2016). The industries where this type of initiative has been implemented are services (26%), transportation (24%), and lodging (19%). Sheswanderful, Womantrip, and La voyagere are other examples of communities that are part of economy sharing among women who want to travel alone, encouraging these women to travel in groups, providing a space to share questions, challenges, dream trips, accommodation, and events in cities to build new relationships.

Regarding the consumer item, the platform tries to reach women who want to travel economically and motivate them to contact other women who enjoy these collaborative experiences. According to an online survey conducted by Voopter (2019) that included 5,419 Brazilian women, the main reasons that prevented them from traveling more often were lack of money (86.6%), lack of time (50.8%), and insecurity/fear (31.9%). Thus, in addition to all fear-related constraints faced by women, they need initiatives to reassure them, but that are also affordable.

The initiative seeks to connect its image to the prospect of uniting women, lowering the obstacles that appear along the way of women who want to travel. Above all, the platform is constantly being modified and adapted to validate its processes and reach new users. The platform is seen as disruptive, representing technological innovation and breaking patterns. First it reaches the "early adopters", meaning those who are willing to try a new product despite taking a risk, which helps the company flourish. With this, the company becomes an option accepted by most people ("mainstreaming"), that is, the features of the initiative are accepted by the users, who are pleased with the platform.

The distribution of the platform occurs through ads, events, and app stores, and the company has partners such as digital influencers, social networks. Its use becomes effective after registration and profile creation, where the entire process goes through a cost-benefit validation, offering collaborative lodging at a fair price. Through these points all the factors that directly influence the performance of the service chosen are checked. It is possible to observe the processes and strategies and also the competitors, who are both in the digital environment and offline. The moment the platform is reached, in terms of its users, is usually called a "Sales Funnel". The goal is to analyze user engagement based on first user access, and from there organize the steps in the client's path to succeed in reaching a previously determined objective (Oliveira, Gois & Goldstein, 2018). This is a strategy adopted by the platform, which creates its Sales Funnel based on the female user process, from the discovery of the brand to demonstration of interest, registration in the platform, and searching, requesting, and confirming a reservation.

CONCLUSION

This study focused on recognizing the needs of a female audience to take independent trips, from the perspective of collaborative tourism and the platform analyzed, discussing topics related to the viability of an exclusive lodging tool for women and examining the relationship between market, platform, and female users.

With regard to the age distribution of the platform users, it is interesting to note the specificities of each age group. The importance of this data is precisely in recognizing the demographic profile of the users, in this way more specifically directing strategies towards the audience in question.

The results of the survey show that the platform comes into existence with the proposition of satisfying the need for security and support among women who are afraid to travel alone. Through the platform, women help each other and deal with the obstacles faced by women every day. The study improved the identification of the female platform users in relation to their sensations and emotion expressions. Considering this information, there is a need to continually introduce tools within the platform itself to provide all of the support these users depend on for traveling.

Among the expressions used by the female users, it is worth mentioning safety, freedom, traveling, support, visiting, meeting, experiences. These expressions depict women who wish to visit new places and meet new people, who seek experiences and support, and who like to feel free and want to achieve self-knowledge. The platform has forums as virtual spaces for dialogue that may encourage women to travel after learning about other experiences in the platform. There is a concern among women who travel about sharing their experiences through the platform.

The tourism accommodation market is constantly changing over the years, which increasingly stimulates the client-company relationship. The development of the platform was observed to occur from the identification of obstacles present in society. Women are more and more interconnected and willing to lessen their constraints through this kind of initiative. Sharing trips and a direct link between ideas were two identified platform attributes. Women who use the platform have similar ideas and share the same needs; the platform itself directly focuses its efforts towards a specific audience, as found from the data presented here.

Through the present analysis, we determined that there is a link between theory and practice, which allows for an examination of how the need for a platform used exclusively by women emerged and how this platform develops from a perspective that includes not only the trip itself, but also a number of expectations and feelings that are exclusive to a female audience. However, the importance of the platform within the Brazilian tourism sector must be asserted, since the platform delivers plans, satisfaction, opportunities, and equality to women travelers.

One limitation of this research was the emerging nature of studies like this, but the objective of the study was achieved and answered. Other studies may be conducted on this topic and on the platform that would be able to examine them from a different point of view. That would create the opportunity for more quantitative research, which could be more focused on women and tourism. In conclusion, the platform is clearly important for strengthening the tourist trade and the understanding of women in the modern world.

REFERENCES

Albertini, G. S. G. G. (2017). *Turismo e economia colaborativa: uma avaliação das plataformas online de serviços vinculados ao turismo* (Undergraduate dissertation). Unidade Federal Fluminense, Rio de Janeiro, Brazil. Retrieved October 2, 2019 from https://app.uff.br/riuff/handle/1/4721

Ansarah, M. G. R., & Panosso Netto, A. (2010). A segmentação dos mercados como objeto de estudo do turismo. *VII Seminário da Associação Brasileira de Pesquisa e Pós-Graduação em Turismo, 20.* Retrieved September 10, 2019 from https://www.anptur.org.br/anais/anais/files/7/25.pdf

Antonioli, F. L. A. (2015). *Viagens no feminino: gênero, turismo e transnacionalidade* (Master's thesis). Universidade Estadual de Campinas, Campinas, Brazil. Retrieved September 10, 2019 from http://www.repositorio.unicamp.br/handle/Reposip/279566

Associação Brasileira de Startups. (2019). *Sobre a abstartups*. Retrieved September 16, 2019, from https://abstartups.com.br/sobre-a-abstartups

Belk, R. (2014). You are what you can access: Sharing and collaborative consumption online. *Journal of Business Research*, *67*(8), 1595–1600. doi:10.1016/j.jbusres.2013.10.001

Berdychevsky, L., Poria, Y., & Uriely, N. (2013). Sexual behavior in women's tourist experiences: Motivations, behaviors, and meanings. *Tourism Management*, *35*, 144–155. doi:10.1016/j.tourman.2012.06.011

Booking.com. (2019). *Mulheres pelo mundo*. Retrieved November 21, 2019, from https://www.mulherespelomundo.com/introducao

Brasil. (1990). Constituição. Código de Defesa do Consumidor nº 8078, date September 11, 1990.

Carvalho, G., Baptista, M. M., & Costa, C. (2015). Mulheres que viajam sozinhas: Reflexões sobre género e experiências turísticas. *Revista Turismo & Desenvolvimento*, *23*, 59–67.

Castells, M. (1999). *A Sociedade em Rede*. Paz e Terra.

Colantuono, A. C. S. (2016). A segmentação da demanda turística no mercado brasileiro e as políticas de fomento ao setor. *Revista Turismo: Estudos e Práticas*, *5*(2), 219–250.

Confederação Nacional de Dirigentes Lojistas. (2019). *Economia compartilhada deixa 89% de seus usuários satisfeitos, revela estudo da CNDL/SPC Brasil*. Retrieved September 29, 2019, from http://site.cndl.org.br/economia-compartilhada-deixa-89-de-seus-usuarios-satisfeitos-revela-estudo-da-cndlspc-brasil-2/

Correia, P. R., Aguiar, J. G., Viana, A. D., & Cabral, G. C. P. (2016). Por que vale a pena usar mapas conceituais no ensino superior? *Revista de Graduação USP*, *1*(1), 41–51. doi:10.11606/issn.2525-376X.v1i1p41-51

Cunha, M. C. G. M. (2018). *Proposta de solução gamificada no e-commerce* (Master's Thesis). Universidade de Lisboa, Lisbon, Portugal.

Da Silveira, L. M., Petrini, M. D., & Santos, A. C. M. Z. (2016). Economia compartilhada e consumo colaborativo: O que estamos pesquisando? *Rege-Revista de Gestão*, *23*(4), 298–305. doi:10.1016/j.rege.2016.09.005

De La Encarnación, A. (2016). El alojamiento colaborativo: Viviendas de uso turístico y plataformas virtuales. *Revista de Estudios de la Administración Local y Autonómica*, 5. Retrieved October 02, 2019 from https://revistasonline.inap.es/index.php/REALA/article/view/10350

De Sousa, J. B. S. (2016). Lean Startups: O sistema de produção enxuta como estratégia competitiva. *Journal of Lean Systems*, *1*(4), 53–65.

ETC/UNWTO. (2008). *Handbook on E-marketing for tourism destinations*. Retrieved November 1, 2019 from https://www.cabdirect.org/cabdirect/abstract/20083275024

Ferreira, F. H. G. (2000). *Segmentação de mercado*. Retrieved October 25, 2019 from https://www.ecrconsultoria.com.br/sites/default/files/Segmentacao_de_Mercado.pdf

Fondo Multilateral De Inversiones. (2016). *Economía colaborativa en América Latina: ¿Cómo se distribuyen geográficamente las iniciativas? ¿Qué antigüedad tienen?* Retrieved October 30, 2019 from https://publications.iadb.org/es/publicacion/15641/economia-colaborativa-en-america-latina

Guimarães, A. S., & Borges, M. P. (2008). *E-Turismo: Internet e negócios do turismo*. Cengage Learning.

Heo, Y. (2016). Sharing economy and prospects in tourism research. *Annals of Tourism Research, 58*, 166–170. doi:10.1016/j.annals.2016.02.002

Huete, R. (2013). El papel de la mujer en el turismo. In *XIII Jornadas Sobre Mujer* (pp. 1–27). Universidad de Alicante.

Kaufman, D. (2010). *Processo de tomada de decisão no ciberespaço: O papel das redes sociais no jogo das escolhas individuais* (Masters thesis). Pontifícia Universidade Católica de São Paulo, São Paulo, Brazil.

Kaufman, D. (2015). *O despertar de Gulliver: os desafios das empresas nas redes digitais* (Doctoral dissertation). Universidade de São Paulo, São Paulo, Brazil.

Kotler, P., & Armstrong, G. (2008). *Princípios de Marketing*. Pearson.

Kotler, P., & Keller, K. L. (2006). *Administração de Marketing: A Bíblia do Marketing*. Prentice Hall Brasil.

Lage, B. H. G. (1992). Segmentação do mercado turístico. *Revista Turismo em Análise, 3*(2), 61–74. doi:10.11606/issn.1984-4867.v3i2p61-74

Maneze, C. A. L., & Pacheco, R. T. B. (2019). La viajera independiente: búsqueda de sí, búsqueda de ser. In Gestión, turismo y género. Thomson Reuters.

Marinho, P. A. S., & Gonçalves, H. S. (2016). Práticas de empoderamento feminino na América Latina. *Revista de Estudios Sociales, 56*, 80–90. doi:10.7440/res56.2016.06

Martins, D. M. G. (2018). *A concetualização da qualidade do serviço no setor do turismo* (Masters thesis). Universidade do Porto, Porto, Portugal.

Moreno-Izquierdo, L. M., Ramón-Rodríguez, A., & Such Devesa, M. J. (2016). Turismo colaborativo: ¿Está AirBnB transformando el sector del alojamiento? *Economistas, 150*, 107–119.

Nogueira, C. (2018). *Mulheres são maioria entre os turistas que buscam passagens aéreas*. Retrieved November 20, 2019, from https://extra.globo.com/noticias/viagem-e-turismo/mulheres-sao-maioria-entre-os-turistas-que-buscam-passagens-aereas-revela-pesquisa-22974058.html

Oliveira, P. H. P., Gois, I. P., & Goldstein, M. C. B. (2018). Aplicação do funil de vendas para melhoria de engajamento em um SVA. *Revista da Universidade Vale do Rio Verde, 16*(2), 1–19. doi:10.5892/ruvrd.v16i2.5020

Perinotto, A. R. C. (2013). *Circulação de imagens turísticas: fotografias de Parnaíba/PI nas mídias* (Doctoral dissertation). Universidade do Vale do Rio dos Sinos, São Leopoldo, Brazil.

Perinotto, A. R. C., & Siqueira, R. A. (2018). As novas tendências do marketing digital para o setor turístico. *Turismo: Estudos & Práticas*, *7*(2), 86–215.

Reis, A. M. (2017). *Mulheres e viagens: insegurança e medo?* (Undergraduate dissertation). Universidade Federal Fluminense, Rio de Janeiro, Brazil. Retrieved November 15, 2019 from https://app.uff.br/riuff/handle/1/4403

Rodrigues, L. de A. (2015). *Mulheres no exterior: um estudo sobre medo e gênero* (Undergraduate dissertation). Universidade Federal do Rio Grande do Sul, Porto Alegre, Brazil.

Sass, C. (2017). *Precisamos falar de turismo!* Retrieved January 3, 2018, from https://www.ecommercebrasil.com.br/artigos/precisamos-falar-de-turismo/

Swarbrooke, J., & Horner, S. (2002). *O Comportamento do Consumidor no Turismo*. Aleph.

Tocquer, G., & Zins, M. (1999). *Marketing do Turismo*. Instituto Piaget.

Tripadvisor. (2015). *Uma a cada quatro brasileiras viaja sozinha*. Retrieved December 01, 2019 from https://tripadvisor.mediaroom.com/2015-03-05-Uma-A-Cada-Quatro-Brasileiras-Viaja-Sozinha

Voopter. (2019). *Viaje, mulher!* Retrieved November 21, 2019, from https://voopter.com.br/viajemulher/

Chapter 12
The Valorization of the Surface Waters on the Inland Regions in Differentiated Tourism Products:
The Offers of Estrela Geopark Territory (Portugal) and the Strategies of Qualifying the Destination

Gonçalo Poeta Fernandes
https://orcid.org/0000-0003-1278-0369
Polytechnic Institute of Guarda, Portugal & Centre for Tourism Research, Development and Innovation (CITUR), Portugal

Hugo Gomes
Associação Geopark Estrela (AGE), Portugal

ABSTRACT

The purpose of this study is to address the importance of water resources in tourism and leisure activities, to establish a framework of activities in inland waters and their meaning in the territory of the Serra da Estrela, namely within the scope of Estrela Geopark. Recognizing tourism as one of the pillars of a UNESCO world geopark, the image and brand of this destination is enhanced by the classification obtained, which drives the strengthening of partnerships; collaboration and cooperation between stackholders; the appreciation and diffusion of its heritage and culture; encouraging increased visitor numbers, consumption of goods and services, and decreasing seasonality of tourism. The attraction of its waters for tourism and leisure activities represents a resource for the strengthening of the economic activities within the region with concrete effects on trade, hotels, and local restaurants, in view of the practices involved, the natural and cultural contexts that the territory offers, and the opportunities for new activities.

DOI: 10.4018/978-1-7998-4855-4.ch012

INTRODUCTION

Tourism in its relationship with the territory and the transformations carried out, conveys the meaning of tourism destination, promoting that its resources become products, which are associated services and generate features for its enjoyment. Forms of ownership should be increasingly carefully and properly planned to ensure their best use and sustainability, particularly in areas of natural relevance.

The surface water resources of the inland regions (rivers, lakes, dams, canals, marshes, among others) have always been attractive to the population for the development of leisure and recreation activities, beyond those considered essential for the human presence, namely the supply water, agriculture and industry, transport and circulation. The existence of water has promoted recreational uses for the riverside communities, or for visitors, who seek in these surface leisure areas, of which bathing, sports activities, fishing, observation of natural ecosystems, among others, are important attractions to associate with tourism. Inland waters acquire increasingly important for leisure and tourism, and various activities that are associated with them and multiple users who seek them throughout the year with different interests and practices.

Local people value tourism activities supported in river environments and realize their relevance to a model of territorial promotion and well-being, fostering interventions to accommodate these activities in safety and the consequent concern for environmental preservation. It is worth mentioning that it is the quality of the waters, the heritage value of the landscape, the maintenance of the characteristics of the river ecosystems, and the own rurality that support the value of these spaces in the tourist activity. In their appropriation, social, economic and environmental interests tend to be divergent, implying the existence of principles and guidelines that promote their proper regulation, establishing limits and conditions for the uses and practices to be developed.

These spaces can take place playful, recreational activities, sports or competition, dependent on geomorphological conditions, existing riparian ecosystems, the size of the water masses and developed infrastructure. Given the environmental value of these resources and the ecosystems they represent, there is a latent fragility that requires proper management and adaptation of spaces to different uses and functions, without compromising the sustainability of ecosystems and user safety.

Taking into account the constant and growing tourism competitiveness, it is essential to open to new markets and focus on non-mass tourism products, supported on quality and differentiation, and result from the optimization of endogenous resources available at local and regional levels, associating services and territorial integration. These tourism and leisure activities supported by local water resources provide socio-economic and cultural impact, which must be valued, but at the same time monitored to ensure its sustainability.

The chapter is structured in three parts, the first corresponding to the framework of the importance of inland waters in the development of tourism and leisure activities, the second promotes an approach to the Estrela Geopark and its potential in terms of recreational and leisure activities to be promoted, given the surface water resources held. Finally, the river beaches of the Geopark territory are analyzed, their characteristics and management forms, developing a perception analysis on their importance for the development of the destination and its communities. The aim is to illustrate the importance of river resources in the promotion of tourism and to assess the recreational and sports potential that the Estrela Geopark holds and analyze the perception of the value of river beaches for local communities and the image of the destination.

The Serra da Estrela, being the largest mountain of the mainland in Portugal, originates from the main portuguese rivers along with a set of lakes and dams of relevant importance for water supply to populations and economic activities, constituting the last two decades as a bathing leisure and sports space. Thus, the forms of management, the entrepreneurial initiatives promoted, and the contributions to the local communities are perceived, in an approach of tourism value and sustainability of the inland waters and the strengthening of the demand and the identity of the destination.

INLAND SURFACE WATERS AND THEIR FUNCTIONS. PERCEPTIONS, NEW USES AND CHANGE CONTEXTS

The management of water resources is a complex issue of collective interest and relevance to the environment, especially when the water sources such as rivers, lakes, reservoirs, canals, marshes, and other aquifers present concern conditions due to the uses and functions that support. They require actions and forms of control that ensure the preservation of their ecosystems and the quality of their own water (Peixoto, 2016; Folgado Fernandez *et al*, 2019). They are public goods and their management should value and protect the interests of the local communities established there and of the economic and social activities developed, many of them ancestral. Over time there has been an increase in water use needs, both in quantity and quality, as a result of population growth and expansion of activities and consumption, which require more water resources and increased control over uses and functions developed (Prideaux *et al*, 2009). The relationship between water and tourism is characterized by a double understanding of this element: as a precious resource and as an attraction. Given that water is a valuable and scarce resource, research should focus on identifying practical and business initiatives through which water-based tourism can be developed without compromising the quality and availability of water resources (Brito-Henriques *et al.*, 2010; Gacia Gonzalez, 2014). In this sense, the relationship between water and tourism is focused on sustainability. On the other hand, as a tourist attraction, water represents a resource with strong potential to attract tourists (Grössling et al., 2015; Cole & Ferguson, 2015). In this scope river tourism contributes significantly to local development. Incentives from local authorities and institutional organizations have favored the emergence of endogenous entrepreneurial strategies (Fachrudin & Lubis, 2016; Norhafiza & Ku Azam, 2017; Violier, 1993)

With regards to Ferreira & Guerreiro de Brito (2009, p. 343), *water is a transversal element and enhancer multiple functions that, assuming value associated with a utility requires that demand and the uses are reconciled with the offer and needs*. The most significant uses of the natural regime of surface water and aquifer systems are, in general terms, those intended for agriculture, industry, energy production and supply to the population, both in terms of consumption and in terms of interventions that promote changes in the flow regime.

Inland waters include river ecosystems and artificial or strongly modified water bodies associated with them (dams and canals), as well as other ecosystems where the aquatic component is of permanent or temporary importance, such as tributary rivers or ponds. In these ecosystems the presence of water (in frequency and/or abundance) is the determining element, leading to the emergence of specific biological components (Ferreira & Guerreiro de Brito,2009). The surface water resources, natural and artificial, are diverse and their conditions have been fostering several uses and activities in terms of leisure and tourism, which gain expression in terms of inland waters, catalyzing attractiveness and recreational sports practices in increasing appreciation (Morar & Pop, 2016). In this context, the main services provided by surface inland waters include:

- Agriculture;
- The production of renewable energy;
- Supply to human populations and sanitation;
- Supply to industrial activities;
- The transportation and movement of goods and raw materials;
- Support for fishing activities and fish production;
- Tourism and leisure activities;

The growing interest in the enhancement and exploitation of inland water resources for activities not related to water supply to communities, industry and agriculture, has been expressing itself in recent decades. Accompanying the growth of leisure and tourism experiences in rural areas, the appreciation of riparian ecosystems is a concern of local entities in the qualification of these resources and in generating new opportunities for local communities (Garcia Gonzales, 2004; Norhafiza & Ku Azam, 2017). The changing sensitivity to water and its importance for health and leisure signal the potential for the development of a social conscience capable of generating new business opportunities that can create value for interior geographic territories with new roles and users. This emerging water culture emphasizes the importance of achieving a harmonious, quality, and sustainable relationship with this element. (Folgado Fernandez et al., 2019)

In this context, fluvial beaches arise from the reuse of water lines/surface of a river channel or reservoir (dam, reservoir or lagoon), constituting a form of creation of a recreational and sporting space, generating a leisure structure for the local communities and an element of tourist attraction (Cole & Ferguson, 2015; Santos, 2016). Its creation requires the development of studies and the evaluation of the appropriate functions, so as not to disturb the natural environment and to create the necessary access and security conditions for its operation. A significant part of the environmental changes result from the pollutant discharges and inability to assimilate them by the water resource, with higher incidence in the summer period, compared to the natural water regime and consequent reduction of flow, increasing the density of pollutants, disturbing and endangering the species and their ecosystems (Cole & Ferguson, 2015; Ferreira & Guerreiro de Brito, 2009). There is a growing concern between the different uses of water resources and the maintenance of species and ecosystems, considering the resulting changes and major conflicts in the interface between use and environmental protection (Ferreira & Guerreiro de Brito, 2009; Hall & Härkönen, 2006).

The direct involvement of the local population in the development and promotion of a tourist attraction is crucial for the recognition of collective interest, opportunities to emerge, and participation in decision making. The expansion that tourism has been experiencing in recent decades has required significant changes in the way business is run, including tourist destinations. Involving public/private leaders in the tourism management process promotes tourism development in a more sustainable way, making it more attractive and competitive, generating entrepreneurial initiatives, innovating, and qualifying the tourism destination.

River Tourism and its Importance for Inland Territories in Portugal

According to the World Tourism Organization (2016, p.10) *Rivers provide solitude, beauty, and interesting history that appeal to local recreationists and tourists. There are also a number of significant direct and indirect roles that rivers play, including location for activities and places of tourist interest.* Like other types of nature-based tourism, river tourism is closely related to the principle of living in harmony with nature, integrating with the environment. It represents an important resource for rural regions, generating new demands and activities related to tourism and leisure, revaluing their perception by local communities and promoting investments in their qualification.

In Portugal the importance of water for leisure and tourism and its development has been analyzed and the target of studies from different perspectives and supported by projects that seek understandings and conceptualizations with increasing interest. (Cavaco & Simões 1998; Costa 2010, Sarmento & Lousada, 2010; Moreira & Santos, 2010, 2016; Santos, 2012, 2016, Moreira, 2012, 2018). *River systems, due to their water characteristics, landscape quality, biological resources, environmental richness, and associated heritage, allow the development of various tourist, recreational and leisure activities, such as river beaches.* (Gouveia, 2009, p.73)

In terms of national fluvial tourism, the rivers, lakes, and reservoirs stand out for their significance as spaces for the reception and development of sports, recreational and recreational activities (Moreira & Santos, 2016). There is an investment in the development of infrastructures and equipment that allow the expansion of tourism and leisure and the creation of the necessary conditions for a qualified performance as recreational spaces, entertainment, and differentiated tourist activities.

Leisure, recreational, sports or competitive activities may take place in these spaces, depending on the geomorphological conditions, the characteristics of existing riparian ecosystems, the size of the water bodies and the infrastructure developed. Due to the environmental value of these resources and the ecosystems that represent a latent fragility that requires the proper management and adaptation of spaces to different uses and functions, without compromising the sustainability of ecosystems and user safety. According to Moreira, C. (2018) diverse and sometimes even conflicting coexistence of socio-economic activities with different interests poses new challenges to the management of rivers and inland waters. Today's dynamics are unprecedented, as rivers have never been the subject of such diverse and intense interests and demand as they are. Its impact on waters and surrounding areas increasingly implies greater control, qualification, and regulation, capable of optimizing and adjusting uses and generating social welfare. (Moreira & Santos, 2016; Vaz, 2008).

It should be noted that the activities to be developed can be considered in the function of the physical conditions of the practitioners, the equipment necessary for their practices and the infrastructures created for their performance, as well as the legislation in force for the respective practices and uses and the established in the instruments of applicable spatial planning and management (Cavaco, 2010; Peixoto, 2016). There is a dependent and consequent relationship between the resource and the territory and the surroundings, in particular with the margins (river or reservoir margin) where the impact of human action is exerted on a large scale by the development of investments in equipment and infrastructure, changes in natural conditions, new uses and associated services (LaVanchy, 2017; Mangorrinha & Pinto, 2010). Thus, and especially in the summer period, conditions and means are required to safeguard environmental values and enable practices that are adjusted to existing conditions. (Moreira et al., 2002; Peixoto, 2016).

There is a disturbance in the structure of surface water resources and their margins, including fragmentation and denaturalization of riparian galleries, as a result of human activities as diverse as leisure and tourism, mining, or the use of adjacent areas for agriculture (Ferreira & Guerreiro de Brito, 2009). As referred by Moreira, C. (2018, p.58) *On the banks there is a vast set of heritage, natural and cultural, built and immaterial, but also equipment and infrastructure that enable and enhance many other hard or soft activities.* This relationship between bank and river resource (rivers, pond, reservoir, canal, among others) requires the creation of appropriate conditions that guarantee safe access, the installation of conditions and means of support, and sustainable management. *River beaches can, therefore, be considered, especially in the interior and rural areas, a new specific segment of the tourist market. They can combine several tourist products, as a result of the specific characteristics they present, the purpose they are intended for, the activities they allow to develop, and the environment in which they operate.. (*Gouveia, 2009, p.32).

The recognition of the tourist value of inland waters, in conjunction with recreational and sporting practices, are factors of social and economic dynamism, with special interest for territories with medium mountain specificity. In this context, the allocation of these resources to tourism, in conjunction with local heritage and lifestyles, makes it possible to promote business initiatives and local investment, expanding the global tourist offer and the well-being of communities, in which we can include:

- Increase in hotel and restaurant services;
- Stimulation to the activity of animation and sport;
- Commercial increase associated with recreational and sports activities;
- Increased demand for personal goods and services;
- Encouraging cultural encounters between tourists and residents;
- Infrastructure development and enhancing local welfare;
- Exchange of experiences, knowledge sharing and ways of life.

Water-based experiences are associated with several positive socioeconomic and cultural impacts, and these comprise benefits to self-identity and esteem through the pursuit of fun, adventure, challenge, fulfillment of belonging needs, and others (Jennings, 2007). It becomes relevant to control and regulate activities, regulating access and practices (activities and behaviors), due to the non-massification of spaces, the perpetuation of natural values, the valorization of local identities and their compatibility in different spaces.

For the development of river tourism, it is necessary to ensure that the central government, local governments, the private sector, and non-governmental organizations work harmoniously within the framework of good governance principles. In these processes, fluvial spaces require investment and qualification so that they are recognized as suitable for specific activities and practices (Gorenak &Trdina, 2017). The diversity of recruiting and sporting activities that inland waters propose and the growing demand for experiences close to nature require new positioning and the development of support infrastructures.

River tourism is an alternative type of tourism that includes nature walks and hikes, swimming, and sports activities such as amateur fishing, rafting, canoeing, kayaking, nautical sports, sailing in different formats, padel, windsurfing, among others. In mountain areas, and the particular case of Estrela Geopark, the possibilities are expanded, local communities perceive the potential and interest and public investments gain meaning, to allow satisfactory conditions of access and use.

ESTRELA GEOPARK AND ITS SUSTAINABILITY OBJECTIVES, ENHANCING TOURISM AND QUALIFYING RESOURCES

Geoparks UNESCO World - Concept, Mission and Commitment to Sustainability

The concept of geopark emerged at the end of the 20th century in Europe, with the creation of the European Network of Geoparks. A geopark is a geographically delimited territory with a sustainable development strategy based on the conservation of the geological and geomorphological heritage in association with the other elements of the natural and cultural heritage and local ways of life (Eder & Patzak, 2004; Frey et al, 2016). It seeks, through the protection and enhancement of the geological heritage, to improve the living conditions of the population, promoting, in an integrated way, the endogenous values and their tourist appropriation (Stueve et al., 2002). The Global Network of Geo-parks (GNG) was created in 2004, with the support of UNESCO (United Nations Educational, Scientific and Cultural Organization), incorporating the definition already established by the European Network of Geo-parks

The creation of geoparks constituted a new paradigm in the way in which a territory is valued, with relevant geological interests, and the geosciences, heritage and ways of life of its communities are disseminated. The management strategy of a geopark is not only geological heritage, but also biodiversity, archeology, and other aspects of cultural heritage. In this way the Geosciences gained public visibility and patrimonial dimension (Zouros & Mc Keever, 2008; Farsani et al., 2011). Thus, the geoparks introduce a great responsibility in the creation of economic, touristic, and social value, i.e. their potentialities can be transversal in helping to create territorial value in a certain geographical area (Hose, 2005; UNESCO, 2015). However, their potentialities can be appropriate to tourism based on the construction of development strategies towards the common good of the community and the integrated valorization of the territory, promoting a sense of belonging and collaborative work.

UNESCO Global Geoparks are single, unified geographical areas where sites and landscapes of international geological significance are managed with a holistic concept of protection, education, and sustainable development. A UNESCO Global Geopark is a geographically delimited territory with a sustainable development strategy based on the conservation of the geological and geomorphological heritage in association with the other elements of the natural and cultural heritage and local ways of life (Eder & Patzak, 2004).

A UNESCO Global Geopark uses its geological heritage, in connection with all other aspects of the area's natural and cultural heritage, to enhance awareness and understanding of key issues facing society, such as using our earth's resources sustainably, mitigating the effects of climate change and reducing natural disasters-related risks. By raising awareness of the importance of the area's geological heritage in history and society today, UNESCO Global Geoparks give local people a sense of pride in their region and strengthen their identification with the area. The creation of innovative local enterprises, new jobs, and high-quality training courses is stimulated as new sources of revenue are generated through geotourism, while the geological resources of the area are protected (Frey *et al.*, 2016).

A Geopark is a well-defined territory, holding a remarkable geological heritage combined with a sustainable development strategy, whose main pillars are Geoconservation, Education for Sustainable Development, and Tourism. Among its objectives, can be included the construction of new infrastructures that promote the conservation of geological heritage, education, and tourism; the development of new local products and services; the encouragement of crafts and local economic growth, and thus the creation of new employment opportunities. Given the three main areas of operation of a Geopark: Conservation of the Geological Heritage, education for Sustainability, Tourism and Local Development; the Global Geopark Network, argues that a geopark should also enhance, promote and preserve Biodiversity, Cultural Heritage, the Gastronomy and Scientific Research (Farsani et al., 2012; Miśkiewicz, 2016).

As UNESCO Global Geoparks are eminently territorial, these are based on territorial approaches such as development, preservation, and promotion of biodiversity, cultural heritage, geology, gastronomy, and scientific research. The apex of action of these spaces are the conservation of geological heritage, the education for sustainability, and therefore a tourist development of the areas where they operate. It is important to note that upstream to a tourism strategy is a regional strategy based on the use of local resources.

Estrela Geoparks and its Contributions to the Geoconservation and Promotion of Sustainable Tourism

Serra da Estrela presents a territory, geographically defined, with geological and geomorphological interest resources, not always duly valorized, based on glacial, periglacial, and granitic forms whose singularity and heritage value bring to this territory an eco-cultural dimension of great relevance that constitutes a "living laboratory" in which the natural resources, the ways of life of the population and the biological and geological diversity blend into cultural landscapes. The Serra da Estrela with 1993 m is the highest mountain range in mainland Portugal and is part of the Iberian Central Cordillera. Bounded by fault scarps, a granite massif occupies the central area forming a summit plateau between ci. 1400 and 2000 m, while surrounding the core area, there is an interplay with schists and greywackes. During the Last Glacial a plateau ice-field and five radiating valley glaciers occupied the highest parts of the mountain with an estimated equilibrium line altitude at 1,650 m asl (Vieira, 2004). The plateau style of the glaciation and the ELA just below the plateau edge made the Estrela very sensitive to climate fluctuations, which has resulted in several terminal moraine complexes that reveal several glacial stages. The central plateau area shows widespread glacial erosion features and an almost complete stripping of the Cenozoic weathering mantle (Vieira, 2004). The non-glaciated plateaus show a rich landscape dominated by granite weathering landforms. The remarkable glacial landscape of the Serra da Estrela when considering its setting in SW Europe, together with other relevant geoheritage such as periglacial, weathering and mass wasting phenomena, tectonic, petrological and hydrogeological features, are at the core of Estrela's UNESCO Global Geopark (Castro et al., 2017)

The Estrela Geopark integrates the Serra da Estrela, from its southwestern limit on the border with the Açor Mountains to the northeast contact with the Iberian Meseta surface, including also the northwest and southeast foothills, where there are millennia, humans live in close relationship with the mountain and what it has to offer. Serra da Estrela is the highest mountain in mainland Portugal, reaching 1993 meters in the Alto da Torre (Figure 1). However, the territory of the Estrela Geopark is a larger area, with its boundaries outlined to include the main geological elements that contributed to the current landscape (Fernandes *et al.*, 2016; Gomes *et al.*, 2017)

The geopark emphasizes the development of a collective strategy, reconciling environmental, social, and economic responsibilities, which articulate for the first time all municipalities of Serra da Estrela are engaged in value action, conservation, and territorial promotion (Castro et al., 2017; Ferrnandes, 2016).

Figure 1. Estrela geopark location
Source: Source: AGE -Associação Geopark Estrela, improved by authors

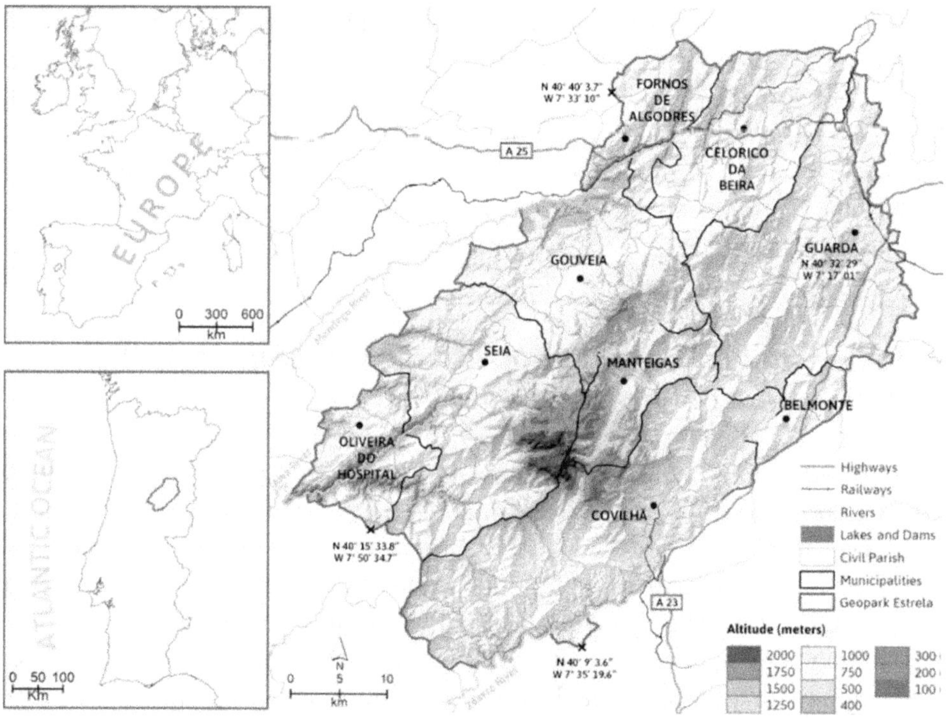

SURFACE WATER IN THE TERRITORY OF ESTRELA GEOPARK AND ASSOCIATED PRODUCTS AND SERVICES

The Serra da Estrela is the main portuguese mountain and its geological, geomorphological, hydrofluvial, biological resources, ecological and landscape and climatic organization give it meaning and its own dynamics. These resources allow the construction of a different tourist offer, either at different times of the year, either in the practices developed or in the profile of tourists. This diversity allows to promote a multifaceted and diversified destination, in the potentialities and opportunities to consider, for the tourism and leisure activities to develop. In this context, we can consider a set of tourism products and activities, which are articulated around resources that, while not being exclusive, assume greater or lesser expression depending on the specificities of the form or the way it is practiced. Thus, we will be able to establish the following tourist products and activities associated with superficial water resources: Rural Tourism, Geotourism, Ecocultural Tourism, Adventure Tourism, Tourism. Scientific, River Tourism, Health, and Wellness Tourism.

The surface waters of the Estrela Geopark territory are composed of river, lakes, and reservoirs, corresponding in total to a surface divided by 53.7 km2 of retained waters (Lagos, reservoirs and RAMSAR site) and 826.8 km of running water, corresponding to rivers, their tributaries and sub-tributaries (Figure 1 and Table 1). The diversity of water lines, and the ecosystems they integrate, together with reservoirs and lagoons, provide the territory with strong potential for the development of water-based tourism and leisure activities.

Figure 2. Map of water resources in the Estrela Geopark territory
Source: AGE - Associação Geopark Estrela

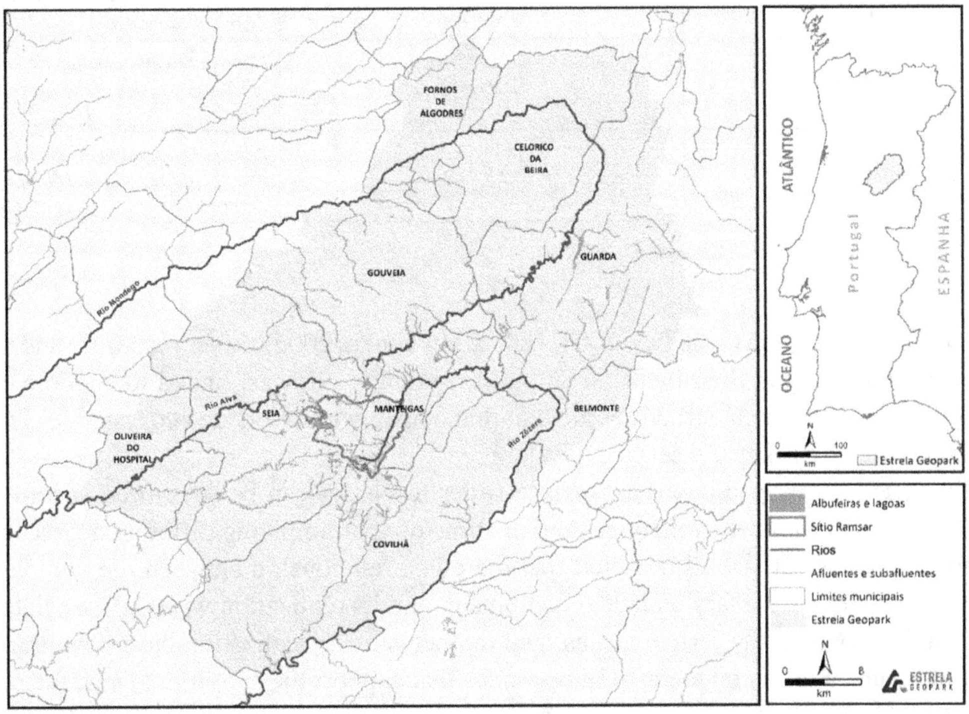

The main activities carried out on these water surfaces, given their characteristics in terms of geomorphology, climate, ecosystems present, and recreational/sports interests of local communities and tourists, can be systematized in Table 2, according to the respective water resources, uses and associated practices and existing restrictions. It appears that it is the practices related to bathing and recreational activities (near water surfaces) that are the most representative. The ease of access and the direct use of nature, particularly in the summer season, together with its reduced demand for equipment or technical support for its realization, enhance its recreational use by local communities and visitors. In this context, the river beaches represent a set of spaces that have been gaining meaning and notoriety, such as leisure equipment, constituting a value offer for the region and its communities, particularly during the summer. In recent years there has been an increase in "new" activities in the region and on surface waters, which imply greater organization and equipment. Of note, on the one hand, is the growth in sport fishing activities, due to the natural conditions and the existence of salmonid populations that attracts practitioners at national and international level. On the other hand, canoeing and padel gaining fans, increasing the attractiveness of the region, for which the water quality and the landscape are decisive arguments. The qualification and expansion of river beaches, as leisure facilities for local communities and visitors, constitute arguments for the qualification of the destination and the expansion of the offer of bathing spaces in mountain regions, conquering a greater number of users and reinforcing the image of nature associated with the destination. The investments made, the qualifications and the provision of services promoted, are perceived to be of value to local communities and to attract visitors. The connection between the community, territory, and tourism is a fundamental process for the promotion of

Table 1. Surface water resources, extension and associated watersheds

Water resources	*Surface / extension*	*Watersheds*
Rivers – Zezere, Mondego, Alva	*826,8 km*	*Tejo and Mondego*
Lakes and dams	*2,9 km2*	*Natural or embalmed lakes in depressions of sub-tributaries of the Zezere and Mondego*
RAMSAR (Zone)	*50,8 km2*	*Zone corresponding to the height Plateau of Serra da Estrela and Upper part of the Zêzere River*

Source: AGE - Associação Geopark Estrela

tourism in rural areas (Garcia Gonzales, 2004; Hall & Harkonen, 2006). River resources and the heritage associated with them, along their banks or in the surrounding villages, linked to economic activities, mobility, or water management structures are simultaneously with riverside ecosystems, elements with the potential to affect tourism and service development.

Tourist products developed based on surface water resources can be structured according to their intensity, the means used for their practice, and the time of year, implying different degrees of physical intensity and regulation. It should be noted that there is a relationship between the complexity of the activity carried out, the necessary technical equipment, and its realization with the associated levels of regulation. In this context, the forms, means, and moments of its realization have been translated into regulation, to reduce impacts on the natural resources that support the activities. Thus, the plans for the organization of dams and lakes and the regulations for activities of animation and sport, are established by the ICNF, based on the principles of organization enshrined in the national planning policy and the guidelines and limitations established in the Serra da Estrela Natural Park plan. However, there is currently reduced signage and information, which allows a better knowledge of the potentialities, the requirements (technical and legal) for their practice, and the adequacy of means that allows the application of the various practices and their proper monitoring.

Given the growing demand for the mountain destination, the development of various tourist products and the perception of the Serra da Estrela as a destination for sports and recreational practices associated with water, it becomes relevant to promote greater attention from local authorities on the potential of these demands and encourage the community to develop services associated with different practices. In this context, it is necessary to consider different types of entrepreneurs, depending on their nature, their positioning in the market, and their technical and material capacities. In terms of the dynamics of rural areas and the problems that face the OECD (2005) identified three specific challenges currently affecting rural areas, which have implications for entrepreneurship, namely:

- Declining employment opportunities in primary industries (mainly agriculture), as a result of structural change, intensified by changes in policy resulting from reviews of the European policies. This emphasizes the need to take steps to stimulate economic activity with employment-generating potential in rural areas;
- An aging population, associated with an outmigration of young people and an in-migration of retired people, in some cases, which in combination affects the supply of potential entrepreneurs;

Table 2. Surface water resources and associated recreational and sport uses

Water resources	*Entertainment uses and sports*	*Restrictions*
Rivers: Zezere and tributaries Mondego and tributaries Alva and tributaries	Recreational (contemplation and picnics) Bathing, Canoeing Rafting Playful and sport fishing	- Only river beaches are properly trained for bathing practices, with means of support, access, and control. - Sportfishing requires licenses for its practice due to its performance in rivers or the protected fishing zone of Serra da Estrela
Dams: Lagoa Comprida; Lagoa Escura; Lagoa do Vale Rossim; Lagoa do Viriato; Lagoa do Covão de Ferro; Albufeira da Barragem do Lagoacho; Lagoa do Covão do Forno; Poços de Loriga; Lagoa Seca Lagoa dos Conchos; Lagoa Redonda; Lagoa da Erva da Fome Lagoa do Covão do Curral; Lagoa do Vale do Conde; Lagoa dos Cântaros	Recreational (contemplation and picnics) Bathing Canoeing Windsurf and Padel Playful and sport fishing Scientific Activities	- Bathing practices limited to lagoons - Use of propelled vessels conditioned; - Fishing subject to a specific calendar and obtaining special licenses for daily fishing in the protected fishing zone in the lagoons of Serra da Estrela
RAMSE zone – Protection zone on the Torre plateau relative to wet areas and small natural lakes	Walking and observation Bathing Scientific activities	Area of great environmental sensitivity, with human activities severely limited in this area of maximum restrictions under the PNSE regulations

Source: Authors elaboration

- Difficulties in maintaining a critical mass of facilities to support economic development, including a range of business services

In this context entrepreneurs play an important role in sustaining rural tourism and the formulation of sustainable strategies being the initiators of the tourism business and the engine of the local development. (Sharif & Ku Azam, 2017). These initiatives can contribute to mitigating the demographic trends cur-

Figure 3. Typologies of tourist and recreational activities in inland waters and physical and regulatory requirements
Source: Authors elaboration

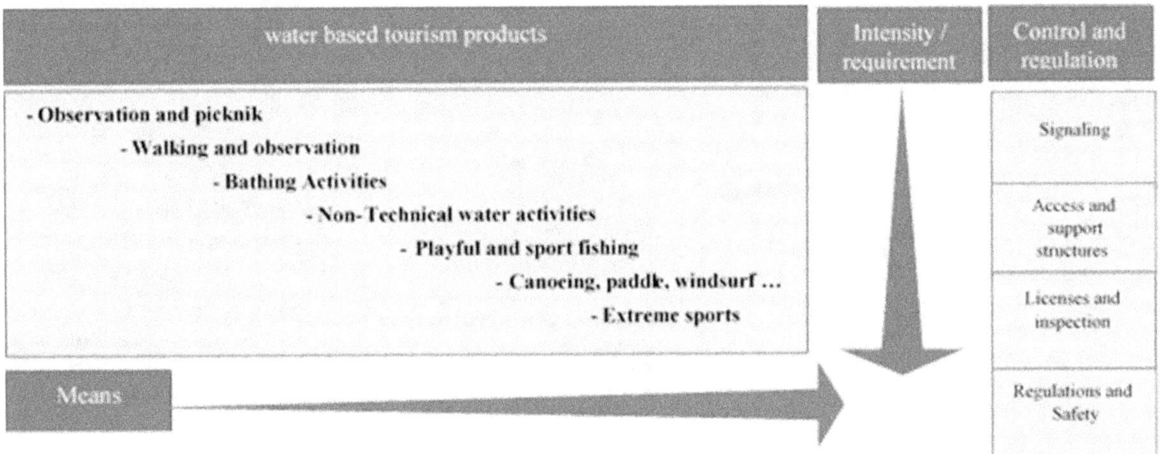

rently observed. Akama and Keiti (2007) and Fernandes and Natariio (2020), proposed several tourism development strategies which aim to achieve sustainable community development. Among them are:

- Increase local employment and self-employment;
- Development opportunities in the informal sector;
- The development of partnerships among public and private sector, NGOs and local communities;
- Improving the social and cultural impact;
- Improve access to infrastructure and services provided for the convenience of tourists;
- Increase the participation of local communities in decision-making;
- Building capacity to promote public participation.

According to Dimitrovski et al., (2012), the rural areas have a unique opportunity to attract tourists through establishing a connection between rural areas and their cultural, historic, ethnic, and geographical roots. The appropriate policy needs to create an entrepreneurial environment that is comfortable for other potential entrepreneurs to start new businesses and simultaneously to provide a process of suitability, between the resources and local communities The development of entrepreneurship acts as a new venture as well as contribute to the development of a destination by creating competition, cooperation, specialization, innovation, investments (Lordkipanidze et. al., 2005) and risk-taking (Bula, 2012).

Figure 4. Collaborative processes of entrepreneurship - public and private - on water-based tourism products
Source: Authors elaboration

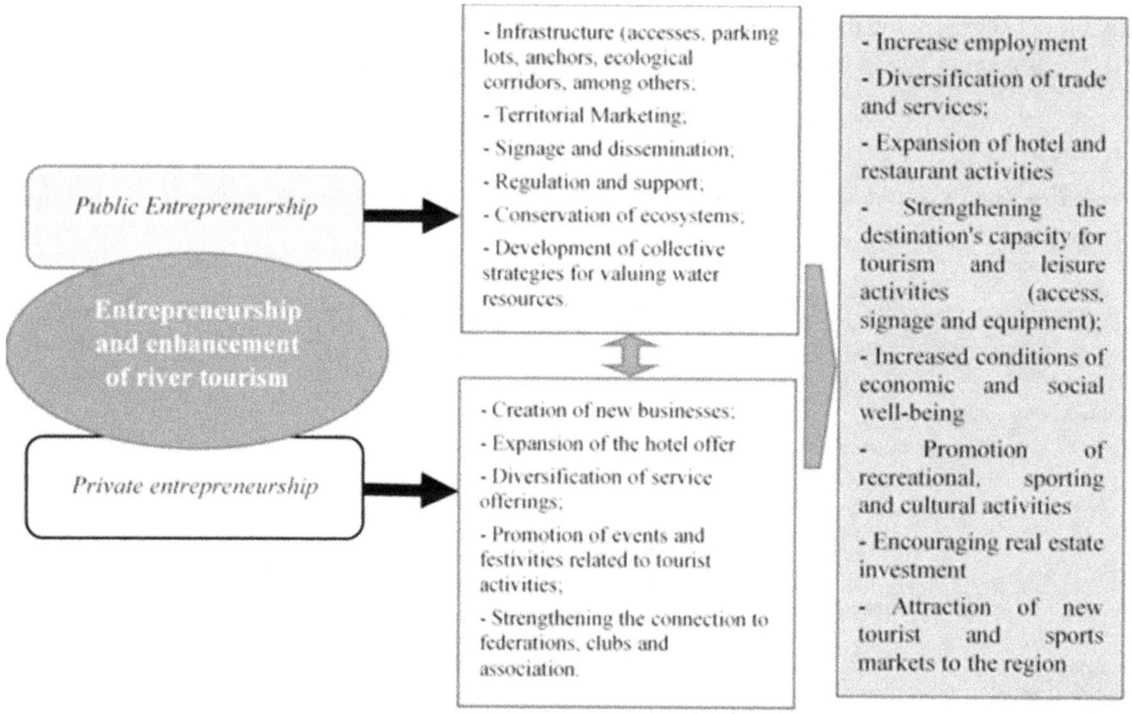

One of the potential disadvantages faced by rural firms in comparison with their urban-based counterparts is the small size of the local market that is typical, associated with low population densities, together with low per capita and household incomes (Smallbone et al., 2005).

Entrepreneurs have a key role to play in contributing to a region's competitiveness by creating new products and services and bringing them to market and keeping one step ahead of their competitors (Norhafiza & Ku Azam, 2017). As a consequence, the performance of (rural) regions with regards to both innovation and entrepreneurship becomes critical to their future development

To achieve the aspiration of sustainable tourism development, it must be managed wisely and require a strong commitment from the government (Lordikipanize et al., 2005; Norhafiza & Ku Azam, 2017) non-governmental organizations and the local community itself to boost social development and generate economic opportunities that are more interesting, in addition to improving the transportation infrastructure in the tourist area to improve the living conditions of local communities (Blancas et al., 2011). In this context, the cooperation and articulation of initiatives (between the public and private sectors) proves to be decisive, in the training of tourist destinations for new tourist practices and the rational and regulated use of the water resource, considering its various functions and weaknesses. The bathing, sporting, recreational, scientific, and cultural potential, should promote differentiated tourist products, leveraged by public actions, and encourage the expansion of economic activities for the community.

CASE STUDY: RIVER BEACHES IN THE TERRITORY OF ESTRELA GEOPARK

The concept of a river beach as a bathing area is relatively recent, but it has been a long time since men have enjoyed the river and its banks for recreation and leisure (family picnics, river bathing, fishing, boating, etc.) (Gouveia, 2009). River beaches can thus be considered, especially in the interior and rural areas, a tourism segment, which combines beach tourism with the three strategic tourism products for the interior areas, namely cultural and landscape touring, nature and gastronomy tourism and wine. (Goldstein, 2011)

According to Goldstein, S. (2011) The development of river beaches was accomplished through the creation of access, parking, support structures (such as snack bar, kiosks selling food and beverages, first aid, grill and camping areas), surveillance of beaches by lifeguard, signage of uses, among others, enhancing the process of sustainable development of regions with good tourism potential.

The appreciation of the water resources of the Estrela Geoparque territory, due to their diversity, environmental value, landscape, and qualifications operated, allow to create conditions for the creation of high-quality bathing spaces. In this way, leisure practices are enhanced with the local communities and at the same time attracting visitors, enhancing the Serra da Estrela tourist destination, and the creation of support services.

In the Estrela Geopark territory we can count globally a set of 38 recreation and recreation spaces supported in inland waters, which can be differentiated in praises Bathing areas and Leisure areas. The identification of inland bathing waters and the classification of fluvial and lacustrine beaches as bathing beaches in the continental territory is defined for each year by ordinance of the competent Ministry (Law No. 44/2004 of 19 August), which identifies for each beach is suitable for bathing and its bathing season (period, fixed annually by administrative determination of the competent authority, during which the obligation to guarantee the assistance to bathers is in force). Recreational and leisure bathing areas are not considered classified beaches, according to Ordinance No. 178/2013, so their use, at the level of bathing, is the sole responsibility of the users. These areas are not monitored and the quality of their waters is not assigned any classification by the Portuguese Environment Agency, IP.

The river beaches of the Estrela Geopark territory (in a total of 11) classified in 2019 (Ordinance No. 141/2019 of 14 May), are divided into 6 municipalities, being in the municipalities of Seia and Oliveira do Hospital where it has the highest expression. Unhais da Serra Valhelhas; Aldeia Viçosa Relva da Reboleira São Gião; São sebastião da Faia; Alvoco da Várzeas e Avô Vila Cova à Coelheira, Loriga, Lapa dos Dinheiros

Table 3. Location of river beaches, bathing areas and recreation areas in the territory Estrela Geopark

Municipalities	Classified Fluvial beach - 2019	Fluvial Beaches in the Territory Estrela Geoaprk	Total
Belmonte		1	1
Celorico da Beira		2	2
Covilhã	1	6	7
Fornos de Algodres		1	1
Guarda	2		5
Gouveia		1	2
Manteigas	1	3	4
Oliveira do Hospital	4	5	8
Seia	3	2	9
Total	11	21	41

Source: Portaria n.º 141/2019 de 14 de maio (DR, 1.ª série -N.º 92) and inventariatio AGE Associação Geopark Estrela

It should be noted that the greater the number of river beaches and swimmers who frequent them, the greater the requirement for the management and control of activities practiced in water bodies and associated equipment. In this context, the responsibility of public agents to monitor and manage bathing areas is crucial, keeping the health risk low in these recreation areas. (Goldstein, S., 2011).

Of the 11 existing river beaches (Table 1), most are managed by public entities, namely parish councils, which represent 75% of the forms of management, with the remainder being municipalities and a beach under a foundation. It should be noted that overall there has been a significant improvement and appreciation of the conditions and equipment available, promoting the qualification of these bathing spaces for wider use, with better access conditions, security, and associated services (Figure 5).

There is a concern for all these spaces with the issues related to the disposal of garbage dumps and its recycling, with the conditions that allow the attendance of publics with reduced mobility, for which access and ramps that allow to enjoy the river beach, the existence of recreation areas associated with specific areas for fire, food preparation, and clean water. In all fluvial beaches there are and are identified parking areas, with non-pedestrian access to the beach very conditioned or only possible by bicycle. It should be noted that in more than 66% of beaches there are camping areas (integrated or adjacent), allowing the overnight stay or stay, thus conditioning the setting of tents in the specific river beach area (Figure 6).

There has been an effective concern for signaling the different spaces associated with river beaches, leading to investments and the development of means that enable better access, identification of differ-

Figure 5. Services held by the river beach/bathing área
Source: Survey on river beach management/concessionaires

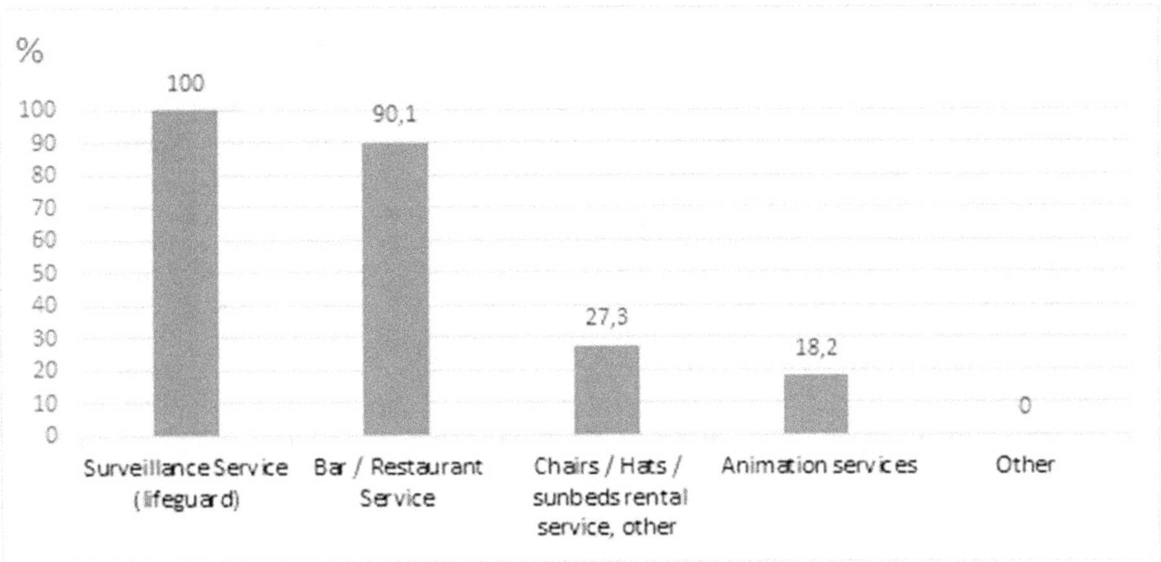

ent beach areas and functions, existing services, rules, and hours of operation and information about the water quality. In this context, there are still some gaps in beaches, related to the signaling/identification of activities not allowed, danger zone signaling, as well as the provision of proximity services and local heritage (Figure 7).

Figure 6. Features of River beach: Access and constraints
Source: Survey on river beach management/concessionaires

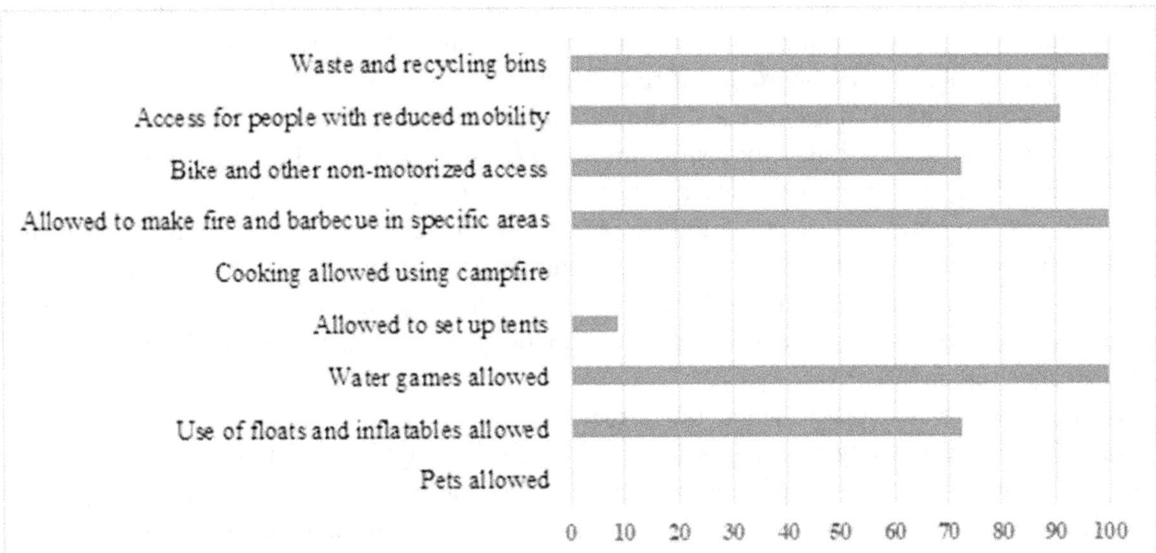

Figure 7. Signals supporting the operation of the river beach
Source: Survey on river beach management/concessionaires

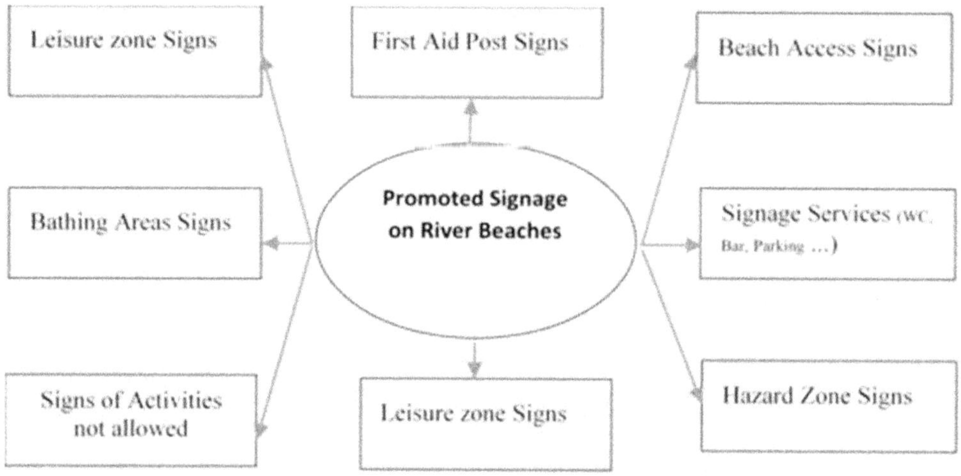

In terms of the contributions that river beaches make to the territory and the respective local communities, there is a strong appreciation/evaluation and a very strong impact on the attraction of tourists, on the improvement of the resident population's quality of life, through the availability of this equipment, presence of larger numbers of visitors and expansion of commerce and services activities. Also in the image of the community and their self-esteem is perceived the importance of river beaches and consequently the disclosure that they associate, in terms of environmental quality, heritage value, and space of tranquility and quality of life.

In general, the river beaches verify an evaluation of very strong importance for the tourist valuation of the Serra da Estrela destination, for the attraction of tourists to the region, for the promotion of experiences, foster a direct relationship with nature and expand the playful conditions for the well-being of local communities (Figure 8). The promotion and dissemination of these tourist facilities, their setting in contexts of relevant natural space, and their contribution to minimizing the attraction effect of the region related to snow (seasonality) and adventure sports, contribute to the valuation as a tourist asset. There is a recognition of the importance of river beaches as a tourist resource to be boosted, broadening the interaction with local heritage and communities, boosting new businesses, and simultaneously constituting a social responsibility for controlling water quality and the activities developed.

The local administrative entities, have made efforts and investments in qualification and dissemination, promoting improvements in accessibility, parking, and increment of services. Local communities recognize its importance and the potential for use outside the summer season, for activities related to sports and outdoor recreation. Some of these structures support during the summer activities of animation, such as concerts and thematic festivities, dissemination of local cuisine, local heritage, and thematic routes over the region.

Water, as well as associated aquatic and terrestrial ecosystems, tend to be subject to multiple uses by a set of users with widely differing and sometimes conflicting interests. In their appropriation the social, economic, and environmental interests tend to be divergent, implying the existence of principles and guidelines that promote their proper regulation, establishing limits and conditions to the developed

Figure 8. Evaluation of the Importance of fluvial beaches for the Community and Destination (assessments of importance according to their impact)
Source: Survey on river beach management/concessionaires

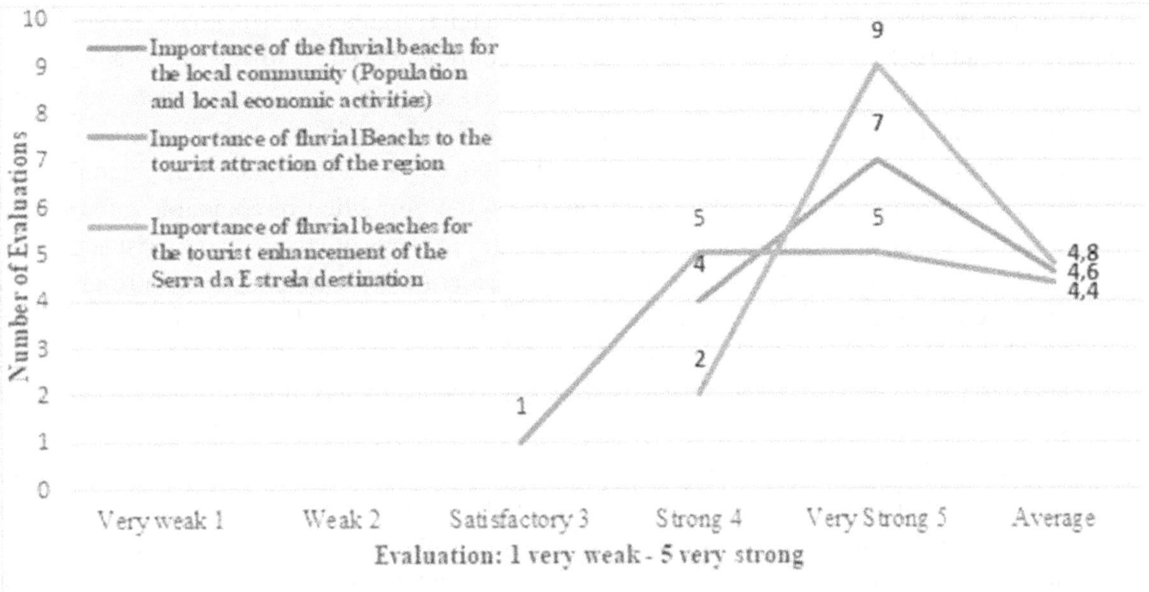

practices and uses. In parallel with the increase in water consumption by various activities and economic sectors, there has been a growing recognition of the value of wetlands and surface waters for tourism and leisure activities (Costa, 2010; Ferreira & Guerreiro de Brito 2009), which record new searches, in particular associated with its bathing, sporting and recreational potential.

Local communities must have a growing involvement in water-based tourism activities, proactive and enterprising intervention, stimulating these activities, and at the same time having a responsible civic attitude, capable of controlling and alerting to situations of incorrect use or practice. Recognizing that the provision of support equipment and services serves to improve the quality of life and economy of local populations.

CONCLUSION

The surface water resources (rivers, lakes, reservoirs, dams, marshes) and the practices associated with them, established relationships with the territory and with the riverside communities that make them a relevant resource from tourism.

The growth of river tourism and leisure activities, particularly active in inland waters, is currently relevant, which advises that inventories of the potential held should be invented, so that the existing and potential suppliers can be known and monitored to better manage it. the different markets looking for this destination.

The diversity of practices and activities that allow development is appreciable, as well as the set of elements that are associated with them and that, must be activated for the benefit of communities and tourism.

Tourism development is vital to the rural communities are not marginalized than urban communities. It is important to ensure the sustainability of the future rural community. The involvement of entrepreneurs in promoting and stimulating tourism in rural areas will provide new perspectives to local communities through a greater openness and interest on their part for the effective recovery of tourist resources. Local administrative entities have a relevant role in the qualification, conservation, and dissemination of these resources, encouraging the involvement of communities in the promotion of economic initiatives and the construction of the image of the destination and its values in terms of leisure and tourist activities. It should be noted that the qualification of rivers, lakes, dams, among other water spaces for river beaches, has been promoted particularly by local administrative entities, creating infrastructures and equipment that promote the best conditions of access and use.

Documenting good practice in river tourism is also critical and should be widely disseminated among relevant stakeholders, fostering entrepreneurial initiatives and the promotion of diverse activities that foster the economy, particularly in inland territories

The image generated and the appreciation given by tourists and visitors of these leisure spaces has broadened their consideration in local tourism strategies and provided more opportunities for these territories. The classification of Serra da Estrela as a Geopark, enhances the notoriety of the destination and recognizes its geoheritage importance, induced growth in demand, and greater responsibility in its management. The quality of the water, the natural value, and the investments made, conciliate collective interests for a progressively more consolidated, collaborative and proactive offer for the region.

REFERENCES

Akama, J. S., & Keiti, D. (2007). Tourism and Socio-Economic Development in Developing Countries: A Case Study of Mombasa Resort in Kenya. *Journal of Sustainable Tourism*, *15*(6), 35–748. doi:10.2167/jost543.0

Blancas, F. J., Oyola, M. L., González, M., Guerrero, F. M., & Caballero, R. (2011). How to Use Sustainability Indicators for Tourism Planning: The Case of Rural Tourism in Andalusia (Spain). *The Science of the Total Environment*, *412*(413), 28–45. doi:10.1016/j.scitotenv.2011.09.066 PMID:22033358

Brito-Henriques, E., Sarmento, J., & Lousada, M. A. (2010). When water meets tourism: An introduction. In E. Brito-Henriques, J. Sarmento, & M. A. Lousada (Eds.), *Water and tourism. Resources management, planning and sustainability* (pp. 13–33). CEG, Universidade de Lisboa.

Bula, H. O. (2012). Evolution and Theories of Entrepreneurship: A Critical Review on the Kenyan Perspective. *International Journal of Business and Commerce*, *1*(11), 81–96.

Castro, E., Vieira, G., Loureiro, F., Patrocínio, F., Firmino, G., Gomes, H. & Fernandes, M. (2017). Managing the Estrela Aspiring Geopark: A framework for promoting geoheritage conservation. *Conference Ciência 2017.*

Cavaco, C. (2010). Água doce: Agricultura versus lazeres e turismo. In E. Brito-Henriques, J. Sarmento, & M. A. Lousada (Eds.), *Water and tourism. Resources management, planning and sustainability* (pp. 223–259). CEG, Universidade de Lisboa.

Cavaco, C., & Simões, J. (1998). *Água, desenvolvimento e bem-estar*. MADRP.

Cole, S., & Ferguson, L. (2015). Towards a gendered political economy of water and tourism. *Tourism Geographies, 17*(4), 511–528. doi:10.1080/14616688.2015.1065509

Costa, C. (2010). O turismo e a água. In CCDRC (Ed.), À beira da água (pp. 261-295). Coimbra: CCDRC.

Dimitrovski, D., Todorovic, A., & Valjarevic, D. (2012). Rural tourism and regional development: Case study of Development of rural tourism in the region of Gruza, Serbia. *Procedia Environmental Sciences, 14*(1), 288–297. doi:10.1016/j.proenv.2012.03.028

Eder, W., & Patzak, M. (2004). Geoparks – geological attractions: A tool for public education, recreation and sustainable economic development. *AUC Geographica, 51*(2), 235–246.

Fachrudin, H. T., & Lubis, M. D. (2016). Planning for Riverside Area as Water Tourism Destination to Improve Quality of Life Local Residents, Case Study: Batuan – Sikambing River, Medan, Indonesia. Procedia - Social and Behavioral Sciences, 234, 434 – 441.

Farsani, N. T., Celeste Coelho, C., Carlos Costa, C., & Carlos Neto, C. (Eds.). (2011). *Geoparks and Geotourism. New Approaches to Sustainability for the 21st Century*. Universal publishers.

Farsani, N. T., Coelho, C., & Costa, C. (2012). Geotourism and geoparks as gateways to sociocultural sustainability in Qeshm rural areas, Iran. *Asia Pacific Journal of Tourism Research, 17*(1), 30–48. doi :10.1080/10941665.2011.610145

Fernandes, G., Castro, E., Firmino, G., & Ferreira, A. (2016). Geoparks and regional and local development: aspiring Estrela Geopark. *European Geoparks Conference*. DOI: 10.13140/RG.2.2.28770.94405

Fernandes, G. P., & Natário, M. M. (2020). Entrepreneurship in Border Territories: Context Costs, Resilience, and Logics of Action of the Companies and Entrepreneurs – Study of the Border Municipalities of Beira Interior in Portugal and Salamanca in Spain. In J. Dantas & L. Carvalho (Eds.), *Handbook of Research on Approaches to Alternative Entrepreneurship Opportunities* (pp. 129–152). IGI Global. doi:10.4018/978-1-7998-1981-3.ch007

Fernandes, S. (2003). *Contributo para a Requalificação Biofísica da Praia Fluvial da Valeta (Arco de Valdevez). Relatório Final de Curso. Bacharelato em Engenharia doAmbiente e dos Recursos Naturais*. Instituto Politécnico de Viana do Castelo. EscolaSuperior Agrária de Ponte de Lima.

Ferreira, M. T., & Guerreiro de Brito, A. (2009). Águas interiores supericiais. In H. Pereira, T. Domingos, L. Vicente, & V. Proença (Eds.), *Ecossistemas e Bem-Estar Humano em Portugal*. Fundação da Faculdade de Ciências da U. L. e Escolar Editora.

Folgado-Fernández, J., Di-Clemente, E., Hernández-Mogollón, J., & Campón-Cerro, A. (2019). Water Tourism: A New Strategy for the Sustainable Management of Water-Based Ecosystems and Landscapes in Extremadura (Spain). *Land Journal, 8*(1), 2. doi:10.3390/land8010002

Frey, M. L., Schafer, K., Buchel, G., & Patzak, M. (2016). Geoparks: A Regional European and Global Policy. Geotourism. Doi:10.1016/B978-0-7506-6215-4.50014-2

García González, L. (2004). Agua y turismo. Nuevos usos de los recursos hídricos en la Península Ibérica. Enfoque integral. *Boletín de la AGE, 37*, 239–255.

Goldstein, S. (2011). *Caracterização ambiental de praias fluviais de Portugal Continental*. Tese de Mestrado, Faculdade de Ciencias/Universidade de Lisboa.

Gomes, H., Fernandes, M., Castro, E., & Vieira, G. &. (2017). The (Serra da) Estrela Aspiring Geopark (Portugal): preserving geoheritage, while promoting science and its links to local communities. In *Geophysical Research Abstracts* (Vol. 19). EGU General Assembly.

Gorenak, M., & Trdina, A. (2017). *Tourism and development 2017: Active & Sports Tourism: Feel the freedom of the water, Book of Abstracts*. University of Maribor Press., doi:10.18690/978-961-286-121-6

Gouveia, V. (2009). *Contributo das praias fluviais para o desenvolvimento regional: a rede de praias fluviais do pinhal interior. Tese de Mestrado*. FCT/UNL.

Grössling, S., Hall, M., & Scott, D. (2015). *Tourism and Water*. Channel View Publications. doi:10.21832/9781845415006

Hall, C., & Härkönen, T. (2006). Lake tourism: an introduction to lacustrine tourism systems. In Lake Tourism: An Integrated Approach to Lacustrine Tourism Systems (pp. 3-26). Channel View Publications. doi:10.21832/9781845410421-004

Hose, T. A. (2005). Geoturismo europeo. Interpretacíon geológica y promoción de la conservación geológica para turistas. In D. Barretino, W. A. P. Winbledon, & E. Gallego (Eds.), *Patrimonio geológico: conservación y gestión*. Istituto Tecnológico Geomineiro de España.

Jennings, G. (2007). *Water-Based Tourism, Sport, Leisure, and Recreation Experiences*. Butterworth-Heinemann.

LaVanchy, G. (2017). When wells run dry: Water and tourism in Nicaragua. *Annals of Tourism Research, 2017*(64), 37–50. doi:10.1016/j.annals.2017.02.006

Lordikipanidze, M., Brezet, H., & Backman, M. (2005). The entrepreneurship factor in sustainable tourism development. *Journal of Cleaner Production, 13*(8), 787–798. doi:10.1016/j.jclepro.2004.02.043

Mangorrinha, J., & Pinto, H. (2010). Water and tourism. Culture heritage and new business opportunities – the legacy of portuguese thermalism. In E. Brito-Henriques, J. Sarmento, & M. A. Lousada (Eds.), *Water and Tourism. Resources management, planning, and sustainability* (pp. 93–106). CEG, Universidade de Lisboa.

Ministério do Ambiente (MA). (1997). *Programa de Valorização das Praias Fluviais*. SERN e IA.

Miśkiewicz, K. (2016). Promoting geoheritage in geoparks as an element of educational tourism. In *Geotourism: organization of the tourism and education in the geoparks in the Middle-Europe Mountains*. University of Business in Wrocław.

Morar, C., & Pop, A. C. (2016). Water, Tourism and Sport. A conceptual approach. *Geo Journal of Tourism and Geosites*, *9*(2), 249–258.

Moreira, C. (2011). Ambientes aquáticos não marítimos no Baixo Mondego: Margens para a recreação, o lazer e o turismo. In *Trunfos de uma geografia activa: Desenvolvimento local, ambiente, ordenamento e tecnologia* (pp. 149–158). IUC. doi:10.14195/978-989-26-0244-8_15

Moreira, C. (2012). Lazer, animação turística e desenvolvimento local. In *Patrimónios, territórios e turismo cultural: Recursos, estratégias e práticas, Coleção Iberografias, 19* (pp. 77–103). CEI, Âncora Editora.

Moreira, C. (2018). *Turismo fluvial, lazeres em águas interiores e desenvolvimento local e regional/River tourism, leisure in inland waterways and local and regional development in Cadernos de Geografia, 18.* CEGOT, Universidade de Coimbra.

Moreira, C., & Santos, N. (2010). New opportunities for water environments: River tourism and water leisure activities. In E. Brito-Henriques, J. Sarmento, & M. A. Lousada (Eds.), *Water and tourism. Resources management, planning and sustainability* (pp. 147–168). CEG, Universidade de Lisboa.

Moreira, C., & Santos, N. (2016). Turismo fluvial em portugal continental: oferta e potencialidades/ river tourism in portugal: touristic offer and potentials. In Territórios de Água | Water Territories, (pp. 255-271). CEGOT, Centro de Estudos de Geografia e Ordenamento do Território, Faculdade de Letras da Universidade de Coimbra.

Moreira, I., Ferreira, M., Cortes, R., Almeida, P. A., & Pinto, P. (Eds.). (2002). Ecossistemas Aquáticos e Ribeirinhos. In Ecologia, Gestão, Conservação. Edições do Instituto da Água.

Norhafiza, M.S., & Ku Azam, T.L. (2017). Sustaining the Entrepreneurship in Rural Tourism Development. *International Journal of Multicultural and Multireligious Understanding*. . doi:10.18415/ijmmu.v4i6.122

OECD. (2005). *Place-Based Policies for Rural Development, Crete, Greece (Case study). Working Party on Territorial Policy in Rural Areas, Public Governance and Territorial Development Directorate*. Territorial Development Policy Committee, OECD.

Peixoto, P. (2016). Os usos sociais dos rios. In P. Peixoto & J. P. Cardielos (Orgs.), *A água como património. Experiências de requalificação das cidades com água e das paisagens fluviais,* (pp. 57-70). Coimbra: IUC.

Prideaux, B., Tomothy, D. J., & Cooper, M. (2009). Introducing River Tourism – Physical, Ecological, and Human Aspects. In River Tourism (pp. 1-22). CABI Publishing, International.

Rodrigues, A., & Rodrigues, A. (2007). *Rede Europeia de Turismo de Aldeia, um Casoe inovação em Turismo rural*. Congresso Internacional de Turismo, Leiria-Oeste.

Santos, N. (2012). A gestão das bacias hidrográficas e a valorização turística dos ambientes fluviais: O rio Mondego. *Revista Geonorte*, *4*(4), 210–219.

Santos, N., & Cunha, L. (2008). Novas oportunidades para o espaço rural. Análise exploratória no Centro de Portugal. In *Lazer: Da libertação do tempo à conquista das práticas* (pp. 209–225). IUC.

Sharif, N. & Ku 'Azam, L. (2017). Sustaining the Entrepreneurship in Rural Tourism Development. *International Journal of Multicultural and Multireligious Understanding, 4*(6), 31-42.

Smallbone, D. (2005). *Improving Access to Business Advices. In Rural Areas: an Evaluation, report commissioned by the Department for Environment, Food and Rural Affairs, Small Business Research Centre*. Kingston University.

Stueve, A. M., Cooks, S., & Drew, D. (2002). *The Geotourism Study: Phase I. Executive Summary*. Travel Industry Association of America.

UNESCO. (2015). *Statutes of the International Geoscience and Geoparks Programme and Operational Guidelines for UNESCO Global Geoparks*. UNESCO.

Vaz, B. (2008). *Contributos para a Avaliação e Gestão de Praias: A Importância da Percepção dos seus Utilizadores*. Dissertação de Mestrado. Faculdade de Ciências Sociais e Humanas. Universidade Nova de Lisboa. Lisboa.

Vieira, G. (2004). *Geomorfologia dos Planaltos e Altos Vales da Serra da Estrela. Ambientes frios do Plistocénico Superior e Dinâmica Atual. PhD Tesis. Universidade de Lisboa*. CEGOT.

Violier, P. (1993). Le tourisme fluvial dans la région Pays de la Loire et la dynamique des espaces et sociétés rurales. Norois, 159, 371-386. doi:10.3406/noroi.1993.6493

World Tourism Organization. (2016). *Mekong River-based Tourism Product Development*. UNWTO.

Zouros, N., & Mc Keever, P. (2008). The European Geoparks Network. *Proceedings. 3rd International UNESCO Conference on Geoparks*, 11-12.

KEY TERMS AND DEFINITIONS

Collaborative Entrepreneurship: A collaborative process of developing formal and informal relationships between organizations that seek, flexibly and innovatively, to qualify for economic performance that generates wealth and social well-being.

Differentiated Touristic Products: Tourist products which, due to their specificities and characteristics, depending on the resources and practices they involve, establish themselves as attractions and promoters of new markets in a given tourist destination, requiring specific guidelines for their management.

Estrela Geopark: Estrela Geopark (integrates since 2019 the UNESCO network of global geoparks) has the mission of contributing to the protection, valorization and dynamization of the natural and cultural heritage, with special emphasis on geological heritage, to deepen and disseminate scientific knowledge, promoting tourism and sustainable development of the Serra da Estrela.

Inland Waters: Inland waters include river ecosystems and artificial or strongly modified water bodies associated with them (dams and canals), as well as other ecosystems where the aquatic component is of permanent or temporary importance, such as tributary rivers or ponds, including its banks, which several ecological, economic and social functions.

Chapter 13
Organizational Culture:
A Key Element for the Development of Mexican Micro and Small Tourist Companies

Vicente Alejandro Jiménez Olivera
Senryakuteki Research and Business Institute, Mexico

Irma Magaña Carrillo
Universidad de Colima, Mexico

ABSTRACT

In this chapter, the authors present a part of an analysis of the organizational culture of micro and small tourist companies in the state of Colima, Mexico. It is the result of mixed research that belongs to an exploratory study carried out from a systemic perspective of tourism organizations. The purpose of the study was to identify elements capable of triggering the productivity, competitiveness, and innovation of Mexican micro and small tourism companies based on their language and organizational structure. The results obtained revealed that the "Clan" type culture is predominant in this type of company in the state of Colima. Hence, a transformation of the mental structure of senior management is necessary and urgent, as it requires a change in an organizational culture focused on the "Market" to achieve its permanence in the sector.

INTRODUCTION

This book chapter is one of the products obtained from a doctoral research entitled: The organizational culture of micro and small tourism businesses in the state of Colima, Mexico. An analysis of innovation and competitiveness in the XXI century. It was a mixed research and exploratory nature. Its overall objective was to analyze the organizational culture of micro and small tourism businesses in the state of Colima to identify the elements that can reach to detonate their productivity, competitiveness and innovation based on their language and organizational structure.

DOI: 10.4018/978-1-7998-4855-4.ch013

The research had the following objectives: a) to diagnose the organizational culture of this type of business, b) to identify the predominant culture of micro and small enterprises in the tourism sector located in the state of Colima, c) to know the meaning psychological stimulus words like 'Tourism', 'Quality', 'Competitiveness', and 'Innovation', d) to identify elements for innovation and competitiveness of micro and small tourism businesses in Colima. To meet these objectives, research techniques were used as natural semantic networks and surveys.

The central theme of the research was the study of the organizational culture of micro and small tourism companies in the state of Colima, Mexico, from a systemic approach. Organizational culture is the essence of the organization, an intangible, that exists within the company and that is in continuous movement, interacting constantly with other parts of the company, forming a series of complex sensitive and open subsystems related to each other (Cleghorn, 2005; Katz & Kahn, 1986).

This chapter book aims to present only part of the diagnosis made of the predominant organizational culture in the 70 micro and small tourism businesses that make up the study. These companies belong to different business line as hotels, restaurants, bars, travel agencies, and cafes. A-part of the study, it is sought to highlight the importance that the type of organizational culture in enterprises has predominant directly related to productivity, competitiveness, and innovation. The chapter first shows a theoretical context and the focus of the investigation in which some concepts related to administration, culture, and organizational culture are generally addressed. Secondly, a brief analysis of tourism in Mexico is presented based on the importance of its micro and small tourism companies. Thirdly, part of the results related to the predominant organizational culture in the micro and small tourist companies studied are disclosed. Fourthly, the role of organizational culture in the fragmentation of the reality of Mexican micro and small tourist companies is analyzed. Finally, some solutions and recommendations are presented, possible future directions of the research, and the conclusion of this chapter.

THEORETICAL CONTEXT AND RESEARCH APPROACH

This first section seeks to substantiate the importance of studying the organizational culture of micro and small tourist companies in Mexico and to present a global vision of the context in which the research was carried out. For this, reference is made to different theoretical concepts related to administration, culture, and organizational culture. Empirical studies, methodologies, and research approaches carried out in different parts of the world such as the United States, Europe, Japan, China, and Latin America are mentioned (Huete & Mantecón, 2017). The objective of this section is to lay the foundations for the study that was carried out. The intention is to have a clear idea of the role played by the organizational culture of the Mexican micro and small companies that belong to the tourism sector in a given territorial space, which in this case was Colima, Mexico.

In a traditional way, administration is understood as an activity that involves conscious and constant human coordination, or as a discipline responsible for studying the various production processes necessary to generate efficiency and effectiveness, with the intention of having, always and at all times, better benefits than are already available (Stoner et al., 1996). These benefits can be of different types, such as: human, technological and of course economic (Hernandez, 2006).Thus, the administration shapes the organization, the organization is made of people and these people have a culture of their own, which may or may not be in communion with the culture of the company. The importance of the study of administration and culture in the company lies in the understanding that the company is the fundamental part of the development of societies, it is everywhere and even if they change over time it is part of the future of the person, an indispensable reason for its study.

To talk about culture is to speak of a controversial and complex term, just as controversial or even more than tourism. It is a topic of broad content, with various variables and meanings that have evolved over the years as well as the identity and characteristics of people and society at large. It is a multipurpose term, complex, imprecise and difficult to conceptualize, but in turn so necessary to understand human reality (Barreto, 2007; Cabedo, 2006; Schein, 1988). Although there are many definitions of the concept of culture, in this book chapter it is agreed with the description proposed by Schein (1988: 24), which mentions that culture is "a learned product of group experience, and therefore something localizable only where a definable group and possessing a significant history there". In the same way, it partakes contributions made by Thevenet (1992: 11) when he says that "culture is not a state but is constantly evolving because it is a continuous learning process of the organization2. Thus, "culture is learned, evolves with new experiences, and can be changed if it understood the dynamics of the learning process" (Schein, 1988: 25).

In recent decades the word culture seems to have become fashionable to talk about, a set of expressions used in different contexts and often with ambiguous meaning. One of these contexts is precisely the company, in which the concept of organizational culture is used since the 70's or 80's in Mexico to refer to both the organization and the person who make up it. Currently it is a topic of importance and relevance due to its presence in each organization and because of its relationship in the individual and collective performance of the workers (Pedraza et al., 2015; Robbins, 2005; Miquilena & Paz, 2008).

Organizational culture is part of classical administrative thinking and aims to respond to the different challenges that the *Gigantism* of organizations brought for the administration, in which economic rationalism prevailed, whose objective was the search for greater productivity and efficiency at all costs. Organizational culture is also understood as the set of vital values for the company which are transmitted implicitly or explicitly to all subsystems of the company. Then it is talk about a system of meanings that differentiates one company from another from another based on its basic characteristics and the internal and external behavior of the organization (Alles, 2012; Robbins, 2005). As you can see the concept of organizational culture is of recent creation and is accompanied by topics such as group standards or organizational climate, among others, being also widely used in the field of psychology in recent decades (French et al., 2007).

The study of culture is relevant, particularly in countries like Mexico, where "there is no tradition of studies on organizational culture, but studies on the working culture and the culture of work stress confrontations between the actors they give inside business organizations" (Zalpa, 2002: 19). It is, therefore, necessary to study the organizational culture of the senior management of the company, which is responsible for the kind of culture that predominates inside these companies.

TOURISM IN MEXICO: AN ANALYSIS FROM ITS MICRO AND SMALL TOURISM BUSINESSES

One of most people's dreams is to travel. Traveling is one of the main activities of tourism. Tourism, on the other hand, is undoubtedly one of the fastest growing sectors of the economy in the world characterized by its continuous expansion and diversification, but in turn with a high degree of sensitivity to political, economic, climatic and currently mobile and health changes. Despite this, tourism represents an important economic force present at a global level, as well as a gigantic economic activity in the world (Nieto et al., 2016; Guerrero & Ramos, 2014; Jafari, 2005). The interest in the study of tourism and travel lies in the understanding that it is an economic activity in which today's societies can fully develop, being in turn a phenomenon capable of transforming the economic and social environment of various specific regions and the dynamics of complete countries. Tourism is currently an interdisciplinary and relatively new field of study that has been studied by different areas and disciplines, such as: social sciences, sociology, humanities, psychology and of course economics, just to mention a few (Ascanio, 2012; Guerrero & Ramos, 2014).

In general, Vallejo (2002: 204) mentions that "tourism arises when people decide to temporarily move from their usual places of residence and work to another place, to enjoy or achieve satisfaction". For this, it was necessary to comply with specific conditions that allowed its realization and that also defined its origins and some basic principles that prevail until now, such as having "disposition, at an individual level, of income and free time", "means of transport that allow spatial mobility in acceptable conditions of time and money" and "the agents (individuals or companies) that facilitate the transfer and reception at the places of destination" (Vallejo, 2002: 204). Therefore, the close relationship that exists between tourism and the emergence of companies focused on meeting the needs generated by said activity, that is, the birth of the now called tourism companies, stands out.

From an economic point of view, tourism is a development option currently being explored by different countries, including Mexico. Mexico has potential as a country focused on micro and small enterprises. However, these companies tend to be studied, considering that their behavior is the same, regardless of the sector to which they belong. While it is true that companies generally exhibit behaviors related "actually their ways of acting respond to diverse factors and dynamics, depending on their size, the extent of which are family businesses, their formalization", etc. (Arechavala, 2014: 13-14). For the World Organization Tourism (UNWTO), the tourism "is a social, cultural and economic phenomenon related to the movement of people who are outside their habitual place of residence for personal or business / professional reasons". Tourism has presented, since World War II, an accelerated and constant growth, generating an unmeasured interest in knowing, understanding and studying it in detail, both as a social phenomenon and an economic activity (Jafari, 2005). Tourism is undoubtedly a fascinating, dynamic and complex concept, however talking about tourism is talking about a young discipline, travel stories, history customs and a phenomenon that has impacted and continues to have significantly impacted various regions and their population (Faraldo & Rodríguez, 2014). Also, generally speaking, tourism is seen by some as the "set of disciplines within science, tourism is mainly related to the following fields set: a) Physical planning, b) Sociology and Psychology, c) Psychology, d) Policy and law, e) anthropology, f) Economy and administration, g) Architecture and planning", but is also part of many others (Ascanio, 2012: 17).

You cannot deny that tourism has been in constant motion and transformation, ceasing to be "a luxury to a well almost necessities, to which from threshold determined income few are willing to give, precisely because manifest a sensible income elasticity of demand in the second half of the twentieth century" (Vallejo, 2002: 204). The tourism "has helped to enrich the cultural heritage of the different regions of the world. In the course of history, tourism has motivated obeying the time, for example, migration, religion, war, commerce, fashion" just to mention a few (Guerrero & Ramos; 2014: 4).

Now the tourism landscape of Mexico worldwide about issues such as innovation, productivity and economic competitiveness, derived from the company, is not favorable if you take into account that the year 2000 to 2010 the country lost on an average place in the tables of competitiveness every three months and 2018 was ranked 51 in the competitiveness Index of the Institute for Management Development (IMD, for its acronym in English). Even, there is evidence that the lag in Mexico is growing, even among Latin American countries regarding indicators of competitiveness, investment in science, technology, and innovation, among other fronts (Arechavala, 2014: 03). Despite this, the tourism, from being an insignificant phenomenon in socio-economic terms, becomes an activity of increasing weight in the international economy and undoubted importance in the regions and receiving countries, as is the case of Mexico, where the international tourist demand in 2018 was approximately 42 million tourists, reaching since 2017 historical figures that position the country as an undisputed power worldwide (SECTUR, 2018; Vallejo, 2002). For all of the above, the population around the world is fascinated with the tourism sector, because it is one of the few activities capable not only of generating but also of exponentially multiplying mainly the economic benefits, such as the generation of jobs. It also promotes the exchange of culture, the international promotion of countries, promotes investment in technology, the breaking of stigmas and paradigms while incentivizing foreign investment (Caiza & Molina, 2012).

Tourism "presupposes the existence of tourist infrastructure, structures, and attractions, which without the intervention of tourism planners they would just be raw resources". The companies should meet the needs of tourists before, during and after their visit (Barreto, 2007: 12). Thus today, we live in a world surrounded by businesses, created for different reasons, which are of different sizes and different objectives; however, it is clear that talking business "is often understood as a modern organization formed by various administrative units and each directed by a person responsible maximum", who for the purpose of this research is referred to senior management (Gil, 2012: 38). Particularly in Mexico, companies are classified into manufacturing companies, trade, or services. These, in turn, are grouped considering the number of employed persons can be micro, if you have 1 to 10 employees, small if you have 11 to 50 employees and medium are those with between 51 and 100 employees toiling.

Table 1. Small and medium-sized Mexican companies, stratification official micro (DOF, 30/06/2009: 1)

Stratification				
Size	**Sector**	**Range number of workers**	**Range of annual sales amount (MDP)**	**combined maximum Top**
Micro	All	Up to 10	Up to $ 4	4.6
Little	Commerce	11 to 30	From $ 4.01 to $ 100	93
	Industry and Services	From 11 to 50	From $ 4.01 to $ 100	95
Median	Commerce	From 31 to 100	From $ 100.01 to $ 250	235
	Services	51 to 100		
	Industry	51 to 250	From $ 100.01 to $ 250	250

Stratification of micro, small and medium previous companies was announced by the agreement published in the Official Gazette Dated June 30, 2009, which is that "micro, small and medium also established companies are a crucial element for the economic development of countries, thus contributing to employment and its contribution to the Gross Domestic Product" (from now on GDP), which in the case of Mexico, according to official data provide 42% of the GDP and 78% of employment in the country (DOF: 2009: 1). This framed within the Law for the Development of the Competitiveness of Small and Medium Enterprises Mexico (Sacramento et al., 2019; Soberanes et al., 2018).

As for the number of Mexican companies and the total employed persons by company size, in 2016 they released the first results of the National Survey on Productivity and Competitiveness of Small and Medium Enterprises (ENAPROCE) 2015, the which aimed to "generate, small and medium enterprises on the managerial skills and entrepreneurship as well as government support they receive national information from micro to contribute to the promotion of an entrepreneurial culture of our country" (INEGI, 2016: 1). The survey revealed that 97.6% of Mexican companies correspond to micro and 2% for small businesses, with only 0.4% considered medium enterprises and even fewer large companies, as shown by the following graphs. The ENAPROCE 2015 allowed to identify some of the characteristics of micro and small enterprises regardless of the sector to which they belong. Here are some highlights.

Table 2. Main characteristics of micro and small Mexican companies (Own elaboration based on the National Survey on Productivity and Competitiveness of Micro, Small and Medium Enterprises (ENAPROCE) 2015)

Micro-enterprise	Small companies
• Sectors in manufacturing, trade, and service, 98 of 100 companies are micro and concentrate more than 75% of the employed personnel. • 17.7% of engaged staff have higher education. • Total personnel are trained, 53% are women. • Just over 40% of businesses say that it does not grow due to a lack of credit and competition from informal firms. • 74% do not use equip computer. 73% do not use the internet.	• 51% of employed personnel have an upper secondary education. • 24.4% indicate that they do not grow due to the excessive number of procedures and high taxes. • 93.4% of companies use computer equipment, and 92.4% use the internet. • Almost 28% of companies monitor 3 to 5 indicators of their performance.

Recently, in November 2018, the field operation began to carry out ENAPROCE 2018 again, at least this was announced by the INEGI in a press release, also indicating that it will be applied to 23,928 companies selected at random from the different sectors, which will allow us to have representative results at the state and national levels by company size.

With the above, one agrees with Sarmiento et al., (2010: 1) when they state that:

Mexico, like some other countries in the world, is a country that has micro and small businesses as its economic backbone. Many Mexicans who are small landowners converge in them, entrepreneurs who were risking many times everything they have launch into the great adventure of seeking to improve living conditions for themselves and their own.

Having made it clear that Mexico is a country of micro and small companies, it is understood that tourism companies also play a fundamental role in the growth and development of this sector. These types of companies and their organizational culture will be discussed immediately, as they were the object of study of the original research (Soberanes et al., 2018).

ORGANIZATIONAL CULTURE OF MICRO AND SMALL TOURIST ENTERPRISES IN THE STATE OF COLIMA, MEXICO

The importance of knowing the organizational culture of micro and small tourism businesses Mexican lies in the fact that, according to the National Institute of Statistics and Geography of Mexico (INEGI, 2016), approximately 99.8.% of the total economic units in Mexico, it is made up of micro, small and medium enterprises, which according to Arechavala (2014: 04) "succumbing daily to the pressures of international competition, and the vast majority of them did not observe changes in behavior that allow providing positive changes down". According to Gil (2012: 40), the 21st century came loaded with challenges for "the business world and society in general, with the revolution that is taking place within the computer industry, information systems, nanotechnology, etc.", and is characterized by the introduction of different forms of organization, management, planning processes, designs and strategies that allow them not only to function but to stand out from the others day by day. Some of the ways to face the current challenges facing micro and small tourism companies are the quality and differentiation of the products and services they offer to the market. For this, tourism companies need to know their customers and satisfy them. However, in most companies, the organizational culture of Senior Management does not consider the productivity or competitiveness of tourism companies as a priority; therefore, it is not part of the organizational culture of the company.

As previously mentioned, the research from which this chapter of the book emerges included a total of 70 micro and small tourist companies distributed in 5 of the ten municipalities of the state of Colima, these being: Manzanillo, Tecomán, Colima, Villa de Álvarez, and Comala. To carry out the diagnosis of organizational culture is predominantly surveyed senior executives of each of the 70 micro and small tourism businesses. It should be noted that according to Naranjo and Calderon (2015) different techniques, tools, and models have been proposed to measure, evaluate and diagnose culture over time. Some of them analyze it from cultural traits, such as those offered by O'Reilly et al. (1991) and Hofstede (1999), and others from cultural typologies that integrate these features. In this particular case, values Framework was used in competition to diagnose the type of predominant culture in micro, and small tourism companies studied.

Similarly, after reviewing different models, it was concluded that the proposal of Cameron and Quinn (1999, 2006) was the most appropriate, considering that without a doubt, the most widespread and widely accepted typology of organizational culture is derived from the well-known model of competing values, based on research by Cameron and Freeman (1991). It should be noted that this model has also been widely applied internationally (Cújar, 2013). Cameron et al., (2016), address the issue of value production, which means that the benefits produced to exceed the cost of providing those benefits. Seen from this perspective, people are value creators in organizations when the value of what they generate is higher than the value of what they consume. Producing more benefits than costs makes people value creators; however, creating value is not easy. The Competing Values Framework has been named as one of the 40 most essential frameworks in business history. It has been studied and tested in organizations for over 30 years by a group of opinion leaders from leading business schools and corporations. It is now one of the most widely used and highly cited frameworks in the world. Their range of application and information processing ranges from neuroscience to financial strategy and organizational culture (Cameron et al., 2014).

According to Cameron and Quinn (1999, 2006), the Competing Value Framework was created to have a high degree of consistency with known and accepted categorical schemes that organize the way people think, their values and assumptions, and the way they process information. About Competing Values Framework, Naranjo and Calderon (2015: 225) mention the following:

The MVC defines four types of culture from two dimensions. The first dimension (stability versus flexibility) refers to whether the organization considers the most crucial order and control (stability) or the dynamism and discretion (flexibility). The second dimension presents two competing values: the company has an internal orientation or has an external direction. Combining these two dimensions or competing values, the model distinguishes four types of culture: clan, adhocracy, hierarchical, and rational.

In general, the dimensions, orientations, and types of culture proposed by Cameron and Quinn (1999, 2006) in their model are as follows:

Figure 1. Fundamental aspects of the competing values framework or competing value framework (Own elaboration based on Cameron and Quinn, 1999, 2006)

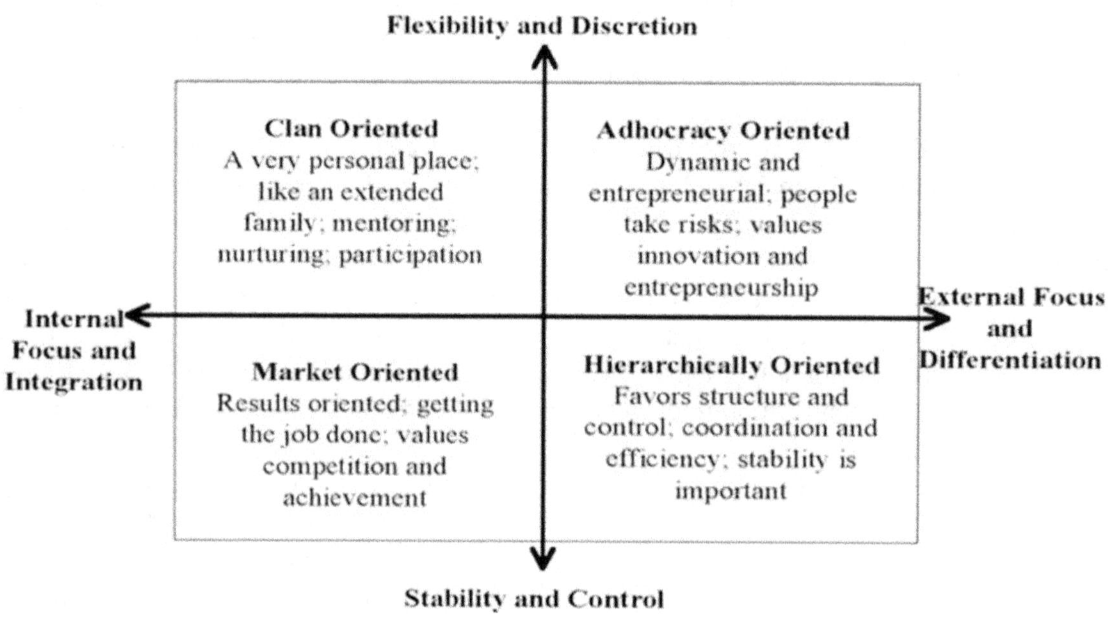

According to Cameron and Quinn (1999: 31), "these organizational effectiveness indicators represent what people value in terms of the performance of the organization. They define what is seen as fair, appropriate, and correct. In other words, set the central values on which judgments are based on the organization". So, you can talk about four major types of organizational culture: Clan Culture (flexibility and internal focus) emphasize human development, teamwork, and employee participation. The adhocratic culture (flexibility and external orientation) mainly appreciates the initiative, creativity, and risk-taking. Intellectual culture, also called market (stability and external direction), power achieving ambitious goals and competitiveness. Finally, in the hierarchical culture, the most important values are efficiency, compliance, and formalization of processes (Cameron & Quinn, 1999; 2006).

Each of them is described in detail below.

Organizational Culture Predominant in Micro and Small Tourist Companies in the State of Colima, Mexico

Table 3. Types of organizational culture according to Cameron and Quinn (Own elaboration based on Cameron and Quinn, 1999, 2006)

Type of culture	Description
Clan: He called clan because of the type of family organization that characterizes this culture.	The organization is a friendly place to work and where people share much with each other. It is generally a family. Leaders or heads of the organization are considered mentors, and perhaps father figures arrive deep inside the institution. The organization is bound by loyalty or tradition. In general, the commitment of its members is high. The organization emphasizes the long-term benefit in the development of human resources and attaches great importance to cohesion and morale. Institutional success is defined in terms of customer satisfaction and consideration of the people. The organization rewards teamwork, participation, and consensus.
Adhocracy: They are highly sensitive to the accelerated changes and turbulences that characterize the organizational world of the 21st century.	Members see the organization as a dynamic place to work, with an entrepreneurial spirit and a creative environment. People, therefore, tend to be creative and take acceptable risks. Leaders are considered innovators and risk-takers.
Market: The term market is not related to the term marketing; preferably, it is applied to an organization that functions as a market.	The market culture operates mainly through the management of transactions with other markets (as interest groups) to generate competitive advantages. It is a results-oriented organization whose primary concern is to get the job done right.
Hierarchized: Bureaucratic organizations.	The organization is a structured and formalized place to work. Procedures govern and tell people what to do daily. The interest of the organization's leaders is to be good coordinators and organizers, maintaining a cohesive organization, where rules and policies play a predominant role.

Of the 70 micro and small tourist companies studied in the state of Colima, 41 are 'Restaurants', 13 are 'Hotels', 9 'Travel Agencies', 4 'Bars', and 3 'Cafes'. As it can be observed, the micro and small tourist companies such as the restaurants predominated the study, in contrast to the bars and cafeterias that ranked 4th and 5th, respectively.

Thanks to the application of the OCAI questionnaire (Organizational Culture Assessment Instrument), it is determined that the type of organizational culture that predominates in the micro and small tourist companies of the state of Colima is the 'Clan' type, followed by the kind of culture 'Hierarchized' and in some cases leaving behind the kinds of culture 'Adhocracy' and 'Market'.

This makes sense considering that Colima is one of the states with the smallest territorial extension in the Mexican Republic, which in large part allows the development of family-type companies rather than the development of large companies that prefer states with a more significant territorial extension and many inhabitants, such as Jalisco and Nuevo León, just to mention a couple of examples.

It was also identified that the type of culture 'Hierarchical' remains in each municipality in second place as the dominant culture in the state, which can be interpreted as a possible change current issues such as globalization require addressing by the companies in general and particularly those micro and small economic sector and as dynamic and changing as tourism economic activity.

Particularly concerning one of the variables studied was the age of the micro and small tourism businesses 27 companies with 'More than 20 years' in the market and secondly 19 with '5 to 10 years' in the market. This indicates that perhaps are the companies that have between '11 and 20 years' in the market, which are at a higher risk of closing before reaching 20 years of age.

The overall results can also be presented graphically, as shown below in order to have an overview of the distribution of the types of culture present in the 70 micro and small tourism businesses in the state of Colima that they analyzed.

Figure 2. Organizational culture is predominant in micro and small tourist companies in the state of Colima (Own elaboration based on field work carried out during the years 2017 and 2018)

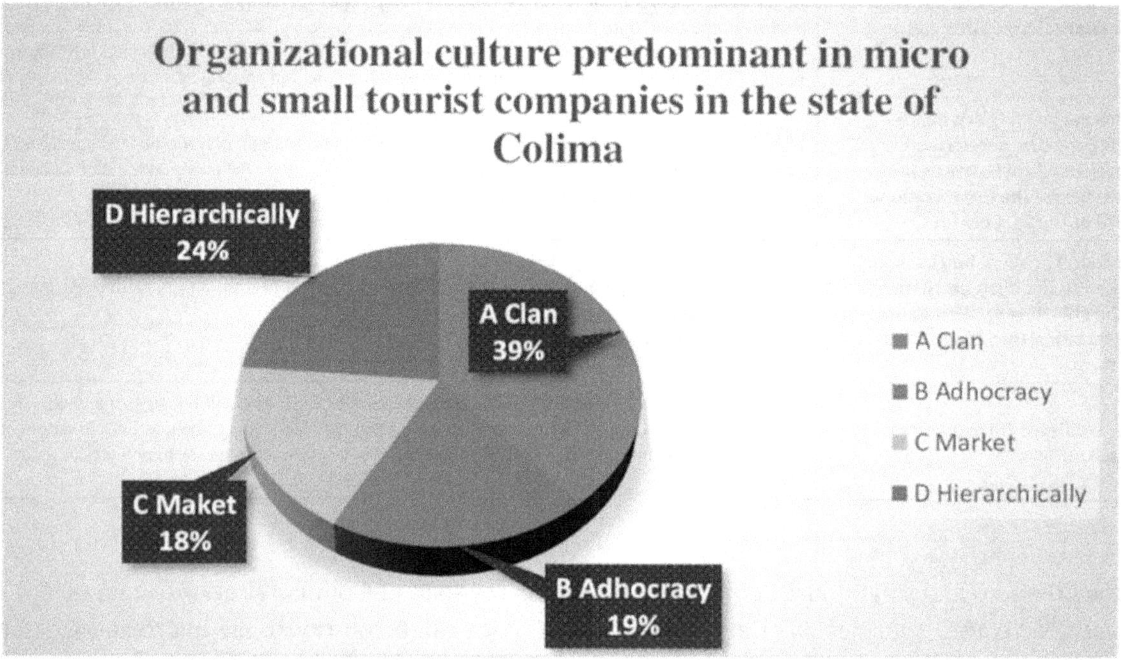

According to the previous graph, the results obtained with the OCAI questionnaire applied to the 70 micro and small tourist companies in the municipalities of Manzanillo, Tecomán, Colima, Villa de Álvarez and Comala, these companies have different types of culture in higher or to a lesser extent, standing out 'Clan culture' with 39% and leaving 'Market culture' lastly focused with just 18%. This is striking since, as the studied municipalities are the ones that receive the most significant number of tourists in the state, there are no micro and small tourist companies that consider a culture more focused on meeting its needs.

The Role of Organizational Culture in the Fragmentation of the Reality of Micro and Small Tourist Companies

The tourist company offers visitors various possibilities to benefit from a cluster of facilities for their enjoyment and recreation. To do this, such facilities and organizations must be both productive and competitive. However, among some of the difficulties that micro and small tourism companies face in a widespread way are the lack of technical financing and training that employees of the tourism organization receive. This results in human capital with a limited level of training. This problem is associated with its reduced participation in the markets, also intervening in aspects related to the organizational culture of the company, transmitted to it by senior management, which directly impacts on the productivity and competitiveness of this type of company in a changing and relatively new sector, as discussed previously (Ascanio, 2012).

98.8% of the total economic units in Mexico are made up of micro, small and medium-sized companies, which face problems such as lack of liquidity, international competition, lack of productivity of their employees, lack of competitiveness, and the definitive closure of its operations, just to mention a few.

Similarly, the lack of work in the mental structure of micro and small entrepreneurs stands out, which seeks to be covered by the training. Such training, when given to the operating managers, is mainly not sufficient. The effects are not lasting or tangible and often have little or no direct-effective relationship with the productivity and competitiveness of the company.

The lack of a systemic vision of companies and the perception of a fragmented reality on the part of entrepreneurs, which begins at a very young age in the school environment, teach young people to see things separately and isolated, resulting in a series of crises in today's society (Cleghorn, 2005). This fragmented view of life that accompanies people from an early age extends to different areas of their lives, with the workplace perhaps being one of the most important, because "we work in departments that are like watertight compartments that isolate each worker in their work area. In this way, it becomes difficult to visualize the system in which we operate, which provides us with a narrow vision, a source of conflict" (Cleghorn, 2005: 23).

The role of micro and small tourism companies in today's world is of vital importance for the development of tourism activity. Its impact on the world economy is undeniable, and on occasions, it has surpassed traditional sectors such as automotive. Without a doubt, tourism, in addition to be a dynamic and exciting phenomenon, is an economic activity that impacts people's lives, their daily lives, and that of the companies that offer their services and products to tourists. The world economic scene changes day by day. Companies must current accordingly, modify, and adapt continuously. This responsibility falls on the Senior Management and will depend on the organizational culture of the latter, which will help or not to overcome the crises through strategies that allow the new challenges to be assumed and overcome. There are several challenges that tourism as an economic activity is continuously facing, making it necessary to remain alert with a willingness to change frequently and attentive to trends and changes in the markets, even if you cannot have a complete guarantee. An example of this is the "UNWTO forecasts, at the beginning of 2008, of annual growth in world demand of 4% until 2020" (Rodríguez & Martínez, 2009: 5). These predictions were abruptly truncated, generating uncertainty and constant turbulence in the markets worldwide.

Finally, it is necessary to understand that the tourism sector is extensive. It goes beyond hotels, restaurants, and bars; therefore, the challenges to face are unknown and increasing every time.

SOLUTIONS AND RECOMMENDATIONS

In the case of Mexico, as in other countries, "tourism is the possibility of creating jobs, increasing markets where small and medium enterprises operate as well as the possibility of preserving a natural and cultural wealth of countries" (PND 2013-2018: 82). According to information provided by Miguel Torruco, current Secretary of Tourism of the Government of Mexico 2019-2020 "they have already taken actions to position Mexico as a competitive tourist power, avant-garde and that this sector is a pillar for development fair and balanced between communities and regions" (SECTUR, 2019).

What is necessary to keep in mind is that people are responsible for working to strengthen the country's tourist activity. The Senior Management of micro and small tourist companies in Mexico are the engine of this possible transformation. Therefore, it is necessary to work on their mental structure and their organizational culture, which they will end up transmitting to their companies and their workers. External efforts such as capital investment in companies are fundamental but destined to fail if the entrepreneur is not mentally prepared to assimilate knowledge and develop effective strategies.

Similarly, it is necessary to recognize that companies in the tourism sector are different from each other, as are we just like people. Therefore, they need to be understood and studied from different perspectives.

Part of their problem lies in the fact that they are considered, studied, and treated generically, when each of them, despite having similarities, also presents significant differences. This makes it necessary to research and understand it globally but also in a particular way, that is, by specific sectors that allow obtaining detailed information on the general situation of the companies that comprise it, as well as the characteristics of its Senior Management.

FUTURE RESEARCH DIRECTIONS

Based on the information presented in this chapter, future research on the organizational culture of micro and small tourism companies is essential. Studies are required that carried out not only from an economic perspective of tourism but also that address the person as a fundamental part of the success or failure of the companies. In particular, the next step is to study in detail the specific characteristics of Senior Management and micro and small tourist companies in Mexico. Currently, there is general information that classifies Mexico as a country of family businesses; however, there is a lack of theoretical and applied research that allows generating theories or replicating techniques, strategies, and procedures that can contribute to business success.

Those working in the tourism sector must have clarity in concepts such as productivity, innovation, and competitiveness. At this point, it must be clear that "the growing foreign competition has taken an increasing toll in terms of several small and medium enterprises that succumb to it. Mexican companies, as a whole, have presented severe problems to adapt to the growing demands of the international competition" and that manifests itself clearly in the low or sometimes zero economic growth in the country compared to other countries of features similar to those of Mexico (Arechavala, 2014: 8).

Also, it is essential to be aware that "the productivity of an economy depends not only on the availability and quality of production inputs but also on how they interact" and that it is "essential to take advantage of the tourism potential of Mexico to generate a more significant economic spillover in the country ... promoting sectorial ordering and transformation ... innovation in supply ... increasing the competitiveness of the tourism sector and improving the flow of investments from it" (PND 2013-2018: 86-87).

Finally, in each study on micro and small companies, it should be borne in mind that the term SME tends to be used as a category of economic units that should have similar behaviors. In reality, their ways of acting respond to very diverse factors and dynamics, depending on their size, the degree to which they are family businesses, so it is known that not all SMEs, for example, seek to grow. Many of them remain satisfied with a level of operations that the owner or founder can personally control ... many arise from defensive self-employment ... others are prey to adverse family dynamics that prevent them from growing (Arechavala, 2014).

CONCLUSION

It is undeniable that the current paradigm of business and tourism is changing. With this, the study and understanding of organizational culture has the opportunity to advance and develop at a higher level of educational, cultural and technological level. The attitude that as a society is taken in the face of this paradigm shift proves to be key for the development of the company, as well as to the growth, development and innovation of sustainable tourism around the world with a greater respect for the environment and landscape (Huete & Mantecón, 2017; Sosa, 2013).The preceding, considering that currently, it is inevitable to be part of a globalized and accelerated environment (everything changes much and more quickly) the maintenance of our organizations is only possible through an adaptive process, the opposite is the end of the organization (Gan & Triginé, 2012).

About the predominant organizational culture, it is concluded that the predominant organizational culture in the micro and small tourist companies of the state of Colima is the clan culture, which indicates that the Senior Management considers that their micro and small company is more like a family, where they work and have friendly treatment regularly. This type of culture "is typical of companies that seek internal control but with flexibility concerned about their employees and showing sensitivity to customers (typically family-owned companies)" (Vivanco & Franco, 2012: 2). In second place is 'Hierarchical culture' and in third and fourth place respectively 'Adhocracy' and 'Mercado', which are perceived as practically non-existent, which can be interpreted as a dynamic in which micro and small Tourism companies are immersed in themselves, looking at all costs to survive the day without having a vision for the future.

At this point, the tourist and the participation of micro and small tourist companies in the state of Colima is limited, and the most worrying thing is that micro and small entrepreneurs did not perceive the need for change, so it is possible to speak of an immense comfort zone that keeps the economic activity of the sector asleep, at least the one that is present in this type of companies.

REFERENCES

Alles, M. (2012). *Comportamiento Organizacional. Cómo lograr un cambio cultural a través de Gestión por competencias*. Granica.

Arechavala, R. (2014). *Twenty years of corporate culture in western Mexico Where are we going?* Universidad de Guadalajara.

Ascanio, A. (2012). *Theory of tourism*. Trillas.

Barreto, M. (2007). *Tourism and Culture. Relations, contradictions, and expectations*. STEPS. Retrieved June 12, 2019, from https://lecturayescrituraunrn.files.wordpress.com/2013/08/turismo-y-cultura-barreto.pdf

Cabedo, M. (2006). *Philosophy and culture of tolerance*. Universidad Jaume I.

Caiza, R., & Molina, E. (2012). Historical analysis of the evolution of tourism in the Ecuadorian territory. *Tourist unit specialties RICIT, 4*, 8-24.

Cameron, K., & Freeman, S. (1991). Cultural congruence, strength, and type: Relationships to effectiveness. *Research in Organizational Change and Development, 5*, 23–58.

Cameron, K. S., & Quinn, R. E. (1999). *Diagnosing and changing organizational culture*. Addison-Wesley.

Cameron, K. S., & Quinn, R. E. (2006). *Diagnosing and changing organizational culture* (Revised Edition). Jossey-Bass.

Cameron, K. S., Quinn, R. E., Degraff, J., & Thakor, A. (2016). *Competing Values Leadership. Creating value in organizations*. New Horizons in Management.

Cleghorn, L. (2005). *Gestión ética para una organización competitiva*. San Pablo.

Cújar, A., Ramos, C., Hernández, H., & López, J. (2013). Cultura organizacional: Evolución en la medición. *Revista Estudios Gerenciales, 29*(128), 350–355. doi:10.1016/j.estger.2013.09.009

Diario Oficial de la Federación (DOF). (2009). *ACUERDO por el que se establece la estratificación de las micro, pequeñas y medianas empresas*. Retrieved June 12, 2019, from https://www.gob.mx/cms/uploads/attachment/file/50882/A539.pdf

Faraldo, J., & Rodríguez, C. (2014). *Introducción a la historia del turismo*. Alianza Editorial.

French, W., Bell, C., & Zawacki, R. (2007). *Desarrollo Organizacional: transformación y administración efectiva del cambio*. McGraw-Hill Interamericana.

Gan, F., & Triginé, J. (2012). *Cultura de empresa y gestión del cambio*. Diaz de Santos.

Gil, M. A. (2012). *Cómo crear y hacer funcionar una empresa*. Alfaomega.

Guerrero, P., & Ramos, R. (2014). *Introducción al turismo*. Patria.

Hofstede, G. (1999). Culturas y organizaciones. El software mental. Madrid: Alianza.

Huete, R., & Mantecón, A. (2017). La clave es el paisaje. Explorando alternativas al turismo de masas. *Arbor, 193*(785), 1–12. doi:10.3989/arbor.2017.785n3003

Instituto Nacional de Estadística y Geografía (INEGI). (2016). *Encuesta Nacional sobre Productividad y Competitividad de las Micro, Pequeñas y Medianas Empresas (ENAPROCE) 2015*. Retrieved June 12, 2019, from https://www.inegi.org.mx/contenidos/programas/enaproce/2015/doc/ENAPROCE_15.pdf

Jafari, J. (2005). El turismo como disciplina científica. *Política y Sociedad, 42*(11), 39–56.

Katz, D., & Kahn, R. (1986). *Psicología Social de la Organizaciones*. Trillas.

Miquilena, D., & Paz, A. (2008). La cultura como fuente de compromiso del capital humano en las organizaciones modernas. *Forum Humanes*, *1*(1), 1–12.

Naranjo, J., & Calderón, G. (2015). Construyendo una cultura de innovación. Una propuesta de transformación cultural. *Revista Estudios Gerenciales*, *31*(135), 223–236. doi:10.1016/j.estger.2014.12.005

Nieto, J. L., Román, I. M., Bonillo, D., & Paulova, N. (2016). El turismo a nivel mundial. International. *Journal of Scientific Management and Tourism*, *2*(1), 129–144.

O'Reilly, C., Chatman, J., & Caldwell, D. (1991). People and organizational culture: A profile comparison approach to assessing person-organization fit. *Academy of Management Journal*, *14*(3), 487–516.

Pedraza, L., Obispo, K., Vásquez, L., & Gómez, L. (2015). Cultura organizacional desde la teoría de Edgar Schein: Estudio fenomenológico. *Revista Clío América*, *9*(17), 17–25.

Robbins, S. P. (2005). *Comportamiento organizacional*. McGraw Hill.

Sacramento, P. M., & Teixeira, R. M. (2019). Innovation and Entrepreneur Learning in the Tourism Sector: Cross-Case Analysis in Small and Medium Enterprises in the city of Aracaju, Sergipe. *Revista Brasileira de Pesquisa em Turismo*, *13*(3), 121–139. doi:10.7784/rbtur.v13i3.1586

Sarmiento, I., García, A., Marcial, F., & Hernández, B. (2010). *Panorama de la micro y pequeña empresa en México*. Retrieved June 12, 2019, from https://www.uaeh.edu.mx/investigacion/productos/6520/panorama_de_la_micro_y_pequena_empresa_en_mexico.pdf

Schein, E. (1988). *Corporate culture and leadership. A dynamic vision*. Plaza y Janes.

Soberanes, M. T. A., Gómez, R. A., & Andrade, M. A. (2018). Opportunity Strategies related to Business Competitiveness in Restaurants in Mexico. Case: Micro-enterprises in the Municipality of Othon P. Blanco, State of Quintana Roo. *Cuadernos Americanos*, *34*(60), 18–29. doi:10.25100/cdea.v34i60.5928

Sosa, A. (2013). *Organizational culture, Japan, the United States, and Mexico: a hypothesis*. Retrieved June 12, 2019, from https://colegios.ingenieria.webunam.mx/sites/default/files/la_cultura_organizacional_japon_estados_unidos_y_mexico_una_hipotesis.pdf?m=1500437373

Stoner, J., Freeman, R., & Gilbert, D. (1996). *Administration* (6th ed.). PEARSON - Prentice Hall.

Thevenet, M. (1992). *Audit of the corporate culture*. Diaz de Santos.

Vallejo, R. (2002). Economy and history of Spanish tourism of the XXI century. *Contemporary History*, *25*, 203–232.

Vivanco, J. S., & Franco, R. (2012). *The types of organizational culture and performance of SMEs in Aguascalientes*. Retrieved June 12, 2019, from http://congreso.investiga.fca.unam.mx/docs/xvii/docs/C28.pdf

Zalpa, G. (2002). Culture in business organizations. *Studies Journal Contemporary Cultures*, *8*(15), 9-33.

KEY TERMS AND DEFINITIONS

Colima: A state in Mexico located in the east region of the country. To the north, it is bordered by the state of Jalisco, to the south by the state of Michoacán and to the west by the Pacific Ocean.

Competitiveness: It is the capacity that the company has to compete in the market in different levels: local, national, international.

Human Capital: Term used in different economic theories which refers to the people the constitute a company, but not just by number, but in their professional training, academic degrees, and quality, including their productivity level, involved personnel in the production process of the company. These personnel generate value to the company.

Innovation: In the global world, it is the capacity that the company has to generate for the consumer the original, the different and the unique.

Micro and Small Companies: Types of companies classified according to the small number of workers involved in the operation of the latter.

Productivity: It is also an economic measure; it has to do with the company's ability to produce. Productivity is directly related to the person and their level of performance within the company.

Senior Management: It is the person or people who have under their responsibility the decision-making in the company. It is responsible for establishing strategic projects, objectives, and goals, designing procedures and methods to develop a comprehensive work involving all the staff that form the company.

Tourism Economic Activity: Activity related to a diversity of commercial actions within the field of tourism.

Chapter 14
Influence of Innovative Processes in the Competitiveness of Tourist Destination

Dilmurad Bekjanov
Urgench State University, Uzbekistan

Bunyod Matyusupov
Urgench State University, Uzbekistan

ABSTRACT

Tourism, with a high economic and social significance for the economy of many countries, is one of the fastest developing fields. A new challenge of transition to an innovative way of developing tourism industry is strongly proposed with the result of existing conditions of fierce rivalry in the field, economic and political instability in many regions, including the Republic of Uzbekistan, the fast-growing needs and demands of tourists. Therefore, destinations, importantly, have to determine their competitiveness to figure out their strengths and weaknesses and thereby develop their future conditions. The authors discuss the issues related to the innovation in tourism destinations and present its main directions in the sector, as well as factors affecting the innovative development of the fieldin Khorezm region of Uzbekistan. Moreover, the chapter illustrates the principles and basic requirements for the system of managing innovative activities in tourism destinations at macro, mesa, and microeconomic levels.

INTRODUCTION

Tourism has become a global economic activity in recent decades. Driven by a relatively strong global economy, a growing middle class in emerging economies, technological advances, new business models, affordable travel costs and visa facilitation, international tourist arrivals rose 5% in 2018 to reach the 1.4 billionmark. This figure was reached two years ahead of UNWTO forecast. Also, export earnings

DOI: 10.4018/978-1-7998-4855-4.ch014

generated by tourism have grown to USD 1.7 trillion in 2018. This makes the sector a true global force for economic growth and development, driving the creation of more and better jobs and serving as catalyst for innovation and entrepreneurship. In short, tourism is helping build better lives for millions of individuals and transforming whole communities. Growth in international tourist arrivals and receipts continues to outpace the world economy and both emerging and advanced economies are benefiting from rising tourism income. During the last seven years, tourism exports grew faster than merchandise exports, reducing trade deficits in many countries (UNWTO, 2019). That's how it works World Tourism Organization UNWTO (2017) from an increase in international arrivals 3.3% per annum to over 1.8 billion by 2030, with the proportion of developing countries, both as destinations and as source markets.

Today, despite the large number of publications published in recent years, there is a growing need for further progress of the topic of the innovative development of tourist destinations with a clear theoretical concept of competitiveness of destinations. Consequently, a consistent methodological derivation of analysis is crucial and enough for use in various economic contexts of destination development.

The main purpose of this work is to discuss the concept of competitiveness in tourism destination and present some adjustable and innovative directions to improve the current conditions. During the recent years in the world, where strong competition is widespread, tourism destination competitiveness owns enormous importance to raise market share. Moreover, in today's saturated markets where competitiveness is highly experienced, one of the most difficult and major tasks fulfilled by sector managers is that of saving and sustaining tourism destination competitiveness. The power of competitiveness in tourism destination increases by the area where all of its alternative and innovative tourism products and attractiveness are conveyed to potential markets in detail.

In order to get the main purpose this research, meanwhile, it is crucial to analyze and determine the conception of implementing innovation in increasing competitiveness within destination management. Recent figures of tourism sector indicate that the market shares of competitive tourism destinations are expected to rise. Because of the recent increase in tourism investments in this region, Khorezm is chosen as the domain of the investigation. Various aspects of destination competitiveness have been studied in the literature, but without reference to the case of the region and a holistic point of view focusing on the conception of innovation. Therefore, this research tries to examine to present some adjustable and innovative directions to improve the current conditions in the region. In the literature review, the concept of innovative management and the definition of innovation in tourism destination are presented together with the factors that define what innovation is and what it is not.

In assessing destination competitiveness, this work follows the method of questionnaires fulfilled with short questions related to the theme, taking a perspective of Khiva as an innovative and competitive destination. During the research, tourism service providers and tourists were surveyed separately. An expressive analysis of the results provides a suggestion regarding the factors that are the most important in identifying the competitiveness of Khiva as a competitive tourism destination. Several tourism service providers and tourists participated in this survey and gave their feedback on tourism issues and providing facilities in the field. Because the research was conducted to find out the most important factors and indicators affecting the competitiveness of Khiva destination by analyzing the suggestions and ideas of both tourists and tourism service providers. The results indicate despite the fact that Khorezm region is rich in historical and cultural resources, other factors such as destination safety, infrastructure and natural resources are also important in enhancing its competitiveness as a tourism destination. Moreover, the results also help identify how Khiva rates in terms of endowed resources, created resources, security issues, situational conditions, supporting factors and tourism destination management and they are essential to encourage the sector to analyze these matters that would finally contribute to the increased competitiveness of Khiva as a competitive destination.

BACKGROUND

According to the data of the last 3 years, it can be said that the tourism industry in the Republic of Uzbekistan is developing rapidly as a result of great attention paid to tourism in the country. In 2019 alone, the country has adopted 15 new normative and legal acts to develop tourism rapidly and make it one of the strategic sectors of the economy. Also, the introduction of a 7-60-day visa-free regime with 86 countries, a transit-visa-free regime with 36 countries and an electronic visa regime with 57 countries has led to a sharp increase in the number of foreign visitors (annual report of the state committee of the Republic of Uzbekistan for tourism development, 2019). As a result of the effective implementation of reforms by the state and the opening of the country to the world community, the number of foreign visitors in 2019 increased by 2.5 times compared to 2017, and exports of tourist services increased by 2.4 times in 2019 compared to the same period (annual report of the state committee of the Republic of Uzbekistan for tourism development, 2019). It might be appropriate to state that the mentioned reforms was carried out to further develop the tourism industry in the country, strengthen its position in the world market of tourist services and increase its competitiveness.

The sector of service is the main productive sector in most organization for economic cooperation and development (OECD) economies. Innovation is nowadays considered as one of the important aspects in developing economic competitiveness of firms. Being one of the most developing industries in the World, tourism caused urgent need to better understand innovation in the field (Carvalho and Costa, 2011). Different scientific articles define assumptions about destination competitiveness and some theoretical and practical matters regarding innovations in tourism sector. They have been investigated in the researches of several scientists all over the World and it is possible to be familiar with their names, suggestions and concepts in the field in this work. In addition, from the scientific point of view, tourism destination competitiveness is studied in accordance with the conditions of Uzbekistan (Navruz-Zoda and Ibragimov, 2017).

Different concepts of, firstly, destination competitiveness in tourism, secondly, the relation of innovation to destination competitiveness are provided in Table 1 and Table 2 respectively.

The term "*innovation*" is still discussed in many recent publications by highlighting its distribution to the competitiveness of the organization, both commercial and industrial. In 1934, Joseph Schumpeter introduced and applied the economic conception of innovation and made some assumptions on linking innovation with economic growth. As a result, other authors in this field started analyzing and discussing the issues related to innovation and its role in economic development (Nelson and Winter, 1982). Concerning to applications of new methods, new processes, new markets, new technologies, new products, new sources of energy and new raw materials, innovation involves changes and enhancements in methods, styles and technologies.

Based on the Schumpeterian idea, the main functions of innovation are, firstly, involvement of new productive forces in the process by reducing the costs and enhancing production efficiency; secondly, improvement of the level of life of every person and society by providing the diversity and quality of products and services that meet the needs of the population; thirdly, adaptation of the structure of pro-

Table 1. The concepts of destination competitiveness

A tourist destination is competitive if it can attract and satisfy potential tourists.	Enright and Newton (2005, 340)
Competitiveness is defined as the qualitative and quantitative superiority of a unit, company or territory above all competitors.	Cracolici and Nijkamp (2008)
Competitiveness influences the quality of life of residents at the destination, assuming that GDP income per capita is a proxy of quality of life.	Croes and Rivera (2010)
City tourism competitiveness reflects the ability of the city to highlight its attractiveness for tourists, provide goods and services for tourists better than other cities do.	Cibinskiene and Snieskiene (2015)
The power of competitiveness posed by a tourism destination increases by the extent to which all of its alternative tourism products and attractiveness are conveyed to potential markets in detail.	Ozen (2017)

duction to the structure of the changing needs by contributing to the maintenance of a balance between supply and demand, between production and consumption (Sandybayev, 2016).

Therefore, in this research, a survey to identify main factors that affect the competitiveness of Khorezm tourism destination was implemented and the results were discussed. Two sides, several tourism service providers and tourists, participated in the survey and their ideas and suggestions were analyzed and discussed thoroughly.

DESTINATION COMPETITIVENESS THEORY

Destination competitiveness plays an important role in tourism because the success of tourism destinations in world markets heavily relies on their competitiveness (Komsic & Dorcic, 2016) and in turn the need for research in the field led to the two most influential models, particularly, the first model was developed by Crouch and Ritchie in 1999 and then improved by themselves in 2003 (described in figure 1) and second model was developed by Dwyer and Kim in 2003 (described in figure 2). This article, in order to observe and discuss the topic of destination competitiveness, investigates Khiva (one of the regions of Uzbekistan, Central Asia)as a competitive destination for tourists by following the one of the

Table 2. The concepts of the relation of innovation to destination competitiveness in tourism

….. the emphasis on the innovation is critical for the destination´s competitiveness. … The innovative destinations are able to foresee the changes in the environment and adequately respond to and predict business opportunities.	Kuscer (2012)
The recognition of the tourism industry is strategically important and priority that provides significant opportunities for the development of innovation activities in this area. After all, innovation is a factor in increasing competitiveness in all fields of management, including tourism.	Sandybayev (2016)
… The increased use and changes in technology hand in hand with the decrease of marketing budgets are forcing destinations to innovate their communications strategies as well. More and more destinations shift their traditional communications strategy based on radio, television, printed media and others towards internet and social media. ….	Kiráľová and Pavlíčeka (2015)
An innovative tourist destination built on an infrastructure of state-of-the-art technology, which guarantees the sustainable development of tourist areas, facilitates the visitor's interaction with and integration into his or her surroundings, increases the quality of the experience at the destination, and improves residents' quality of life.	Lopez de Avila (2015)

two aforementioned models which is the model developed by Dwyer and Kim (2003). In general, these two models, taking into account the importance of each factor that affect destination competitiveness, enable the comparison between countries and tourism sector industries in terms of their competitiveness. A descriptive analysis of the results indicates the factors which are the most important in determining the competitiveness of Khiva as a competitive tourism destination.

Kuscer (2012) defines that the aforementioned two destination competitiveness models proved mainly useful tools in the tourism sector. Authors of these models applied the idea of the competitiveness of nations on the tourism destination level. Enhancing the model, developed by Crouch and Ritchie, Dwyer and Kim added demand conditions as a determinant of destination competitiveness and recognized that destination competitiveness is not the final purpose in itself, but the intermediate purpose towards the goal of regional or national economic development. The destination competitiveness model is a mechanism used by the tourism sector for the analysis, prediction, planning, and communication of competitiveness strategies. It is based on the concept of comparative and competitive advantages of the destination (Kuscer, 2012).

According to the model developed by Crouch and Ritchie (2003), first, comparative advantages include several resource endowments such as human resources, physical resources, knowledge resources, capital resources, infrastructure and tourism superstructure, historical and cultural resources and the size of economy, second, competitive advantages consist of resource deployment such as audit and inventory, maintenance, growth and development, efficiency and effectiveness. Briefly, it can be said that the amount of the available socio-cultural and natural resources is the comparative advantage of a destination and the ability of the destination to ensure the long-term use of these resources is the competitive advantage. Although natural resources are not the product of labor and cannot be created by human beings in equal quality, their quality may be degraded by them. The proper management of resources and the consensus between the stakeholders regarding tourism development strategies at the destination are extremely important for the long-term competitiveness of a tourism destination and for the proper planning of the suitable destination development. Kuscer (2012) mentions that the destination riches in resources can be less competitive than the destination with much fewer resources that are more efficiently used; the efficiency of resource usage ensures the competitive advantage of the destination (Crouch and Ritchie, 2000).

In order to fully understand and describe the destination competitiveness, the consideration of only comparative and competitive advantages is not enough and it is crucial to mention all the factors of competitiveness and understand the connections among them even though the perception of comparative and competitive advantages is a theoretical foundation for the advance of the destination competitiveness model.

Thoroughly discussing comparative and competitive advantages in their model (figure 1), Crouch and Ritchie also recognized two types of environments that influence the elements at the destination: the competitive (micro) environment and the global (macro) environment. The first environment, the competitive (micro) environment, consists of factors developed different private and public sector institutions at the destination while the second one affects the destination but managers have no control over this. Moreover, some other factors within these environments which are the supporting factors and resources, the core resources and attractors, destination management, destination planning, policy and research and finally qualifying and amplifying determinants were set by authors considering them as the factors that enable destination competitiveness and sustainability. Analyzing and enhancing the model,

Figure 1. Crouch and Ritchie Conceptual Model of Destination Competitiveness
Source: Ritchie & Crouch, 2003

developed by Ritchie and Crouch, Dwyer and Kim achieved a result that led to their own integrated model of destination competitiveness.

To describe the difference between these two models, it is important to show several changes done by Dwyer and Kim. Firstly, the integrated model comprises some types of determinants missing from the Ritchie-Crouch model. The explicitly drawn difference between inherited (endowed) and created resources, "market ties" that included among the Supporting factors and resources not under Core resources and attractors, Demand Conditions that new-added determinant by being recognized as an important determinant of destination competitiveness and explicitly recognition that destination competitiveness is not the final purpose in itself, but the intermediate purpose towards the goal of regional or national economic development should be highlighted to describe new determinants that added by Dwyer and Kim. Secondly, Destination, Policy and Development, a separate determinant included several factors that provided in the Ritchie-Crouch model, was subsumed under Destination Management which included other several dimensions such as Destination Management Organization, Destination Marketing Management, Human Resources Development and Environmental Management in the integrated model. Thirdly, there is a category called "Situational Conditions" in the integrated model created by relabeling the category of "Qualifying and Amplifying Determinants" in the Crouch-Ritchie model. It consists of, similar to "Qualifying and Amplifying Determinants", Destination Location, Safety and Price competitiveness. Moreover, "Situational Conditions" includes Competitive (micro) and Global (macro) environment by excluding interdependencies which included in "Qualifying and Amplifying Determinants" in the Crouch-Ritchie model. Interdependencies are recognized as differently, and therefore come into view as Accessibility of Destination under Supporting Factors and Resources in the integrated model.

Figure 2. Model of Destination Competitiveness
Source: Dwyer & Kim, 2003

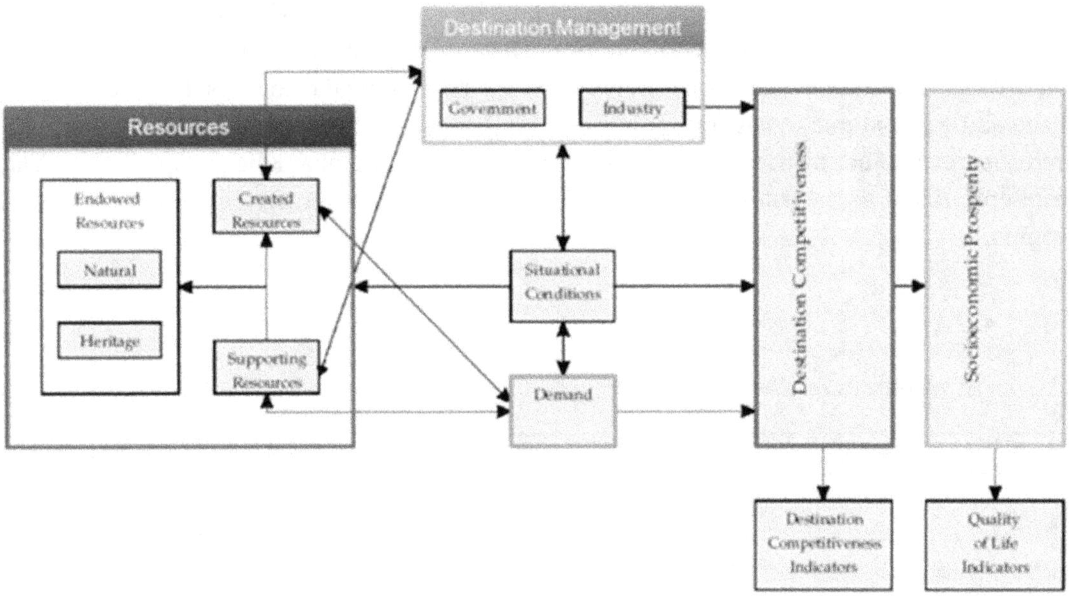

Another change is that tourist awareness about destination is included as a Demand Condition in the integrated model. Next, enterprise is included under Supporting Factoring and Resources in the Crouch-Ritchie model while the model developed by Dwyer and Kim includes it within elements of Competitive (micro) Environment. Also, tourism infrastructure and general infrastructure are described separately in the integrated model while they are included under the label "superstructure" among the Core Resources and Attractors in the Crouch-Ritchie model. Additionally, being recognized as a major attraction, shopping is included under "Created Resources" in the integrated model.

Taking into account the importance of the aforementioned two models in carrying out this research, they have been described in detail above. Meanwhile, in this study, the Dwyer and Kim model was used to study the competitiveness of Khiva destination through questionnaires. It analyzed how various factors affect the competitiveness of the destination and made recommendations for future priorities.

ANALYSIS AND DISCUSSION

General Information About Khorezm

Being located in the center of the famous "Great Silk Road", Khorezm, one of the regions of Uzbekistan, is considered as one of the most popular tourist destinations that possessing high tourist potential in our country. Khiva city located in Khorezm is a museum city under the open sky and the inclusion of the famous Khorezm national dance Lazgi in the UNESCO World Heritage list will help to attract more tourists to this destination and strengthen its position in the global tourism market.

There are more than 255 historical and cultural heritages in Khorezm region, dating from the ancient Neolithic period up to the 19th century. Of them, 18 are archeological, 133 are architectural, 66 are sculptural monuments, 6 are sightseeing places and 32 are sacred sites. The main parts of these tourist resources are located in Khiva and Urgench cities. They are described in Table 3. Moreover, there are also a wide range of resources such as natural resources, archeological resources, historical monuments, pilgrimage and gastronomic tourism resources, cultural and entertainment tourism resources and agro tourism resources in other districts of the region such as Bagat, Khanka, Khazarasp, Shovot, Yangiariq and Kushkupir. All of these factors have a great contribution to the attractiveness and competitiveness of the region.

Table 3. Tourist resources of Khorezm region

Tourist resources	Structure	Geographic location
Natural Resources and Biological Diversity	Nature of the desert, surface relief, flora and fauna	Throughout the region.
The Amu Darya and its Basin	The recreational resources of the river and its banks, the steppes and the forests of the river, the flora.	In Bagat, Khanka, Urgench and Khazarasp districts.
Archeological resources	Golden rings of the capitals of the ancient state of Khorezm, ancient rings, archaeological findings	Districts of Khiva, Urgench, Shovot, Bagat, Khazarasp and Kushkupir.
Resources associated with historical monuments	Historical monuments	Cities of Khiva and Urgench, Districts of Shovot, Bagat, Khazarasp and Yangiarik.
Pilgrimage tourism	Shrines of Sultan Uvays, Yusuf Hamadoni, Pahlavon Mahmud, Shaikhmutovali, Norimjonbobo and other places of worship	Districts of Khiva, Urgench, Bagat, Yangiarik
Gastronomic tourism resources	Tukhumbarak, shivitoshi (the green lagman), special minced meat in Khiva called Itzhan (Khiva ijjoni) and other national dishes exclusively for Khorezm region	Throughout the region
Cultural and entertainment tourism resources	National holidays celebrated in the region and the ever-expanding traditional dance festival "Raqs sehri" (Magic of dance)	Cities of Khiva and Urgench
Agro tourism resources	Covering the whole territory of the region the melon festival traditionally held called "Gurvak – the symbol of Khorezm melon" and four special Agritourism communities.	Districts of Khiva, Urgench, Bagat, Yangiarik
The resources of the other special types of Tourism	Craftsmanship, custom and traditions, national arts and games	Throughout the region

Source: annual report of the territorial administration of Khorezm region for tourism development (2019)

The Analysis of Current Condition of Tourism in Khorezm Region

The conducted analyses are showing the number of tourists who are visiting to Khorezm region is rising year by year. The number of foreign and domestic tourists in the total number of tourists is significant (Table 4).

Table 4. The number of tourists who visited to Khorezm region during 2016-2019

Year	2016	2017	2018	2019
Total	754 843	1 160 600	1 653 665	2 646 597
Foreign tourists	46 614	60 600	123 210	420 828
Domestic tourists	708 229	1 100 000	1 530 455	2 225 769

Source: annual report of the territorial administration of Khorezm region for tourism development (2019)

Such positive changes in the development of tourism are, first of all, connected with the globalization of tourism in our country and the strong growth in the global tourism market. At the same time, several factors such as a number of undertaken activities to develop the services sector by the government under the auspices of the President, regular representation of the tourism potential of our country at the world tourism fairs, regular annual fair "Tourism along the Silk Road" in Tashkent, improved road infrastructure linking Khorezm oasis with other tourist centers, the increased quality of tourism services and other such kind of factors have a positive impact on the sustainable development of tourism in the region. The results of the analysis show that Khorezm region takes the first place, in the country, in terms of the density of tourist excursion resources and the number of historical and architectural objects, second place in terms of technological reserves of excursions (Table 5).

From the one side comparative advantage comprises the resources available to a destination, competitive advantage on the other side relates to the proper management of resources and the efficiency of resource usage over the long-term (Crouch and Ritchie, 1999, p.143; Bobirca and Cristureanu, 2008). According to Borbica et al. (2008), competitive factors can be produced and improved by tourist firms or governmental policy. It is, therefore, crucial to mention that inefficient usage of resources may affect the competitiveness of a destination endowed with a wealth of resources and it can be less competitive than a destination lacking in resources. Despite the great potential, in terms of the number of hotels and tourism service providers and the quality of services, Khorezm region lags far behind Samarkand and Bukhara regions. Because of this, the number of tourists visiting the region, hotel employment and tourism exports are also low. Having been observed since 2004, this trend is indicative of the poor resolution of marketing and promotion activities, interregional transport and other organizational issues in Khorezm region.

Organizations Providing Tourism Services

Despite the high tourism potential of Khorezm region, the number of tourist organizations (tour operators) operating in the region is very low (Table 6). This is due to the lack of specialists capable of providing quality tourism services. As a result, the proceeds from tourists arriving in Khorezm will be transferred to tourist organizations in other regions.

The numbers of tourist organizations in the region and the inadequate qualification have a negative impact on the tourism export potential of Khorezm. In Khorezm region, export revenue from tourism is mainly derived from hotel activities, where tourism operators have a very low share.

Table 5. Main indicators of tourist centers of Uzbekistan as of 01.01.2020

Indicators		Khorezm	Samar kand	Bukhara	Tashkent city
Technological reserves of the resources of excursions		75.9	75.4	163.2	-
The density of tourist excursion resources		16.9	3.1	1.1	-
Hotels	Total	88	148	152	186
	Established in 2019	28	30	26	8
Tour operators	Total	58	170	100	816
	Established in 2019	24	43	54	169
Tourist transport enterprises	Total	42	24	21	38
	Established in 2019	25	8	8	7
Entertainment parks	Total	13	19	10	18

Source: annual report of the state committee of the Republic of Uzbekistan for tourism development (2019)

Global Information Resources

It should be noted that the tourism potential of the region on global internet resources is also inadequate. For instance, www.booking.com is a popular tourism resource where the number of hotels available for booking is 285 in Tashkent, 168 in Samarkand, 194 in Bukhara and 98 in Khorezm. This can be seen by foreign tourists as the evidence that tourism services and infrastructure in Khorezm are poorly developed compared to other historic cities in Uzbekistan. In fact, this is the result of poor information exchange facilities in the tourism sector in Khorezm region and it has bad impact on its online reputation. It is important to mention about the relationship between the online reputation and the competitiveness of a tourism destination that reputation is highly useful to improve competitiveness (Pechlaner, Smeral, & Matzier, 2002). According to authors such as Caldwell and Freire (2004) or De Moya & Jain (2013), this reputation partly explains final competitiveness and competitiveness and reputation mutually influence one another. Likewise, Marchiori and Cantoni (2011) stressed that visits to a destination could be influenced by online reviews and the subjective experiences of travelers. Finally, the relevance of this reputation for tourist satisfaction and destination competitiveness is emphasized in the related literature

Table 6. The number of organizations providing tourism services (tour operators) in tourist centers of Uzbekistan as of 01.01.2020

The names of places	The number of tour operators	Share (%)
Total in the Republic of Uzbekistan	1380	100.0
Tashkent city	816	59.1
Samarkand region	170	12.3
Bukhara region	100	7.2
Khorezm region	58	4.2

Source: annual report of the state committee of the Republic of Uzbekistan for tourism development (2019)

reviews conducted by various authors such as Gallarza, Gil-Saura, and García (2002), Pike (2002), or Zhang, Fu, Cai, and Lu (2014).

The analyses show that foreign tourists sometimes do not plan to visit Khorezm because they do not have enough information about the advantages available here. At the same time, foreign tourists say that their impressions during their visit to Khorezm are not inferior to those of Samarkand and Bukhara. In addition, international tourism media reported that cooking of different nations is well developed only in Tashkent, with relatively few places for foreign tourists, particularly in Khorezm.

Cultural, Natural and Recreational Resources

Crouch and Ritchie (1999) and Dwyer and Kim (2003) believe that being the important characteristics of a destination that attract visitors and creating the main foundations of sustainable tourism, tourism resources and attractors should be recognized as the basis of destination competitiveness. Khorezm region occupies an important place in the world in terms of cultural and natural resources. The available resources will allow the region to deeply diversify the tourism opportunities and extend the tourist season.

Moreover, Khorezm is known for being one of the centers of science in history. The share of young people aged under 30 to total population is 59 percent as of 01.01.2020 (annual report of the state committee of the Republic of Uzbekistan on statistics, 2019), which means that the region has great potential in human resources. However, at the same time, there is a lack of qualified personnel in the tourism sector. From the lowest levels of tourist service to the management level, there is a great need for qualified personnel in this area. There is also the possibility of hiring from other neighboring regions or foreign countries for qualified labor to enhance competition.

Khiva tourism destination is a combination of such historical and cultural monuments as well as archeological and gastronomic resources. By informing the world population, it will be possible to increase the attractiveness of tourist destinations and increase the flow of foreign tourists. Nowadays, the main tourist destination in Khorezm for tourists from foreign countries is Khiva. It is possible to increase the tourism income by increasing the flow of tourists and the tourist night stay.

Measuring the indicators of tourist destination competitiveness is done with the help of the model developed by Dwyer and Kim. This method was used by researches from the Austrian Management Center Innsbruck Tourismus Destination Research Lab to study the tourist destinations of Tirol (MCI Tourismus Destination Research Lab 2019). The list of indicators for determining the competitiveness was initially modified based on the specifics of Khiva destination. Indicators, in this, such as swimming on the beach, seaside recreation and mountain climbing have been excluded from the list. The reason for that, all these kind of indicators are not relevant to Khiva destination because of its location long far away from any sea or mountains.

In the research work, the factors influencing the competitiveness of tourist destination in various levels are studied through a survey of tourists and the implementation of the model of Dwyer and Kim. As the main factors of competitiveness, these factors include natural, created and supporting resources (formation of motivation in tourists, supported resources and external environment), as well as factors that determine the attractiveness of the destination. The views of the tourism service providers and tourists in this process are different. This research relies on a survey on opinions and attitudes of not only tourists but also tourism service providers towards the above factors. Comparing the views of these two sides, this study examines the specific factors that characterize the activities of the proposed economic entities and the satisfaction of tourists with them. These are achieved by finding answers to the following scientific questions:

- What are the opportunities for the application of the model of competitiveness of tourist destinations, developed by Dwyer and Kim, in Khorezm region?
- What factors are considered as the most important factors in the competitiveness of Khiva tourism destination from the point of view of tourism service providers?
- To what extent is the attractiveness of tourist services actually provided in Khiva tourism destination assessed by tourists on specific factors?
- What is the difference between the factors that are considered important from the point of view of the providers and the tourist services actually provided to the tourists?
- What factors of the competitiveness and attractiveness of the destination affect the satisfaction of tourists with the activities of tourism service providers?

In this research, like Austrian scientists, the questionnaire method was used as the research method. At the same time, some questions were changed based on the specifics of the tourist destination. Each indicator that represents the competitiveness and attractiveness of tourist destinations, according to significance, is used by Austrian scientists on a 7-point Likert scale. We rated it on a 5-point scale (1=”no”, 5=“very high”) for the convenience of the analysis, because this survey was conducted only on one tourism destination (Khiva). Our survey was conducted in collaboration with Urgench State University and the Territorial administration of Khorezm region for tourism development from March to September in 2019 in Khiva, the main tourist center of Khorezm region, and Urgench, main center of the region. The study, conducted in the Alpine tourist destinations in Tyrol, was conducted in a form of an online survey. However, this survey was carried out on the basis of a face-to-face interview in the framework of the project “Improvement of competitiveness and innovative approaches to tourism destination” of the Department of Tourism, Faculty of Tourism and Economics, Urgench State University. The representatives of 52 tourism companies and 480 tourists attended in the survey.

Tourism service providers and tourists were selected to describe the competitiveness and attractiveness of Khiva tourism destination. The information about the tourist companies participated in the survey is described in Table7.

Table 7. The information about the tourism service providers participated in the survey during May – September, 2019

Tourism service providers	Quantity
Hotel and other accommodation facilities	16
Tour operators	6
Transport enterprises	4
Restaurant and other such kind of establishments	11
Entrepreneurs operating on the territory of the State Museum “Ichan Kala”	9
Craftsmen	6

The tourism service providers attended in the survey operate especially in Khiva city, since tourists who visited to tourist destination mainly come to see the historical and cultural heritages situated in Khiva and it can be explained with the location of "Ichan Kala" state museum there (Figure 3).

According to the survey conducted to get some opinions from tourists in Khiva, the average age of tourists from the far foreign countries was 47, and the average age of the tourists from the Commonwealth of Independent States (CIS) was 39, and the average age of domestic tourists from other parts of the country was 42. These respondents were identified by the following categories of gender:

- far foreign countries: 50.6% male and 50.4% women;
- the CIS countries: 62.2% male and 37.8% female;
- Uzbekistan: 51.8% male and 49.2% female.

Figure 3. The main purposes of travel of respondents attended in the survey (%)

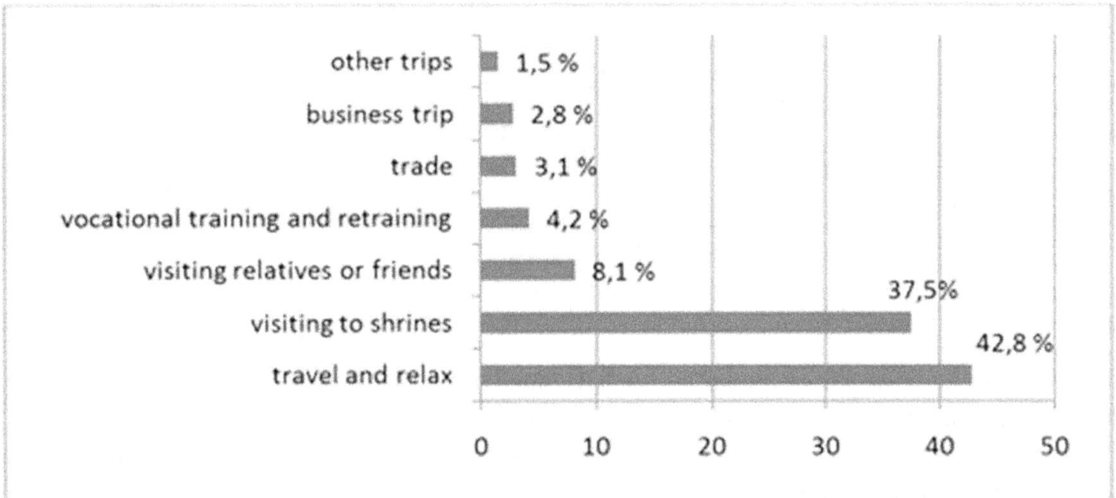

Foreign tourists participated the survey especially visited to Khiva to travel and relax while domestic tourists especially visited to religious and pilgrimage tourism (pilgrimage tourism means to visit to shrines).

During the survey, the results obtained by averaging the discretionary value of the most important competitors and attractiveness indicators were selected, and were used by the groups to analyze the factor. As a result of discretionary evaluation of the average quantity of the obtained results, the most important indicators of competitiveness and attractiveness were selected and divided into mutually compatible groups to analyze with factors.

From the point of view of tourism service providers, nature activity, historical and cultural excursions, destination safety, environmental cleanliness, satisfaction of tourists with the quality of services provided, the quality and variety of regional products, the quality and variety of tourism infrastructure,

Table 8. The assessment of 18 basic indicators that represent the competitiveness of Khiva tourism destination from the point of view of tourism service providers (n=52)

Nº	Indicators	Assessment results (1="no"; 5="very high"
1	Nature activity	4.27
2	Historical and cultural tourism	4.25
3	Destination safety	4.22
4	Environmental cleanliness	4.21
5	Satisfaction with service quality	4.1
6	Product quality and variety	4
7	Quality of tourist infrastructure	4
8	Quality-price relationship of the product	3.9
9	Electricity and Gas Supply	3.9
10	Quality of provided services	3.8
11	Information about tourists	3.8
12	Climate	3.5
13	Information technology (Wi-Fi, mobile)	3.45
14	Implementation of innovations	3.42
15	The number of hotels and service quality	3.42
16	Barrier-free access to destination	3.4
17	The quality and diversity of national cuisine	3.4
18	Quality of food	3.38

quality-price relationship of the product were determined as the most important indicators for the competitiveness of Khorezm tourism destination (Table 8).

Tourism service providers cited electricity and gas supply, the quality of the services provided, tourist destination climate, the number of guests and quality of service, the possibility of barrier-free access to destination, the degree of the development of information technologies, the literacy of the population on tourism and the quality and variety of foodstuffs as other indicators that were important for the competitiveness of tourist destinations.

At the next stage, from the point of view of the tourism service providers, the most important factors affecting the competitiveness of Khorezm tourism destination are evaluated.

From the point of view of tourism service providers, destination safety, historical-cultural heritage, natural resources, general infrastructure, the possibility of trade in tourism destination and others were determined as the most important factors that represent the competitiveness of the tourism destination

To improve the competitiveness of tourism destination, tourism service providers are required not only to study the indicators that are important for themselves, but also learn and analyze the suggestions and ideas of tourists who are customers of tourism products and services to provide the variety of tourism products from the point of view of customers and make an appropriate conclusion.

Table 9. The assessment of the main 20 factors represent the competitiveness and attractiveness of Khiva tourism destination from the point of view of the tourism service providers (n=52)

N°	Factors	Assessment results (1-"no"; 5-"very high"
1	Historical and cultural resources	4.85
2	Natural resources	4.75
3	Destination safety	4.74
4	Tourism activity	4.72
5	Hospitality	4.7
6	General infrastructure	4.68
7	Opportunity to trade	4.62
8	Quality of services	4.6
9	Knowledge and skills of the staff	4.5
10	Globalization	4.4
11	Destination Tourism Policy	4.38
12	The competitive environment	4.35
13	Destination marketing	4.3
14	Cost competitiveness	4.25
15	Geographical location of Destination	4.2
16	Destination Management	4.1
17	Special events and festivals	3.9
18	Free entry to Destination	3.8
19	Business relations	3.7
20	Protection of the environment	3.5

Important factors for increasing the competitiveness of tourist destinations by tourism service providers are highlighted in the form of ratings based on the results of the survey. The most important factors are the historical and cultural resources of the tourist destination, natural resources and safety in the destination. The main reasons for this are the fact that Khiva tourism destination is a famous historical tourist center located on the Great Silk Road, the unique nature of Khorezm oasis and the safety of tourism service providers in the tourist destination. The inviolability of property, free conduct of business and state support of tourism service providers operating in Khiva destination are considered as a positive factor in increasing the competitiveness of the tourist destination.

Now the indicators and factors symbolize competitiveness and attractiveness of tourism destination are analyzed from the point of view of tourists.

It is important to note that more noticeable changes happened during the evaluation process of the competitiveness of tourism destination from the point of view of tourists than that of tourism service providers. Here tourists determined the historical-cultural tourism as the most important indicator because of that tourism destination was rich in historical-cultural resources. Moreover, it should be highlighted that tourist activity regarding nature, safety, satisfaction of tourists on providing services and climate

Table 10. The assessment of the main 18 indicators which represent the competitiveness of Khiva tourism destination from the point of view of the tourists (n=480)

Nº	Indicators	Assessment results (1-"no"; 5-"very high"
1	Historical and cultural tourism	4.55
2	Nature activity	4.45
3	Destination safety	4.4
4	Environmental cleanliness	4.35
5	Satisfaction of tourists with the quality of services provided	4.25
6	Climate	4.22
7	Electricity and Gas Supply	4.2
8	The quality and variety of regional products	4.15
9	Quality of provided services	4.12
10	The number of hotels and service quality	4.1
11	Diversity of national cuisine	4
12	The literacy of the population on tourism	3.9
13	Adventurous activity	3.8
14	Quality and variety of tourist infrastructure	3.8
15	Information technology (Wi-Fi, mobile)	3.6
16	Quality-price relationship of the product	3.5
17	Tourist information	3.5
18	Quality and diversity of food products	3.3

were considered as important indicators that affecting the competitiveness of tourism destination (Table 9 and Table 10).

From the point of view of tourists, the historical-cultural resources of the tourism destination, safety, natural resources, the variety of tourist activities, tourism and transport infrastructure were determined as the most important factors represent the attractiveness of tourism destination (Table 11).

From the point of view of tourists, the richness of historical and cultural resources and their concentration in one place, as well as the fact that Khiva destination is represented as the legendary "Museum under the open sky" are important factors for the attractiveness of the destination. The tourist destination has not experienced any natural disasters, various infectious diseases, terrorist threats and criminal cases against tourists so far, moreover owns enough conditions for tourists to feel free and safe during the trip and enjoy the trip. These factors also play an important role to increase the attractiveness of tourist destination.

The assessment of the tourists on the attractiveness of Khiva tourism destination, in terms of the overall results, is better and more positive than that of the tourism service providers. For tourism service providers, the importance of competitiveness of tourist destinations is reflected in the overall results. The research focused on the importance of the opinions expressed by tourists, as their opinions are interpreted as an important tool in making changes in the future activities of tourism service providers. It is possible to conclude that the 5 most important factors such as historical and cultural resources, se-

Table 11. The assessment of the main 12 factors which represent the attractiveness of Khiva tourism destination from the point of view of the tourists (n=480)

N°	Factors	Assessment results (1-"no"; 5-"very high"
1	Historical and cultural resources	4.65
2	Destination safety	4.55
3	Natural resources	4.3
4	Tourism activity	4.1
5	Tourism infrastructure	3.9
6	Transport infrastructure	3.8
7	Quality of services	3.7
8	Opportunity to trade	3.5
9	External price competition	3.45
10	Events and festivals	3.4
11	Active communication to locals	3.35
12	Internal price competition	3.3

curity in destination, natural resources, tourist activity and infrastructure identified in the analysis of the survey will be the basis for Khiva destination to further improve its position in the tourist markets and become more competitive in the future. Tourists often approached the assessment of factors affecting the competitiveness of a tourist destination from a seasonal perspective. Therefore, they are difficult to use in destination management processes. There are very few factors such as historical, cultural and natural resources, trade opportunities, infrastructure and quality of services provided that affect the satisfaction of tourists with the travel process and their positive impact on their lifestyle. It is crucial to take these factors into account in the process of creating a tourism product and make sure that they have a positive impact on the lives of tourists.

OPPORTUNITIES TO DEVELOP TOURISM IN KHOREZM REGION UNTIL 2030

The matter of developing tourism in Khorezm region depends on a number of factors related to economy, demography, ecology, technology and others. They consist of the followings:

- All over the World, International tourism that has been developing steadily for over sixty years, also increasing demand for tourism services in many developing countries, such as China, India, Russia, the Middle East and East Asia, close to Uzbekistan;
- Traveling around the Great Silk Road and countries that reflect Eastern culture has become an interesting destination for many Western countries, popularizing historical and cultural tourism;
- Increasing interest in personally visiting areas that are particularly vulnerable to environmental damage, in order to gain a deeper understanding of environmental issues among the World's population, especially tourists;

- Increasing interest in ethnographic, cultural and national cuisine (gastronomic tourism) of different nations;
- Increasing interest in youth for active leisure activities (sports, hunting, extreme tourism, etc.);
- New requirements made by people of all ages for family travel around the World, and others.

As the experience of the developed countries in terms of tourism industry shows, the proper infrastructure and a number of essential conditions for tourism planning should be created, taking into account the long-term trends. In this sense, the main objective of this strategy is to ensure the accelerated development of tourism in Khorezm region and to increase the competitiveness of the region at national and international levels. Successful development of the region, as defined in the strategy of tourism, depends on the active involvement of regional government, the territorial administration of Khorezm region for tourism development and a number of other responsible organizations and local tourism businesses representatives.

Due to the main objective of the strategy, one of the long-term objectives is to increase tourism level in Khorezm region to the level of Samarkand and Bukhara regions. In order to reach this purpose, it is crucial that Khorezm should be one of the leading centers in the global tourism market and a significant increase in the number of tourists to visit should be achieved. Moreover, it is crucial to pay attention for the potential of Khiva to be smart destination with the help of its characteristics. Because, many historical and cultural resources, tourism service providers and information-communication infrastructure objects are situated quite close to one another in Khiva. Therefore, it is easy and enjoyable to make a trip by walking there and there is no need for driving cars during the travel. It may enable Khiva have a chance to be smart destination. In turn, developed Khiva destination will contribute to the raise of tourism level of Khorezm region.

CONCLUSION

According to the survey, while the metrics of tourist destination attractiveness are better and more positively rated by tourists, the importance of competitiveness indicators for tourism service providers was highlighted. Tourists also focused on seasonality in covering competitiveness factors in the survey, which is often overlooked by tourist destination managers. As a result, the number of tourists who want to travel during the season is high, but due to the availability of tourist infrastructure, not all tourists can travel. Therefore, they travel to other tourist destinations. Through the development of information and communication technology infrastructure in tourist destination, the task is to maintain the motivation of potential tourists who are unable to travel this season to travel there. This can be achieved by creating new websites that promote the tourist destination and make it accessible to tourists virtually, opening new pages on social networks and creating virtual excursions of museums.

Tourism service providers considered several factors such as establishment of transport routes in different directions, the image of tourist destination, knowledge and skills of personnel, access to major tourist markets, the level of development of information and communication infrastructure, interaction of the subjects operating in the tourist destination and government support measures as important not only in the development of their activities, but also in increasing the well-being of the local population living here.

The importance of the competitiveness of the tourism destination is expressed in general results for tourism service providers. Within the scientific research the importance of the suggestions and ideas of tourists has been paid attention, because their suggestions and ideas are important tools to make required changes in the activities of tourism service providers. The most important five factors defined due to the survey results are historical-cultural resources, destination safety, natural resources, tourist activity and infrastructure should be stated as a conclusion that they are the bases for Khiva destination to improve its reputation and be competitive in the tourist markets.

Tourists often approach seasonality in assessing factors affecting the destination competitiveness. Therefore, it is quiet difficult to apply them in the process of destination management. Very few factors such as historical-cultural and natural resources, trade opportunity and high quality of infrastructure and provided services positively influence on tourists to make great satisfaction with travelling and their lifestyle. It is crucial to take these factors into account in the process of making tourism products and achieve a positive influence on the lives of tourists.

REFERENCES

Annual report of the state committee of the Republic of Uzbekistan for tourism development, 2019,

Annual report of the state committee of the Republic of Uzbekistan on statistics, 2019

Annual report of the territorial administration of Khorezm region for tourism development, 2019

Bobirca, A., & Cristureanu, C. (2008). Analyzing Romania's competitiveness as a tourism destination. *Advances in Hospitality Leisure*, (4), 75-99.

Caldwell, N., & Freire, R. J. (2004). The differences between branding a country, a region and a city: Applying the brand box model. *Journal of Brand Management*, *12*(1), 50–61. doi:10.1057/palgrave.bm.2540201

Carvalho, L., & Costa, T. (2011). Tourism innovation – a literature review complemented by case study research. *International conference on tourism and management studies proceedings*.

Cibinskienea, A., & Snieskieneb, G. (2015). Evaluation of city tourism competitiveness. *20th International Scientific Conference Economics and Management - 2015 (ICEM-2015).*

Cracolici, M., & Nijkamp, P. (2008). The attractiveness and competitiveness of tourist destinations: A study of Southern Italian regions. *Tourism Management*, *30*(3), 336–344. doi:10.1016/j.tourman.2008.07.006

Croes, R., & Rivera, M. (2010). Testing the empirical link between tourism and competitiveness: Evidence from Puerto Rico. *Tourism Economics*, *16*(1), 217–234. doi:10.5367/000000010790872114

Crouch, G. I., & Ritchie, J. R. B. (1999). Tourism, competitiveness, and societal prosperity. *Journal of Business Research*, *44*(3), 137–152. doi:10.1016/S0148-2963(97)00196-3

Crouch, G. I., & Ritchie, J. R. B. (2000). The competitive destination: A sustainability perspective. *Tourism Management*, *21*(1), 1–7.

Crouch, G. I., & Ritchie, J. R. B. (2003). *The Competitive Destination: A Sustainable Tourism Perspective*. CABI Publishers.

De Moya, M., & Jain, R. (2013). When tourists are your "friends": Exploring the brand personality of Mexico and Brazil on Facebook. *Public Relations Review, 39*(1), 23–29. doi:10.1016/j.pubrev.2012.09.004

Dwyer, L., & Kim, C. (2003). Destination competitiveness: Determinants and indicators. *Current Issues in Tourism, 6*(5), 369–414. doi:10.1080/13683500308667962

Enright, M. J., & Newton, J. (2004). Tourism destination competitiveness: A quantitative approach. *Tourism Management, 25*(6), 777–788. doi:10.1016/j.tourman.2004.06.008

Gallarza, M. G., Gil-Saura, I., & García, H. C. (2002). Destination image: Towards a conceptual framework. *Annals of Tourism Research, 29*(1), 56–78. doi:10.1016/S0160-7383(01)00031-7

Kiráľová, A., & Pavlíčeka, A. (2015). Development of Social Media Strategies in Tourism Destination. *Procedia: Social and Behavioral Sciences, 175*, 358–366. doi:10.1016/j.sbspro.2015.01.1211

Komsic, J., & Dorcic, J. (2016). Tourism destination competitiveness and online reputation: Conceptualization and literature framework analysis. In *Tourism & Hospitality Industry 2016, Congress Proceedings* (pp. 144–157). Opatua, Croatia: University of Rueka.

Kuscer, K. (2012). *Modelling mountain tourism destination development with focus on innovativeness* (Doctoral dissertation). Ljubljana.

Lopez de Avila, A. (2015). Smart destinations: XXI century tourism. ENTER2015 conference on information and communication technologies in tourism, 4–6.

Marchiori, E., & Cantoni, L. (2011). The online reputation construct: Does it matter for the tourism domain? A literature review on destinations' online reputation. *Information Technology & Tourism, 13*(3), 139–159. doi:10.3727/109830512X13283928066715

Navruz-Zoda, B.N., Ibragimov, N.S., Navruz-Zoda, Z.B., & Navruz-Zoda, Sh.B. (2017). *Tourism destination competitiveness, a monograph*. Bukhara, Uzbekistan: "Sadriddin Salim Buxoriy" Durdona Publishing.

Nelson, R. R., & Winter, S. G. (1982). *An Evolutionary Theory of Economic Change, Mass*. Harvard University Press.

Ozen, K.Y. (2017). Competitiveness of Istanbul as A Tourism Destination for Luxury Market. *Journal of Tourismology, 3*(2).

Pechlaner, H., Smeral, E., & Matzier, K. (2002). Customer value management as a determinant of the competitive position of tourism destinations. *Tourism Review, 57*(4), 15–22. doi:10.1108/eb058390

Pike, S. (2002). Destination image analysis - a review of 142 papers from 1973 to 2000. *Tourism Management, 23*(5), 541–549. doi:10.1016/S0261-5177(02)00005-5

Sandybayev A. (2016). Strategic Innovation in Tourism. A Conceptual and Review Approach. *International Journal of Research in Tourism and Hospitality, 2*(4), 5-10.

UNWTO Statistics and Tourism Satellite Account Programme. (2019). *Measuring Sustainable Tourism: Developing a statistical framework for sustainable tourism*. UNWTO.

Tourismusforschung Tirol–Executive Summary. Forschungsbereich 1. Destinations management. MCI Tourismus Destination Research Lab. (2019). https://www.ttr.tirol/sites/default/files/2019-07/2.1%20MCI%20Tourismus%20Destination%20Research%20Lab.pdf

UNWTO Measuring Sustainable Tourism. (2017). *A Call for Action, Report of the 6th International Conference on Tourism Statistics*. Manila, Philippines: UNWTO.

Zhang, H., Fu, X., Cai, L. A., & Lu, L. (2014). Destination image and tourist loyalty: A meta-analysis. *Tourism Management*, *40*, 213–223. doi:10.1016/j.tourman.2013.06.006

Chapter 15
What We (Do Not) Know About Research on Innovation in Hotels:
A Bibliometric Analysis

Catarina Fernandes
Instituto Politécnico de Bragança, Portugal & Centro de Economia e Finanças, Universidade do Porto, Portugal

Rui Alexandre R. Pires
https://orcid.org/0000-0001-8217-8459
Instituto Politécnico de Bragança, Portugal

ABSTRACT

This chapter performs a bibliometric analysis on innovation in hotels field. More specifically, it maps the current research front on innovation in hotels field based on 287 papers published in refereed journals indexed to the Scopus database. Using the VOSviewer software and the bibliographic coupling method, it was possible to identify and systematize the main research streams of this field: 1) technological innovation, 2) sustainable innovation, 3) innovative work behavior, 4) market orientation' role in promoting innovation, 5) knowledge-based resources as a driver of innovation and performance as an outcome of innovation, 6) service innovation, 7) knowledge sharing and management as a key factor for innovation, 8) innovation ambidexterity, 9) innovation in time of crisis, and 10) learning approach as a source of innovation. The findings presented in this chapter will enable future authors studying innovation in hotels to focus their studies more effectively.

DOI: 10.4018/978-1-7998-4855-4.ch015

INTRODUCTION

Innovation, although not being a new phenomenon, has not always received the required attention. However, in recent years the research on the role of innovation is proliferating, since it is increasingly regarded as crucial for competitiveness and performance. In fact, "innovation has been widely accepted as a critical factor for hospitality and tourism enterprises, organizations and destinations, and is recognized as a strategic issue to achieve growth and long-term oriented success" (Pikkemaat, Peters, & Bichler, 2019, p. 184).

Tourism firms, namely hotels, are pressured to innovate continually because of the speed and ease with which their competitors can copy successful ideas (Gomezelj, 2016; Vila, Enz, & Costa, 2012). Thus, hospitality firms, such as hotels, are shown as an ideal example of a market which could benefit from the implementation of service innovation (Victorino, Verma, Plaschka, & Dev, 2005). From a customer's perspective, the hospitality market is continually inundated by many similar, often easily substitutable service offerings, which makes it more difficult to differentiate a hotel from its competitors (Reid & Sandler, 1992). One solution to this challenge may be to offer innovative features to customers, with focus on quality, customization, and technological interfaces. Service innovation does matter when customers are selecting a hotel (Victorino et al., 2005).

For Orfila-Sintes, Crespí-Cladera, and Martínez-Ros (2005), there are three main characteristics of the hotel industry that make different from other service activities and justify a separate analysis. Hotel-specific characteristics are category, governance, and chain structure. "The ownership structure and the degree of dependence on other organizational structures influence the framework of management innovation decisions" (p. 854). Some hotels are independently managed while others belong to a hotel chain or to a more diversified company conglomerate. "The hotel category classification indicates the level and complexity of services provided. Their influence over the innovation propensity is simultaneously an endogenous and exogenous outcome" (p. 854).

The main purpose of this study is to map the current research front on innovation in hotels field by performing a bibliometric analysis. This approach aims to study the research on the field (Gutiérrez-Salcedo, Martínez, Moral-Munoz, Herrera-Viedma, & Cobo, 2018) and its use has growing in tourism research in recent years (e.g., Durán-Sánchez, Álvarez-García, del Río-Rama, & Rosado-Cebrián, 2019; Okumus, Köseoglu, Putra, Dogan, & Yildiz, 2019). Although Durán-Sánchez et al. (2019) and Okumus et al. (2019) carry out bibliometric analyses on tourism innovation and lodging-context research, respectively, to best of the authors' knowledge, until this moment, no bibliometric analysis on innovation in hotels field was performed. Moreover, Okumus et al. (2019) suggest the use of advanced bibliometric analyses in the lodging-context research. In this sense, this study contributes to deepen knowledge on the field and helps future researchers studying innovation in hotels to focus their studies more effectively. Furthermore, it helps hotels' managers and employees to know the main types and features of innovation in hotels, their determinants and outcomes, and the role of customers, managers, and employees in this process.

Innovation is not only a driver of hotels' value creation and long term success but also a source of tourism destinations economic development and growth and the role of the local community cannot be ignored.

The remainder of the chapter is structured as follows. Section 2 presents an overview on innovativeness and innovation, namely in tourism and hotel industries. Section 3 describes the methodological procedures followed. Section 4 presents a brief sample description and performance analysis, specifically publications' evolution, journals, authors, and most cited papers, and maps the field on innovation in hotels using the bibliographic coupling method. Section 5 provides some research gaps which represent opportunities for further studies. Finally, section 6 provides the conclusion, directions for future research, and limitations.

INNOVATIVENESS AND INNOVATION: AN OVERVIEW

Innovativeness versus Innovation: An Analysis

The word innovation is derived from the Latin word "innovare" meaning to renew, to make new, or to change. One of the most important authors on innovation was Joseph Schumpeter (1883-1950), often called "the "godfather" of innovation studies" (Tidd & Bessant, 2018, p. 8), and, even, "the father of the economic theory of innovation" (Divisekera & Nguyen, 2018, p. 158).

In much of the literature, innovation and innovativeness are used as synonym concepts. As a result of inconsistency in defining both terms[1] and, thus, the interchangeable use of them, there have been studies which present inconsistent and sometimes conflicting findings (Kamaruddeen, Yusof, & Said, 2010). On the other hand, "while the concept of innovation is widely analyzed in the literature, the definition of "innovativeness" is rarely discussed and there are many interpretations and approaches, leading to ambiguity and confusion in practice" (Tajeddini, Trueman, & Larsen, 2006, p. 530). For these reasons, it is critical that researchers recognize the difference between innovativeness and innovation.[2]

Different definitions of innovativeness were found in the literature. Several researchers establish a relationship between innovativeness and the degree of newness (e.g., Garcia & Calantone, 2002; Hurley & Hult, 1998), while other view innovativeness as an enduring organizational characteristic (e.g., Subramanian & Nilakanta, 1996) and an innovative capability (e.g., Kahn, 2018; Wang & Ahmed, 2004). Still, other studies regard innovativeness as the willingness to change (e.g., Calantone, Cavusgil, & Zhao, 2002; Tajeddini et al., 2006). Thus, the definition of innovativeness includes "newness", "orientation toward innovation", "innovative behavior", "innovative capability", and "willingness and ability to adopt innovations". Various factors of the firm's culture, such as participative decision making support, affect whether the firm has an innovative orientation. Also, a temporal dimension is considered in the definition of innovativeness, in particular the time of adoption and the consistency over time.

In the hotel industry, innovativeness "embraces a broad spectrum of activities such as developing appropriate strategies, new technologies, supportive leadership, improved services, safety, and environmental friendly issues as well as information and communication technologies interaction" (Tajeddini, 2010, p. 223).

"Innovation is a commonly used term, in various settings and with different meanings" (Wikhamn, Armbrecht, & Wikhamn, 2018, p. 2493). It is defined differently "depending on the research focus" (Pikkemaat et al., 2019, p. 184). Nevertheless, innovation "is often associated with aspects such as creativity, novelty, value creation and economic growth" (Wikhamn et al., 2018, p. 2493). "Common to all definitions of innovation is newness, as innovation is strongly associated with something new. In tourism research, the definitions by Schumpeter (1934) and the OECD/Eurostat (2018) are frequently used" (Pikkemaat et al., 2019, p. 184).

For Shumpeter (1934, p. 66) innovation is "the carrying out of new combinations" and covers: 1) the introduction of a new or improved (new quality) good; 2) the introduction of a new method of production; 3) the opening of a new market; 4) the conquest of a new source of supply; and 5) the carrying out of a new form of organization. Thompson (1965, p. 2), also, emphasizes the link between innovation and "the new" and considers the innovation as a set of sequential phases, given that innovation is defined as "the generation, acceptance and implementation of new ideas, processes products or services." On the contrary, some authors only focus on the adoption/implementation of "the new", presenting a narrow definition of innovation, as is the case of Daft (1978) and Amabile, Conti, Coon, Lazenby, and Herron (1996). They define innovation, respectively, as "the adoption of a new idea or behavior by an organization" (p. 197) and "the successful implementation of creative ideas within an organization" (p. 1155). Although with a more restricted scope, such definitions include a central element of innovation: implementation. "A common feature of an innovation is that it must have been implemented" (OECD/Eurostat, 2005, p. 47). Gault (2018, p. 619) stresses that "an innovation is the implementation of a new or significantly changed product or process." In fact, "innovation does not occur when a new idea is generated, but rather when that new idea is put into use" (Damanpour, 1987, p. 676).

"Innovation is more than ideas and creating something new; execution in terms of getting the offering into the hands of consumers, having purposeful use, and achieving market acceptance is an essential part of innovation" (Kahn, 2018, p. 457). Edquist (1997, p. 1) defines innovation as "new creations with economic significance". The concept of economic significance means that innovation should create value. It should be disseminated and accepted by the market in order to result in commercial transactions. Gault (2018) following the earlier work (Gault, 2012, 2016) proposes a more general definition than that of the Oslo Manual (OECD/Eurostat, 2005)[3], replacing "introduced on the market" for product innovation by "made available to potential users" and "the firm's operations" by "in the operation of the institutional unit, including the making of product available to potential users." Thus, the proposed general definition is the following: "An innovation is the implementation of a new or significantly changed product or process. A product is a good or a service. Process includes production or delivery, organisation, or marketing processes. A new or significantly changed product is implemented when it is made available to potential users. New or significantly changed processes are implemented when they are brought into actual use in the operation of the institutional unit, including the making of product available to potential users" (Gault, 2018, p. 619). The change "introduced on the market" to "made available to potential users" create the possibility of firms innovating by making free products available to potential users, in addition to placing products on the market at market prices. In accordance with Gault (2018), the Oslo Manual (OECD/Eurostat, 2018, p. 20) states that "an innovation is a new or improved product or process (or combination thereof) that differs significantly from the unit´s previous products or processes and that has been made available to potential users (products) or brought into use by the unit (process)."

Moreover, several authors directly relate knowledge and innovation stressing the importance of knowledge creation, combination, diffusion, and management as a vital element for the innovation process to occur (e.g., Edquist, 1997; Feldman, 1994; Plessis, 2007). Feldman (1994, p. 1) argues that "innovation perhaps more than other economic activity, depends on knowledge." Moreover, Plessis (2007, p. 21), for instance, defines innovation as "the creation of new knowledge and ideas to facilitate new business outcomes, aimed at improving internal business processes and structures and to create market driven products and services."

In an attempt to develop a multidisciplinary and integrative concept of innovation Baregheh, Rowley, and Sambrook (2009) surveyed sixty definitions. Through a content analysis, they identify six attributes of innovation definitions, which form the basis for an integrative definition: 1) nature of innovation; 2) type of innovation; 3) stages of innovation; 4) social context; 5) means of innovation; and 6) aim of innovation. Thus, Baregheh et al. (2009, p. 1334) propose the following multidisciplinary and integrative definition of innovation: "Innovation is the multi-stage process whereby organizations transform ideas into new/improved products, service or processes, in order to advance, compete and differentiate themselves successfully in their marketplace." Innovation is, therefore, a process and not a discrete act. More recently, Kahn (2018) argues that innovation is an outcome, a process, and a mindset. Innovation as an outcome addresses the end the organization intends to achieve. Process includes the ways and means through which innovation can occur. Mindset aligns employees and reveals the culture needed for innovation to happen.

In short, when comparing innovation with innovativeness "innovation seems to incorporate the adoption or/and implementation of "new" defined rather in subjective ways, whereas innovativeness appears to embody some kind of measurement contingent on an organization's proclivity towards innovation" (Salavou, 2004, p. 33). Innovation can been seen as the result of innovativeness or, in other words, the outcome of being innovative: if people in an organization are not innovative or not allowed to think and work innovatively then innovation cannot happen. Thus being innovative results in innovations.

Importance and Types of Innovation on Tourism

Innovation plays an important role both at the countries/tourism destinations (as a whole) level and at the individual firms level – tourism firms and other firms belonging to a destination. Innovation has been widely accepted as a vital factor for hospitality and tourism firms, other firms, and destinations and is recognized as a strategic issue to achieve growth and long-term oriented success (Pikkemaat et al., 2019). Tourism destinations comprise local actors (community, public organizations, tourism firms and other private organizations,) and links to external actors whose strategies and policies influence the destination's management and development.

Innovations are functions of "creative destruction" (Schumpeter, 1934)[4], which challenges market equilibriums and provide new opportunities for the existing and the new firms (Wikhamn et al., 2018). "Innovation is a first-magnitude competitive force" (Camisón & Monfort-Mir, 2012, p. 776). Due to the globalization, stiff competition in the global market, and an explosion of technology in recent years, innovation is considered as a need for every firm (Tajeddini & Trueman, 2008). As more firms compete globally, there is increasingly more focus on innovation as a way to have a competitive advantage in the market (Ratten, Ferreira, & Fernandes, 2017). "The expected output of the innovation process is an enhanced competitiveness and/or improvement of performance" (Orfila-Sintes & Mattsson, 2009). Therefore, according to the existing literature, innovation is crucial for survival, sustained success, competitiveness/competitive advantages, growth, and performance (e.g., Orfila-Sintes & Mattsson, 2009; Ratten et al., 2017; Sundbo, Orfila-Sintes, & Sørensen, 2007; Úbeda-García, Claver-Cortés, Marco-Lajara, García-Lillo, & Zaragoza-Sáez, 2018).

In an increasingly global environment, innovation is also a key factor to maintain competitiveness in tourism destinations (Pikkemaat et al., 2019). Also, the "increasing globalization and great rivalry among national and international destinations imply that radical innovation has become a critical factor for the competitiveness of firms in the tourism industry" (Martínez-Pérez, Elche, & García-Villaverde, 2019, p. 80). On the one hand, the image of a destination is co-create by tourism firms (Saraniemi & Kylänen, 2011). Thus, continuous innovation is vital not only for firms competing within the destination but also (collaborating) to compete with other destinations to attract more tourists (Hjalager, 2010a, 2010b). Also, Martínez-Pérez et al. (2019) argue that if tourism firms are able to generate a higher level of radical innovation, the destination will be considered innovative, resulting in a competitive advantage and a better image compared to other destinations with identical characteristics. On the other hand, when a destination is widely recognized and has a good image, all the tourism firms located in this destination will benefit (Martínez-Pérez et al., 2019). Innovation can occasionally be seen exclusively as an internal activity in individual firms. "The destination as such is a repository of competence and knowledge, and parts of this knowledge are unique and inimitable, and crucial for the development of products and services" (Hjalager, 2010a, p. 6). In fact, "it is widely recognized that tourism enterprises can seldom be understood entirely as solitary units; tourist's consumption patterns link them together in bundles, either in formal, informal or in de facto structures at the destination or in other environments" (Hjalager, 2010a, p. 7). Thus, tourism firm innovation is related to both firm-level factors (such as professionalism) and firms' participation in local and non-local networks or in other words, networks within and beyond destinations (Sundbo et al., 2007). Sundbo et al. (2007, p. 88) highlight that "tourism firms operate in a competitive sector where innovating is often a condition for survival." In fact, "developing and applying new ideas that add value to a service is no easy task, particularly when hotel competitors seek to duplicate any innovation they detect" (Vila et al., 2012, p. 75). Thus, it is important to identify innovations that are difficult for competitors to copy (Vila et al., 2012). Similarly, for Reid and Sandler (1992, p. 68) "in the hotel industry, product differentiation has become increasingly difficult, as there is increased segmentation, diverse and overlapping options, and increased competition. The customer is provided a wide range of choices in most markets." By increasing the capacity for innovation, a tourism supplier can become increasingly exclusive (Gomezelj, 2016). Competitiveness in tourism is particularly dependent on innovation for achieving lower costs and higher quality outputs (Ottenbacher & Gnoth, 2005; Sundbo et al., 2007) and meet the demand requirements of potential customers (Sundbo et al., 2007). As the complexity of the tourism experience is determined by interrelated services and products as well as by the interaction among several different actors: tourists, tourism firms, and other destination firms and the local community, innovation occurs at various levels with diverse stakeholders and generates distinct forms of innovation (Pikkemaat et al., 2019).

Innovation can take a variety of forms. In fact, there are different types of innovations, characterized by having a distinct nature, distinct determinants and impacts. A typology of innovations offers a good orientation for research and practice (Hjalager, 2002). However, as noted in OECD/Eurostat (2005, p. 53), Oslo Manual, "many innovations may have characteristics that span more than one type of innovation. It can be both difficult and misleading, in terms of types of innovation activities undertaken by firms, to categorize these innovations as a single type." Thus, distinguishing innovation types is not necessarily easy as innovations are frequently bundled: innovation in one field leads to subsequent innovations in other(s) field(s). Often, reciprocity between innovations exists.

In tourism, customer experience is comprised of interwoven and networked single tourism products and services (Pikkemaat et al., 2019). Martin-Rios and Ciobanu (2019) for instance, recommend that hospitality firms should consider implementing a combination of different innovation types that lead to increased performance. "Specifically, hospitality professionals should focus on bundles of innovative initiatives, for example, new or improved supporting activities for processes in combination with new methods for service placement and sales channels" (p. 227).

Schumpeter's (1934) categories of innovations have been applied to some extent in tourism studies (Hjalager, 2010a) and inspired subsequent research as is the case of Hjalager (1997, 2002, 2010a) and OECD and Eurostat's Oslo Manual (2005). Specifically, Schumpeter divides innovation into the following categories: 1) the introduction of a new good or of a new quality of a good (**product innovation**); 2) the introduction of a new method of production (**process innovation**); 3) the opening of a new market (**marketing innovation**); 4) the conquest of a new source of supply/new sources in production (**input innovation**); and (5) the creation of a new form of organization (**organizational innovation**).

Close to Schumpeter's definition, OECD and Eurostat's Oslo Manual (2005) classifies innovations into the four following types: 1) **product innovation** – introduction of a new or significantly improved good or service with respect to its characteristics or intended uses; 2) **process innovation** – implementation of a new or significantly improved production or delivery method, including significant changes in techniques, equipment, and/or software; 3) **marketing innovation** – implementation of a new marketing method involving significant changes in product design or packaging, product placement, product promotion, or pricing; and 4) **organizational innovation** – implementation of a new organizational method in a firm's business practices, workplace organization, or external relations.

Several authors, such as Orfila-Sintes and Mattsson (2009), Vila et al. (2012), and Nicolau and Santa-María (2013), study innovation in the tourism/hotel industry using typologies of innovation. To analyze the innovation behavior in the Balearic hotel industry, Orfila-Sintes and Mattsson (2009) develop and test a model which relates four types of innovation (i.e., management, external communication, service scope, and back-office) to the service provider characteristics, customer competences, and the market drivers.

Later, Hjalager (2010a) proposed five different types of innovations in tourism, including a new category of innovation – **institutional innovation** – which is specific to tourism (Pikkemaat et al., 2019). The categories of innovation are: 1) **product or service innovations** – changes directly observed by the customer/tourist and considered as new, new in the sense of never seen before or new to the specific firm or destination; 2) **process innovations** – typically backstage initiatives which intend to increase efficiency, productivity, and flow; information and communication technology (ICT) constitutes a main factor for process innovations; 3) **managerial innovations** – new ways of organizing internal collaboration and developing human resource practices such as directing and empowering staff, compensating work with financial or nonfinancial benefits building careers, improving workplace satisfaction, fostering internal knowledge, and building team spirit. A main challenge for many tourism firms is to develop methods to retain staff, keep flexibility, and control costs. Methods to retain employees are extremely valuable in the tourism sector, as it is highly labor-intensive; 4) **marketing innovations** – new marketing concepts that change the way that overall communication to, and with, customers is undertaken and how relationships between them and the service provider are created and retain (e.g., loyalty programs and co-production of brands); and 5) **institutional innovations** – new embracing collaborative/organizational structures or legal frameworks that develop the business in certain fields of tourism further (e.g., networks and alliances). Also, regarding tourism innovation, Vila et al. (2012) in their research of the innovative practices in the Spanish hotel industry considered (only) four types of innovations: 1) **product innovations** – products, services, and their attributes; 2) **processes innovations** – operational processes; 3) **knowledge of the market** – distribution channels, web-based communication, customer loyalty, information sharing, and marketing innovations; and 4) **management innovations** – changes in organizational structures, policies, nonoperational processes, and the informal culture.

More recently, OECD and Eurostat's Oslo Manual (2018) presents the taxonomies of innovation, in general, considering: 1) innovation types by object, product, and business process innovations; and 2) innovation types according to novelty and impacts.

METHODOLOGICAL PROCEDURES

As stated above, the main purpose of this chapter is to identify the more recent contributions and detect the current research trends on innovation in hotels. To carry out this study, instead of a traditional narrative review, a bibliometric approach is used, which complements traditional methods of review (Zupic & Čater, 2015). This approach increases the objectivity and quality of reviews by adopting a systematic, transparent, and reproducible review process, avoiding subjectivity and introducing quantitative rigor (Zupic & Čater, 2015). In this way, a bibliometric analysis allows to study or measure a research field through the analysis of the scientific publications indexed in bibliographic databases (Gutiérrez-Salcedo et al., 2018), such as Clarivate Analytics' Web of Science, Elsevier' Scopus, and Google Scholar.

In performing a bibliometric analysis, alternative, but complementary, approaches can be adopted. For instance, Köseoglu, Rahimi, Okumus, and Liu (2016), and Mehraliyev, Choi, and Köseoglu (2019) refer two approaches: evaluative bibliometrics and relational bibliometrics. Gutiérrez-Salcedo et al. (2018) also highlight two main approaches: performance analysis and science mapping. On the one hand, evaluative bibliometrics or performance analysis evaluate the productivity and activity's impact of individuals and groups of scientific actors (e.g., researchers, universities, and countries) using production indicators (e.g., number of published papers in a period of time), impact indicators based on received citations (e.g., total number of citations and H-index), and indicators based on the impact of the journal (Gutiérrez-Salcedo et al., 2018). One the other hand, relational bibliometrics or science mapping reveal intellectual, social, and conceptual structures of research fields based on bibliographic networks using the methods of co-citation, bibliographic coupling, co-author, and co-word (Gutiérrez-Salcedo et al., 2018; Köseoglu et al., 2016; Mehraliyev et al., 2019). This study focuses on both evaluative bibliometrics or performance analysis and relational bibliometrics or science mapping, while it uses science mapping, more precisely intellectual structure, to achieve its main purpose.

Although various workflows are used in the literature to carry out a bibliometric analysis (e.g., Cobo, López-Herrera, Herrera-Viedma, & Herrera, 2011; Zupic & Čater, 2015), this study adopts the five-step procedures defined by Zupic and Čater (2015) for conducting science mapping analyses.

In accordance with these steps, in January 2020 a search was made on Scopus database. Other databases, such as the Web of Science, could also have been considered. Nevertheless, according to Durán-Sánchez et al. (2019) and Mongeon and Paul-Hus (2016), the Scopus database covers more indexed publications in this field of research, being used, also, in similar studies (e.g., Jiang, Ritchie, & Benckendorff, 2019).

The search was made using as keywords "innovat*" AND "hotel*"[5] in three alternative fields: title, keywords, and abstract. The sample was restricted to articles published in international journals by the end of 2019, written in English, and allocated to the subject area of "Business, Management and Accounting". To ensure the homogeneity of the sample other publications such as books, chapter books, conference papers, and reports were excluded. This search process resulted in a total of 619 papers.

The bibliographic data extracted from Scopus database was then exported to Elsevier's software Mendeley 1.19 to manage the papers selected. In this process (i.e., data cleaning) 332 papers were excluded because they were out of the scope of this study. Therefore, the final sample includes a total of 287 papers.

In order to explore and analyze the selected papers and to build the bibliometric network three different procedures were followed. Firstly, the bibliographic data of the final sample was exported and analyzed using Microsoft Excel 2016. In this phase a graphic description was elaborated to examine publications' evolution, main journals and authors, and most cited papers. Secondly, the VOSviewer 1.6.13 software was used to identify the words (and terms) most quoted in the title and abstract of the papers and the WordClouds software to represent them. Thirdly, the VOSviewer 1.6.13 software was used to map the current research front on innovation in hotels field.

To reach the goal of mapping the current research front on innovation in hotels field two bibliometric methods could be used (Vogel & Güttel, 2013; Zupic & Čater, 2015): co-citation and bibliographic coupling. Nevertheless, co-citation is not optimal to mapping research fronts (Zupic & Čater, 2015) whereas bibliographic coupling allows to detect the current research trends (Vogel & Güttel, 2013). According to Vogel and Güttel (2013, p. 429), co-citation neglects "the publication dynamics at the forefront of research" while bibliographic coupling "captures more recent contributions, including the classics of tomorrow." Additionally, co-citation is used in most bibliometric studies in management and organization, which represents an opportunity for the use of bibliographic coupling (Zupic & Čater, 2015).

The bibliographic coupling method analyses the extent to which two publications are related by virtue of them both citing a third publication. This is, publications that cite the same references are related to each other, and that the greater the number of shared references between two publications, the more related they are (Vogel & Güttel, 2013; Zupic & Čater, 2015). In this way, bibliographic coupling links publications, authors, and journals that reference the same set of cited publications (Boyack & Klavans, 2010).

RESULTS

This section contains two parts. The first one presents a brief description of the sample and performance analysis, exploring the issues such as publications' evolution, journals, and authors. The second one comprises the science mapping on innovation in hotels field using the bibliographic coupling analysis of papers, which enabled to identify the main research streams.

Sample Description and Performance Analysis

Figure 1 shows the publications' evolution on innovation in hotels from 1982, year of the first publication founded, to 2019. As can be seen the number of papers published until 2004 is small (14 papers, 4.88% of the sample). However, this trend has changed and there is a significant increase of the publications until 2019, which reveals a growing interest of the scientific community in the field. The data also show that the publications in the last five years (i.e., form 2015 to 2019) represent more than half of the sample (159 papers, 55.40% of the sample) and 2019 was the most productive year, with 44 papers. These findings are in accordance with those founded by Durán-Sánchez et al. (2019), showing an exponential growth of publications on innovation in tourism.

The 287 papers were published in 107 journals. The majority was published in journals from the field of hospitality and tourism but some of them also in several journals from different fields, such as service business, marketing, and strategic management. As the papers were published in a wide variety of journals, Figure 2 presents only the journals with four or more publications.[6] Among these, International Journal of Contemporary Hospitality Management (IJCHM) ($N = 35$, 12.20%), International Journal of Hospitality Management (IJHM) ($N = 35$, 12.20%), and Tourism Management (TM) ($N = 19$, 6.62%) appear to be the major outlets to publications on innovation in hotels. In fact, these journals are the most productive and leading journals in publishing research on lodging-context, namely the IJHM and IJCHM (Okumus et al., 2019). This information is relevant for researchers from the field given that

Figure 1. Number of publications by year

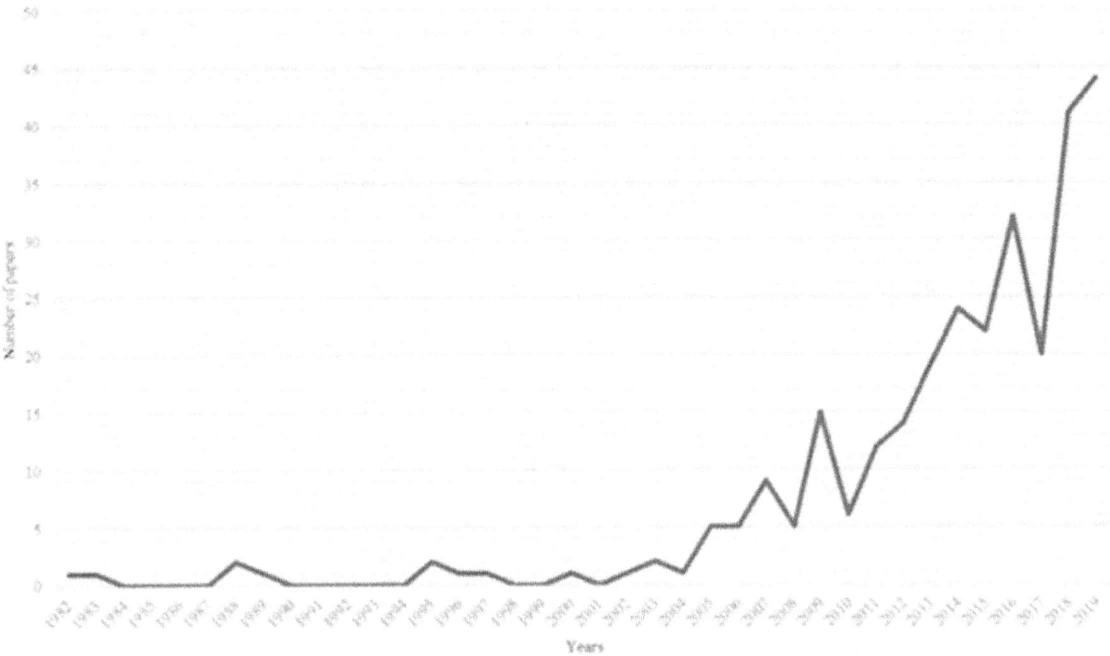

Figure 2. Journals with four or more publications

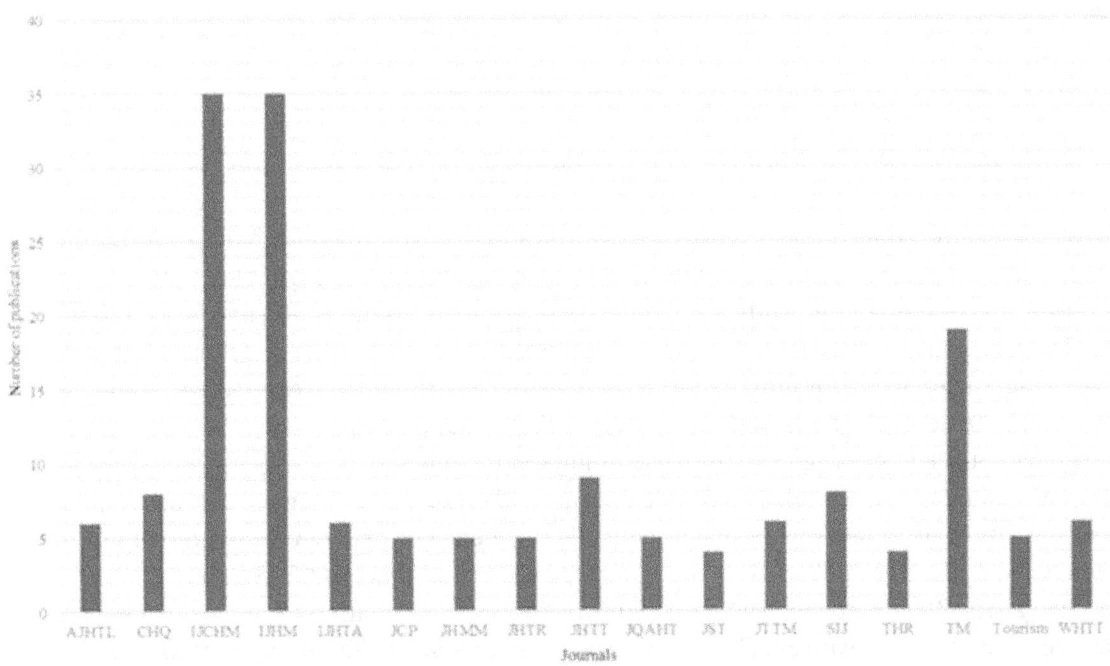

these journals should be considered when choosing where to publish the outcomes of their research (Durán-Sánchez et al., 2019).

Regarding the papers' authorship, there is a wide dispersion, since the 287 papers were produced by 596 authors. This reveals the interest of the field, which has attracted the attention of the scientific community. In this way, Figure 3 only shows the authors with three or more publications. Julia Nieves leads the ranking of the most productive authors with seven papers, followed by Francina Orfila-Sintes with six papers, and Ahmed Hassanien, Kayan Tajeddini, and Ta-Wei Tang with five papers each one.

Figure 3. Authors with three or more publications

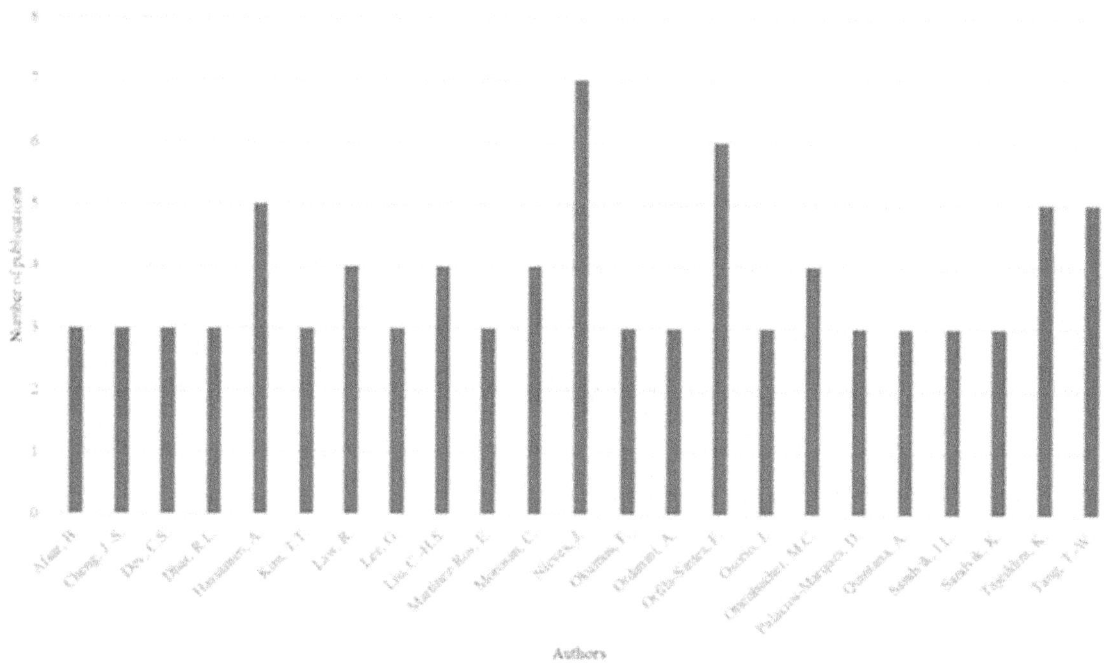

Given that some publications assume seminal roles in the evolution of the field (Furrer, Thomas, & Goussevskaia, 2008), Figure 4 presents the 10 most cited papers. The papers of Ordanini and Parasuraman (2011) and Agarwal, Erramili, and Dev (2003) stand out with a total of 302 and 289 citations, respectively. The first paper, published in the Journal of Service Research (JSR), explores some determinants and outcomes of service innovation's volume and radicalness in Italian luxury hotels. The second paper, published in the Journal of Services Marketing (JSM), examines the mediating effect of innovation between market orientation and performance in international hotels.

The most cited papers examine some antecedents and consequents of innovativeness and innovation in hotels. Most of them investigate, specifically, the influence of market orientation on innovativeness and innovation, and their influence on performance (Agarwal et al., 2003; Sandvik & Sandvik, 2003; Tajeddini, 2010; Zhou, Brown, & Dev, 2009). Market orientation allows to take advantage of market opportunities and has a positive influence on innovativeness and different types of innovation. In addi-

Figure 4. The 10 most cited papers

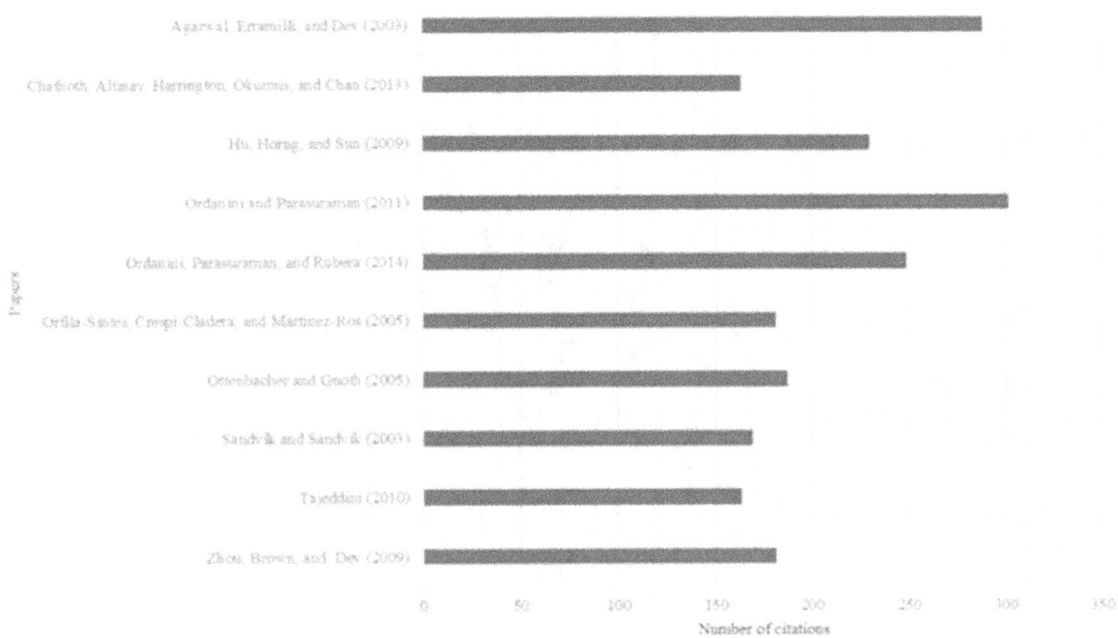

tion, Hu, Horng, and Sun (2009) show the relevance of knowledge sharing behavior and team culture to service innovation performance (i.e., employee service innovation behavior and new service development) and Ordanini and Parasuraman (2011) highlight collaborative competences, dynamic capabilities of customer orientation, and knowledge interfaces as determinants of service innovation's volume and radicalness. The majority of the 10 most cited papers reveals that innovation has a positive influence on performance (e.g., Ordanini & Parasuraman, 2011; Tajeddini, 2010; Zhou et al., 2009), mainly on market performance (e.g., customer satisfaction and service quality), which shows the relevance of innovation as a critical factor for hotels' survival and success.

Finally, using the VOSviewer software the words (and terms) most quoted in the title and abstract of the papers were identified. Figure 5 reveals the results and represents the 70 words (and terms), from 83, quoted more than 19 times in the papers' title and abstract. 13 words (and terms), such as article, author, data, and paper, were excluded because they are quoted in most studies independently of their focus. This is, these words (and terms) are not specific of the field under study in this chapter.

As Figure 5 shows, the words (and terms) that most stand out are hotel and innovation, which match the goal of this study. These words (and terms) appear interrelated with other words (and terms), with less emphasis in the Figure previously mentioned, such as effect, model, relationship, and performance. Figure 5 also highlights words (and terms) related with the process of innovativeness and innovation (e.g., capability, knowledge, and innovative behavior) and its main actors (i.e., customers, employees, and managers).

Figure 5. Representation of the 73 words (and terms) most quoted in the title and abstract of the papers analyzed

Intellectual Structure Based on Bibliographic Coupling of Papers

Using the VOSviewer software and the bibliographic coupling of papers, which reveals clusters of studies with similar sources (Mehraliyev et al., 2019), it was obtained the intellectual structure of the research on innovation in hotels. The network visualization, considering a minimum threshold of five citations, is presented in Figure 6.

Based on both the network diagram and the papers' content analysis, the authors systemized the literature, identifying ten clusters, corresponding to ten main research streams (not mutually exclusive), and indicating the main references, summarized in Table 2: 1) technological innovation; 2) sustainable innovation; 3) innovative work behavior; 4) market orientation' role in promoting innovation; 5) knowledge-based resources as a driver of innovation, and performance as an outcome of innovation; 6) service innovation; 7) knowledge sharing and management as a key factor for innovation; 8) innovation ambidexterity; 9) innovation in time of crisis; and 10) learning approach as a source of innovation.

The cluster "technological innovation", the red one, includes papers that examine the process of adoption and implementation of several types of technological innovations as well as their role and importance in the hotel industry activity such as: mobile technologies – mobile apps and mobile devices (Lei et al.,

Figure 6. Network diagram of the bibliographic coupling of papers

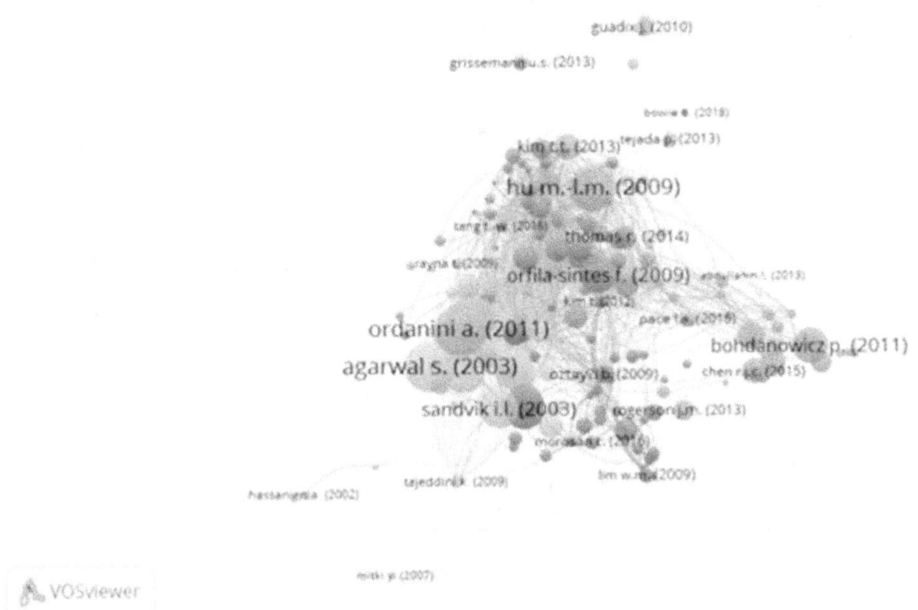

2019; Morosan, 2018; Morosan & DeFranco, 2016; Sarmah et al., 2017); radio frequency identification (RFID) (Aluri & Palakurthi, 2011; Cobos et al., 2016; Öztayşi et al., 2009); internet technologies (Lim, 2009; Palacios-Marqués et al., 2015); in room-technologies (Bilgihan et al., 2016); biometric technologies (Kim & Bernhard, 2014); and ICT (Karadag & Dumanoglu, 2009).

On the one hand, the literature, in general, stresses that the technology is increasingly becoming a vital source of sustainable competitive advantage in the hotel industry given that it enhances service quality, customer satisfaction, and performance (e.g., Karadag & Dumanoglu, 2009; Öztayşi et al., 2009). Sarmah et al. (2017) and Morosan (2018), for example, highlight the role of mobile technologies in the process of co-creation to customize services and, consequently, to fulfil customer needs and expectations, generating satisfaction and profitability. On the other hand, the research, as is the case of Lam and Law (2019), also emphasizes the need for hotels to have the right approach to technology to reach a deeper customer knowledge and, consequently, to attain higher customer satisfaction, loyalty, and usage. In this context, Lei et al. (2019) examine how hoteliers leverage mobile technologies to shape services that allow customers to create their own unique and personalized experiences. Also, Lim (2009) analyses the antecedents affecting technology adoption as well as the hoteliers' intensity of technology adoption. Regarding the "customer side", Aluri and Palakurthi (2011) and Morosan and DeFranco (2016) investigate the factors influencing consumers' intentions to use technologies and, additionally, Aluri and Palakurthi (2011) explore the influence of demographic factors (i.e., age, gender, education, income) on consumer attitudes to use them.

The cluster "sustainable innovation", the green one, contains papers that analyze the effects of eco-innovation practices/green strategies (e.g., recycle materials and waste and save water and energy) on customer perceptions and selection decision (i.e., on customers attraction), customer experience, post-

Table 2. Main research streams on innovation in hotels and main references

Research Stream	Main References
Cluster 1 – Technological innovation	Aluri and Palakurthi (2011), Bilgihan, Smith, Ricci, and Bujisic (2016), Cobos, Mejia, Ozturk, and Wang (2016), Karadag and Dumanoglu (2009), Kim and Bernhard (2014), Lam and Law (2019), Lei, Wang, and Law (2019), Lim (2009), Morosan (2018), Morosan and DeFranco (2016), Öztayşi, Baysan, and Akpinar (2009), Palacios-Marqués, Merigó, and Soto Acosta (2015), and Sarmah, Kamboj, and Rahman (2017).
Cluster 2 – Sustainable innovation	Aboelmaged (2018), Bohdanowicz, Zientara, and Novotna (2011), Chan (2013), Chen (2015), García-Pozo, Sánchez-Ollero, and Ons-Cappa (2016), Horng, Liu, Chou, Tsai, and Chung (2017), and Pace (2016).
Cluster 3 – Innovative work behavior	Bani-Melhem, Zeffane, and Albaity (2018), Dhar (2015, 2016), Grissemann, Pikkemaat, and Weger (2013), Jung and Yoon (2018), Kim and Koo (2017), and Li and Hsu (2016).
Cluster 4 – Market orientation' role in promoting innovation	Agarwal et al. (2003), Grissemann, Plank, and Brunner-Sperdin (2013), Jalilvand (2017), Sandvik and Sandvik (2003), Tajeddini (2010), and Zhou et al. (2009).
Cluster 5 – Knowledge-based resources as a driver of innovation, and performance as an outcome of innovation	Hernández-Perlines, Ariza-Montes, Han, and Law (2019), Mattsson and Orfila-Sintes (2014), Nicolau and Santa-María (2013), Nieves and Diaz-Meneses (2016, 2018), Orfila-Sintes and Mattsson (2009), Tang (2016), and Thomas and Wood (2014).
Cluster 6 – Service innovation	Beshr and Hossan (2018), Hassanien and Eid (2007), Ordanini and Parasuraman (2011), Ordanini et al. (2014), and Ottenbacher and Harrington (2010).
Cluster 7 – Knowledge sharing and management as a key factor for innovation	Hu et al. (2009), Kim and Lee (2013), Salem (2014), and Tang, Wang, and Tang (2013).
Cluster 8 – Innovation ambidexterity	Cheng, Tang, Shih, and Wang (2016), Tang (2014), Úbeda-García et al. (2018), and Wang, Tang, and Cheng (2018).
Cluster 9 – Innovation in time of crisis	Pappas (2015a, 2015b).
Cluster 10 – Learning approach as a source of innovation	Mitki and Herstein (2007), Tajjedini (2009), and Vuković, Gagić, Terzić, Petrović, and Radovanović (2018).

experience and loyalty, competitive advantage, labor productivity, and hotel performance. Focusing on sustainability hotels are responding to the demands of environmentally conscious clients (Horng et al., 2017). Bohdanowicz et al. (2011) specifically analyze Hilton's we care! programme for enhancing the environmental performance of the 70 Hilton Worldwide hotels. In turn, Chen (2015) examines how sustainable-effort-related factors have an impact on consumer purchasing behaviors, experiences, and revisit intention, concluding that among the most significant attributes are the use of recycled materials, energy-efficient lighting, and water-saving. Pace (2016) focuses, in particular, on the innovation for sustainable energy consumption. His findings show that hotels that build internal capabilities for adoption of energy efficient measures and technologies are more likely to co-produce innovative energy efficient solutions.

The cluster "innovative work behavior", the blue one, encompasses papers that investigate what actually drives innovative behavior or, in other words, the set of factors/determinants of such behavior, with focus on employees. In fact, employee innovative behavior is a key factor for business success (e.g., performance and long-term survival), particularly for service firm (Li & Hsu, 2016). Thus, identifying potential drivers of employee innovative behavior remains an important research topic, especially in the hotel industry (Bani-Melhem et al., 2018). The research on this specific topic identifies several determinants, firm-internal determinants and firm-external determinants, such as: employee engagement (Grissemann, Pikkemaat, et al., 2013); development of employee skills, active participation in the institutional decision making process, employee mobility, training for professional development, and motivation (Dhar, 2015); ethical leadership (Dhar, 2016); information and emotional interactions between customer and employees (Li & Hsu, 2016); leaders' role in shaping employees' work attitudes and performance (Kim & Koo, 2017); conflict management climate (Jung & Yoon, 2018); and workplace happiness, job stress, and coworker support (Bani-Melhem et al., 2018).

Concerning firm-internal factors, for instance, Grissemann, Pikkemaat, et al. (2013) find that employee engagement (i.e., permanent training and empowerment of employees) fosters innovation activities. "Highly skilled and open-minded employees are needed to create an innovation-friendly culture within the firm and help to transform ideas into successful innovations (p. 18). Also, Jung and Yoon (2018) highlight that when frontline hotel employees can overcome conflicts their levels of engagement can increase and innovative behavior may appear.

The cluster "market orientation's role in promoting innovation", the "brownish green" one, includes papers that investigate the influence of market orientation on hotel's performance through innovation (e.g., Agarwal et al., 2003; Grissemann, Plank, et al., 2013; Sandvik & Sandvik, 2003). The main purpose of the majority of the existing papers is, thus, to assess the mediating role of innovation in the relationship between market orientation and performance or, in other words, innovation is used as mediator of the influence of market orientation on business performance. In particular, Agarwal et al. (2003) find that the immediate impact of market orientation is to spur innovation, which, in turn, enhances performance. Similarly, Sandvik and Sandvik (2003) show that market orientation through product innovation contributes positively to superior market and financial performance. However, two exceptions to the prevailing research should be mentioned: Tajeddini (2010) and Jalilvand (2017). Both authors analyze the direct relationship between customer orientation, innovativeness, and performance, this is, without mediating effects.

The cluster "knowledge-based resources as a driver of innovation, and performance as an outcome of innovation", the "purple" one, comprises papers that explore the role of knowledge in the innovation (e.g., Nieves & Diaz-Meneses, 2016, 2018; Tang, 2016) and papers that examine the influence of innovation on performance (e.g., Hernández-Perlines et al., 2019; Mattsson & Orfila-Sintes, 2014; Tseng, Kuo, & Chou, 2008). The first set of papers highlights, amongst other aspects, the relevance of the acquisition (potential absorptive capacity) and, specially, the use (realized absorptive capacity) of external knowledge to promote innovative behavior (Thomas & Wood, 2014). In particular, Tang (2016) finds that both the environment scanning (i.e., the acquisition and exploitation of information regarding events and external relationships) and the social capital mediate the relationship between managers' proactive personality and service innovation as well as service improvement. Nieves and Diaz-Meneses (2018), in turn, show that local and non-local knowledge sources and intra-organizational collaboration determine incremental innovations but, contrary to the expected, non-local knowledge sources do not influence radical innovations. The second set of papers, shows that different innovation types and different combinations of these types influence performance in a different manner. Mattsson and Orfila-Sintes (2014), for instance, find that full innovation (i.e., management, external communications, service scope, and back-office innovations) and the combination of service and back-office innovations positively influence both the occupancy rate and the profitability. Nicolau and Santa-María (2013) report that process and marketing innovations have a higher positive impact on market value than product and organization innovations. Furthermore, Hernández-Perlines et al. (2019) stress the relevance of innovation capacity in maximizing hotel performance, highlighting, also, the role of quality certification in this process. In accordance, innovation capacity has a positive influence on performance and this influence is higher in hotels that have quality certification.

The cluster "service innovation", the "light blue" one, contains papers that investigate the environment and facets (i.e., volume and radicalness) of new service development (i.e., service innovation). Service innovation in the hotel industry is critical to satisfy customer requirements, build sustainable competitive advantage, and promote performance (Beshr & Hossan, 2018; Ordanini & Parasuraman, 2011). Hassanien and Eid (2007), for instance, explore how Egyptian hoteliers use the new service development as a marketing tool, revealing that only 5 star hotels integrate service development policies within their marketing plans. Ottenbacher and Harrington (2010) show the key success factors for radical and incremental service innovations, and Ordanini et al. (2014) stress that service innovation adoption depends on the fit between new service attributes and coproduction requirements. More recently, Beshr and Hossan (2018) highlight that organizational support for innovation, organizational trust, and work engagement are the key drivers of service innovation in the Unit Arab Emirates' hotel industry.

The cluster "knowledge sharing and management as a key factor for innovation", the "orange" one, encompasses papers that analyze the importance of knowledge management and, in particular, knowledge sharing in supporting innovation. In fact, knowledge is the key of any business' success (Hu et al., 2009), given that it enables to improve creativity, innovation, and productivity. For instance, Kim and Lee (2013) and Tang et al. (2013) explore, respectively, the role of goal orientations (i.e., learning goal orientation and performance goal orientation) and social capital as antecedents of knowledge sharing and how knowledge sharing influences innovation. Kim and Lee (2013) find that employees with higher levels of learning goal orientation reveal higher levels of knowledge sharing while employees with higher levels of performance goal orientation reveal lower levels of knowledge sharing. In turn, Tang et al. (2013) show that capital social enhances knowledge sharing. Both studies report that it promotes innovation, particularly employee service innovation behavior (Kim & Lee, 2013) and service innovation capability (Tang et al., 2013). Additionally, Hu e al. (2009) reveal that team culture promotes the relationship between knowledge sharing and service innovation performance.

The cluster "innovation ambidexterity", the "brown" one, contains papers that investigate, in particular, how hotels concurrently develop service improvement (i.e., exploitative innovation) and service innovation (i.e., explorative innovation) capabilities for continually creating customer value. In particular, Tang (2014) shows that customer orientation positively influences both service improvement and service innovation capabilities which, in turn, have a positive influence on market performance (i.e., competitive advantage, customer satisfaction, and service quality). Also, Cheng et al. (2016) explore, through a case study, how a hotel can adopt art as a management strategy and become an ambidextrous hotel improving the art knowledge and performance techniques of its service employees (i.e., service improvement capabilities) and providing new art-oriented services by providing them with new skills (i.e., service innovation capabilities). The authors highlight as main support resources a customer-based culture and high-performance human resource practices. In fact, Úbeda-García et al. (2018) show that high-performance work systems influences ambidexterity innovation not only directly, but also indirectly through the ambidextrous organizational culture. Moreover, they find that ambidexterity innovation has a positive influence on Spanish hotels' performance. Thus, these papers highlight that ambidextrous hotels, balancing two contradictory service capabilities, simultaneously meet customer needs and develop new services to create new needs. In this way, hotels improve their performance.

The cluster "innovation in time of crisis", the "pink" one, comprises papers that examine the importance of innovation, respectively, to strengthen competitive advantages (Pappas, 2015a) and as a practice against crisis (Pappas, 2015b) of Greek hotels during a period of recession and political instability. Further, García-Pozo et al. (2016) assess the economic crisis' influence on the relationship between eco-innovation and labor productivity in the Andalusian hotel industry, showing that the introduction of eco-innovation measures has a positive effect on labor productivity, although the economic crisis reduce that effect. These papers highlight that, particularly during crisis periods, a firm's productivity and competitiveness depend on their ability to change and adapt to the new reality. In addition, Campo, Díaz, and Yagüe (2014) explore the influence of the international economic and financial crisis on the relationship between innovation and performance, revealing the importance of the adoption of innovations in periods of crisis.

Finally, the cluster "learning approach as a source for innovation", the "light pink" one, includes papers that focus on the role of the learning process within an organization, which affects its ability to innovate, and thus to compete on the global market and be successful. In particular, Mitki and Herstein (2007) describe an innovative training approach, based on employees' participation and commitment, which transforms hotels into learning organizations that enable them "to maintain a high level of flexibility and to adapt to changes in their environment, while identifying needs and opportunities in advance" (p. 175). Also, Tajeddini (2009) shows that learning orientation (i.e., commitment to learning, open-mindedness, and shared vision) positively influences the development of new service which, in turn, has a positive influence on performance in the Iranian hotel industry. In fact, such orientation is reflected in increased employees' efforts to expand their knowledge and skills, and learn new and better ways of interacting with customers. Other studies confirm that organizational learning has a positive influence on innovativeness and several innovation types which, in turn, positively influence performance (e.g., Nair, 2019).

WHAT WE DO NOT KNOW AND WOULD LIKE TO KNOW ABOUT INNOVATION IN HOTELS

The chapter identifies and systematizes the main research streams on innovation in hotels. The findings show that the existing research successfully covers different topics related to innovation in the specific context of hotels and, also, that technological innovation and sustainable innovation have been gained increased attention. However, some research gaps, which reflects what it is not (sufficiently) known about the specific topic analyzed, emerged and provide promising directions for future research. First, tourism firms, and hotels are no exception, are mainly micro and small firms. Thus, more research is needed to explore the role of family ties/dynamics in innovation in such context (e.g., assess the relation between family-related values and beliefs and innovation). Second, future research needs to examine the interdependencies between innovation at the hotel level and at the destination level and explore the potential of open innovation. The role of policy-makers, politicians, and public authorities (policy actions) to support innovation should be detailed analyzed. Third, regarding methodological approaches, the vast majority of sample studies is cross-sectional and uses quantitative approaches based on questionnaires to collect the data. Further research should adopt longitudinal approaches in order to examine the effect of time on the influence of many internal and external factors on innovation and the influence of innovation on performance. Also, future research should use qualitative approaches to investigate extensively these influences. Longitudinal and qualitative approaches will allow to understand deeply the process of innovation within hotels and the role of internal (e.g., managers and employees) and external actors (e.g., customers, competitors, and business partners) in this process.

CONCLUSION

This study sought to identify the more recent contributions and detect the current research trends on innovation in hotels performing a bibliometric analysis. In fact, the use of bibliometric analysis allowed to carry out a performance analysis, obtaining relevant information on the publications' evolution, journals that publish on the field, most productive authors, and highly cited papers, and to reveal the intellectual structure on innovation in hotels based on science mapping. Thus, the findings of this study provide important contributions to research and practice about innovation in such context.

First, the academic community's interest about innovation in hotels appears to emerge in 1982, with a growing number of publications since 2004, being 2019 the most productive year, which reflects the growing interest of the field and enables to foresee that this increasing tendency will continue in the coming years. Regarding the journals and authors' productivity, IJCHM, IJHM, and TM are the journals with more publications on innovation in hotels in the period considered, and Julia Nieves is the most prolific author, according with the number of appearances.

Second, the intellectual structure obtained using the bibliographic coupling method revealed ten clusters of papers, corresponding to ten main research streams (not mutually exclusive): 1) technological innovation; 2) sustainable innovation; 3) innovative work behavior; 4) market orientation' role in promoting innovation; 5) knowledge-based resources as a driver of innovation, and performance as an outcome of innovation; 6) service innovation; 7) knowledge sharing and management as a key factor for innovation; 8) innovation ambidexterity; 9) innovation in time of crisis; and 10) learning approach as a source of innovation.

This study obviously has some limitations that should be taken into account when interpreting the conclusions. First, the data for this study were exclusively collected from Scopus database. Additionally, the selected papers only are allocated to the subject area of "Business, Management and Accounting". Future studies could collect data from other databases, such as Web of Science database and publications from other subject areas. Second, this study only reveals the intellectual structure of the innovation in hotels field. However, authors' papers in progress will analyze social and conceptual structures from this field. Finally, the study only used papers as unit of analysis. Nevertheless, the bibliographic coupling method can also use journals and authors as unit of analysis. Thus, in the future development of this research such units of analysis will be considered.

REFERENCES

Aboelmaged, M. (2018). Direct and indirect effects of eco-innovation, environmental orientation and supplier collaboration on hotel performance: An empirical study. *Journal of Cleaner Production, 184*, 537–549. doi:10.1016/j.jclepro.2018.02.192

Agarwal, S., Erramilli, M. K., & Dev, C. S. (2003). Market orientation and performance in service firms: Role of innovation. *Journal of Services Marketing, 17*(1), 68–82. doi:10.1108/08876040310461282

Aluri, A., & Palakurthi, R. (2011). The influence of demographic factors on consumer attitudes and intentions to use RFID technologies in the US hotel industry. *Journal of Hospitality and Tourism Technology, 2*(3), 188–203. doi:10.1108/17579881111173749

Amabile, T. M., Conti, R., Coon, H., Lazenby, J., & Herron, M. (1996). Assessing the work environment for creativity. *Academy of Management Journal, 39*(5), 1154–1184. doi:10.2307/256995

Bani-Melhem, S., Zeffane, R., & Albaity, M. (2018). Determinants of employees' innovative behavior. *International Journal of Contemporary Hospitality Management, 30*(3), 1601–1620. doi:10.1108/IJCHM-02-2017-0079

Baregheh, A., Rowley, J., & Sambrook, S. (2009). Towards a multidisciplinary definition of innovation. *Management Decision, 47*(8), 1323–1339. doi:10.1108/00251740910984578

Beshr, H. A., & Hossan, C. G. (2018). Key drivers for service innovation in the UAE's hotel sector. *International Journal of Entrepreneurship and Innovation Management, 22*(3), 251–268. doi:10.1504/IJEIM.2018.091769

Bilgihan, A., Smith, S., Ricci, P., & Bujisic, M. (2016). Hotel guest preferences of in-room technology amenities. *Journal of Hospitality and Tourism Technology, 7*(2), 118–134. doi:10.1108/JHTT-02-2016-0008

Bohdanowicz, P., Zientara, P., & Novotna, E. (2011). International hotel chains and environmental protection: An analysis of Hilton's we care! programme (Europe, 2006-2008). *Journal of Sustainable Tourism, 19*(7), 797–816. doi:10.1080/09669582.2010.549566

Boyack, K. W., & Klavans, R. (2010). Co-citation analysis, bibliographic coupling, and direct citation: Which citation approach represents the research front most accurately? *Journal of the American Society for Information Science and Technology, 61*(12), 2389–2404. doi:10.1002/asi.21419

Calantone, R. J., Cavusgil, S. T., & Zhao, Y. (2002). Learning orientation, firm innovation capability, and firm performance. *Industrial Marketing Management, 31*(6), 515–524. doi:10.1016/S0019-8501(01)00203-6

Camisón, C., & Monfort-Mir, V. M. (2012). Measuring innovation in tourism from the Schumpeterian and the dynamic-capabilities perspectives. *Tourism Management, 33*(4), 776–789. doi:10.1016/j.tourman.2011.08.012

Campo, S., Díaz, A. M., & Yagüe, M. J. (2014). Hotel innovation and performance in times of crisis. *International Journal of Contemporary Hospitality Management, 26*(8), 1292–1311. doi:10.1108/IJCHM-08-2013-0373

Chan, E. S. W. (2013). Gap analysis of green hotel marketing. *International Journal of Contemporary Hospitality Management, 25*(7), 1017–1048. doi:10.1108/IJCHM-09-2012-0156

Chen, R. J. C. (2015). From sustainability to customer loyalty: A case of full service hotels' guests. *Journal of Retailing and Consumer Services, 22*, 261–265. doi:10.1016/j.jretconser.2014.08.007

Cheng, J.-S., Tang, T.-W., Shih, H.-Y., & Wang, T.-C. (2016). Designing lifestyle hotels. *International Journal of Hospitality Management, 58*, 95–106. doi:10.1016/j.ijhm.2016.06.010

Cobo, M. J., López-Herrera, A. G., Herrera-Viedma, E., & Herrera, F. (2011). Science mapping software tools: Review, analysis, and cooperative study among tools. *Journal of the American Society for Information Science and Technology, 62*(7), 1382–1402. doi:10.1002/asi.21525

Cobos, L. M., Mejia, C., Ozturk, A. B., & Wang, Y. (2016). A technology adoption and implementation process in an independent hotel chain. *International Journal of Hospitality Management, 57*, 93–105. doi:10.1016/j.ijhm.2016.06.005

Daft, R. L. (1978). A dual-core model of organizational innovation. *Academy of Management Journal, 21*(2), 193–210. doi:10.2307/255754

Damanpour, F. (1987). The adoption of technological, administrative, and ancillary innovations: Impact of organizational factors. *Journal of Management, 13*(4), 675–688. doi:10.1177/014920638701300408

Dhar, R. L. (2015). The effects of high performance human resource practices on service innovative behaviour. *International Journal of Hospitality Management, 51*, 67–75. doi:10.1016/j.ijhm.2015.09.002

Dhar, R. L. (2016). Ethical leadership and its impact on service innovative behavior: The role of LMX and job autonomy. *Tourism Management, 57*, 139–148. doi:10.1016/j.tourman.2016.05.011

Divisekera, S., & Nguyen, V. K. (2018). Determinants of innovation in tourism evidence from Australia. *Tourism Management, 67*, 157–167. doi:10.1016/j.tourman.2018.01.010

Durán-Sánchez, A., Álvarez-García, J., del Río-Rama, M. de la C., & Rosado-Cebrián, B. (2019). Science mapping of the knowledge base on tourism innovation. *Sustainability, 11*(12), 1–17. doi:10.3390u11123352

Edquist, C. (1997). *Systems of innovation: Technologies, institutions and organizations*. Pinter Publishers/Cassell Academic.

Feldman, M. P. (1994). *The geography of innovation* (Vol. 2). Springer Netherlands. doi:10.1007/978-94-017-3333-5

Furrer, O., Thomas, H., & Goussevskaia, A. (2008). The structure and evolution of the strategic management field: A content analysis of 26 years of strategic management research. *International Journal of Management Reviews, 10*(1), 1–23. doi:10.1111/j.1468-2370.2007.00217.x

Garcia, R., & Calantone, R. J. (2002). A critical look at technological innovation typology and innovativeness terminology: A literature review. *Journal of Product Innovation Management, 19*(2), 110–132. doi:10.1111/1540-5885.1920110

García-Pozo, A., Sánchez-Ollero, J. L., & Ons-Cappa, M. (2016). Eco-innovation and economic crisis: A comparative analysis of environmental good practices and labour productivity in the Spanish hotel industry. *Journal of Cleaner Production, 138*, 131–138. doi:10.1016/j.jclepro.2016.01.011

Gault, F. (2012). User innovation and the market. *Science & Public Policy, 39*(1), 118–128. doi:10.1093cipolcs005

Gault, F. (2016). Defining and measuring innovation in all sectors of the economy: Policy relevance. *OECD Blue Sky Forum III.*

Gault, F. (2018). Defining and measuring innovation in all sectors of the economy. *Research Policy, 47*(3), 617–622. doi:10.1016/j.respol.2018.01.007

Gomezelj, D. O. (2016). A systematic review of research on innovation in hospitality and tourism. *International Journal of Contemporary Hospitality Management, 28*(3), 516–558. doi:10.1108/IJCHM-10-2014-0510

Grissemann, U. S., Pikkemaat, B., & Weger, C. (2013). Antecedents of innovation activities in tourism: An empirical investigation of the Alpine hospitality industry. *Tourism (Zagreb), 61*(1), 7–27.

Grissemann, U. S., Plank, A., & Brunner-Sperdin, A. (2013). Enhancing business performance of hotels: The role of innovation and customer orientation. *International Journal of Hospitality Management, 33*(1), 347–356. doi:10.1016/j.ijhm.2012.10.005

Gutiérrez-Salcedo, M., Martínez, M. Á., Moral-Munoz, J. A., Herrera-Viedma, E., & Cobo, M. J. (2018). Some bibliometric procedures for analyzing and evaluating research fields. *Applied Intelligence, 48*(5), 1275–1287. doi:10.100710489-017-1105-y

Hassanien, A., & Eid, R. (2007). Developing new products in the hospitality industry: A case of Egypt. *Journal of Hospitality & Leisure Marketing, 15*(2), 33–53. doi:10.1300/J150v15n02_03

Hernández-Perlines, F., Ariza-Montes, A., Han, H., & Law, R. (2019). Innovative capacity, quality certification and performance in the hotel sector. *International Journal of Hospitality Management, 82*, 220–230. doi:10.1016/j.ijhm.2019.04.027

Hjalager, A.-M. (1997). Innovation patterns in sustainable tourism: An analytical typology. *Tourism Management, 18*(1), 35–41. doi:10.1016/S0261-5177(96)00096-9

Hjalager, A.-M. (2002). Repairing innovation defectiveness in tourism. *Tourism Management, 23*(5), 465–474. doi:10.1016/S0261-5177(02)00013-4

Hjalager, A.-M. (2010a). A review of innovation research in tourism. *Tourism Management, 31*(1), 1–12. doi:10.1016/j.tourman.2009.08.012

Hjalager, A.-M. (2010b). Regional innovation systems: The case of angling tourism. *Tourism Geographies, 12*(2), 192–216. doi:10.1080/14616681003725201

Horng, J.-S., Liu, C.-H. S., Chou, S.-F., Tsai, C.-Y., & Chung, Y.-C. (2017). From innovation to sustainability: Sustainability innovations of eco-friendly hotels in Taiwan. *International Journal of Hospitality Management, 63*, 44–52. doi:10.1016/j.ijhm.2017.02.005

Hu, M.-L. M., Horng, J.-S., & Sun, Y.-H. C. (2009). Hospitality teams: Knowledge sharing and service innovation performance. *Tourism Management, 30*(1), 41–50. doi:10.1016/j.tourman.2008.04.009

Hurley, R. F., & Hult, G. T. M. (1998). Innovation, market orientation, and organizational learning: An integration and empirical examination. *Journal of Marketing, 62*(3), 42–54. doi:10.1177/002224299806200303

Jalilvand, M. R. (2017). The effect of innovativeness and customer-oriented systems on performance in the hotel industry of Iran. *Journal of Science and Technology Policy Management, 8*(1), 43–61. doi:10.1108/JSTPM-08-2016-0018

Jiang, Y., Ritchie, B. W., & Benckendorff, P. (2019). Bibliometric visualisation: An application in tourism crisis and disaster management research. *Current Issues in Tourism, 22*(16), 1925–1957. doi:10.1080/13683500.2017.1408574

Jung, H. S., & Yoon, H. H. (2018). Improving frontline service employees' innovative behavior using conflict management in the hospitality industry: The mediating role of engagement. *Tourism Management, 69*, 498–507. doi:10.1016/j.tourman.2018.06.035

Kahn, K. B. (2018). Understanding innovation. *Business Horizons, 61*(3), 453–460. doi:10.1016/j.bushor.2018.01.011

Kamaruddeen, A. M., Yusof, N. A., & Said, I. (2010). Innovation and innovativeness: Difference and antecedent relationship. *The Icfai University Journal of Architecture, 2*(1), 66–78.

Karadag, E., & Dumanoglu, S. (2009). The productivity and competency of information technology in upscale hotels: The perception of hotel managers in Turkey. *International Journal of Contemporary Hospitality Management, 21*(4), 479–490. doi:10.1108/09596110910955712

Kim, J. S., & Bernhard, B. (2014). Factors influencing hotel customers' intention to use a fingerprint system. *Journal of Hospitality and Tourism Technology, 5*(2), 98–125. doi:10.1108/JHTT-11-2013-0031

Kim, M.-S., & Koo, D.-W. (2017). Linking LMX, engagement, innovative behavior, and job performance in hotel employees. *International Journal of Contemporary Hospitality Management, 29*(12), 3044–3062. doi:10.1108/IJCHM-06-2016-0319

Kim, T. T., & Lee, G. (2013). Hospitality employee knowledge-sharing behaviors in the relationship between goal orientations and service innovative behavior. *International Journal of Hospitality Management, 34*(1), 324–337. doi:10.1016/j.ijhm.2013.04.009

Köseoglu, M. A., Rahimi, R., Okumus, F., & Liu, J. (2016). Bibliometric studies in tourism. *Annals of Tourism Research, 61*, 180–198. doi:10.1016/j.annals.2016.10.006

Lam, C., & Law, R. (2019). Readiness of upscale and luxury-branded hotels for digital transformation. *International Journal of Hospitality Management, 79*, 60–69. doi:10.1016/j.ijhm.2018.12.015

Lei, S. I., Wang, D., & Law, R. (2019). Hoteliers' service design for mobile-based value co-creation. *International Journal of Contemporary Hospitality Management, 31*(11), 4338–4356. doi:10.1108/IJCHM-03-2018-0249

Li, M., & Hsu, C. H. C. (2016). Linking customer-employee exchange and employee innovative behavior. *International Journal of Hospitality Management, 56*, 87–97. doi:10.1016/j.ijhm.2016.04.015

Lim, W. M. (2009). Alternative models framing UK independent hoteliers' adoption of technology. *International Journal of Contemporary Hospitality Management, 21*(5), 610–618. doi:10.1108/09596110910967836

Martin-Rios, C., & Ciobanu, T. (2019). Hospitality innovation strategies: An analysis of success factors and challenges. *Tourism Management, 70*(July), 218–229. doi:10.1016/j.tourman.2018.08.018

Martínez-Pérez, Á., Elche, D., & García-Villaverde, P. M. (2019). From diversity of interorganizational relationships to radical innovation in tourism destination: The role of knowledge exploration. *Journal of Destination Marketing & Management, 11*, 80–88. doi:10.1016/j.jdmm.2018.12.002

Mattsson, J., & Orfila-Sintes, F. (2014). Hotel innovation and its effect on business performance. *International Journal of Tourism Research, 16*(4), 388–398. doi:10.1002/jtr.1933

Mehraliyev, F., Choi, Y., & Köseoglu, M. A. (2019). Social structure of social media research in tourism and hospitality. *Tourism Recreation Research, 44*(4), 451–465. doi:10.1080/02508281.2019.1609229

Mitki, Y., & Herstein, R. (2007). Innovative training in designing corporate identity. *Industrial and Commercial Training, 39*(3), 174–179. doi:10.1108/00197850710742289

Mongeon, P., & Paul-Hus, A. (2016). The journal coverage of Web of Science and Scopus: A comparative analysis. *Scientometrics, 106*(1), 213–228. doi:10.100711192-015-1765-5

Morosan, C. (2018). An empirical analysis of intentions to cocreate value in hotels using mobile devices. *Journal of Hospitality & Tourism Research (Washington, D.C.), 42*(4), 528–562. doi:10.1177/1096348015597034

Morosan, C., & DeFranco, A. (2016). Modeling guests' intentions to use mobile apps in hotels: The roles of personalization, privacy, and involvement. *International Journal of Contemporary Hospitality Management, 28*(9), 1968–1991. doi:10.1108/IJCHM-07-2015-0349

Nair, G. K. (2019). Dynamics of learning orientation, innovativeness, and financial performance of the hotel industry. *The Journal of Hospitality Financial Management, 27*(1), 27–38. doi:10.7275/8832-cr71

Nicolau, J. L., & Santa-María, M. J. (2013). The effect of innovation on hotel market value. *International Journal of Hospitality Management, 32*(1), 71–79. doi:10.1016/j.ijhm.2012.04.005

Nieves, J., & Diaz-Meneses, G. (2016). Antecedents and outcomes of marketing innovation: An empirical analysis in the hotel industry. *International Journal of Contemporary Hospitality Management, 28*(8), 1554–1576. doi:10.1108/IJCHM-11-2014-0589

Nieves, J., & Diaz-Meneses, G. (2018). Knowledge sources and innovation in the hotel industry: Empirical analysis on Gran Canaria Island, a mature mass-tourism destination. *International Journal of Contemporary Hospitality Management, 30*(6), 2537–2561. doi:10.1108/IJCHM-07-2016-0341

OECD/Eurostat. (2005). *Oslo manual: Guidelines for collecting and interpreting innovation data* (3rd ed.). Paris: OECD Publishing; doi:10.1787/9789264013100-

OECD/Eurostat. (2018). *Oslo manual: Guidelines for collecting, reporting and using data on innovation* (4th ed.). Paris: OECD Publishing; doi:10.1787/9789264304604-

Okumus, F., Köseoglu, M. A., Putra, E. D., Dogan, I. C., & Yildiz, M. (2019). A bibliometric analysis of lodging-context research from 1990 to 2016. *Journal of Hospitality & Tourism Research (Washington, D.C.), 43*(2), 210–225. doi:10.1177/1096348018765321

Ordanini, A., & Parasuraman, A. (2011). Service innovation viewed through a service-dominant logic lens: A conceptual framework and empirical analysis. *Journal of Service Research*, *14*(1), 3–23. doi:10.1177/1094670510385332

Ordanini, A., Parasuraman, A., & Rubera, G. (2014). When the recipe is more important than the ingredients: A qualitative comparative analysis (QCA) of service innovation configurations. *Journal of Service Research*, *17*(2), 134–149. doi:10.1177/1094670513513337

Orfila-Sintes, F., Crespí-Cladera, R., & Martínez-Ros, E. (2005). Innovation activity in the hotel industry: Evidence from Balearic Islands. *Tourism Management*, *26*(6), 851–865. doi:10.1016/j.tourman.2004.05.005

Orfila-Sintes, F., & Mattsson, J. (2009). Innovation behavior in the hotel industry. *Omega*, *37*(2), 380–394. doi:10.1016/j.omega.2007.04.002

Ottenbacher, M. C., & Gnoth, J. (2005). How to develop successful hospitality innovation. *The Cornell Hotel and Restaurant Administration Quarterly*, *46*(2), 205–222. doi:10.1177/0010880404271097

Ottenbacher, M. C., & Harrington, R. J. (2010). Strategies for achieving success for innovative versus incremental new services. *Journal of Services Marketing*, *24*(1), 3–15. doi:10.1108/08876041011017853

Öztayşi, B., Baysan, S., & Akpinar, F. (2009). Radio frequency identification (RFID) in hospitality. *Technovation*, *29*(9), 618–624. doi:10.1016/j.technovation.2009.05.014

Pace, L. A. (2016). How do tourism firms innovate for sustainable energy consumption? A capabilities perspective on the adoption of energy efficiency in tourism accommodation establishments. *Journal of Cleaner Production*, *111*, 409–420. doi:10.1016/j.jclepro.2015.01.095

Palacios-Marqués, D., Merigó, J. M., & Soto-Acosta, P. (2015). Online social networks as an enabler of innovation in organizations. *Management Decision*, *53*(9), 1906–1920. doi:10.1108/MD-06-2014-0406

Pappas, N. (2015a). Achieving competitiveness in Greek accommodation establishments during recession. *International Journal of Tourism Research*, *17*(4), 375–387. doi:10.1002/jtr.1995

Pappas, N. (2015b). Marketing hospitality industry in an era of crisis. *Tourism Planning & Development*, *12*(3), 333–349. doi:10.1080/21568316.2014.979226

Pikkemaat, B., Peters, M., & Bichler, B. F. (2019). Innovation research in tourism: Research streams and actions for the future. *Journal of Hospitality and Tourism Management*, *41*, 184–196. doi:10.1016/j.jhtm.2019.10.007

Plessis, M. (2007). The role of knowledge management in innovation. *Journal of Knowledge Management*, *11*(4), 20–29. doi:10.1108/13673270710762684

Ratten, V., Ferreira, J. J., & Fernandes, C. I. (2017). Innovation management - Current trends and future directions. *International Journal of Innovation and Learning*, *22*(2), 135–155. doi:10.1504/IJIL.2017.085916

Reid, R. D., & Sandler, M. (1992). The use of technology to improve service quality. *The Cornell Hotel and Restaurant Administration Quarterly*, *33*(3), 68–73. doi:10.1177/001088049203300319

Salavou, H. (2004). The concept of innovativeness: Should we need to focus? *European Journal of Innovation Management*, *7*(1), 33–44. doi:10.1108/14601060410515628

Salem, I. E.-B. (2014). Toward better understanding of knowledge management: Correlation to hotel performance and innovation in five-star chain hotels in Egypt. *Tourism and Hospitality Research*, *14*(4), 176–196. doi:10.1177/1467358414542265

Sandvik, I. L., & Sandvik, K. (2003). The impact of market orientation on product innovativeness and business performance. *International Journal of Research in Marketing*, *20*(4), 355–376. doi:10.1016/j.ijresmar.2003.02.002

Saraniemi, S., & Kylänen, M. (2011). Problematizing the concept of tourism destination: An analysis of different theoretical approaches. *Journal of Travel Research*, *50*(2), 133–143. doi:10.1177/0047287510362775

Sarmah, B., Kamboj, S., & Rahman, Z. (2017). Co-creation in hotel service innovation using smart phone apps: An empirical study. *International Journal of Contemporary Hospitality Management*, *29*(10), 2647–2667. doi:10.1108/IJCHM-12-2015-0681

Schumpeter, J. A. (1934). *The theory of economic development: An inquiry into profits, capital, credit, interest and the business cycle*. Harvard University Press.

Subramanian, A., & Nilakanta, S. (1996). Organizational innovativeness: Exploring the relationship between organizational determinants of innovation, types of innovations, and measures of organizational performance. *Omega*, *24*(6), 631–647. doi:10.1016/S0305-0483(96)00031-X

Sundbo, J., Orfila-Sintes, F., & Sørensen, F. (2007). The innovative behaviour of tourism firms-Comparative studies of Denmark and Spain. *Research Policy*, *36*(1), 88–106. doi:10.1016/j.respol.2006.08.004

Tajeddini, K. (2009). The impact of learning orientation on NSD and hotel performance: Evidence from the hotel industry in Iran. *Education, Business and Society*, *2*(4), 262–275. doi:10.1108/17537980911001099

Tajeddini, K. (2010). Effect of customer orientation and entrepreneurial orientation on innovativeness: Evidence from the hotel industry in Switzerland. *Tourism Management*, *31*(2), 221–231. doi:10.1016/j.tourman.2009.02.013

Tajeddini, K., & Trueman, M. (2008). Effect of innovativeness, risk taking, self confidence and the need for achievement on venture performance: Evidence from Switzerland effect of customer orientation and innovativeness on business performance: A study of small-sized service retailers. *International Journal of Entrepreneurship and Small Business*, *6*(2), 280–295. doi:10.1504/IJESB.2008.018633

Tajeddini, K., Trueman, M., & Larsen, G. (2006). Examining the effect of market orientation on innovativeness. *Journal of Marketing Management*, *22*(5–6), 529–551. doi:10.1362/026725706777978640

Tang, T.-W. (2014). Becoming an ambidextrous hotel: The role of customer orientation. *International Journal of Hospitality Management*, *39*, 1–10. doi:10.1016/j.ijhm.2014.01.008

Tang, T.-W. (2016). Making innovation happen through building social capital and scanning environment. *International Journal of Hospitality Management*, *56*, 56–65. doi:10.1016/j.ijhm.2016.04.002

Tang, T.-W., Wang, M. C.-H., & Tang, Y.-Y. (2013). Developing service innovation capability in the hotel industry. *Service Business*, *9*(1), 97–113. doi:10.100711628-013-0220-z

Thomas, R., & Wood, E. (2014). Innovation in tourism: Re-conceptualising and measuring the absorptive capacity of the hotel sector. *Tourism Management*, *45*, 39–48. doi:10.1016/j.tourman.2014.03.012

Thompson, V. A. (1965). Bureaucracy and innovation. *Administrative Science Quarterly*, *10*(1), 1–20. doi:10.2307/2391646

Tidd, J., & Bessant, J. R. (2018). *Managing innovation: Integrating technological, market and organizational change* (6th ed.). Willey.

Tseng, C.-Y., Kuo, H.-Y., & Chou, S.-S. (2008). Configuration of innovation and performance in the service industry: Evidence from the Taiwanese hotel industry. *Service Industries Journal*, *28*(7), 1015–1028. doi:10.1080/02642060701882080

Úbeda-García, M., Claver-Cortés, E., Marco-Lajara, B., García-Lillo, F., & Zaragoza-Sáez, P. C. (2018). Continuous innovation in the hotel industry: The development of organizational ambidexterity through human capital and organizational culture in Spanish hotels. *International Journal of Contemporary Hospitality Management*, *30*(12), 3609–3631. doi:10.1108/IJCHM-06-2017-0326

Victorino, L., Verma, R., Plaschka, G., & Dev, C. S. (2005). Service innovation and customer choices in the hospitality industry. *Managing Service Quality*, *15*(6), 555–576. doi:10.1108/09604520510634023

Vila, M., Enz, C. A., & Costa, G. (2012). Innovative practices in the Spanish hotel industry. *Cornell Hospitality Quarterly*, *53*(1), 75–85. doi:10.1177/1938965511426562

Vogel, R., & Güttel, W. H. (2013). The dynamic capability view in strategic management: A bibliometric review. *International Journal of Management Reviews*, *15*(4), 426–446. doi:10.1111/ijmr.12000

Vuković, A. J., Gagić, S., Terzić, A., Petrović, M. D., & Radovanović, M. (2018). The impact of organisational learning on innovation: Case study of the serbian hotel industry. *Journal for East European Management Studies*, *23*(4), 673–692. doi:10.5771/0949-6181-2018-4-673

Wang, C. L., & Ahmed, P. K. (2004, December 1). The development and validation of the organisational innovativeness construct using confirmatory factor analysis. *European Journal of Innovation Management*, *7*(4), 303–313. Advance online publication. doi:10.1108/14601060410565056

Wang, T.-C., Tang, T.-W., & Cheng, J.-S. (2018). Art-oriented model of hotel service innovation. *International Journal of Contemporary Hospitality Management*, *30*(1), 160–177. doi:10.1108/IJCHM-02-2016-0059

Wikhamn, W., Armbrecht, J., & Wikhamn, B. R. (2018). Innovation in Swedish hotels. *International Journal of Contemporary Hospitality Management*, *30*(6), 2481–2498. doi:10.1108/IJCHM-06-2017-0323

Zhou, K. Z., Brown, J. R., & Dev, C. S. (2009). Market orientation, competitive advantage, and performance: A demand-based perspective. *Journal of Business Research*, *62*(11), 1063–1070. doi:10.1016/j.jbusres.2008.10.001

Zupic, I., & Čater, T. (2015). Bibliometric methods in management and organization. *Organizational Research Methods*, *18*(3), 429–472. doi:10.1177/1094428114562629

ENDNOTES

1 The inconsistency has resulted from overlapping conceptualizations of innovativeness and innovation or from the absence of a clear distinction between them.

2 Some definitions of innovativeness focuses on the individual and not on the organizational level. However, the authors should note that, in this chapter, the unit of analysis is the organization.

3 According to the paragraphs 146 and 150 of the Oslo Manual (OECD/Eurostat, 2005, para. 47): "146. An innovation is the implementation of a new or significantly improved product (good or service), or process, a new marketing method, or a new organisational method in business practices, workplace organization or external relations.

150. A common feature of an innovation is that it must have been implemented. A new or improved product is implemented when it is introduced on the market. New processes, marketing methods or organizational methods are implemented when they are brought into actual use in the firm's operations."

4 According to Schumpeter (1934), entrepreneurs represent a major contribution to innovative dynamics, who with their concepts and ideas set new standards and change the taste and preferences of their customers. Also, for Hjalager (2010a, p. 4) "entrepreneurship is a crucial factor in the evolutionary redirection of tourism products and increasing competitiveness."

5 The use of asterisk allows to include in the search the words containing its root as well as in singular and plural form. For example, "innovat*" can include innovation, innovations, innovative, and innovativeness.

6 African Journal of Hospitality and Tourism Administration (AJHTA), Cornell Hospitality Quarterly (CHQ), International Journal of Contemporary Hospitality Management (IJCHM), International Journal of Hospitality Management (IJHM), International Journal of Hospitality and Tourism Administration (IJHTA), Journal of Cleaner Production (JCP), Journal of Hospitality Marketing and Management (JHMM), Journal of Hospitality and Tourism Research (JHTR), Journal of Hospitality and Tourism Technology (JHTT), Journal of Quality Assurance in Hospitality and Tourism (JQAHT), Journal of Sustainable Tourism (JST), Journal of Travel and Tourism Marketing (JTTM), Service Industries Journal (SIJ), Tourism and Hospitality Research (THR), Tourism Management (TM), Worldwide Hospitality and Tourism Themes (WHTT).

Chapter 16
Innovation Applied to Community Tourism:
Bibliometric Analysis

Eunice Ramos Lopes
Instituto Politécnico de Tomar, Portugal

Joao Tomaz Simoes
https://orcid.org/0000-0003-3923-5555
Instituto Politécnico de Tomar, Portugal

Maria Rita Nunes
Instituto Politécnico de Tomar, Portugal

ABSTRACT

This chapter intended to perceive the knowledge produced by the scientific community about tourism, innovation, and community tourism. In order to achieve this objective, a search was carried out in the SCOPUS database, which provided with more than 30 thousand articles from the year 1857 to 2019. It is noticed, however, that the publications that report community tourism and innovation are scarcer, which demonstrates the lack of research within the three areas. It is noticed that for future works it is interesting to insert other sources of research, such as Sicielo and Latinex, and give more visibility to productions in other languages, which gives a broader understand of other publications outside the English-speaking world.

DOI: 10.4018/978-1-7998-4855-4.ch016

INTRODUCTION

Community tourism or community-based tourism (CBT) comes within the scope of world tourist phenomenon as an alternative to mass tourism. Despite difficulties encountered in assigning a widely accepted definition of what would come CBT (Castro & Pinto, 2013), this type of tourism has as its premise the collective social, economic and cultural development, based on the host community.

It is the host community that provides, manages and profits from tourism. There is a development based on sharing the wealth that tourism and its flow causes (Pearce, 1992). In the CBT the host has or at least intends to have control the entire tourism production chain.

According to Irving (2009), CBT favors the social bond and the collective meaning of life in society. There is, therefore, a need for community tourism to be sustained by endogenous motivation, because although there is a demand for this type of experience on the part of tourists, if the community is not willing to do so, the tourist phenomenon in that location will not occur. There is also a dependence on the involvement of local residents and entrepreneurs (Blackstock, 2005) in the construction of an attractive destination that promotes remarkable experiences supported by local values.

Thus, it is clear that community empowerment is one of the great premises for the realization of the CBT. This empowerment is even seen in the ability to not carry out tourism as it will define if the activity takes place and, if implemented, how it will be implemented. It is very common within tourism studies that the profitability of tourism is for the few and that the biggest disadvantaged in the development of mass tourism, is the indigenous population (Gomez, *et al.*, 2015). In this way, CBT emerges as a response to this reality, where not only adverse effects are socialized, but also the contributions of tourism, be it social, cultural or environmental and, above all, economic (Giampiccoli & Saayman, 2016).

Nonetheless, innovation is imperative for business, also it is for CBT. This article aims to understand the interest and growth of research in the areas, namely, tourism, innovation and, more specifically, innovation linked to community tourism by a bibliometrically analysis from the Scopus ® software data base.

BACKGROUND

CBT is widely discussed within the literature about alternative tourism, whether some definitions of community-based tourism are considered more relevant to its correct understanding (See Table I).

It is noted, from the concepts presented, a close connection between the principles of sustainability and community tourism. The necessary understanding is that in tourism community based there is the empowerment of the community, the alleviation of the harmful effects of tourism, such as massification, cultural and environmental degradation, for greater cooperation between the community and the tourist in order to achieve an activity capable to provide development and growth.

However, some authors such as Blackstock, (2005); Giampiccoli (2015) and Castro & Pinto (2013) point out that the lack of technical knowledge in product management (accommodation, routes or itineraries, agency service, etc.) conditions the success of its tourism practices to only a small number of communities.

According to Valente & Mielke (2017), to overcome this reality of failure and even distrust in the management capacity of the CBT, innovation is necessary. One of the ways to innovate within community tourism, according to Giampiccoli & Saayman (2016), is to make this tourism practice well known. According to the authors, it is necessary to move from the local scale to an International. Mak-

Table 1. Concepts on community-based tourism

Authors	CBT concepts
(Giampiccoli 2016, p.1).	Community-based tourism (CBT) is a type of tourism rooted in the alternative development approach.
(Jorge, Mielke, & Pegas, 2013, p. 173)	Community-based tourism projects should have as their basic principle the improvement of the living conditions of people living in regions where, as a rule, they are excluded from the economic system.
(Lopez-guzman, Borges, & Cerezo, 2016, p. 1608)	Community-based tourism is being presented as an alternative to traditional mass tourism in developing countries. This kind of tourism allows for greater contact with local residents and generates greater economic and environmental benefits for the local community.
(Castro & Pinto, 2013, p. 216)	Community tourism can stimulate local participation to management and the use of local power for tourism development, in the sense of the enhancement of local power, according to a slow and intense transformation, having as its central axis the political part of the subjects involved in the process, therefore their empowerment through an inclusive and potentially self-managed governance.
(Fabrino & Costa, 2012, p. 547)	TBC is considered a model of development for tourism activity centered on community resources, being the main agent in the design, development, and management.
(Gomez, Falcão, Castilho, Correia, & Oliveira, 2015, p. 1218)	Community-based Tourism (TBC) has been promulgated as a "new" way of doing tourism that opposes mass tourism, and as an initiative that has potential for social transformation and local sustainable development. Such initiatives have as main characteristic, the search for quality of life for those involved, the search for social and/or environmental gains in the local context, and for initiatives for the benefit of the collectivity
(Blackstock, 2005, p. 39)	Community based tourism (CBT) centers on the involvement of the host community in planning and maintaining tourism development in order to create a more sustainable industry.
(Nelson, 2004, p. 8)	Community-based tourism as tourism activities or enterprises in which local communities participate, occurring on their lands, and scaffolding on their cultural heritage and natural attractions and asset.

Source: (Blackstock, 2005; Castro, Pinto, Luisa, & Castro, 2013; Fabrino & Costa, 2012; Giampiccoli & Mtapuri, 2012; Gomez et al., 2015; Jorge et al., 2013; Lopez-guzman et al., 2016; Nelson, 2004)

ing community-based tourism known worldwide to promote its practice and make it much more viable and interesting.

CBT Experiences

In the prerogative for a better understanding about the practices of community-based tourism, it will be analyzed some experiences of the CBT. Such practices report the need for an integration between local governments, society and, above all, with tourists. CBT's premise is sustainable practice, with a focus on smallholders, local communities and their cultures, so the experiences presented here will be an alternative to mass tourism. It will not be possible to find a tourism that manages big profits and that changes cities and places, as the authors have seen through the great massification of territories. Here, will be presented territories that have the practice of tourism, but not only. They present a diversity of activities that generate and revolve the local economy. A tourist monoculture is not perceived (Krippendorf, 2003), on the contrary, the plurality of activities is encouraged, as a form of diversification and elimination of dependence on tourism.

In Latin America, CBT experiences and initiatives are identified in almost all countries, from Mexico to the extreme south of Argentina (Irving, Cristina, & Mendonça, 2018). This fact is possible due to the organization and associations to foster and encourage the practice of CBT. Fact that resulted in the creation of a network, the Tourism Network Community of Latin America (REDTURS), stimulated by the International Organization Labor Organization (ILO). A prominent case in Latin America is perceived in Brazil, to quote, namely, the case of Prainha do Canto Verde in Ceará do Brasil. The Prainha do Canto Verde has the Residents' Association which they created through a partnership with Cooperativa de Artesanato e Turismo - Coopecantur, a social fund that turns the region's economy for tourism. It has four hostels and receives around thousand tourists (Nunes, Santos, & Azevedo, 2014).

In Europe, there is an emphasis on practices in Italy, detected by Chiodo, Adriani, Navarro and Salvatore (2019). According to the authors, in the 1990s, in that country, there was a spontaneous appearance of "villages of rural tourism", focusing on gastronomy and enology in the interior regions. By the growth of tourist demand and in the process to protection for cultural heritage material and immaterial of the villages, several associative initiatives were created, committed to the objective of promoting a quality offer for tourists as well as protecting the environment and local culture.

In Africa, the experience narrated by Lopez-guzman *et al.* (2016) in Cape Verde, on the island of Santiago. There, is a marked growth of tourist activity and this growth is due to the engagement of the local community in the activity. Most tourists who arrive there stay, eat and do some tourist animation in ventures of resident families, which work in the tourist activity, not only as a work force, but as owners of their own business, which develops the community.

The studies by Joo, Choi & Kim (2019) in Korea, demonstrates the government's attempt to insert community-based tourism into the country. The "Tour Dure" as the CBT government initiative was called, was created from the Korean cultural valorization, called "Dure". The Tour, with that name, values Korean culture, encourages local entrepreneurship and economy practices collaborative, aiming, most importantly, to increase social capital and innovation in Korean communities.

With the practices presented here, the imperative of an integrated community is perceived, empowered and aware of the need to act in the territories as protagonists. The CBT, only comes into existence when there is cohesion between the elements of society, as well as a civil organization, whether spontaneous as found in Italy or even encouraged by governments as presented in Korea. A society with little or no social capital (Nunes *et al.*, 2014), is not very efficient in the development of alternative tourism practices and territorial improvement.

Innovation: A Dynamic Approach

Schumpeter (*apud* Simões, 2014) indicates that there are five types of innovation: *i*) introduction of a new good; *ii*) introduction of a new production method; *iii*) opening a new market; *iv*) new way of acquiring raw materials (new source of) and *v*) implementation of a new form of organization (structuring).

Schumpeter also states that innovation is distinct from invention in that the innovation is possible without an invention, and an invention does not necessarily induce innovation (Mota & Oliveira, 2014), because inventions are not exclusively for industrial use while innovations are amplified by inventions or just brilliant general ideas for making them useful products (Hjalager, 2002) (See Figure 1).

On the other hand, Higgins (1996), states that innovation arises basically in two ways: *i*) the first being the result of an innovative process; *ii*) the second is that it is a process that creates new products and services, or improves existing ones, or creation of organizational methods (processes) that promote a significant impact in people, groups, organizations, industry or society. The author defends that "the process of creating innovation is based on creativity, considering it a key element" (Higgins, 1996).

Figure 1.
Source: Higgins (1996, p. 371)

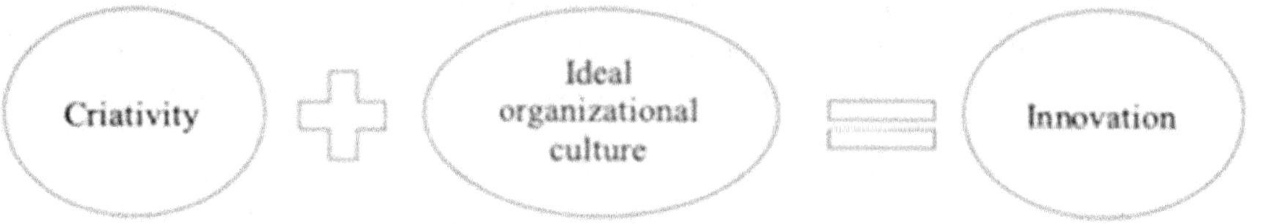

Companies currently implement information technologies in their innovation processes, according to Lindič *et al.*, (2011, p.189) "Technology can play an important role to support and enable the management of knowledge all along the innovation process, particularly when the number of ideas get overwhelming 'idea overload'". The technology allows the renewal of methods, structural development and ideal organizational cultural creativity, cyclical innovation and rapid evolution, updating and subsequent improvement the conditions of production, development and design of goods.

Abernathy and Clark (1985 *apud* Hjalager, 2002), present a characteristic approach of innovation in the automotive industry that can fit into tourism (See Figure 2)

The Model presents four types of innovation - Regular, niche, revolutionary and architectural, basically consisting of two axes. The horizontal indicates whether a certain innovation breaks the process/item making it obsolete or, if on the other hand, strengthens existing ones. The vertical axis indicates knowledge and skills necessary for the creation of a certain product or service, in some cases, ideas skills and qualifications are out of date and need to be replaced while that others prove more relevant when they are adapted and later developed.

Innovation plays an increasingly central role in our society, not only in pursuit of economic sustainability as well as social development. "In step with the increased importance of innovation as a means of sustaining and developing competitive positions, firms increasingly seek to develop their creative capacity" (Mota & Oliveira, 2014, p.1). Although, in the tourism industry, as stated by Hjalager (2002, p.469): "In some important respects, tourism differs from the types of businesses usually targeted for the dissemination of research-based knowledge".

According to Lagowski (1990), these words are generally used to represent something good or better, which leads us to discuss the concept of "best". As he says: "What does the observation that an innovation produces a better result in one environment mean to a second environment? "Better" is one of those slippery, anthropocentric words like "truth" or "beauty." (p.449).

Community Territorial Empowerment

The perception of the development space of the territories as a form of empowerment territorial development, reveals itself in the form of economic, social, cultural and environmental. This resizing aims to community development, through community activation, the basis of which spans impulses for active participation local actors in a territory, leading the community itself to force endogenous resources, transforming them into innovative tourism products.

Figure 2.
Source: Hjalager, 2002, p. 467

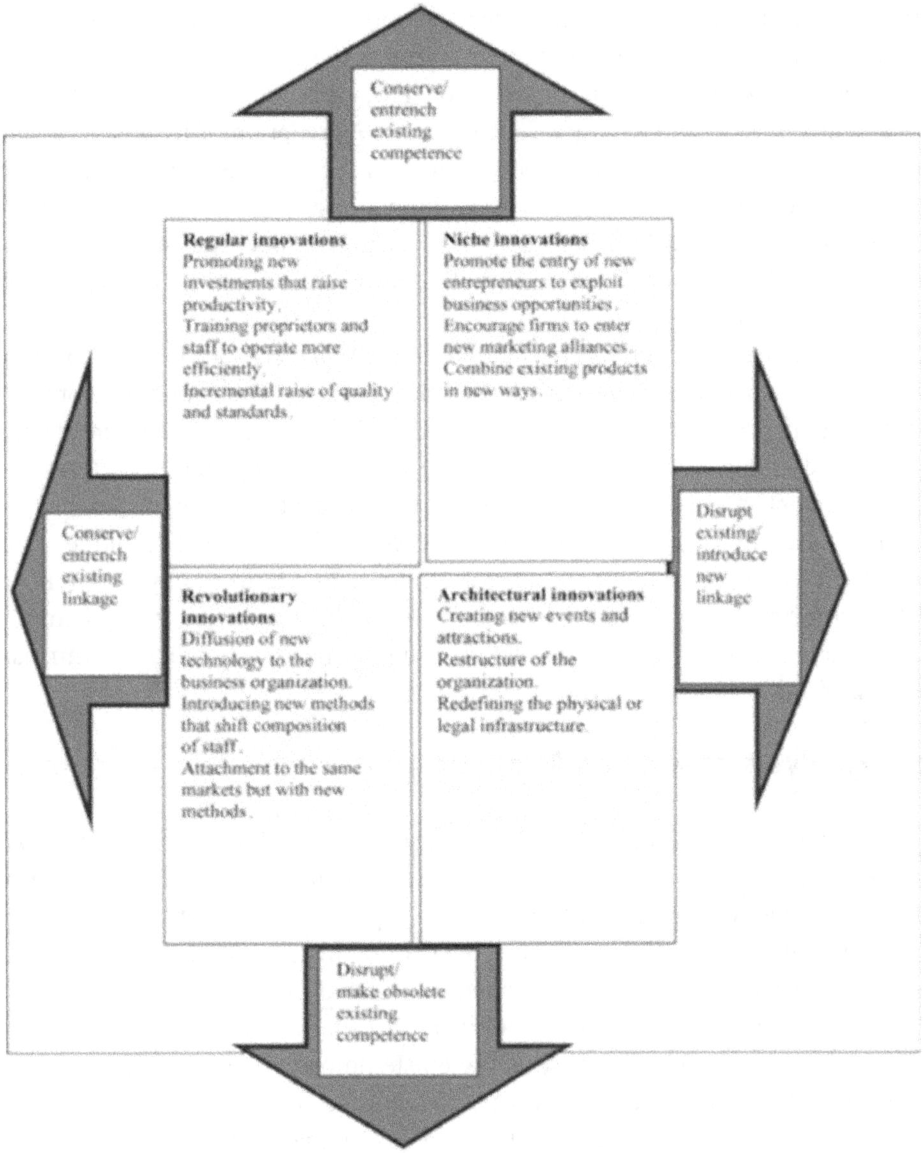

Community development and affirmation is a constructive process where the local actors' communities have a preponderant role in the identification and territorial empowerment, if necessary, to know the territorial particularities through the patrimonial resources giving space for community activation, sustainable building of these resources endogenous factors that can promote the promotion of identity (re)appropriation activities territorial.

In this context, citizen participation is essential, it increases the power of individuals and communities to make individual and collective decisions. A participation active, favors the sustainability of actions, strengthens the approach to the community, allows innovation, enhances the dissemination of good practices and partnership results, favors information sharing, strengthens networking and peer mediation in resolving disagreements and conflicts (Almeida, 2017).

The identification of the participative strategic ideal understood as the main mechanism promoter of innovation, change and diversification of initiatives, through the exercise of citizenship and community development, by raising awareness of the existence of rights and duties (Nunes, 2002).

The future attractiveness of the territory depends on the perception of its potential value in a world globalization, its ability to generate and attract savings and reproductive investment that values "territorial inheritance" and generates sustainable development based on cultural and natural heritage. The sustainable enhancement of cultural heritage, both material and immaterial, has a fundamental role in the rehabilitation and revitalization of territories, as well as promoting social participation and the exercise of citizenship (PNPOT, 2018).

The relationship between the community and tourism with the territory is complex, requiring instruments of territorial organization to proceed with its management. The economic contribution and the imposition that the tourist dynamics exert on the territorial structure and even on the receiving communities lack an increasingly strategic and equitable territorial management.

The very involvement of the community and the local physical environment in the tourist development in social terms calls for policies and planning that enable the preservation of the local identity and the enhancement of the heritage (cultural and natural), this action aims to improve living conditions.

Community territorial empowerment can contribute to a social fabric strengthened by the social interactions and relationships it promotes. Adding the importance attributed to its dimensions of social and territorial cohesion.

Bibliometric Analysis of Articles Applied to Innovation and Tourism Community

The chosen approach was considered so that, in a first stage, through the analysis quantitative, it could ascertain the interest and growth of research in these areas, namely, tourism, innovation and more specifically, innovation linked to community tourism. Subsequently, a qualitative analysis was performed as a way of presenting the data collected, namely, the geographical location of the authors, newspapers, magazines academic articles and articles most cited and related to the theme.

For this purpose, and using the Scopus ® software, the entire research was carried out in English except for the term 'tourism', which because of the epistemologically derived French *tourisme*, it will be in this language that the first articles, more specifically, the first registrant entitled "Le tourisme et la houille blanche devant la chambre des députés, by Paul Bougault, dated and published by the Cour d'Appel"(Lopes & Simões, 2019, p. 136). As for the word 'innovation', has its first appearance in an 1857 letter entitled "John Carter, FSA, author of the "Pursuits of architectural innovation", by JGN (1857). This investigation was structured around 8 researches (see Table 2):

With the application of the first criterion, the aim is to determine the total number of publications and the annual distribution of this total, containing at least once the terms presented in Table 2. In this second criterion, the search will result in publications closer to search terms, as they must be present in the title, abstract or keywords of the article. The third criterion will present the articles that relate exactly to the terms the authors intend to search for, requesting the software that present publications that list two or more words in their study, at least that the authors define a distance of five words between them, in the title.

Table 2. Research criteria

Term Searched	1st Criterion (in any part of the text)	2nd Criterion (title, abstract and keywords)	3rd Criterion (title)	4th Criterion (a distance of five words in title)
'Tourism'	x		x	
'Innovation'	x		x	
'Innovation + Community Tourism	x	x	No data	No data

Source: author's elaboration, 2019

For a more reliable research on the general theme of this work - Innovation and Community Tourism, the authors will apply all four criteria to these two terms and the exception when mentioned, this investigation will have as a final reference the year 2018 that is closed, with 2019 data not yet complete.

Results Presentation

After applying the first search criterion to the terms: 'tourism' and 'innovation' (separately), 263 617 documents containing at least one the word 'tourism' in any part of the text, and 1 809 501 containing the word 'innovation' (See Figures 3 and 4).

Looking at the previous graphs, the authors verified a non-uniform distribution, in publication of documents, but of continuous growth, concentrating essentially on recent years, which validates the statement about the recent interest in these areas. The outstanding authors are Law, Hall, Ryan, Han, Wall, Saayman, Prideaux, Lee,Song, Uysal, Rogerson and Scott, among others, as shown in the next table: Author search "Tourism"(Graphic 3), Authors search "Innovation" (see Figure 5).

Applying the third search criterion to the same terms, the authors predictably obtain the publications that focus on the topics covered. In this case, 36 528 documents emerged about tourism and 97 417 with the word 'innovation' in the title (See Figure 6 and 7).

Through the above-mentioned criterion, essentially the same authors and newspapers / magazines predominate scientific research that stand out the most are, in the case of 'tourism', Tourism Management with 1540, Annals of Tourism Research with 1503, Tourism Recreation Research with 747, Journal of Sustainable Tourism with 735 and Journal of Travel Research with 588, and In the case of 'innovation', Research Policy stands out with 783, Technological Forecasting and Social Change with 533, Technovation with 478, the Journal of Product Innovation Management with 416 and, the International Journal Of Technology Management with exactly 400 documents published to date (See Figure 8 and 9).

'Innovation' + 'Community Tourism', that is, there is no record of documents that have these words together in the title, the authors decide to focus on the results obtained with the 1st and 2nd criteria. However, the authors verify the existence of works, namely theses of PhDs that approach community tourism as a factor of innovation, outside the Scopus ® platform. In any case, the scarcity of scientific articles related to this problem may generate two conditions, an opportunity, and a difficulty. Indeed, researchers may "be more impressed by the few publications in prestigious magazines, than many in more popular magazines, and this is not uncommon in selection exams, for example. However, quality science can also appear in journals with different impact factors " (Kirchhof & Lacerda, 2012, p. 189).

Figure 3.
Source: Scopus, 2019

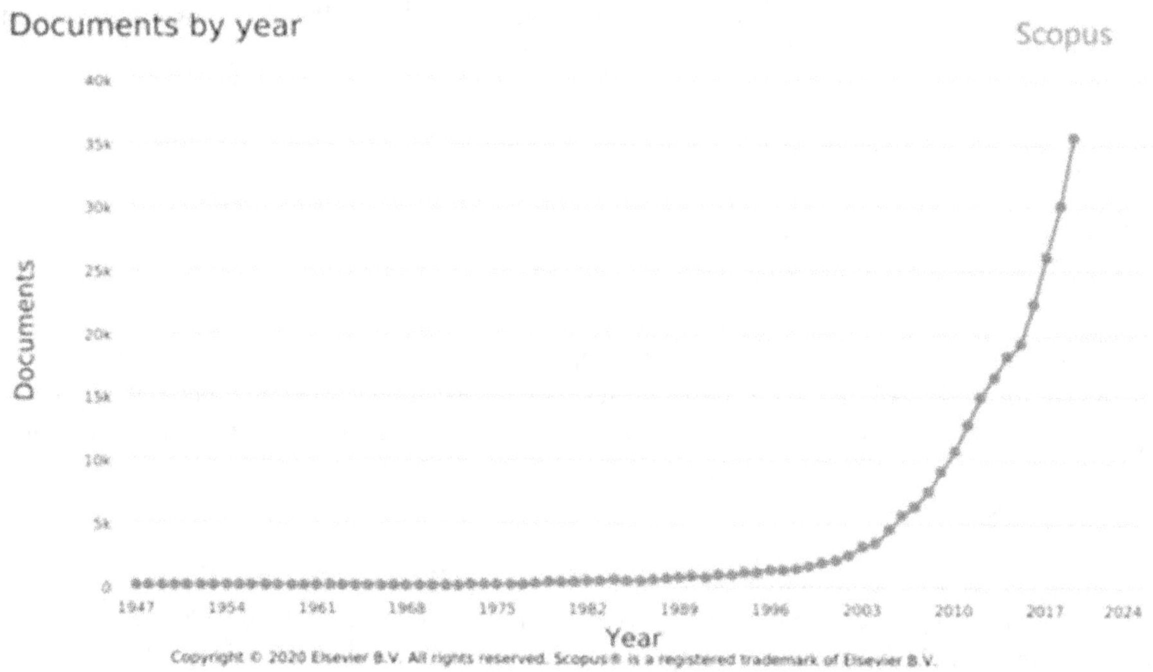

Figure 4.
Source: Scopus, 2019

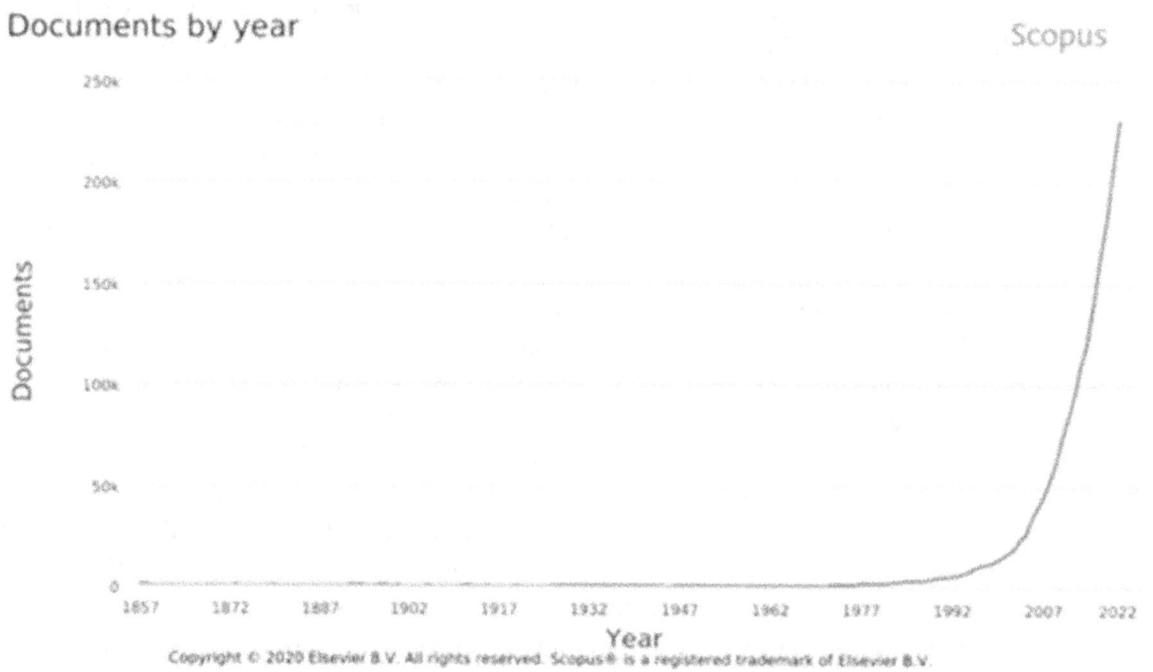

Figure 5.
Source: Scopus, 2019

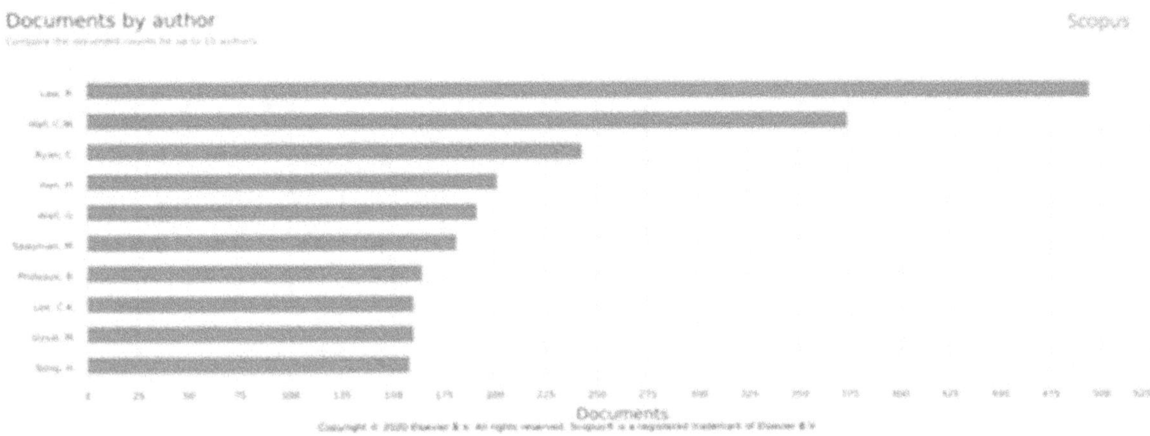

Figure 6.
Source: Scopus, 2019

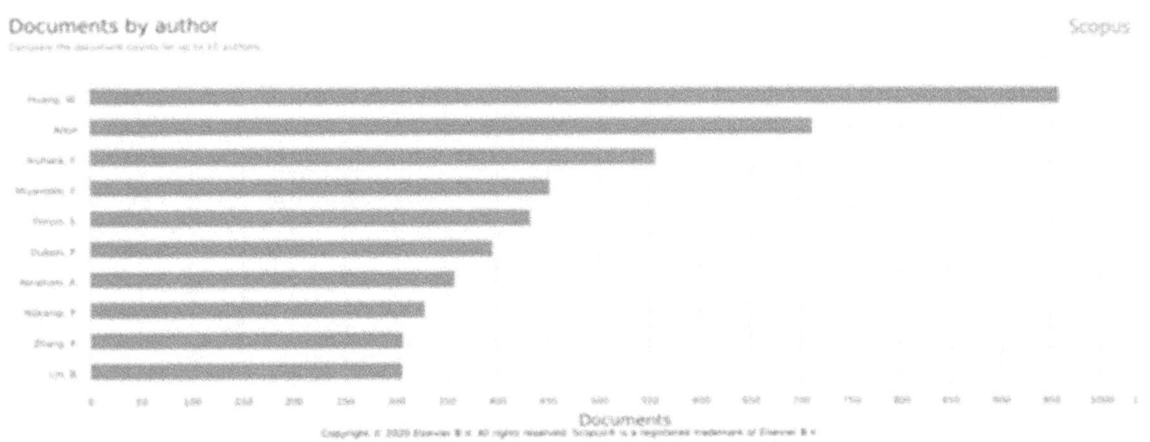

The second condition could be the lack of an efficient guideline, such as surveys and validated data, studies and scarcity of bibliography in general the relationship between 'innovation' and 'community tourism' is established in 943 documents applying the first criterion, the first of which arose in 1986 through an article titled " Tourism and computers: Quo vadis ?" by Var T., *et al.*, (1986) published in the Annals of Tourism Research, 13 (1), pp. 109-117 (Scopus, 2019) (See Figure 10 and 11).

These documents exist in several formats, but the one that stands out most is the Article with 71.8% of publications, followed by book chapters with 12%, and the books themselves with 6.8%. As regards the geographical area, the United Kingdom leads, followed by the United States of America and Australia. Highlight for Italy followed by Spain who lead European countries.

Applying the second criterion (title, abstract and keywords), 12 results were obtained, four of which were published in the current year of 2019 and another is repeated (See Table 3). As for its format, the Article continues to predominate with 83.3%, books with 8.3% and, the notes with 8.3% also.

Figure 7.
Source: Scopus, 2019

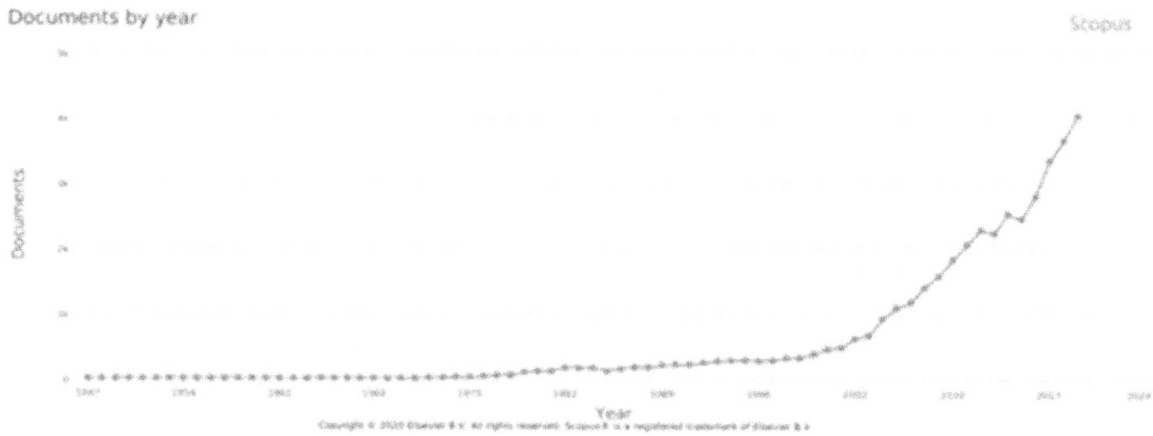

Figure 8.
Source: Scopus, 2019

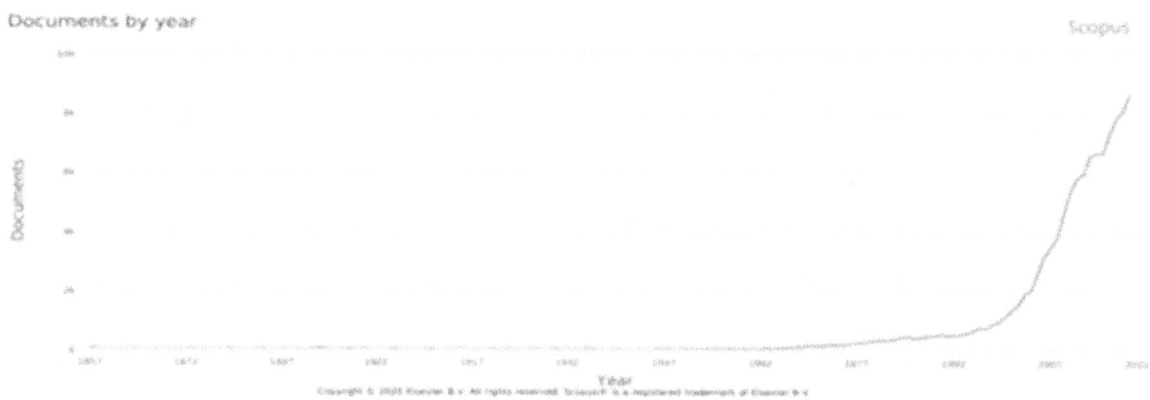

In the previous table is shown that this bibliography is very recent and, therefore, the margin progression on the topic also. The following table shows the publications taking into account your country or territory of origin. The largest number of published articles was in the United Kingdom, followed by Italy and South Africa immediately afterwards and, finally, Argentina, Australia and Canada (See Figure 12 and 13).

It appears that the size of this sample is quite small, namely for the number publications containing the expressions community tourism and innovation in the same document. With regard to major areas of interest, these publications are divided into the sciences with 37%, followed immediately by economics, management and finance with 29.6% and environmental sciences with 14.8% (See Figure 14).

Figure 9.
Source: Scopus, 2019

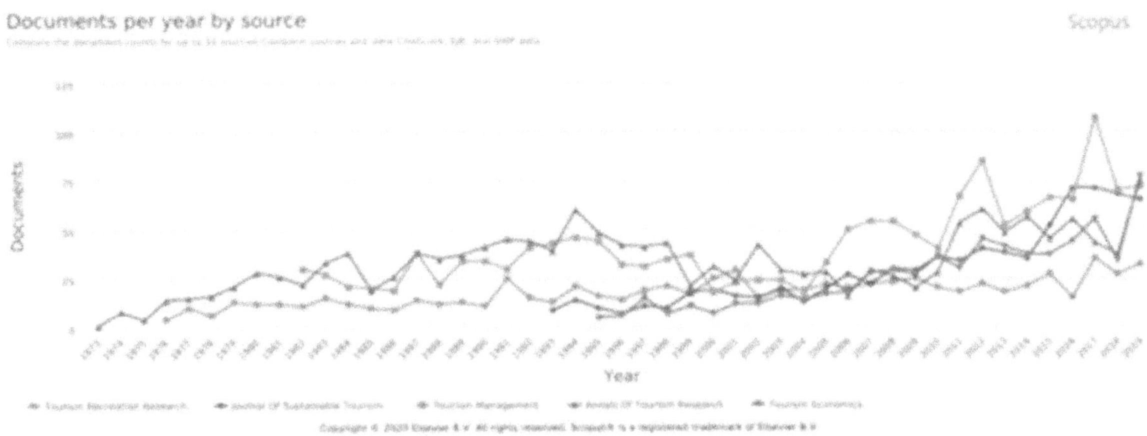

Figure 10.
Source: Scopus, 2019

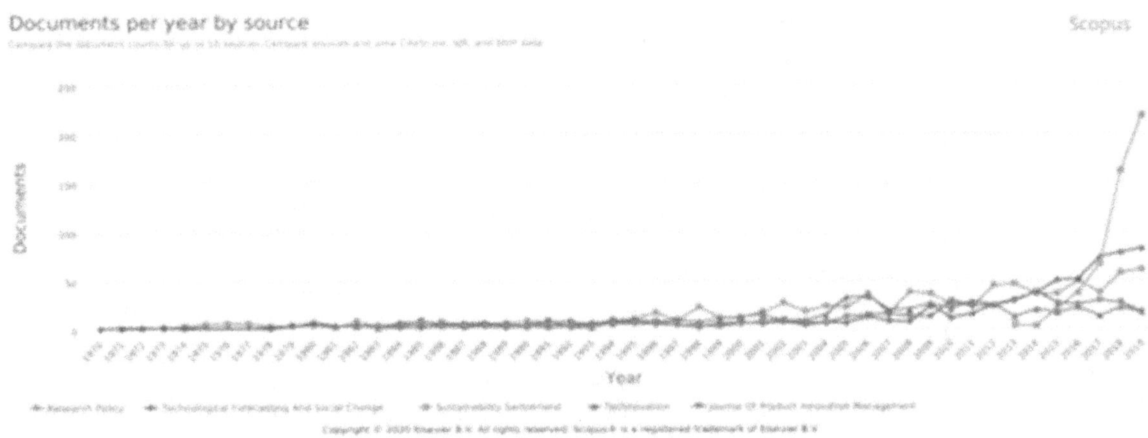

FUTURE RESEARCH DIRECTIONS

For future investigations and research, it is necessary to diversify the databases to bibliometric research. It is also necessary to expand the languages used to carry out the study. In this way, it would be better understood and closer to reality.

CONCLUSION

Community-based tourism (CBT) is a plausible alternative to mass tourism and above all, empowerment of local communities. It is understood that there is a need to address the local problems of the communities for a cultural change that is beneficial to all who integrate these local communities.

Figure 11.
Source: Scopus, 2019

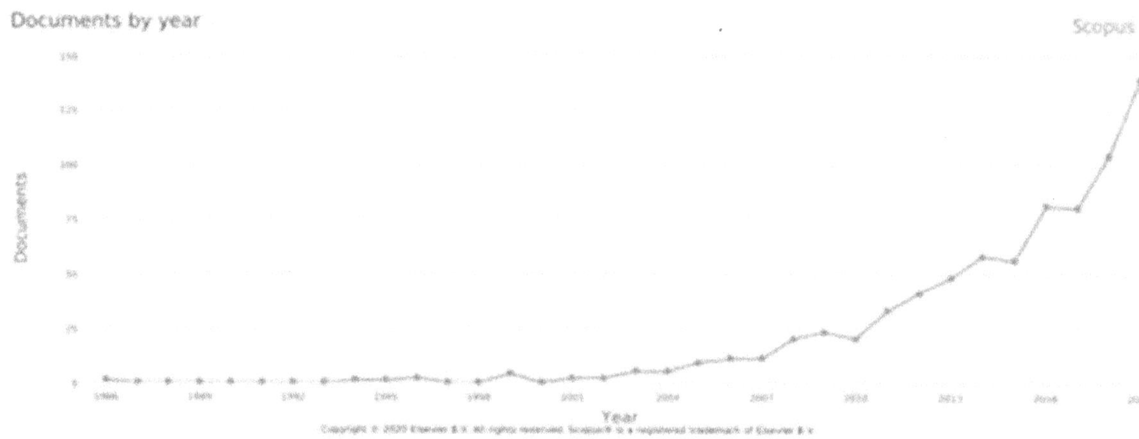

Table 3. Results after applying the second research criterion

Year	Title	Author(es)	Kind
2019	Examining roles of tour dure producers for social capital and innovativeness in community-based tourism	Joo, J., Choi,J-J., Kim, N.	Article
2019	Collaborative processes and collective impact in tourist rural villages-Insights from a comparative analysis between Argentinian and Italian cases	Chiodo, E., Adriani, H.L., Navarro, F.P., Salvatore, R.	Article
2019	Network Structure in Coastal and Marine Tourism: Diving into the Three Clusters	Yustika, B.P., Goni, J.I.C.	Article at press
2019	Thoughts from the think tank: lessons learned from the sustainable Indigenous tourism symposium	Graci, S., Maher, P.T., Peterson, B., Hardy, A., Vaugeois, N.	Article at press
2019	South African community-based tourism operational guidelines: Analysis and critical review	Giampiccoli, A., Saayman, M	Article
2016	Logics and interlocking directorships in a multi-stakeholder system	Beritelli, P., Buffa, F., Martini, U	Article
2015	Integrating Communities into Tourism Planning Through Social Innovation	Malek, A., Costa, C.	Article
2015	Innovative practices in education to empower rural communities: A case study of a "non-formal education" school in Thailand	Rawat, K., Bouchon, F., Nair, V	Article
2012	Responsible tourism: Critical issues for conservation and development	Spenceley, A.	Book
2011	Community networks and sustainable livelihoods in tourism: The role of entrepreneurial innovation	Kokkranikal, J., Morrison, A.	Article
2008	Developing woodlands for tourism: Concepts, connections and challenges	Martin, S.	Article

Source: Scopus - Citation overview, 2019

Figure 12.
Source: Scopus, 2019

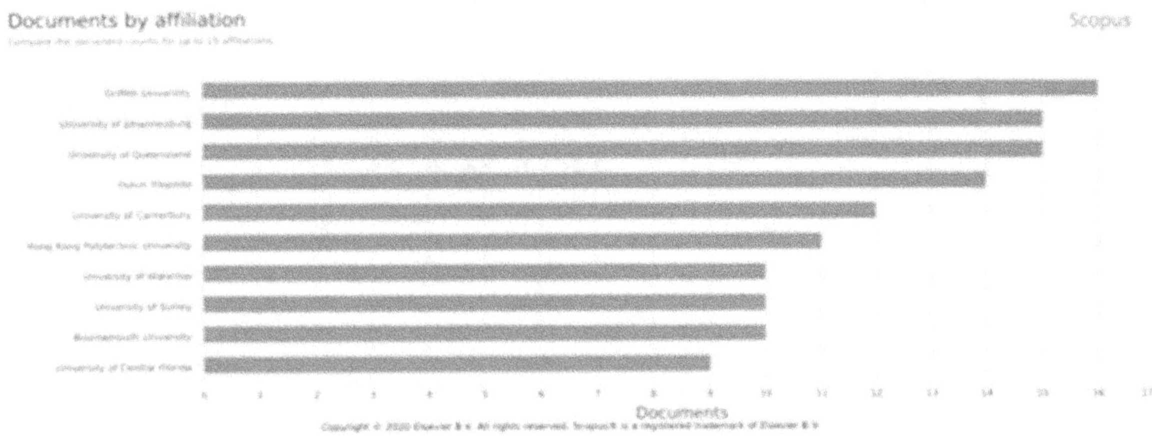

Figure 13.
Source: Scopus, 2019

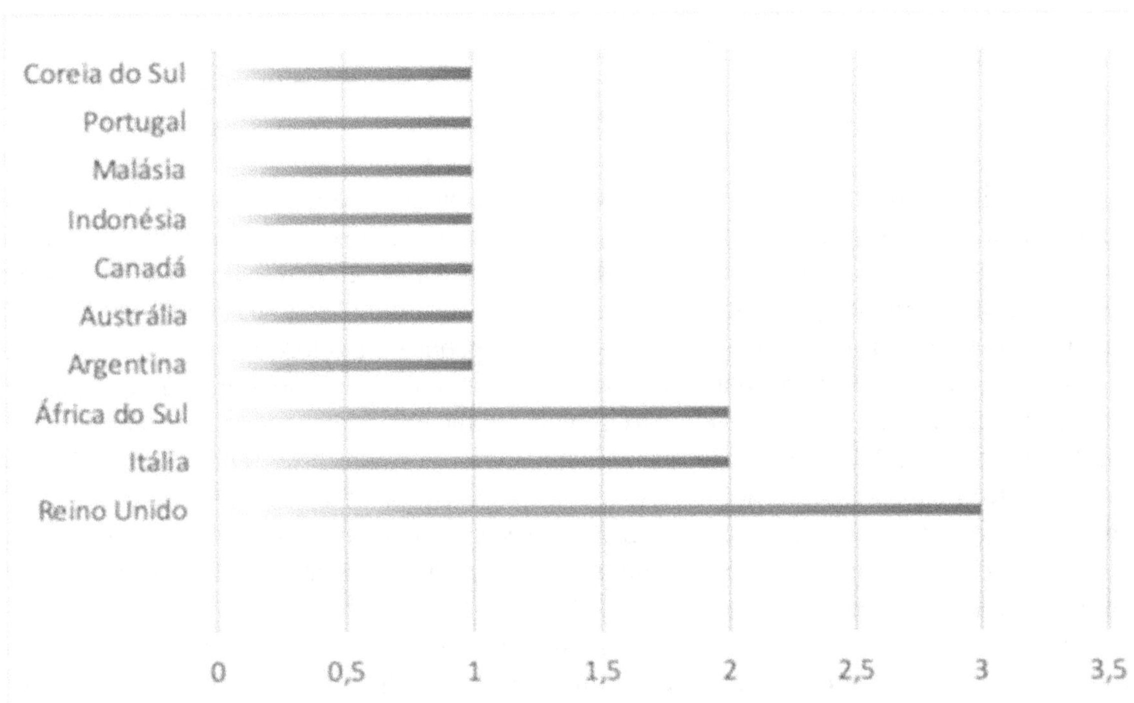

With this, it is possible to envision a less predatory and socially fairer tourism, where the possibilities of growth goals are not limited only to the large owners of hotel chains or to the profit holders of the leisure industry. However, due to the strong competition existing in the tourist territories, one of the alternatives for growth and positioning of CBT would be the capacity for innovation.

Figure 14.
Source: Scopus, 2019

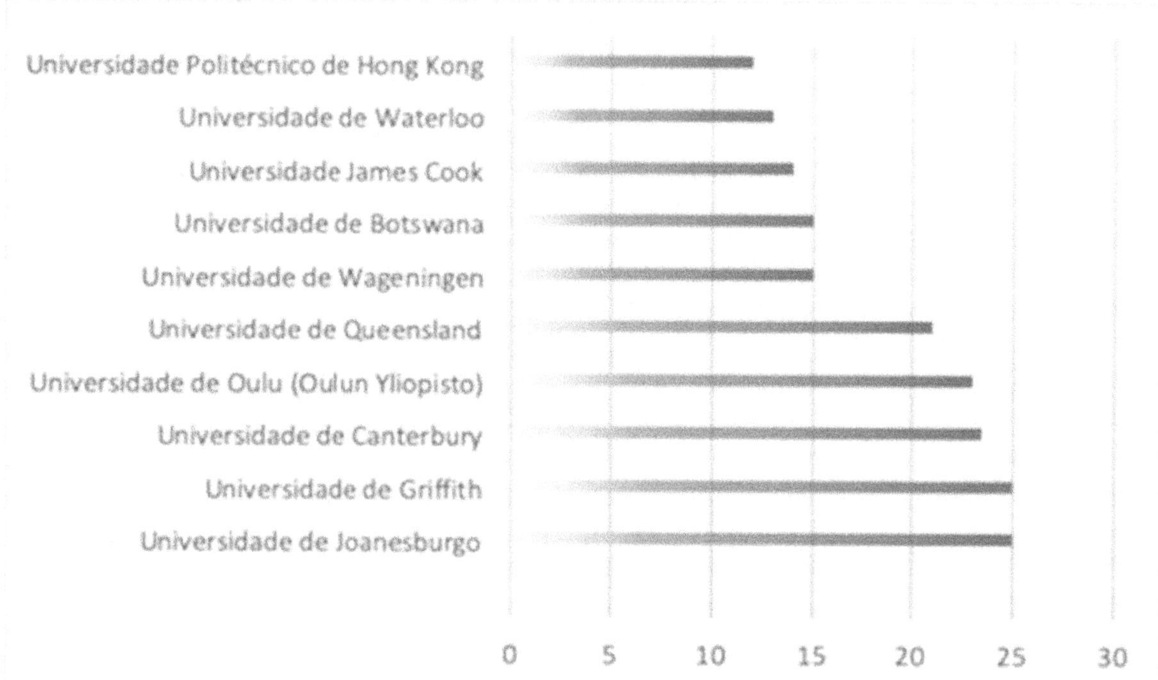

The main objective of this study was to understand the interest and growth of research in the areas, namely, tourism, innovation and, more specifically, innovation linked to community tourism. It is concluded that the practices of innovations in this type of tourism are little researched in the academy. As observed in the methodologies applied to this research, regarding bibliometric analysis, both the terminologies on which this research focused, as well as the publications closest to the search terms, are still scarce in international reference journals. On the one hand, there is a non-uniform distribution in the publication of documents, but of continuous growth. On the other hand, the terms searched for 'Innovation' combined with 'Community Tourism' do not appear in any records in documents with those words in the title of the publications.

This fact has still fragile consequences for an urgent need in community tourism to address the local problems of the communities as examples of innovating for a cultural change for the development and improvement of activities for the communities.

The almost inexistence of scientific articles that research or attest to the importance of innovation in CBT may reveal two conditions, nonetheless, little interest in the topic or difficulties in studying fieldwork, being restricted to European countries. In any case, an opportunity to develop works related to the problem of this study in the future, which were presented in greater depth and linked to concrete examples from local communities.

Thus, it is necessary to encourage production on the subject, to contribute to a more competitive and innovative CBT practice. It is understood that the innovation component can be associated with a community territorial empowerment strategy. Empowerment can be applied to a resilience recognition strategy of local communities justified by a process of formation, perception, reflection and interpretation of social reality, individual and/or collective, allowing to develop significant changes for the construction of a more participative community in a cultural and social engagement process.

In this understanding, the notion of community territorial empowerment can be innovative as an active process for the development of local communities. This study allowed us to conclude that the connection between community tourism and innovation plays an increasingly central role in our society, not only in terms of economic sustainability, but also of social development. Combining empowerment with community tourism represents innovation, although a continuous incentive is needed to deepen these dynamics for a CBT practice, reinforcing them with examples through case studies and experiences emanated by the local communities themselves.

REFERENCES

Almeida, H. (2017 in press). *Participação como eixo estratégico renovado na intervenção social.* Universidade de Coimbra.

Blackstock, K. (2005). A critical look at community based tourism. *Community Development Journal: An International Forum*, *40*(May), 39–49. doi:10.1093/cdj/bsi005

Castro, L., & Pinto, R. (2013). Sustentabilidade e turismo comunitário : Aspectos teóricos-conceituais. *Caderno Virtual de Turismo*, *13*(2), 213–226.

Castro, L., Pinto, R., Luisa, L., & Castro, C. (2013). Sustentabilidade e turismo comunitário : Aspectos teóricos-conceituais. *Caderno Virtual de Turismo*, *13*(2), 213–226.

Chiodo, E., Adriani, H., Navarro, F., & Salvatore, R. (2019). Collaborative Processes and Collective Impact in Tourist Rural Villages — Insights from a Comparative Analysis between Argentinian and Italian Cases. *Susta, 11*(2).

Fabrino, N. H., Costa, H. A., & Nascimento, E. P. (2012). Turismo de Base Comunitária (TBC): Elementos chaves para aferir seu desempenho na perspectiva da sustentabilidade. *Revista Brasileira de Ecoturismo*, *5*(3), 546–559. doi:10.34024/rbecotur.2012.v5.6224

Giampiccoli, A., & Mtapuri, O. (2012). Community-based tourism : An exploration of the concept (s) from a political perspective. *Tourism Review International*, *16*(1), 29–43. doi:10.3727/154427212X13431568321500

Giampiccoli, A., & Saayman, M. (2016). Community-based tourism : From a local to a global push. *Acta Commercii - Independent Research Journal in the Management Sciences, 1*, 1–10.

Gomez, C., Falcão, M., Castilho, L., Correia, S., & Oliveira, V. (2015). Turismo de Base Comunitária como Inovação Social : Congruência entre os constructos. *PASOS Revista de Turismo y Patrimonio Cultural, 13*(5), 1213–1227. doi:10.25145/j.pasos.2015.13.082

Irving, M. D. A., Cristina, T., & Mendonça, M. (2018). *Turismo de base comunitária na América Latina : uma estratégia em rede.* Academic Press.

Joo, J., Choi, J., & Kim, N. (2019). *Examining Roles of Tour Dure Producers for Social Capital and Innovativeness in Community-Based Tourism.* Academic Press.

Jorge, E., Mielke, C., & Pegas, F. V. (2013). Turismo de Base Comunitária no Brasil. *Insustentabilidade é uma Questão de Gestão, 24*, 170–189.

Kirchhof, A. L. C., & Lacerda, M. R. (2012). Desafios e perspectivas para a publicação de artigos - uma reflexão a partir de autores e editores TT - Challenges and prospects for publishing articles - consideration based on statements from authors and publishers TT - Retos y perspectivas de la publ. *Texto & Contexto Enferm, 21*(1), 185–193. doi:10.1590/S0104-07072012000100021

Krippendorf, J. (2003). *Sociologia do Turismo: para uma nova compreensão do lazer e das viagens* (3rd ed.).

Lagowski, J. J. (1990). *What's So Great about Innovation?* Academic Press.

Lindič, J., Baloh, P., Ribire, V. M., & Desouza, K. C. (2011). Deploying information technologies for organizational innovation: Lessons from case studies. *International Journal of Information Management, 31*(2), 183–188. doi:10.1016/j.ijinfomgt.2010.12.004

Lopez-guzman, T., Borges, O., & Cerezo, J. (2016). Community-based tourism and local socio-economic development : A case study in Cape Verde. *African Journal of Business Management, 5*, 1608–1617.

Nelson, F. (2004). *The evolution and impacts of community-based ecotourism in northern Tanzania.* Academic Press.

Nunes, M., Santos, K., & Azevedo, F. (2014). Turismo e capital social: uma aproximação teórica e conceitual. *Revista de Turismo y Patrimonio Cultural, 12*(2008), 443–452.

Valente, T., & Mielke, E. (2017). Evolução e Potencialidades de Inovação do Turismo de Base Comunitária no Município de Guaraqueçaba – PR (Brasil). *Turismo e Sociedade, 10*(3), 1–26.

KEY TERMS AND DEFINITIONS

Bibliometric: Measuring written documents through some criteria.

Creativity: It is a human process (from conception to interpretation) influenced and framed by the surrounding environment over time, and which is therefore manifested and is always understood differently, not limited to a single definition as indicated by Robinson (2008 *apud* Richards 2011, p. 1226), " the lack of a single widely accepted definition of creativity is undoubtedly due to the wide range of views on the function of creativity".

Culture: The identity of a civilization, a nation, a society. Their values, tradition, heritage, and ways.

Empowerment: Give the human and technical tools to promote the knowledge of individuals and institutions, so they can develop in the most sustainable way possible.

Host Community: The locals who receive, welcome, and provide the visitors leisure, cultural knowledge thru experiences and activities.

Hotel Chain: Composition of all hotels a company owns.

Mass Tourism: This term is usually used as a consequence of the standard, planning and formatted kind of product that general tourists buy. Usually cheaper and an all-in-one solution (accommodation, food, and beverage).

Compilation of References

Ab Karim, S., & Chi, C. (2010). Culinary tourism as a destination attraction: An empirical examination of destinations' food image. *Journal of Hospitality Marketing & Management*, *19*(6), 531–555. doi:10.1080/19368623.2010.493064

Aboelmaged, M. (2018). Direct and indirect effects of eco-innovation, environmental orientation and supplier collaboration on hotel performance: An empirical study. *Journal of Cleaner Production*, *184*, 537–549. doi:10.1016/j.jclepro.2018.02.192

Abrahamsson, A. (2008). Sustainopreneurship –business with a cause: the promise of creative business organizing for sustainability. In Bridging the functional and territorial views on regional entrepreneurship and development (pp. 137-155). Swedish Foundation for Small Business Research.

Agarwal, S., Erramilli, M. K., & Dev, C. S. (2003). Market orientation and performance in service firms: Role of innovation. *Journal of Services Marketing*, *17*(1), 68–82. doi:10.1108/08876040310461282

Akama, J. S., & Keiti, D. (2007). Tourism and Socio-Economic Development in Developing Countries: A Case Study of Mombasa Resort in Kenya. *Journal of Sustainable Tourism*, *15*(6), 35–748. doi:10.2167/jost543.0

Alario Trigueros, M., & Morales Prieto, E. (2016). Iniciativas de las mujeres: Emprendimiento y oportunidades en el espacio rural de Castilla y León. *Documents d'Analisi Geografica*, *62*(3), 613–637. doi:10.5565/rev/dag.369

Albanian Development Fund. (2018). *Market research and value chain assessment*. Horwath HTL.

Albertini, G. S. G. G. (2017). *Turismo e economia colaborativa: uma avaliação das plataformas online de serviços vinculados ao turismo* (Undergraduate dissertation). Unidade Federal Fluminense, Rio de Janeiro, Brazil. Retrieved October 2, 2019 from https://app.uff.br/riuff/handle/1/4721

Alles, M. (2012). *Comportamiento Organizacional. Cómo lograr un cambio cultural a través de Gestión por competencias*. Granica.

Almeida-García, F., Peláez-Fernández, M. Á., Balbuena-Vázquez, A., & Cortés-Macias, R. (2016). Residents' perceptions of tourism development in Benalmádena (Spain). *Tourism Management*, *54*, 259–274. doi:10.1016/j.tourman.2015.11.007

Almeida, H. (2017 in press). *Participação como eixo estratégico renovado na intervenção social*. Universidade de Coimbra.

Alonso, A., & Nortchote, J. (2010). The Development of olive tourism in Western Australia: A case study of an emerging tourism industry. *International Journal of Tourism Research*, *12*(6), 696–708. doi:10.1002/jtr.786

Aluri, A., & Palakurthi, R. (2011). The influence of demographic factors on consumer attitudes and intentions to use RFID technologies in the US hotel industry. *Journal of Hospitality and Tourism Technology*, *2*(3), 188–203. doi:10.1108/17579881111173749

Amabile, T. M., Conti, R., Coon, H., Lazenby, J., & Herron, M. (1996). Assessing the work environment for creativity. *Academy of Management Journal*, *39*(5), 1154–1184. doi:10.2307/256995

Andrade, A. (2010). *O Nascimento do Windsurf no Ceará.* Retrieved November 21, 2019, from http://blogs.diariodonordeste.com.br/manobraradical/windsurf/o-nascimento-do-windsurf-no-ceara/

André, I., & Abreu, A. (2006). Dimensions and spaces of social innovation. *Finisterra, 41*(81), 121–141. Retrieved November 2, 2019, from https://revistas.rcaap.pt/finisterra/article/view/1465

Annual report of the state committee of the Republic of Uzbekistan for tourism development, 2019,

Annual report of the state committee of the Republic of Uzbekistan on statistics, 2019

Annual report of the territorial administration of Khorezm region for tourism development, 2019

Ansarah, M. G. R., & Panosso Netto, A. (2010). A segmentação dos mercados como objeto de estudo do turismo. *VII Seminário da Associação Brasileira de Pesquisa e Pós-Graduação em Turismo, 20*. Retrieved September 10, 2019 from https://www.anptur.org.br/anais/anais/files/7/25.pdf

Antonioli, F. L. A. (2015). *Viagens no feminino: gênero, turismo e transnacionalidade* (Master's thesis). Universidade Estadual de Campinas, Campinas, Brazil. Retrieved September 10, 2019 from http://www.repositorio.unicamp.br/handle/Reposip/279566

Arboleya, J., Garcia-Quiroga, M., Lasa, D., Oliva, O., & Luis-Aduriz, A. (2014). Effect of highly aerated food on expected satiety. *International Journal of Gastronomy and Food Science, 2*(1), 14–21. doi:10.1016/j.ijgfs.2013.12.002

Arechavala, R. (2014). *Twenty years of corporate culture in western Mexico Where are we going?* Universidad de Guadalajara.

Arnott, M. (1975). *The Anthropology of Food and Food Habits*. Mouton.

Arruda, D. O., & Gonçalves, J. P. (2020). Limits and possibilities in the development of Community-Based Tourism strategies in quilombola territory. *Interações (Campo Grande), 1*(21), 107–123. doi:10.20435/inter.v21i1.1968

Ascanio, A. (2012). *Theory of tourism*. Trillas.

Associação Brasileira de Startups. (2019). *Sobre a abstartups*. Retrieved September 16, 2019, from https://abstartups.com.br/sobre-a-abstartups

Ateljevic, I., & Doorne, S. (2000). Staying within the fence: Lifestyle entrepreneurship in tourism. *Journal of Sustainable Tourism, 8*(5), 378–392. doi:10.1080/09669580008667374

Ayora-Diaz, S. (2019). Making and Changing Yucatecan Taste in Yucatán: Innovation and Persistence in Yucatecan Gastronomy. In S. Ayora-Diaz (Ed.), *Taste, Politics, and Identities in Mexican Food* (pp. 101–115). Bloomsbury. doi:10.5040/9781350066700.ch-007

Bâc, D. (2012). *The impacts of tourism on society*. University of Oradea. Retrieved from http://steconomiceuoradea.ro/anale/volume/2012/n1/072.pdf

Baker, M., & Kim, K. (2019). Heritage and Authenticity in Gastronomic Tourism. In S. Dixit (Ed.), *The Routledge Handbook of Gastronomic Tourism* (pp. 252–260). Routledge. doi:10.4324/9781315147628-29

Bak-Geller Corona, S., & Matta, R. (2020). Las cocinas mestizas en México y Perú. Claves para interpretar el multiculturalismo en América Latina. *Antipoda: Revista de Antropologia y Arqueologia, 39*(39), 69–93. doi:10.7440/antipoda39.2020.04

Bani-Melhem, S., Zeffane, R., & Albaity, M. (2018). Determinants of employees' innovative behavior. *International Journal of Contemporary Hospitality Management, 30*(3), 1601–1620. doi:10.1108/IJCHM-02-2017-0079

Baregheh, A., Rowley, J., & Sambrook, S. (2009). Towards a multidisciplinary definition of innovation. *Management Decision*, *47*(8), 1323–1339. doi:10.1108/00251740910984578

Barreto, M. (2007). *Tourism and Culture. Relations, contradictions, and expectations*. STEPS. Retrieved June 12, 2019, from https://lecturayescrituraunrn.files.wordpress.com/2013/08/turismo-y-cultura-barreto.pdf

Barreto, M. (1991). Planejamento e Organização. In *Turismo*. Papirus Editora.

Barrios, A., & Pantoja, C. (2016). Comida de la calle en el Centro Histórico. *Tlamati Sabiduria*, *7*(2), 1–11.

Bautista, J., & Smit, M. (2012). Sustentabilidad y agricultura en la" región del mezcal" de Oaxaca. *Revista Mexicana de Ciencias Agrícolas*, *3*(1), 5–20. doi:10.29312/remexca.v3i1.1477

Beauvoir, S. (2009). *O segundo sexo (2nd ed.)*. Nova Fronteira.

Belisle, F. (1983). Tourism and food production in the Caribbean. *Annals of Tourism Research*, *10*(4), 497–513. doi:10.1016/0160-7383(83)90005-1

Belk, R. (2014). You are what you can access: Sharing and collaborative consumption online. *Journal of Business Research*, *67*(8), 1595–1600. doi:10.1016/j.jbusres.2013.10.001

Beni, M. C. (1998). *Análise Estrutur.al do Turismo*. Editora Senac.

Beni, M. C. (1999). Política e Estratégia do Desenvolvimento Regional: Planejamento I. *Turismo em Análise*, *1*(1), 7–17. doi:10.11606/issn.1984-4867.v10i1p7-17

Beni, M. C. (2006). *Tourism policy and planning in Brazil*. Aleph.

BEPA (Bureau of European Policy Advisers). (2011). Empoweringpeople, drivingchange - social innovation in the European Union. Publications Office of the European Union.

Berdychevsky, L., Poria, Y., & Uriely, N. (2013). Sexual behavior in women's tourist experiences: Motivations, behaviors, and meanings. *Tourism Management*, *35*, 144–155. doi:10.1016/j.tourman.2012.06.011

Beritelli, P. (2011). Cooperation among prominent actors in a tourist. *Annals of Tourism Research*, *38*(2), 607–629. doi:10.1016/j.annals.2010.11.015

Bertella, G. (2011). Knowledge in food tourism: The case of Lofoten and Maremma Toscana. *Current Issues in Tourism*, *14*(4), 355–371. doi:10.1080/13683500.2010.489638

Beshr, H. A., & Hossan, C. G. (2018). Key drivers for service innovation in the UAE's hotel sector. *International Journal of Entrepreneurship and Innovation Management*, *22*(3), 251–268. doi:10.1504/IJEIM.2018.091769

Besteiro, B. (2004). El desarrollo del Turismo Náutico en Galicia. *Cuadernos de Turismo*, *13*, 145–163.

Bielinski, M., Jr., Disconzi, C. M. D. G., Knoll, K. R. H., & Fortunato, A. (2016). *Desenvolvimento de Roteiros Turísticos na Cidade de Santana Do Livramento/RS: Um Estudo sobre a Ferradura dos Vinhedo*. XVI Mostra de Iniciação Científica, Pós-graduação, Pesquisa e Extensão. UC.S.

Bilgihan, A., Smith, S., Ricci, P., & Bujisic, M. (2016). Hotel guest preferences of in-room technology amenities. *Journal of Hospitality and Tourism Technology*, *7*(2), 118–134. doi:10.1108/JHTT-02-2016-0008

Blackstock, K. (2005). A critical look at community based tourism. *Community Development Journal: An International Forum*, *40*(May), 39–49. doi:10.1093/cdj/bsi005

Blancas, F. J., Oyola, M. L., González, M., Guerrero, F. M., & Caballero, R. (2011). How to Use Sustainability Indicators for Tourism Planning: The Case of Rural Tourism in Andalusia (Spain). *The Science of the Total Environment*, *412*(413), 28–45. doi:10.1016/j.scitotenv.2011.09.066 PMID:22033358

Bobirca, A., & Cristureanu, C. (2008). Analyzing Romania's competitiveness as a tourism destination. *Advances in Hospitality Leisure*, (4), 75-99.

Boesen, S., Sundbo, D., & Sundbo, J. (2017). Local food and tourism: An entrepreneurial network approach. *Scandinavian Journal of Hospitality and Tourism*, *17*(1), 76–91. doi:10.1080/15022250.2016.1218629

Bohdanowicz, P., Zientara, P., & Novotna, E. (2011). International hotel chains and environmental protection: An analysis of Hilton's we care! programme (Europe, 2006-2008). *Journal of Sustainable Tourism*, *19*(7), 797–816. doi:10.1080/09669582.2010.549566

Boka, M., & Torluccio, G. (2013). Informal Economy in Albania. *Academic Journal of Interdisciplinary Studies*, *2*(8), 2012.

Booking.com. (2019). *Mulheres pelo mundo*. Retrieved November 21, 2019, from https://www.mulherespelomundo.com/introducao

Borges, C. I. O. S. (2011). Community tourism in traditional communities of the coastal zone of Ceará: focus on the experience of the Red Tucum (Master in Geography). Universidad Estatal de Ceará, Fortaleza.

Bosworth, G., & Farrell, H. (2011). Tourism entrepreneurs in Northumberland. *Annals of Tourism Research*, *38*(4), 1474–1494. doi:10.1016/j.annals.2011.03.015

Boyack, K. W., & Klavans, R. (2010). Co-citation analysis, bibliographic coupling, and direct citation: Which citation approach represents the research front most accurately? *Journal of the American Society for Information Science and Technology*, *61*(12), 2389–2404. doi:10.1002/asi.21419

Boyne, S., Williams, F., & Hall, D. (2002). The Isle of Arran taste trail. In A. Hjalager & G. Richards (Eds.), *Tourism and Gastronomy* (pp. 91–114). Routledge.

Bramwell, B., & Lane, B. (2011). Critical research on the governance of tourism and sustainability. *Journal of Sustainable Tourism*, *19*(4/5), 411–421. doi:10.1080/09669582.2011.580586

Brasil. (1990). Constituição. Código de Defesa do Consumidor nº 8078, date September 11, 1990.

Brillat-Savarin, J. (1994). The Physiology of Taste (A. Drayton, Trans.). Harmondsworth: Penguin. (Original publication 1825)

Brito-Henriques, E., Sarmento, J., & Lousada, M. A. (2010). When water meets tourism: An introduction. In E. Brito-Henriques, J. Sarmento, & M. A. Lousada (Eds.), *Water and tourism. Resources management, planning and sustainability* (pp. 13–33). CEG, Universidade de Lisboa.

Bula, H. O. (2012). Evolution and Theories of Entrepreneurship: A Critical Review on the Kenyan Perspective. *International Journal of Business and Commerce*, *1*(11), 81–96.

Bunghez, C. L. (2016). The importance of tourism to a destination's economy. *Journal of Eastern Europe Research in Business & Economics*, *2016*, 1–9. Advance online publication. doi:10.5171/2016.143495

Burgos, A., & Mertens, F. (2016). Collaborative networks in community-based tourism: Implications for participatory management. *Tourism & Management Studies*, *12*(2), 18–27. doi:10.18089/tms.2016.12203

Burgos-Doria, R. (2016). El turismo comunitario como iniciativa de desarrollo local. Caso localidades de Ciudad Bolívar y Usme zona rural de Bogotá. *Hallazgos*, *13*(26), 193–214. doi:10.153321794-3841.2016.0026.08

Bustos, J., & Galván, A. (2015). A Taste of Oaxaca: A Chef, a Recipe, and a Cookbook. *Diálogo, 18*(1), 177–178. doi:10.1353/dlg.2015.0057

Butler. (2006). *The tourism area life cycle.* Channel View Publications.

Butler, J. (2015). *Problemas de gênero: feminismo e subversão da identidade. Col.Sujeito & História* (8th ed.). Civilização Brasileira.

Butler, R. W. (1980). The concept of a tourist area cycle of evolution: Implications for management of resources. *The Canadian Geographer. Geographe Canadien, 24*(1), 5–12. doi:10.1111/j.1541-0064.1980.tb00970.x

Buttler, R. (1990). Alternative tourism: Pious hope or Trojan horse? *Journal of Travel Research, 31*(4), 40–45. Advance online publication. doi:10.1177/004728759002800310

Cabanilla, E. (2017). *Configuración socio-espacial del turismo comunitario: caso República del Ecuador* (Tesis Doctoral). Universidad Nacional del Sur, Argentina. Accesado el 1 de mazo 2019 de:http://repositoriodigital.uns.edu.ar/handle/123456789/3430

Cabedo, M. (2006). *Philosophy and culture of tolerance*. Universidad Jaume I.

Caiza, R., & Molina, E. (2012). Historical analysis of the evolution of tourism in the Ecuadorian territory. *Tourist unit specialties RICIT, 4*, 8-24.

Calantone, R. J., Cavusgil, S. T., & Zhao, Y. (2002). Learning orientation, firm innovation capability, and firm performance. *Industrial Marketing Management, 31*(6), 515–524. doi:10.1016/S0019-8501(01)00203-6

Caldwell, N., & Freire, R. J. (2004). The differences between branding a country, a region and a city: Applying the brand box model. *Journal of Brand Management, 12*(1), 50–61. doi:10.1057/palgrave.bm.2540201

Cameron, K. S., & Quinn, R. E. (1999). *Diagnosing and changing organizational culture*. Addison-Wesley.

Cameron, K. S., Quinn, R. E., Degraff, J., & Thakor, A. (2016). *Competing Values Leadership. Creating value in organizations*. New Horizons in Management.

Cameron, K., & Freeman, S. (1991). Cultural congruence, strength, and type: Relationships to effectiveness. *Research in Organizational Change and Development, 5*, 23–58.

Camisón, C., & Monfort-Mir, V. M. (2012). Measuring innovation in tourism from the Schumpeterian and the dynamic-capabilities perspectives. *Tourism Management, 33*(4), 776–789. doi:10.1016/j.tourman.2011.08.012

Campo, S., Díaz, A. M., & Yagüe, M. J. (2014). Hotel innovation and performance in times of crisis. *International Journal of Contemporary Hospitality Management, 26*(8), 1292–1311. doi:10.1108/IJCHM-08-2013-0373

Carlsen, J., Morrison, A., & Weber, P. (2008). Life style oriented small tourism firms. *Tourism Recreation Research, 33*(3), 255–263. doi:10.1080/02508281.2008.11081549

Carmo, H. (Ed.). (2014). Desenvolvimento comunitário (3rd ed.). Lisboa: Universidade Aberta.

Caro-Montero, E. (2018). *Desarrollo de valores organizacionales para la innovación social. Una mirada desde el emprendimiento en el contexto cubano actual*. Cuba: Universidad de la Habana.

Carvalho, L., & Costa, T. (2011). Tourism innovation – a literature review complemented by case study research. *International conference on tourism and management studies proceedings*.

Carvalho, G., Baptista, M. M., & Costa, C. (2015). Mulheres que viajam sozinhas: Reflexões sobre género e experiências turísticas. *Revista Turismo & Desenvolvimento, 23*, 59–67.

Castells, M. (1999). *A Sociedade em Rede*. Paz e Terra.

Castro, E., Vieira, G., Loureiro, F., Patrocínio, F., Firmino, G., Gomes, H. & Fernandes, M. (2017). Managing the Estrela Aspiring Geopark: A framework for promoting geoheritage conservation. *Conference Ciência 2017.*

Castro, L., & Pinto, R. (2013). Sustentabilidade e turismo comunitário : Aspectos teóricos-conceituais. *Caderno Virtual de Turismo, 13*(2), 213–226.

Cavaco, C. (2010). Água doce: Agricultura versus lazeres e turismo. In E. Brito-Henriques, J. Sarmento, & M. A. Lousada (Eds.), *Water and tourism. Resources management, planning and sustainability* (pp. 223–259). CEG, Universidade de Lisboa.

Cavaco, C., & Simões, J. (1998). *Água, desenvolvimento e bem-estar.* MADRP.

Cavicchi, A., & Ciampi Stancova, K. (2016). *Food and gastronomy as elements of regional innovation strategies.* European Commission, Joint Research Centre, Institute for Prospective Technological Studies.

Çela, E., Metani, I., & Çeliku, E. (2019). *Analiza dhe zhvillimet strukturore të ekonomisë Shqiptare: përqasje sipas matricave të kontabilitetit social.* Bank of Albania.

Çeliku, E., & Metani, I. (2011). *Pozicioni konkurrues i ekonomisë shqiptare në terma të produktivitetit e të kostos së punës.* Bank of Albania.

CEMEFI. (n.d.). *El concepto de Responsabilidad Social Empresarial.* https://www.cemefi.org/esr/images/stories/pdf/esr/concepto_esr.pdf

Centro Tecnológico del Mar (CETMAR). (2014). *Análisis de la participación de la mujer en la actividad pesquera y acuícola.* Ministerio de Agricultura, Alimentación y Medioambiente.

Chambers, R., & Conway, R. G. (1992). *Sustainable rural livelihoods: Practical concepts for the 21st century.* IDS.

Chan, E. S. W. (2013). Gap analysis of green hotel marketing. *International Journal of Contemporary Hospitality Management, 25*(7), 1017–1048. doi:10.1108/IJCHM-09-2012-0156

Che, D. (2006). Select Michigan: Local food production, food safety, culinary heritage and branding in Michigan agritourism. *Tourism Review International, 9*(4), 349–363. doi:10.3727/154427206776330616

Chek Sulaiman, F., Jaini, N., Rouzee Jamaluddin, E., & Idzhainee Hashim, N. (2018). The impact of marine tourism towards the local community at Palau Pangkor, Perak. *Asian Journal of Quality of Life, 3*(13), 168–176. doi:10.21834/ajqol.v3i13.174

Chen, C. (2010). Diversifying fisheries into tourism in Taiwan: Experiences and prospects. *Ocean and Coastal Management, 53*(8), 487–492. doi:10.1016/j.ocecoaman.2010.06.003

Cheng, J.-S., Tang, T.-W., Shih, H.-Y., & Wang, T.-C. (2016). Designing lifestyle hotels. *International Journal of Hospitality Management, 58*, 95–106. doi:10.1016/j.ijhm.2016.06.010

Chen, R. J. C. (2015). From sustainability to customer loyalty: A case of full service hotels' guests. *Journal of Retailing and Consumer Services, 22*, 261–265. doi:10.1016/j.jretconser.2014.08.007

Chiodo, E., Adriani, H., Navarro, F., & Salvatore, R. (2019). Collaborative Processes and Collective Impact in Tourist Rural Villages — Insights from a Comparative Analysis between Argentinian and Italian Cases. *Susta, 11*(2).

Choi, H. C., & Murray, I. (2010). Resident attitudes toward sustainable community tourism. *Journal of Sustainable Tourism, 18*(4), 575–594. doi:10.1080/09669580903524852

Cibinskienea, A., & Snieskieneb, G. (2015). Evaluation of city tourism competitiveness. *20th International Scientific Conference Economics and Management - 2015 (ICEM-2015).*

Ciro, A. (2020). *Entrepreneurial Eco-systems for sustainable tourism development in Liminal Spaces.* IDAUP Pubblication.

Ciro, A., Toska, M., & Nientied, P. (2018). Social innovation and sustainable economic development: participatory torusim destination management. In M. Jaššo, M. Husar, & M. Finka (Eds.), *The role of public sector in innovative local economic and territorial development.* Springer.

Clavé, S. A. & Calabuig, J. (2005). *Planificación y gestión de destinos turísticos.* Master en Dirección y Gestión Turística. Madrid, IUP – Grupo Santillana (edición electrónica).

Cleghorn, L. (2005). *Gestión ética para una organización competitiva.* San Pablo.

Cloutier, J. (2003). Qu'est-ce que l'innovationsociale? Cahier de recherchedu CRISES (n. ET0314). Montreal: UQAM.

Cobo, M. J., López-Herrera, A. G., Herrera-Viedma, E., & Herrera, F. (2011). Science mapping software tools: Review, analysis, and cooperative study among tools. *Journal of the American Society for Information Science and Technology, 62*(7), 1382–1402. doi:10.1002/asi.21525

Cobos, L. M., Mejia, C., Ozturk, A. B., & Wang, Y. (2016). A technology adoption and implementation process in an independent hotel chain. *International Journal of Hospitality Management, 57*, 93–105. doi:10.1016/j.ijhm.2016.06.005

Cocking, L. (2016). *Oaxaca's Guelaguetza Festival Should Be on Everyone's Bucket List.* Culture trip. Available on: https://theculturetrip.com/north-america/mexico/articles/oaxacas-guelaguetza-festival-should-be-on-everyones-bucket-list/

Coelho, M. H. P., & Sakowski, P. A. M. (2014). *1938: Texto para discussão - Perfil da mão de obra do Turismo no Brasil nas atividades características do Turismo e em ocupações. Instituto de Pesquisa Econômica Aplicada.* Ipea.

Cohen, E., & Avieli, N. (2004). Food in Tourism, Attraction and Impediment. *Annals of Tourism Research, 31*(4), 755–778. doi:10.1016/j.annals.2004.02.003

Colantuono, A. C. S. (2016). A segmentação da demanda turística no mercado brasileiro e as políticas de fomento ao setor. *Revista Turismo: Estudos e Práticas, 5*(2), 219–250.

Cole, S., & Ferguson, L. (2015). Towards a gendered political economy of water and tourism. *Tourism Geographies, 17*(4), 511–528. doi:10.1080/14616688.2015.1065509

Confederação Nacional de Dirigentes Lojistas. (2019). *Economia compartilhada deixa 89% de seus usuários satisfeitos, revela estudo da CNDL/SPC Brasil.* Retrieved September 29, 2019, from http://site.cndl.org.br/economia-compartilhada-deixa-89-de-seus-usuarios-satisfeitos-revela-estudo-da-cndlspc-brasil-2/

Coriolano, L. N. (2003). *Tourism of Inclusion and Local Development.* Fortaleza: Funece.

Coriolano, L. N. (2012). Community tourism in Brazil. In L. N. Coriolano & C. A. C. Sampaio (Eds.), *Territories of Latin American Solidarity and Community Tourism in the Reduction of Transnational Mega Enterprises* (pp. 27–41). Ed. UECE.

Correia, É., Oliveira, V., & Gomez, C. (2016). Dimensions of social innovation and the roles of organizational actor: the proposition of a framework. RAM. *Revista de Administração Mackenzie, 17*(6), 102-133. doi:10.1590/1678-69712016/administracao.v17n6p102-133

Correia, P. R., Aguiar, J. G., Viana, A. D., & Cabral, G. C. P. (2016). Por que vale a pena usar mapas conceituais no ensino superior? *Revista de Graduação USP, 1*(1), 41–51. doi:10.11606/issn.2525-376X.v1i1p41-51

Costa, C. (2010). O turismo e a água. In CCDRC (Ed.), À beira da água (pp. 261-295). Coimbra: CCDRC.

Courtine, J. B. (Ed.). (1996). *Larousse Gastronomique*. Lewis Esson.

Cracolici, M., & Nijkamp, P. (2008). The attractiveness and competitiveness of tourist destinations: A study of Southern Italian regions. *Tourism Management, 30*(3), 336–344. doi:10.1016/j.tourman.2008.07.006

Creaco, S., & Querini, G. (2003). The role of tourism in sustainbale development. In *43rd Congress of the European Regional Science Association: "Peripheries, Centres, and Spatial Development in the New Europe"*. Jyväskylä, Finland: European Regional Science Association (ERSA). Retrieved from http://hdl.handle.net/10419/115956

Croes, R., & Rivera, M. (2010). Testing the empirical link between tourism and competitiveness: Evidence from Puerto Rico. *Tourism Economics, 16*(1), 217–234. doi:10.5367/000000010790872114

Crouch, G. I., & Ritchie, J. R. B. (1999). Tourism, competitiveness, and societal prosperity. *Journal of Business Research, 44*(3), 137–152. doi:10.1016/S0148-2963(97)00196-3

Crouch, G. I., & Ritchie, J. R. B. (2000). The competitive destination: A sustainability perspective. *Tourism Management, 21*(1), 1–7.

Crouch, G. I., & Ritchie, J. R. B. (2003). *The Competitive Destination: A Sustainable Tourism Perspective*. CABI Publishers.

Csurgo, B., Hindley, C., & Smith, M. (2019). The role of gastronomic tourism in rural development. In S. Dixit (Ed.), *The Routledge Handbook of Gastronomic Tourism* (pp. 62–70). Routledge. doi:10.4324/9781315147628-8

Cújar, A., Ramos, C., Hernández, H., & López, J. (2013). Cultura organizacional: Evolución en la medición. *Revista Estudios Gerenciales, 29*(128), 350–355. doi:10.1016/j.estger.2013.09.009

Cunha, M. C. G. M. (2018). *Proposta de solução gamificada no e-commerce* (Master's Thesis). Universidade de Lisboa, Lisbon, Portugal.

Da Silva, R. G. O. (2014). *O papel dos serviços prestados pelos bugueiros na composição e no diferencial do produto turístico potiguar de sol e praia: um estudo na APCBA/RN* (Unpublished master's thesis). Universidade Federal do Rio Grande do Norte, Natal, RN.

Da Silveira, L. M., Petrini, M. D., & Santos, A. C. M. Z. (2016). Economia compartilhada e consumo colaborativo: O que estamos pesquisando? *Rege-Revista de Gestão, 23*(4), 298–305. doi:10.1016/j.rege.2016.09.005

Daft, R. L. (1978). A dual-core model of organizational innovation. *Academy of Management Journal, 21*(2), 193–210. doi:10.2307/255754

Damanpour, F. (1987). The adoption of technological, administrative, and ancillary innovations: Impact of organizational factors. *Journal of Management, 13*(4), 675–688. doi:10.1177/014920638701300408

Daterra. (2016). *Círculo de bananeiras*. Retrieved January 5, 2020, from http://www.ambientaldaterra.com.br/circulo-de-bananeiras/

Datta, P. B., & Gailey, R. (2012). Empowering women through social entrepreneurship: Case study of a women's cooperative in India. *Entrepreneurship Theory and Practice, 36*(3), 569–587. doi:10.1111/j.1540-6520.2012.00505.x

Davis, D., Allen, J., & Cosenza, R. M. (1988). Segmenting local residents by their attitudes, interests, and opinions toward tourism. *Journal of Travel Research, 27*(2), 2–8. doi:10.1177/004728758802700201

De La Encarnación, A. (2016). El alojamiento colaborativo: Viviendas de uso turístico y plataformas virtuales. *Revista de Estudios de la Administración Local y Autonómica*, 5. Retrieved October 02, 2019 from https://revistasonline.inap.es/index.php/REALA/article/view/10350

De Masi, D. (2000). *O Ócio criativo*. Sexante.

De Moraes, E. M., Irving, M. A., & Mendonça, T. C. M. (2018). Turismo de base comunitária na América Latina: uma estratégia em rede. *Revista Tur., visão e ação, 20*(2), 249-265.

De Moya, M., & Jain, R. (2013). When tourists are your "friends": Exploring the brand personality of Mexico and Brazil on Facebook. *Public Relations Review, 39*(1), 23–29. doi:10.1016/j.pubrev.2012.09.004

De Sousa, J. B. S. (2016). Lean Startups: O sistema de produção enxuta como estratégia competitiva. *Journal of Lean Systems, 1*(4), 53–65.

Dhar, R. L. (2015). The effects of high performance human resource practices on service innovative behaviour. *International Journal of Hospitality Management, 51*, 67–75. doi:10.1016/j.ijhm.2015.09.002

Dhar, R. L. (2016). Ethical leadership and its impact on service innovative behavior: The role of LMX and job autonomy. *Tourism Management, 57*, 139–148. doi:10.1016/j.tourman.2016.05.011

Diario Oficial de la Federación (DOF). (2009). *ACUERDO por el que se establece la estratificación de las micro, pequeñas y medianas empresas*. Retrieved June 12, 2019, from https://www.gob.mx/cms/uploads/attachment/file/50882/A539.pdf

Díaz-Carrión, I.A., Verján-Quiñones, R. & Castrejón-Alonso, B.E. (2014). Estereotipos de género y turismo: análisis cualitativo del "suelo pegajoso" en hoteles de 4 y 5 estrellas en Tijuana (México). *Turydes Revista Turismo y Desarrollo Revista Turydes: Turismo y Desarrollo, 17.*

Diedrich, A., & García-Buades, E. (2009). Local perceptions of tourism as indicators of destination decline. *Tourism Management, 30*(4), 512–521. doi:10.1016/j.tourman.2008.10.009

Dimitrovski, D., Todorovic, A., & Valjarevic, D. (2012). Rural tourism and regional development: Case study of Development of rural tourism in the region of Gruza, Serbia. *Procedia Environmental Sciences, 14*(1), 288–297. doi:10.1016/j.proenv.2012.03.028

Divisekera, S., & Nguyen, V. K. (2018). Determinants of innovation in tourism evidence from Australia. *Tourism Management, 67*, 157–167. doi:10.1016/j.tourman.2018.01.010

Dodds, R., Ali, A., & Galaski, K. (2018). Mobilizing knowledge: Determining key elements for success and pitfalls in developing community-based tourism. *Current Issues in Tourism, 21*(13), 1547–1568. doi:10.1080/13683500.2016.1150257

Domínguez Orquendo, E. M. (2019), Innovación social: nociones de organismos multilaterales y concepciones universitarias. *Revista Virtual Universidad Católica del Norte,* (56), 72-88.

Dreamstime. (2020). *Señales, lugares de visita turístico de excursión en el mapa de Suramérica (141183564).* Academic Press.

du Rand, G., & Heath, E. (2006). Towards a framework for food tourism as an element of destination marketing. *Current Issues in Tourism, 9*(3), 206–234. doi:10.2167/cit/226.0

Duarte, C. (2002). *Céu de Quixadá está na rota do vôo livre*. Retrieved March 2, 2020, from https://www1.folha.uol.com.br/fsp/turismo/fx0306200222.htm

Duarte, A. R. F. (2006). Betty Friedan: Morre a feminista que estremeceu a América. *Revista de Estudos Feministas, 14*(1), 287–293. doi:10.1590/S0104-026X2006000100015

Durán-Sánchez, A., Álvarez-García, J., del Río-Rama, M. de la C., & Rosado-Cebrián, B. (2019). Science mapping of the knowledge base on tourism innovation. *Sustainability, 11*(12), 1–17. doi:10.3390u11123352

Dwyer, L., & Kim, C. (2003). Destination competitiveness: Determinants and indicators. *Current Issues in Tourism*, *6*(5), 369–414. doi:10.1080/13683500308667962

Earle, R. (2010). "If You Eat Their Food ...": Diets and Bodies in Early Colonial Spanish America. *The American Historical Review*, *115*(3), 668–713. doi:10.1086/ahr.115.3.688

EBST. (2016). *A escola: identificação*. Retrieved March 20, 2019, from http://ebstitanzinho.blogspot.com/p/a-escola.html

Echeverria, O. (2006). Europa Y Su Percepción Del Nuevo Mundo A Través De Las Especies Comestibles Y Los Espacios Americanos En El Siglo XVI. *Historia (Wiesbaden, Germany)*, *39*(2), 531–556.

Eder, W., & Patzak, M. (2004). Geoparks – geological attractions: A tool for public education, recreation and sustainable economic development. *AUC Geographica*, *51*(2), 235–246.

Edquist, C. (1997). *Systems of innovation: Technologies, institutions and organizations*. Pinter Publishers/Cassell Academic.

Ellis, A., Park, E., Kim, S., & Yeoman, I. (2018). What is food tourism? *Tourism Management*, *68*, 250–263. doi:10.1016/j.tourman.2018.03.025

Ellul. D & Schlüter R. (2008). Gastronomía y turismo en Argentina polo gastronómico Tomás Jofré. *Pasos: Revista de turismo y Patrimonio Cultural*, *6*(2), 249-268.

Emmendoerfer, M. L., Moraes, W. V., & Fraga, B. O. (2016). Turismo Criativo e Turismo de Base Comunitária: congruências e peculiaridades. *El periplo sustentable*, (31), 1-18.

Enright, M. J., & Newton, J. (2004). Tourism destination competitiveness: A quantitative approach. *Tourism Management*, *25*(6), 777–788. doi:10.1016/j.tourman.2004.06.008

Eshliki, S., & Kaboudi, M. (2012). Community Perception of Tourism Impacts and Their Participation in Tourism Planning: A Case Study of Ramsar, Iran. *Procedia: Social and Behavioral Sciences*, *36*, 333–341. doi:10.1016/j.sbspro.2012.03.037

Espinoza-López, P.C., Moreno-Dena, J.M., Robles-Parra, J. M., Borbón-Morales, C. G., & Salazar-Solano, V. (2019). Innovation processes for micro, small and medium tourism enterprises development in the Sonora River region. *Revista de alimentación contemporánea y desarrollo regional*, *29*(53), 1-23.

ETC/UNWTO. (2008). *Handbook on E-marketing for tourism destinations*. Retrieved November 1, 2019 from https://www.cabdirect.org/cabdirect/abstract/20083275024

Everett, S. (2009). Beyond the visual gaze? The pursuit of an embodied experience through food tourism. *Tourist Studies*, *8*(3), 337–358. doi:10.1177/1468797608100594

Everett, S. (2012). Production places or consumption spaces? The place-making agency of food tourism in Ireland and Scotland. *Tourism Geographies*, *14*(4), 535–554. doi:10.1080/14616688.2012.647321

Everett, S. (2016). *Food and Drink Tourism*. Routledge.

Everett, S., & Aitchison, C. (2008). The role of food tourism in sustaining regional identity: A case study of Cornwall, South West England. *Journal of Sustainable Tourism*, *16*(2), 150–167. doi:10.2167/jost696.0

Everett, S., & Slocum, S. (2013). Food and tourism: An effective partnership? A UK-Based review. *Journal of Sustainable Tourism*, *21*(6), 789–809. doi:10.1080/09669582.2012.741601

Fabrino, N. H., Costa, H. A., & Nascimento, E. P. (2012). Turismo de Base Comunitária (TBC): Elementos chaves para aferir seu desempenho na perspectiva da sustentabilidade. *Revista Brasileira de Ecoturismo*, *5*(3), 546–559. doi:10.34024/rbecotur.2012.v5.6224

Fachrudin, H. T., & Lubis, M. D. (2016). Planning for Riverside Area as Water Tourism Destination to Improve Quality of Life Local Residents, Case Study: Batuan – Sikambing River, Medan, Indonesia. Procedia - Social and Behavioral Sciences, 234, 434 – 441.

Faraldo, J., & Rodríguez, C. (2014). *Introducción a la historia del turismo*. Alianza Editorial.

FARNET. (2011). *Guía 5 Diversificación de las zonas de pesca.* Retrieved from: https://www.mapa.gob.es/es/pesca/temas/diversificacion/analisiseconomomicodeladiversifiacionpesquerayacuicola_tcm30-290383.pdf

FARNET. (2013). *Uniendo la pesca con la economía del turismo*. Retrieved from: https://webgate.ec.europa.eu/fpfis/cms/farnet/files/documents/FARNET_Magazine_09_ES_0.pdf

Farsani, N. T., Celeste Coelho, C., Carlos Costa, C., & Carlos Neto, C. (Eds.). (2011). *Geoparks and Geotourism. New Approaches to Sustainability for the 21st Century*. Universal publishers.

Farsani, N. T., Coelho, C., & Costa, C. (2012). Geotourism and geoparks as gateways to sociocultural sustainability in Qeshm rural areas, Iran. *Asia Pacific Journal of Tourism Research*, *17*(1), 30–48. doi:10.1080/10941665.2011.610145

Feldman, M. P. (1994). *The geography of innovation* (Vol. 2). Springer Netherlands. doi:10.1007/978-94-017-3333-5

Felicidades-García, J., & Piñeiro-Antelo, M. A. (2017). Competitividad, Sostenibilidad y Diversificación en la Fachada Atlántica de la Península Ibérica: La Gestión Local del Fondo Europeo de Pesca (2007-2013). *Revista de Estudios Andaluces*, *34*(1), 350–379. doi:10.12795/rea.2017.i34.12

Férias Brasil. (2019). *Esportes náuticos*. Retrieved January 20, 2020, from https://www.feriasbrasil.com.br/ce/fortaleza/esportesnauticos.cfm

Fernandes, T. S., Lopes, G. S.C., Watanabe, M., Yamaguchi, C. K., & Godoi, C. K. (2016). Dimensões do empoderamento feminino: autonomia ou dependência? *Revista Alcance -Eletrônica*, *23*(3), 391-413.

Fernandes, G. P., & Natário, M. M. (2020). Entrepreneurship in Border Territories: Context Costs, Resilience, and Logics of Action of the Companies and Entrepreneurs – Study of the Border Municipalities of Beira Interior in Portugal and Salamanca in Spain. In J. Dantas & L. Carvalho (Eds.), *Handbook of Research on Approaches to Alternative Entrepreneurship Opportunities* (pp. 129–152). IGI Global. doi:10.4018/978-1-7998-1981-3.ch007

Fernandes, G., Castro, E., Firmino, G., & Ferreira, A. (2016). Geoparks and regional and local development: aspiring Estrela Geopark. *European Geoparks Conference*. DOI: 10.13140/RG.2.2.28770.94405

Fernandes, L. M. M. (2014). O Ceará turístico: Política de regionalização e governança nos destinos indutores. *Revista GeoUECE*, *3*(5), 213–214.

Fernandes, L. M. M., Soares, J. R. R., & Coriolano, L. N. M. T. (2020. (in press). Governance in the regionalization tourism policy of the state of Ceará/Brazil. *Revista Portuguesa de Estudos Regionais*.

Fernandes, S. (2003). *Contributo para a Requalificação Biofísica da Praia Fluvial da Valeta (Arco de Valdevez). Relatório Final de Curso. Bacharelato em Engenharia doAmbiente e dos Recursos Naturais*. Instituto Politécnico de Viana do Castelo. EscolaSuperior Agrária de Ponte de Lima.

Ferreira, F. H. G. (2000). *Segmentação de mercado*. Retrieved October 25, 2019 from https://www.ecrconsultoria.com.br/sites/default/files/Segmentacao_de_Mercado.pdf

Ferreira, M. T., & Guerreiro de Brito, A. (2009). Águas interiores supericiais. In H. Pereira, T. Domingos, L. Vicente, & V. Proença (Eds.), *Ecossistemas e Bem-Estar Humano em Portugal*. Fundação da Faculdade de Ciências da U. L. e Escolar Editora.

Filimonau, V., & De Coteau, D. (2019). Tourism resilience in the context of integrated destination and disaster management (DM 2). *International Journal of Tourism Research*. Advance online publication. doi:10.1002/jtr.2329

Flecha da Mata. (2018). *A permacultura em vivência comunitária*. Retrieved January 5, 2020, from http://flechadamata.com/#rt-utility

Folgado-Fernández, J., Di-Clemente, E., Hernández-Mogollón, J., & Campón-Cerro, A. (2019). Water Tourism: A New Strategy for the Sustainable Management of Water-Based Ecosystems and Landscapes in Extremadura (Spain). *Land Journal*, *8*(1), 2. doi:10.3390/land8010002

Fondo Multilateral De Inversiones. (2016). *Economía colaborativa en América Latina: ¿Cómo se distribuyen geográficamente las iniciativas? ¿Qué antigüedad tienen?* Retrieved October 30, 2019 from https://publications.iadb.org/es/publicacion/15641/economia-colaborativa-en-america-latina

Forsberg, G., & Lindgren, G. (2015). Regional policy, social networks and informal structures. *European Urban and Regional Studies*, *22*(4), 368–382. doi:10.1177/0969776412474663

Foucault, M. (1998). A história da sexualidade: a vontade de saber (18th ed.). Graal.

Foundation for Puerto Rico. (2018A). *Bottom up destination recovery: Orocovis case study*. Retrieved from https://www.foundationforpuertorico.org/research

Foundation for Puerto Rico. (2018B). *Small Business Cash Grant Program*. Retrieved from https://www.foundationforpuertorico.org/research

French, W., Bell, C., & Zawacki, R. (2007). *Desarrollo Organizacional: transformación y administración efectiva del cambio*. McGraw-Hill Interamericana.

Frenzel, F., & Koens, K. (2012). Slum tourism: Developments in a young field of interdisciplinary tourism research. *Tourism Geographies*, *14*(2), 195–212. doi:10.1080/14616688.2012.633222

Frey, M. L., Schafer, K., Buchel, G., & Patzak, M. (2016). Geoparks: A Regional European and Global Policy. Geotourism. Doi:10.1016/B978-0-7506-6215-4.50014-2

Friedan, B. (1971). *A mística feminia*. Editora Vozes Limitada.

Fuentes García, F. J., & Sánchez Cañizares, S. M. (2010). Análisis del perfil emprendedor: Una perspectiva de género. *Estudios de Economía Aplicada*, *28*(3), 1–28.

Funceme. Fundação Cearense de Meteorologia e Recursos Hídricos. (2019). *Com temporada de ventos mais fortes, Guaramiranga registra rajada de quase 60 km/h*. Retrieved March 28, 2020, from: http://www.funceme.br/?p=5481

Fundación Cotec. (2015). *Innovación*. Retrieved February 2, 2020, from https://cotec.es/quienes-somos/presentacion/

Furrer, O., Thomas, H., & Goussevskaia, A. (2008). The structure and evolution of the strategic management field: A content analysis of 26 years of strategic management research. *International Journal of Management Reviews*, *10*(1), 1–23. doi:10.1111/j.1468-2370.2007.00217.x

Furtado, C. (1974). *Essential*. Sao Paulo: Company of letters.

Furtado, C. (1961). *Development and underdevelopment*. Culture Fund.

Furtado, C. (1964). *Dialectic of Development*. Culture Fund.

Furtado, C. (1984). *Culture and development in times of crisis*. Peace and Earth.

Furtado, C. (2002). *In search of a new model: reflections on the contemporary crisis.* Peace and Earth.

Gallarza, M. G., Gil-Saura, I., & García, H. C. (2002). Destination image: Towards a conceptual framework. *Annals of Tourism Research, 29*(1), 56–78. doi:10.1016/S0160-7383(01)00031-7

Gan, F., & Triginé, J. (2012). *Cultura de empresa y gestión del cambio.* Diaz de Santos.

García González, L. (2004). Agua y turismo. Nuevos usos de los recursos hídricos en la Península Ibérica. Enfoque integral. *Boletín de la AGE, 37*, 239–255.

García Negro, M.C. & Zotes, Y.N. (2006). El trabajo de las mujeres en el sector pesquero gallego. Análisis de los problemas relacionados con su tratamiento estadístico. *Revista galega de economía, 15*(1), 1-25.

García Ramón, M. D., & Baylina Ferré, M. (2000). *El nuevo papel de las mujeres en el desarrollo rural.* Oikos-tau.

García, L., Sarabia, M., & Mendoza, A. (2017). Turismo comunitario y responsabilidad Social en Santa Elena: Un modelo para armar. *Estudios Turísticos, 211-212*, 171–187.

García-Pozo, A., Sánchez-Ollero, J. L., & Ons-Cappa, M. (2016). Eco-innovation and economic crisis: A comparative analysis of environmental good practices and labour productivity in the Spanish hotel industry. *Journal of Cleaner Production, 138*, 131–138. doi:10.1016/j.jclepro.2016.01.011

Garcia, R., & Calantone, R. (2002). A critical look at technological innovation typology and innovativeness terminology: A literature review. *Journal of Product Innovation Management, 19*(2), 110–132. doi:10.1111/1540-5885.1920110

Gault, F. (2016). Defining and measuring innovation in all sectors of the economy: Policy relevance. *OECD Blue Sky Forum III.*

Gault, F. (2012). User innovation and the market. *Science & Public Policy, 39*(1), 118–128. doi:10.1093cipolcs005

Gault, F. (2018). Defining and measuring innovation in all sectors of the economy. *Research Policy, 47*(3), 617–622. doi:10.1016/j.respol.2018.01.007

Georgiadou, A., Gonzalez-Perez, M., & Olivas-Lujan, M. (2019). *Diversity within diversity management-Community Based Perspective.* Emerald Publishing Limited.

Gessa Perera, A. & Toledano. (2011). Turismo, emprendimiento y sostenibilidad en los espacios naturales protegidos. El caso de Andalucía (España). *Estudios y perspectivas en turismo, 20*(5), 1154-1174. Retrieved from: https://dialnet.unirioja.es/servlet/articulo?codigo=3739754

Getz, D., Robinson, R., Andersson, T., & Vujicic, S. (2014). *Foodies & food tourism.* Goodfellow Publishers.

Giampiccoli, A., & Saayman, M. (2016). Community-based tourism : From a local to a global push. *Acta Commercii - Independent Research Journal in the Management Sciences, 1*, 1–10.

Giampiccoli, A., & Mtapuri, O. (2012). Community-based tourism : An exploration of the concept (s) from a political perspective. *Tourism Review International, 16*(1), 29–43. doi:10.3727/154427212X13431568321500

Giampiccoli, A., & Mtapuri, O. (2014). The role of the international cooperation in community-based tourism. *Mediterranean Journal of Social Sciences, 5*(4), 638–644. doi:10.5901/mjss.2014.v5n4p638

Gillespie, C. (2002). *European Gastronomy into the 21st Century.* Butterworth-Heinemann.

Gil, M. A. (2012). *Cómo crear y hacer funcionar una empresa.* Alfaomega.

Gilmore, J., & Pine, B. J. (2007). *Authenticity: What Consumers Really Want?* Harvard Business School Press.

Global Entrepreneurship Monitor. (2018). *Informe Global 2018.* Retrieved from: https://www.gemconsortium.org/report

Gobierno de Puerto Rico. (2018). *Transformación e innovación luego de la devastación. Un plan de recuperación económica y de desastres para Puerto Rico.* Retrieved from https://recovery.pr/document-library

Goeldner, C., & Ritchie, B. (2012). *Tourism, principles, practices, philosophies.* John Wiley & Sons.

Goldstein, S. (2011). *Caracterização ambiental de praias fluviais de Portugal Continental.* Tese de Mestrado, Faculdade de Ciencias/Universidade de Lisboa.

Gomes, A. O., & Vieira Neto, J. P. (2010). *Historian of Curral Velho (Acaraú / CE). Terramar.* Institute and the Cearense Community Tourism Network.

Gomes, H., Fernandes, M., Castro, E., & Vieira, G. &. (2017). The (Serra da) Estrela Aspiring Geopark (Portugal): preserving geoheritage, while promoting science and its links to local communities. In *Geophysical Research Abstracts* (Vol. 19). EGU General Assembly.

Gomez, C., Falcão, M., Castilho, L., Correia, S., & Oliveira, V. (2015). Turismo de Base Comunitária como Inovação Social : Congruência entre os constructos. *PASOS Revista de Turismo y Patrimonio Cultural, 13*(5), 1213–1227. doi:10.25145/j.pasos.2015.13.082

Gomezelj, D. O. (2016). A systematic review of research on innovation in hospitality and tourism. *International Journal of Contemporary Hospitality Management, 28*(3), 516–558. doi:10.1108/IJCHM-10-2014-0510

Gómez-Urrutia, V., & Jiménez Figueroa, A. (2015). Family co-responsability and the work-family balance: Ways to improve gender equity *Polis. Revista Latinoamericana, 14*(40), 377–396.

Gonzalez, M., Husted, B., & Aigner, D. (2017). Opportunity discovery and creation in social entrepreneurship: An exploratory study in Mexico. *Journal of Business Research, 81*, 212–220. doi:10.1016/j.jbusres.2016.10.032

Goodnee, W. (2016). Corporate Social Responsibility as a Mechanism of Community Development: A Study of the Nigeria Liquefied Natural Gas Limited's CSR for the Bonny Kingdom, Niger Delta, Nigeria. *Journal of Corporate Responsibility and Leadership, 3*, 50–68.

Gorenak, M., & Trdina, A. (2017). *Tourism and development 2017: Active & Sports Tourism: Feel the freedom of the water, Book of Abstracts.* University of Maribor Press., doi:10.18690/978-961-286-121-6

Gouveia, V. (2009). *Contributo das praias fluviais para o desenvolvimento regional: a rede de praias fluviais do pinhal interior. Tese de Mestrado.* FCT/UNL.

Graciano, P. F., & de Holanda, L. A. (2020). Bibliometric analysis of the scientific literature on community-based tourism from 2013 to 2018. *Revista Brasileira de Pesquisa em Turismo, 14*(1), 161–179. doi:10.7784/rbtur.v14i1.1736

Grebel, T., Pyka, A., & Hanusch, H. (2003). An evolutionary approach to the theory of entrepreneurship. *Industry and Innovation, 10*(4), 493–514. doi:10.1080/1366271032000163702

Grissemann, U. S., Pikkemaat, B., & Weger, C. (2013). Antecedents of innovation activities in tourism: An empirical investigation of the Alpine hospitality industry. *Tourism (Zagreb), 61*(1), 7–27.

Grissemann, U. S., Plank, A., & Brunner-Sperdin, A. (2013). Enhancing business performance of hotels: The role of innovation and customer orientation. *International Journal of Hospitality Management, 33*(1), 347–356. doi:10.1016/j.ijhm.2012.10.005

Grössling, S., Hall, M., & Scott, D. (2015). *Tourism and Water.* Channel View Publications. doi:10.21832/9781845415006

Gross, S., & Grimm, B. (2018). Sustainable mode of transport choices at the destination–public transport at German destinations. *Tourism Review*, *73*(3), 401–420. doi:10.1108/TR-11-2017-0177

Grünewald, R. A. (2003). Turismo e etnicidade. *Horizontes Antropológicos, Porto Alegre*, *20*, 140–160.

Guerrero, P., & Ramos, R. (2014). *Introducción al turismo*. Patria.

Guimarães, A. S., & Borges, M. P. (2008). *E-Turismo: Internet e negócios do turismo*. Cengage Learning.

Gursoy, D., Jurowski, C., & Uysal, M. (2002). Resident attitudes: A structural modeling approach. *Annals of Tourism Research*, *29*(1), 79–105. doi:10.1016/S0160-7383(01)00028-7

Gutiérrez-Salcedo, M., Martínez, M. Á., Moral-Munoz, J. A., Herrera-Viedma, E., & Cobo, M. J. (2018). Some bibliometric procedures for analyzing and evaluating research fields. *Applied Intelligence*, *48*(5), 1275–1287. doi:10.100710489-017-1105-y

Guzmán-López, T., Borges, O., & Castillo, A. M. (2011). Desarrollo económico local y turismo comunitario en países en vías de desarrollo. Un estudio de caso. *Revista de Ciencias Sociales*, *17*(3), 432–446. doi:10.31876/rcs.v17i3.25574

Hall, C., & Härkönen, T. (2006). Lake tourism: an introduction to lacustrine tourism systems. In Lake Tourism: An Integrated Approach to Lacustrine Tourism Systems (pp. 3-26). Channel View Publications. doi:10.21832/9781845410421-004

Hall, C., & Mitchell, R. (2005). Gastronomic tourism: comparing food and wine tourism experiences. In Niche Tourism: Contemporary Issues, Trends and Cases (pp. 89-100). Oxford: Butterworth-Heinemann.

Hall, C. M. (2019). Constructing sustainable tourism development: The 2030 agenda and the managerial ecology of sustainable tourism. *Journal of Sustainable Tourism*, *27*(7), 1–17. doi:10.1080/09669582.2018.1560456

Hall, C. M., Sharples, L., Mitchell, E., Macionis, N., & Cambourne, B. (Eds.). (2003). *Food Tourism Around The World*. Buttertworth-Heinemann.

Hall, C. M., & Williams, A. (2008). *Tourism and Innovation*. Routledge. doi:10.4324/9780203938430

Hall, C., & Mitchell, R. (2001). Wine and food tourism. In N. Douglas & R. Derrett (Eds.), *Special Interest Tourism* (pp. 307–329). Wiley.

Hall, M. (2009b). The public policy context of tourism entrepreneurship. In I. Ateljevic & S. J. Page (Eds.), *Tourism and entrepreneurship* (pp. 243–263). Butterworth-Heinemann.

Hashemkhani Zolfani, S., Sedaghat, M., Maknoon, R., & Zavadskas, E. K. (2015). "Sustainable tourism: A comprehensive literature review on frameworks and applications", *Economic research-Ekonomska. Istrazˇivanja*, *28*(1), 1–30.

Hashimoto, A., & Telfer, D. (2006). Selling Canadian culinary tourism: Branding the global and regional product. *Tourism Geographies*, *8*(1), 31–55. doi:10.1080/14616680500392465

Hassanien, A., & Eid, R. (2007). Developing new products in the hospitality industry: A case of Egypt. *Journal of Hospitality & Leisure Marketing*, *15*(2), 33–53. doi:10.1300/J150v15n02_03

Heesup, H., Taeyeon, E., Al-Ansi, A., Hyungseo, R. B., & Wansoo, K. (2019). Community-based tourism as a sustainable direction in destination development: An empirical examination of visitor behaviors. *Sustainability*, *11*. Advance online publication. doi:10.3390u11102864

Hegarty, J., & O'Mahony, B. (2001). Gastronomy: A phenomenon of cultural expressionism and aesthetic for living. *International Journal of Hospitality Management*, *20*(1), 3–13. doi:10.1016/S0278-4319(00)00028-1

Heldt Cassel, S., & Pettersson, K. (2015). Performing gender and rurality in Swedish farm tourism. *Scandinavian Journal of Hospitality and Tourism*, *15*(1), 138–151. doi:10.1080/15022250.2015.1005333

Henderson, J. (2009). Food Tourism reviewed. *British Food Journal*, *111*(4), 317–326. doi:10.1108/00070700910951470

Heo, Y. (2016). Sharing economy and prospects in tourism research. *Annals of Tourism Research*, *58*, 166–170. doi:10.1016/j.annals.2016.02.002

Hernández-Ascanio, J., & Rich-Ruiz, M. (2020). Caracterización de organizaciones del Tercer Sector desde el enfoque de la innovación social. Discusión a partir de un estudio de caso. *Innovar*, *30*(*75*), 71–86. doi:10.15446/innovar.v30n75.83258

Hernández-Ascanio, J., Tirado-Valencia, P., & Ariza-Montes, A. (2016). El concepto de innovación social: Ámbitos, definiciones y alcances teóricos. *C.I.R.I.E.C. España*.

Hernández-Perlines, F., Ariza-Montes, A., Han, H., & Law, R. (2019). Innovative capacity, quality certification and performance in the hotel sector. *International Journal of Hospitality Management*, *82*, 220–230. doi:10.1016/j.ijhm.2019.04.027

Hernández, R. (2015). *La complejidad del concepto de desarrollo. Curso Planificación, Gobierno y Desarrollo*. ILPES/CEPAL.

Herrera-Racionero, P., Miret-Pastor, L., & Lizcano, E. (2018). Viajar con la tradición: Los pescadores artesanales ante la pesca-turismo en la Comunidad Valenciana (España). *Cuadernos de Turismo*, *41*(41), 279–293. doi:10.6018/turismo.41.327031

Hinojosa, M. del C. (2009). México y sus regiones gastronómicas. *Hospitalidad ESDAI*, (15), 109-128.

Hirata, H., & Kergoat, D. (2007). Novas configurações da divisão sexual do trabalho (Tradução: Fátima Murad). *Cadernos de Pesquisa*, *37*(132), 595–609. doi:10.1590/S0100-15742007000300005

Hirata, H., Laborie, F., Le Doaré, H., & Senotier, D. (2009). *(Orgs.). Dicionário Crítico do Feminismo*. Editora UNESP.

Hjalager, A. (2002). A typology of gastronomy tourism. In A. Hjalager & G. Richards (Eds.), *Tourism and Gastronomy* (pp. 21–36). Routledge.

Hjalager, A. M. (2002). Repairing innovation defectiveness in tourism. *Tourism Management*, *23*(5), 465–474. doi:10.1016/S0261-5177(02)00013-4

Hjalager, A. M. (2010). A review of innovation research in tourism. *Tourism Management*, *31*(1), 1–12. doi:10.1016/j.tourman.2009.08.012

Hjalager, A.-M. (1997). Innovation patterns in sustainbale tourism. An analytical typology. *Tourism Management*, *18*(1), 35–41. doi:10.1016/S0261-5177(96)00096-9

Hjalager, A.-M. (2010b). Regional innovation systems: The case of angling tourism. *Tourism Geographies*, *12*(2), 192–216. doi:10.1080/14616681003725201

Hofstede, G. (1999). Culturas y organizaciones. El software mental. Madrid: Alianza.

Horng, J., Liu, C., Chou, H., & Tsai, C. (2012). Understanding the impact of culinary brand equity and destination familiarity on travel intentions. *Tourism Management*, *33*(4), 815–824. doi:10.1016/j.tourman.2011.09.004

Horng, J.-S., Liu, C.-H. S., Chou, S.-F., Tsai, C.-Y., & Chung, Y.-C. (2017). From innovation to sustainability: Sustainability innovations of eco-friendly hotels in Taiwan. *International Journal of Hospitality Management*, *63*, 44–52. doi:10.1016/j.ijhm.2017.02.005

Horng, J., & Tsai, C. (2012). Culinary tourism strategic development: An Asia-Pacific perspective. *International Journal of Tourism Research, 14*(1), 40–55. doi:10.1002/jtr.834

Hose, T. A. (2005). Geoturismo europeo. Interpretacíon geológica y promoción de la conservación geológica para turistas. In D. Barretino, W. A. P. Winbledon, & E. Gallego (Eds.), *Patrimonio geológico: conservación y gestión*. Istituto Tecnológico Geomineiro de España.

Hristov, D. (2015). Tourism versus the Visitor Economy and the shifting landscape of Destination Management. *Tourismos: An International Multidisciplinary Journal of Tourism, 10*(1), 219–234.

Huete, R. (2013). El papel de la mujer en el turismo. In *XIII Jornadas Sobre Mujer* (pp. 1–27). Universidad de Alicante.

Huete, R., Brotons, M., & Sigüenza, M. C. (2016). La desigualdad entre mujeres y hombres en el sector hotelero español. *Estudios y Perspectivas en Turismo, 25*, 73–87.

Huete, R., & Mantecón, A. (2017). Landscape is the key. Exploring mass tourism alternatives. *Arbor, 193*(785), a397. doi:10.3989/arbor.2017.785n3003

Huh, C., & Vogt, C. A. (2008). Changes in residents' attitudes toward tourism over time: A cohort analytical approach. *Journal of Travel Research, 46*(4), 446–455. doi:10.1177/0047287507308327

Hu, M.-L. M., Horng, J.-S., & Sun, Y.-H. C. (2009). Hospitality teams: Knowledge sharing and service innovation performance. *Tourism Management, 30*(1), 41–50. doi:10.1016/j.tourman.2008.04.009

Hunt, C., & Stronza, A. (2014). Stage-based tourism models and resident attitudes towards tourism in an emerging destination in the developing world. *Journal of Sustainable Tourism, 22*(2), 279–298. doi:10.1080/09669582.2013.815761

Hurley, R. F., & Hult, G. T. M. (1998). Innovation, market orientation, and organizational learning: An integration and empirical examination. *Journal of Marketing, 62*(3), 42–54. doi:10.1177/002224299806200303

Ignatov, E., & Smith, S. (2006). Segmenting Canadian Culinary Tourists. *Current Issues in Tourism, 9*(3), 235–255. doi:10.2167/cit/229.0

Institute of Statistics. (2019). *Tourism in figures.* Tiranë, Albania: Institute of Statistics.

Instituto Nacional de Estadística y Geografía (INEGI). (2016). *Encuesta Nacional sobre Productividad y Competitividad de las Micro, Pequeñas y Medianas Empresas (ENAPROCE) 2015.* Retrieved June 12, 2019, from https://www.inegi.org.mx/contenidos/programas/enaproce/2015/doc/ENAPROCE_15.pdf

Ioannides, D., & Debbage, K. (1997). Post-Fordism and flexibility: The travel industry polyglot. *Tourism Management, 18*(4), 229–241. doi:10.1016/S0261-5177(97)00019-8

Irving, M. A. (2009). Reinventing reflection on community tourism, is innovation possible? In Community tourism: diversity of perspectives and Brazilian experiences. Rio de Janeiro: Letter and Image.

Irving, M. D. A., Cristina, T., & Mendonça, M. (2018). *Turismo de base comunitária na América Latina : uma estratégia em rede*. Academic Press.

ISO 26000:2010. Guidance on social responsibility. International Organization for Standardization. Geneva, Switzerland.

Jaafar, M., Rasoolimanesh, S. M., & Lonik, K. A. T. (2015). Tourism growth and entrepreneurship: Empirical analysis of development of rural highlands. *Tourism Management Perspectives, 14*, 17–24. doi:10.1016/j.tmp.2015.02.001

Jafari, J. (2005). El turismo como disciplina científica. *Política y Sociedad, 42*(11), 39–56.

Jalilvand, M. R. (2017). The effect of innovativeness and customer-oriented systems on performance in the hotel industry of Iran. *Journal of Science and Technology Policy Management*, *8*(1), 43–61. doi:10.1108/JSTPM-08-2016-0018

Jantunen, A. (2005). Knowledge-processing capabilities and innovative performance: A empirical study. *European Journal of Innovation Management*, *8*(3), 336–349. doi:10.1108/14601060510610199

Jennings, G. (2007). *Water-Based Tourism, Sport, Leisure, and Recreation Experiences*. Butterworth-Heinemann.

Jiang, Y., Ritchie, B. W., & Benckendorff, P. (2019). Bibliometric visualisation: An application in tourism crisis and disaster management research. *Current Issues in Tourism*, *22*(16), 1925–1957. doi:10.1080/13683500.2017.1408574

Jiménez Medina, L., & Escalante Conde, N. (2009). Miradas y reflexiones antropológicas sobre el desfile del 5 de mayo en la ciudad de Puebla. *Cuicuilco*, *16*(45), 227–245.

Johns, N., & Mattson, J. (2005). Destination development through entrepreneurship: A comparison of two cases. *Tourism Management*, *26*(4), 605–616. doi:10.1016/j.tourman.2004.02.017

Joo, J., Choi, J., & Kim, N. (2019). *Examining Roles of Tour Dure Producers for Social Capital and Innovativeness in Community-Based Tourism*. Academic Press.

Jorge, E., Mielke, C., & Pegas, F. V. (2013). Turismo de Base Comunitária no Brasil. *Insustentabilidade é uma Questão de Gestão, 24*, 170–189.

Jung, H. S., & Yoon, H. H. (2018). Improving frontline service employees' innovative behavior using conflict management in the hospitality industry: The mediating role of engagement. *Tourism Management*, *69*, 498–507. doi:10.1016/j.tourman.2018.06.035

Kahn, K. B. (2018). Understanding innovation. *Business Horizons*, *61*(3), 453–460. doi:10.1016/j.bushor.2018.01.011

Kamaruddeen, A. M., Yusof, N. A., & Said, I. (2010). Innovation and innovativeness: Difference and antecedent relationship. *The Icfai University Journal of Architecture*, *2*(1), 66–78.

Karadag, E., & Dumanoglu, S. (2009). The productivity and competency of information technology in upscale hotels: The perception of hotel managers in Turkey. *International Journal of Contemporary Hospitality Management*, *21*(4), 479–490. doi:10.1108/09596110910955712

Karagiannis, D., & Metaxas, T. (2017). *Local Authorities' contribution on Gastronomy Tourism Development: the case of Ontario, Canada, MPRA Paper 79325*. University Library of Munich.

Kasapi, I., & Koc, M. (2013). Changing tourism consumer behaviour: the impacts on tourism demand in Albania. *Creative and Knowledge Society, 2*(2), 16-34. doi:10.2478/v10212-011-0023-3

Katz, D., & Kahn, R. (1986). *Psicología Social de la Organizaciones*. Trillas.

Kaufman, D. (2010). *Processo de tomada de decisão no ciberespaço: O papel das redes sociais no jogo das escolhas individuais* (Masters thesis). Pontifícia Universidade Católica de São Paulo, São Paulo, Brazil.

Kaufman, D. (2015). *O despertar de Gulliver: os desafios das empresas nas redes digitais* (Doctoral dissertation). Universidade de São Paulo, São Paulo, Brazil.

Kearney, M. (1970). The Social Meaning of Food Sharing in Mexico. *Kroeber Anthropological Society Papers*, *43*, 32–41.

Kimbu, A. N., & Ngoasong, M. Z. (2016). Women as vectors of social entrepreneurship. *Annals of Tourism Research*, *60*(1), 63–79. doi:10.1016/j.annals.2016.06.002

Kim, J. S., & Bernhard, B. (2014). Factors influencing hotel customers' intention to use a fingerprint system. *Journal of Hospitality and Tourism Technology*, *5*(2), 98–125. doi:10.1108/JHTT-11-2013-0031

Kim, K., Uysal, M., & Sirgy, M. J. (2013). How does tourism in a community impact the quality of life of community residents? *Tourism Management*, *36*, 527–540. doi:10.1016/j.tourman.2012.09.005

Kim, M.-S., & Koo, D.-W. (2017). Linking LMX, engagement, innovative behavior, and job performance in hotel employees. *International Journal of Contemporary Hospitality Management*, *29*(12), 3044–3062. doi:10.1108/IJCHM-06-2016-0319

Kim, S., & Ellis, A. (2015). Noodle production and consumption: From agriculture to food tourism in Japan. *Tourism Geographies*, *17*(1), 151–167. doi:10.1080/14616688.2014.978812

Kim, T. T., & Lee, G. (2013). Hospitality employee knowledge-sharing behaviors in the relationship between goal orientations and service innovative behavior. *International Journal of Hospitality Management*, *34*(1), 324–337. doi:10.1016/j.ijhm.2013.04.009

Kim, W., & Kim, H. (2013). Regional Development Strategy for Increasing - Cultural Tourism Business in South Korea. *Asia Pacific Journal of Tourism Research*, *18*(6), 534–548. doi:10.1080/10941665.2012.695285

Kim, Y., Yuan, J., Goh, B., & Antun, J. (2009). Web Marketing in Food Tourism: A Content Analysis of Web Sites in West Texas. *Journal of Culinary Science & Technology*, *7*(1), 52–64. doi:10.1080/15428050902788352

Kiralova, A., & Hamarneh, I. (2017). Local gastronomy as a prerequisite of food tourism development in the Czech Republic. *Marketing and Management of Innovations*, (2), 15–25.

Kiráľová, A., & Pavlíčeka, A. (2015). Development of Social Media Strategies in Tourism Destination. *Procedia: Social and Behavioral Sciences*, *175*, 358–366. doi:10.1016/j.sbspro.2015.01.1211

Kirchhof, A. L. C., & Lacerda, M. R. (2012). Desafios e perspectivas para a publicação de artigos - uma reflexão a partir de autores e editores TT - Challenges and prospects for publishing articles - consideration based on statements from authors and publishers TT - Retos y perspectivas de la publ. *Texto & Contexto Enferm*, *21*(1), 185–193. doi:10.1590/S0104-07072012000100021

Kivela, J., & Crotts, J. (2006). Tourism and gastronomy: Gastronomy's influence on how tourists experience a destination. *Journal of Hospitality & Tourism Research (Washington, D.C.)*, *30*(3), 354–377. doi:10.1177/1096348006286797

Klein, L., & Dodds, R. (2018). Blue flag beach certification: An environmental management tool or tourism promotional tool? *Tourism Recreation Research*, *43*(1), 39–51. doi:10.1080/02508281.2017.1356984

Kline, C., Hao, H., Alderman, D., Kleckley, J. w., & Gray, S. (2004). A spatial analysis of tourism, entrepreneurship and the entrepreneurial ecosystem in North Carolina USA. *Tourism Planning & Development*, 305–316. doi:10.1080/21568316.214.890127

Kline, C., & Milburn, L. A. (2010). Ten categories of entrepreneurial climate to encourage rural tourism development. *Annals of Leisure Research*, *13*(1/2), 320–348. doi:10.1080/11745398.2010.9686850

Koens, K., Steinbrink, M., & Rogerson, C. M. (2015). Slum tourism: State of the art. *Tourism Review International*, *18*(4), 237–252. doi:10.3727/154427215X14230549904017

Koh, K., & Hatten, T. (2002). The tourism entrepreneur: The overlooked player in tourism development studies. *International Journal of Hospitality & Tourism Administration*, *3*(1), 21–48. doi:10.1300/J149v03n01_02

Kokkranikal, J., & Morrison, A. (2011). Community networks and sustainable livelihoods in tourism: The role of entrepreneurial innovation. *Tourism Planning & Development*, *8*(2), 137–156. doi:10.1080/21568316.2011.573914

Komppula, R. (2014). Role of individual entrepreneurs in the development of competitiveness for a rural tourism destination – A case study. *Tourism Management*, *40*, 361–371. doi:10.1016/j.tourman.2013.07.007

Komsic, J., & Dorcic, J. (2016). Tourism destination competitiveness and online reputation: Conceptualization and literature framework analysis. In *Tourism & Hospitality Industry 2016, Congress Proceedings* (pp. 144–157). Opatua, Croatia: University of Rueka.

Köseoglu, M. A., Rahimi, R., Okumus, F., & Liu, J. (2016). Bibliometric studies in tourism. *Annals of Tourism Research*, *61*, 180–198. doi:10.1016/j.annals.2016.10.006

Kotler, P., & Armstrong, G. (2008). *Princípios de Marketing*. Pearson.

Kotler, P., & Keller, K. L. (2006). *Administração de Marketing: A Bíblia do Marketing*. Prentice Hall Brasil.

Krippendorf, J. (2003). *Sociologia do Turismo: para uma nova compreensão do lazer e das viagens* (3rd ed.).

Kristo, I. (2013). *Tourism, culture, hospitality, and agrotourism.* Academic Press.

Kuscer, K. (2012). *Modelling mountain tourism destination development with focus on innovativeness* (Doctoral dissertation). Ljubljana.

Lage, B. H. G. (1992). Segmentação do mercado turístico. *Revista Turismo em Análise*, *3*(2), 61–74. doi:10.11606/issn.1984-4867.v3i2p61-74

Lagowski, J. J. (1990). *What's So Great about Innovation?* Academic Press.

Lam, C., & Law, R. (2019). Readiness of upscale and luxury-branded hotels for digital transformation. *International Journal of Hospitality Management*, *79*, 60–69. doi:10.1016/j.ijhm.2018.12.015

Lane, B., & Kastenholz, E. (2015). Rural Tourism: The evolution of practice and research approaches-towars a new generation concept? *Journal of Sustainable Tourism*, *23*(8-9), 1133–1156. doi:10.1080/09669582.2015.1083997

Lane, E., Jones, R., Jones, A., & Mitchelmore, S. (2016). Exploring the potential of local food and drink entrepreneurship in rural Wales. *Local Economy: The Journal of the Local Economy Policy Unit*, *31*(5), 602–618. doi:10.1177/0269094216654669

Lashley, C., & Rowson, B. (2010). Lifestyle businesses: Insights into Blackpool's hotel sector. *International Journal of Hospitality Management*, *29*(3), 511–519. doi:10.1016/j.ijhm.2009.10.027

LaVanchy, G. (2017). When wells run dry: Water and tourism in Nicaragua. *Annals of Tourism Research*, *2017*(64), 37–50. doi:10.1016/j.annals.2017.02.006

Lee, I., & Arcodia, C. (2011). The role of regional food and festivals for destination branding. *International Journal of Tourism Research*, *13*(4), 355–367. doi:10.1002/jtr.852

Lee, T. H. (2009b). A structural model for examining how destination image and interpretation service affect future behavior: A case study of taiwan's Taomi eco-village. *Journal of Sustainable Tourism*, *17*(6), 727–745. doi:10.1080/09669580902999204

Lee, T. H. (2013). Influence analysis of community resident support for sustainable tourism development. *Tourism Management*, *34*, 37–46. doi:10.1016/j.tourman.2012.03.007

Lee, T. H., & Hsieh, H. P. (2016). Indicators of sustainable tourism: A case study from a taiwan's wetland. *Ecological Indicators*, *67*, 779–787. doi:10.1016/j.ecolind.2016.03.023

Lei, S. I., Wang, D., & Law, R. (2019). Hoteliers' service design for mobile-based value co-creation. *International Journal of Contemporary Hospitality Management*, *31*(11), 4338–4356. doi:10.1108/IJCHM-03-2018-0249

León, Baptista, & Contreras. (2012). La innovación social en el contexto de la responsabilidad social empresarial. *Forum Empresarial, 17*(1), 31-63. https://www.redalyc.org/articulo.oa?id=631/63124039002

Lepp, A. (2007). Residents' attitudes towards tourism in Bigodi village, Uganda. *Tourism Management, 28*(3), 876–885. doi:10.1016/j.tourman.2006.03.004

Li, M., & Hsu, C. H. C. (2016). Linking customer-employee exchange and employee innovative behavior. *International Journal of Hospitality Management, 56*, 87–97. doi:10.1016/j.ijhm.2016.04.015

Lim, W. M. (2009). Alternative models framing UK independent hoteliers' adoption of technology. *International Journal of Contemporary Hospitality Management, 21*(5), 610–618. doi:10.1108/09596110910967836

Lindič, J., Baloh, P., Ribire, V. M., & Desouza, K. C. (2011). Deploying information technologies for organizational innovation: Lessons from case studies. *International Journal of Information Management, 31*(2), 183–188. doi:10.1016/j.ijinfomgt.2010.12.004

Lin, Y., Pearson, T., & Cai, L. (2011). Food as a form of destination identity: A tourism destination perspective. *Tourism and Hospitality Research, 11*(1), 30–48. doi:10.1057/thr.2010.22

Liu, A., & Wall, G. (2006). Planning tourism employment: A developing country perspective. *Tourism Management, 27*, 159–170.

Lohmann, G., & Netto, A. P. (2017). *Tourism theory: concepts, models and systems*. CABI. doi:10.1079/9781780647159.0000

Long, P., Perdue, R., & Allen, L. (1990). Rural resident tourism perceptions and attitudes by community level of tourism. *Journal of Travel Research, 28*(3), 3–9.

Lopez de Avila, A. (2015). Smart destinations: XXI century tourism. ENTER2015 conference on information and communication technologies in tourism, 4–6.

López-Guzmán, López, Gálvez, & Vélez. (2019). Gastronomy, Culture and Tourism in Ecuador. In The Routledge Handbook of Gastronomic Tourism (pp. 39-47). London: Routledge.

Lopez-guzman, T., Borges, O., & Cerezo, J. (2016). Community-based tourism and local socio-economic development : A case study in Cape Verde. *African Journal of Business Management, 5*, 1608–1617.

López-Guzmán, T., Lotero, C., Gálvez, J., & Rivera, I. (2017). Gastronomic festivals: Attitude, motivation and satisfaction of the tourist. *British Food Journal, 119*(2), 267–283. doi:10.1108/BFJ-06-2016-0246

López-Guzmán, T., & Sánchez Cañizares, S. M. (2009a). Desarrollo socioeconómico de las zonas rurales con base en el turismo comunitario. Un estudio de caso de Nicaragua. *Cuadernos de Desarrollo Rural, 6*(62), 81–97.

Lordikipanidze, M., Brezet, H., & Backman, M. (2005). The entrepreneurship factor in sustainable tourism development. *Journal of Cleaner Production, 13*(8), 787–798. doi:10.1016/j.jclepro.2004.02.043

Loureiro, C. V. (2013). Community tourism as an alternative for the preservation of coastal ecosystems: the case of the Curral Velho Community, Acaraú-CE-Brazil. *Revista de Turismo Contemporâneo, 1*(1), 1 - 17.

Louro. (2007). *O corpo educado*. Autêntica.

Lugosi, P. (2012). Food, drink and identity. In D. Sloan (Ed.), *Food and Drink: The Cultural Context* (pp. 20–50). Goodfellow Publishers Limited.

Lundberg, E. (2015). *The level of tourism development and resident attitudes: A comparative case study of coastal destinations*. Academic Press.

Mak, A., Lumbers, M., & Eves, A. (2012). Globalisation and Food Consumption in Tourism. *Annals of Tourism Research, 39*(1), 171–196. doi:10.1016/j.annals.2011.05.010

Maldonado, C. (2009). Community rural tourism in Latin America: genesis, characteristics and policies. In R. Bartholo (Ed.), Community-based tourism: diversity of Brazilian perspectives and experiences. Academic Press.

Maneze, C. A. L., & Pacheco, R. T. B. (2019). La viajera independiente: búsqueda de sí, búsqueda de ser. In Gestión, turismo y género. Thomson Reuters.

Mangorrinha, J., & Pinto, H. (2010). Water and tourism. Culture heritage and new business opportunities – the legacy of portuguese thermalism. In E. Brito-Henriques, J. Sarmento, & M. A. Lousada (Eds.), *Water and Tourism. Resources management, planning, and sustainability* (pp. 93–106). CEG, Universidade de Lisboa.

Mannise, R. (2012). *Casas de "superadobe", ¿Qué son y como se hacen?* Retrieved January 5, 2020, from https://ecocosas.com/construccion/casas-de-superadobe-como-hacen/?cn-reloaded=1

Marchiori, E., & Cantoni, L. (2011). The online reputation construct: Does it matter for the tourism domain? A literature review on destinations' online reputation. *Information Technology & Tourism, 13*(3), 139–159. doi:10.3727/109830512X13283928066715

Marinho, P. A. S., & Gonçalves, H. S. (2016). Práticas de empoderamento feminino na América Latina. *Revista de Estudios Sociales, 56*, 80–90. doi:10.7440/res56.2016.06

Martínez-Pérez, Á., Elche, D., & García-Villaverde, P. M. (2019). From diversity of interorganizational relationships to radical innovation in tourism destination: The role of knowledge exploration. *Journal of Destination Marketing & Management, 11*, 80–88.

Martínez García, P. (2017). Mirando al mar desde los feminismos. Comunidad y supervivencia en torno a las trabajadoras de la pesca. *Prima Social, 19*, 183-214. Retrieved from: http://hdl.handle.net/10251/61538

Martínez-Pérez, Á., Elche, D., & García-Villaverde, P. M. (2019). From diversity of interorganizational relationships to radical innovation in tourism destination: The role of knowledge exploration. *Journal of Destination Marketing & Management, 11*, 80–88. doi:10.1016/j.jdmm.2018.12.002

Martin-Rios, C., & Ciobanu, T. (2019). Hospitality innovation strategies: An analysis of success factors and challenges. *Tourism Management, 70*(July), 218–229. doi:10.1016/j.tourman.2018.08.018

Martins, D. M. G. (2018). *A concetualização da qualidade do serviço no setor do turismo* (Masters thesis). Universidade do Porto, Porto, Portugal.

Mattsson, J., & Orfila-Sintes, F. (2014). Hotel innovation and its effect on business performance. *International Journal of Tourism Research, 16*(4), 388–398. doi:10.1002/jtr.1933

Mbakogu, I. A. (2004). Is There Really a Relationship Between Culture and Development? *Anthropologist, 6*(1), 37–43. doi:10.1080/09720073.2004.11890827

McKenzie, R., & Brackfield, D. (2008). *The OECD system of unit labor cost and related indicators*. OECD Statistics Directorate. doi:10.1787/18152031

McKercher, B., Okumus, F., & Okumus, B. (2008). Food tourism as a viable market segment: It's all how you cook the numbers. *Journal of Travel & Tourism Marketing, 25*(2), 137–148. doi:10.1080/10548400802402404

Mead, M. (2015). Sexo e Temperamento em Três Sociedades Primitivas (5th ed.). Perspectiva.

Medeiros, C. P. (2017). Inovação social além da tecnologia social constructos em discussão. *Race: revista de administração, contabilidade e economia, 16*(3), 957-982.

Medina, L., & Schneider, F. (2018). *Shadow economies around the world: what did we learn over the last 20 years?* IMF Working Paper.

Medina, F., & Tresserras, J. (2018). Introduction. Gastronomy and Tourism: Social and Cultural perspectives. In F. Medina & J. Tresseras (Eds.), *Food, Gastronomy and Tourism. Social and Cultural Perspectives.* Universidad de Guadalajara.

Mehraliyev, F., Choi, Y., & Köseoglu, M. A. (2019). Social structure of social media research in tourism and hospitality. *Tourism Recreation Research, 44*(4), 451–465. doi:10.1080/02508281.2019.1609229

Mejía Vázquez, R., Serrano-Barquín, R. C., López Carré, E., & Serrano-Barquín, H. P. (2017). Participación de las mujeres en empresas de hospedaje del municipio de Toluca, México. Análisis de su empoderamiento. *Investigaciones Turísticas, 14*(14), 21–44. doi:10.14198/INTURI2017.14.02

Melo, K. J. P., & Soares, J. R. R. (2019). La participación de las mujeres en los emprendimientos turísticos: Una mirada sobre la realidad brasileña. In Turismo y Género.Thomson Reuters.

Mendoza, Q. (2012). *Gastronomia novoandina del distrito de Puno, una alternativa para el turismo.* Universidad Nacional del Altiplano.

Menguc, B., & Auh, S. (2006). Creating a firm-level dynamic capability through capitalizing on market orientation and innovativeness. *Journal of the Academy of Marketing Science, 34*(1), 63–73. doi:10.1177/0092070305281090

Meschkank, J. (2011). Investigations into slum tourism in mumbai: Poverty tourism and the tensions between different constructions of reality. *GeoJournal, 76*(1), 47–62. doi:10.100710708-010-9401-7

Metaxas, T., & Karagiannis, D. (2016). Culinary tourism in Greece: Can the past define the future? Dimensions of Innovation, Entrepreneurship and Regional Development. *Journal of Developmental Entrepreneurship, 21*(3), 1–20. doi:10.1142/S1084946716500187

Metro-Roland, M. (2013). Goulash nationalism: The culinary identity of a nation. *Journal of Heritage Tourism, 82*(2-3), 172–181. doi:10.1080/1743873X.2013.767814

Mexconnect. (2020). *Cuisine.* Mexconnect. Available on: https://www.mexconnect.com/cuisine

Mezcal National Board. (2020). *Unique Spirits.* Mezcal National Board. Available on: http://www.sientemezcal.com/2009/eng/espiritu_unico.php

Miller, G. L., & Lugo, A. E. (2009). *Guide to the ecological systems of Puerto Rico.* Gen. Tech. Rep. IITF-GTR-35. San Juan, PR: U.S. Department of Agriculture, Forest Service, International Institute of Tropical Forestry.

Minayo, M. C. de S. (2001). Social research: theory, method and creativity. Rio de Janeiro: Vozes.

Minayo, M. C. de S. (2004). The knowledge challenge: qualitative health research (8th ed.). São Paulo: HUCITEC-BRASCO.

Ministerio de Turismo de Ecuador (MINTUR). (2017). *Boletín de Estadísticas Turísticas 2011-2015.* Quito Accsesada el 20 de febrero del 2020 https://servicios.turismo.gob.ec/descargas/Turismo-cifras/AnuarioEstadistico/Boletin-Estadisticas-Turisticas-2011-2015.pdf

Ministério do Ambiente (MA). (1997). *Programa de Valorização das Praias Fluviais.* SERN e IA.

Ministério do Turismo. (2013). *National Tourism Plan 2013-2016: Tourism giving much more in Brazil.* Retrieved October 28, 2017, from: http://www.Turismo.gov.br/images/pdf/plano_nacional_2013.pdf

Ministério do Turismo. (2018). *PRODETUR*. Retrieved January 04, 2018, from: http://www.Turismo.gov.br/programas/5066-prodetur.html

MINTUR. (2010). *Reglamento para los centros turísticos comunitarios, Pub. L. No. Acuerdo Ministerial 16 Ecuador.* Recuperado el 8 de marzo 2020: http://www.quito-turismo.gob.ec/descargas/lotaip2013diciembre/BASELEGAL/REGLAMENTO DE CENTROS DE TURISMO COMUNITARIO.pdf

Miquilena, D., & Paz, A. (2008). La cultura como fuente de compromiso del capital humano en las organizaciones modernas. *Forum Humanes*, *1*(1), 1–12.

Miranda, J. M. G. de. (2003). *Improvement of Clusters: Experiences from Asia and Latin America - "Evaluation and monitoring of APL Projects in SEBRAE: method and indicators"*. Term of Reference Sebrae.

Miret-Pastor, L., Muñoz Zamora, C., Herrera-Racionero, C., & Martínez Novo, R. (2015). Análisis regional del turismo pesquero en España. *Revista de Análisis Turístico*, *20*, 23–28.

Miśkiewicz, K. (2016). Promoting geoheritage in geoparks as an element of educational tourism. In *Geotourism: organization of the tourism and education in the geoparks in the Middle-Europe Mountains*. University of Business in Wrocław.

Mitki, Y., & Herstein, R. (2007). Innovative training in designing corporate identity. *Industrial and Commercial Training*, *39*(3), 174–179. doi:10.1108/00197850710742289

Molina, A., González, J. F., & García, C. (2010). El turismo pesquero como instrumento de apoyo al desarrollo sostenible en zonas litorales: la experiencia del proyecto Sagital. *XIV International Congress on Project Engineering*.

Mongeon, P., & Paul-Hus, A. (2016). The journal coverage of Web of Science and Scopus: A comparative analysis. *Scientometrics*, *106*(1), 213–228. doi:10.100711192-015-1765-5

Montanari, A. (2009). Geography of taste and local development in Abruzzo (Italy): Project to establish a training and research centre for the promotion of enogastronomic culture and tourism. *Journal of Heritage Tourism*, *4*(2), 91–103. doi:10.1080/17438730802366482

Montaño Armendáriz, A., Trejo Briseño, N., Pérez Conche, J. C., & López Torres, V. G. (2018). Emprendiminetos de turismo de naturaleza. Una vía para propiciar el empoderamiento de de la mujer em la zona rural de Los Cabos, Baja California Sur. In *Turismo y género. Una mirada desde Iberoamérica* (pp. 75–100). Universidad Autónoma del Estado de México.

Moraes, E. M. de., Irving, M. A., & Mendonça, T. C. M. (2018). Community-based tourism in Latin America: a network strategy. *Revista Tur., Vision and Action, 20*(2), 249-265.

Morar, C., & Pop, A. C. (2016). Water, Tourism and Sport. A conceptual approach. *Geo Journal of Tourism and Geosites*, *9*(2), 249–258.

Moreira, C., & Santos, N. (2016). Turismo fluvial em portugal continental: oferta e potencialidades/river tourism in portugal: touristic offer and potentials. In Territórios de Água | Water Territories, (pp. 255-271). CEGOT, Centro de Estudos de Geografia e Ordenamento do Território, Faculdade de Letras da Universidade de Coimbra.

Moreira, I., Ferreira, M., Cortes, R., Almeida, P. A., & Pinto, P. (Eds.). (2002). Ecossistemas Aquáticos e Ribeirinhos. In Ecologia, Gestão, Conservação. Edições do Instituto da Água.

Moreira, P., & Urriolagoitia, L. (2011). El emprendimiento social. *Revista española del tercer sector*, (17), 17-40. Retrieved from: http://www.injuve.es/sites/default/files/2018/29/publicaciones/5._emprendimiento_social.pdf

Moreira, C. (2011). Ambientes aquáticos não marítimos no Baixo Mondego: Margens para a recreação, o lazer e o turismo. In *Trunfos de uma geografia activa: Desenvolvimento local, ambiente, ordenamento e tecnologia* (pp. 149–158). IUC. doi:10.14195/978-989-26-0244-8_15

Moreira, C. (2012). Lazer, animação turística e desenvolvimento local. In *Patrimónios, territórios e turismo cultural: Recursos, estratégias e práticas, Coleção Iberografias, 19* (pp. 77–103). CEI, Âncora Editora.

Moreira, C. (2018). *Turismo fluvial, lazeres em águas interiores e desenvolvimento local e regional/River tourism, leisure in inland waterways and local and regional development in Cadernos de Geografia, 18*. CEGOT, Universidade de Coimbra.

Moreira, C., & Santos, N. (2010). New opportunities for water environments: River tourism and water leisure activities. In E. Brito-Henriques, J. Sarmento, & M. A. Lousada (Eds.), *Water and tourism. Resources management, planning and sustainability* (pp. 147–168). CEG, Universidade de Lisboa.

Moreno Muñoz, D. (2018). Aportación a los conceptos de turismo marinero/pesquero y pesca-turismo. *Cuadernos de Turismo*, *42*(42), 385–396. doi:10.6018/turismo.42.17

Moreno-Izquierdo, L. M., Ramón-Rodríguez, A., & Such Devesa, M. J. (2016). Turismo colaborativo: ¿Está AirBnB transformando el sector del alojamiento? *Economistas*, *150*, 107–119.

Morosan, C. (2018). An empirical analysis of intentions to cocreate value in hotels using mobile devices. *Journal of Hospitality & Tourism Research (Washington, D.C.)*, *42*(4), 528–562. doi:10.1177/1096348015597034

Morosan, C., & DeFranco, A. (2016). Modeling guests' intentions to use mobile apps in hotels: The roles of personalization, privacy, and involvement. *International Journal of Contemporary Hospitality Management*, *28*(9), 1968–1991. doi:10.1108/IJCHM-07-2015-0349

Mottiara, Z., Bolukb, K., & Klinec, C. (2018). The roles of social entrepreneurs in rural destination development. *Annals of Tourism Research*, *68*, 77–88. doi:10.1016/j.annals.2017.12.001

Mottiar, Z. (2015). The importance of local area as a motivation for cooperation among rural tourism entrepreneur. *Tourism Planning & Development*, *12*, 1–16.

Movono, A., & Dahles, H. (2017). Female empowerment and tourism: A focus on businesses in a Fijian village. *Asia Pacific Journal of Tourism Research*, *22*(6), 681–692. doi:10.1080/10941665.2017.1308397

MTUR. (2010). *Turismo de Sol e Praia: Orientações Básicas*. Retrieved November 11, 2019, from http://www.turismo.gov.br

Mulcahy, J. (2019). Historical evolution of gastronomic tourism. In S. Dixit (Ed.), *The Routledge Handbook of Gastronomic Tourism* (pp. 24–32). Routledge. doi:10.4324/9781315147628-3

Muler Gonzalez, V., Coromina, L., & Galí, N. (2018). Overtourism: Residents' perceptions of tourismimpact as an indicator of resident social carrying capacity-case study of a Spanish heritage town. *Tourism Review*, *73*(3), 277–296. doi:10.1108/TR-08-2017-0138

Mullo, E., Vera, V. M., & Guillén, S. (2019). El desarrollo del turismo comunitario en ecuador: Reflexiones necesarias. *Universidad y Sociedad*, *11*(2), 178–183. http://scielo.sld.cu/scielo.php?script=sci_arttext&pid=S2218-36202019000200178

Murayama, M. (2012). Promoting Japanese food culture and products. In *Food and The Tourism Experience: The OECD-Korea Workshop* (pp. 147–153). OECD Publishing.

Murphy, P. (1985). *Tourism: A community approach*. Methuen.

Murphy, P. E. (2012). *Tourism: A Community Approach* (4th ed.). Routledge.

Muthuri, J. N. (2007). Corporate Citizenship and Sustainable Community Development. *Journal of Corporate Citizenship, 2007*(28), 73–84. doi:10.9774/GLEAF.4700.2007.wi.00008

Nair, G. K. (2019). Dynamics of learning orientation, innovativeness, and financial performance of the hotel industry. *The Journal of Hospitality Financial Management, 27*(1), 27–38. doi:10.7275/8832-cr71

Naranjo, J., & Calderón, G. (2015). Construyendo una cultura de innovación. Una propuesta de transformación cultural. *Revista Estudios Gerenciales, 31*(135), 223–236. doi:10.1016/j.estger.2014.12.005

Narvaz, M. G., & Koller, S. H. (2006). Metodologias feministas e estudos de gênero: Articulando pesquisa, clínica e política. *Psicologia em Estudo, 11*(3), 647–654. doi:10.1590/S1413-73722006000300021

National Geographic. (2018). *Asiste al festival gastronómico más importante de Mérida*. National Geographic. Available on: https://www.ngenespanol.com/traveler/asiste-al-festival-gastronomico-mas-importante-de-merida/

Naumov, N. (2019). *Food Tourism as a Cultural Experience*. Paper presented at FoodTrex Thessaloniki, Thessaloniki, Greece.

Navarro, M. (2019, January 9). Why Puerto Rico Is No. 1 on our Places to Go list. *The New York Times*. Retrieved from https://www.nytimes.com/2019/01/09/travel/visit-puerto-rico.html

Navruz-Zoda, B.N., Ibragimov, N.S., Navruz-Zoda, Z.B., & Navruz-Zoda, Sh.B. (2017). *Tourism destination competitiveness, a monograph*. Bukhara, Uzbekistan: "Sadriddin Salim Buxoriy" Durdona Publishing.

Nelson, F. (2004). *The evolution and impacts of community-based ecotourism in northern Tanzania*. Academic Press.

Nelson, R. R., & Winter, S. G. (1982). *An Evolutionary Theory of Economic Change, Mass*. Harvard University Press.

Neto, J. A. (2016). *Voo livre colore o céu de Quixadá*. Retrieved March 28, 2020, from https://diariodonordeste.verdesmares.com.br/editorias/regiao/voo-livre-colore-o-ceu-de-quixada-1.1630237

Ngo, T., Lohmann, G., & Hales, R. (2018). Collaborative marketing for the sustainable development of community-based tourism enterprises: Voices from the field. *Journal of Sustainable Tourism, 26*(8), 1325–1343. doi:10.1080/09669582.2018.1443114

Nicholas, L. N., Thapa, B., & Ko, Y. J. (2009). Residents' perspectives of a world heritage site: The pitons management area, st. Lucia. *Annals of Tourism Research, 36*(3), 390–412. doi:10.1016/j.annals.2009.03.005

Nicolau, J. L., & Santa-María, M. J. (2013). The effect of innovation on hotel market value. *International Journal of Hospitality Management, 32*(1), 71–79. doi:10.1016/j.ijhm.2012.04.005

Nieto, J. L., Román, I. M., Bonillo, D., & Paulova, N. (2016). El turismo a nivel mundial. International. *Journal of Scientific Management and Tourism, 2*(1), 129–144.

Nieves, J., & Diaz-Meneses, G. (2016). Antecedents and outcomes of marketing innovation: An empirical analysis in the hotel industry. *International Journal of Contemporary Hospitality Management, 28*(8), 1554–1576. doi:10.1108/IJCHM-11-2014-0589

Nieves, J., & Diaz-Meneses, G. (2018). Knowledge sources and innovation in the hotel industry: Empirical analysis on Gran Canaria Island, a mature mass-tourism destination. *International Journal of Contemporary Hospitality Management, 30*(6), 2537–2561. doi:10.1108/IJCHM-07-2016-0341

Nogueira, C. (2018). *Mulheres são maioria entre os turistas que buscam passagens aéreas*. Retrieved November 20, 2019, from https://extra.globo.com/noticias/viagem-e-turismo/mulheres-sao-maioria-entre-os-turistas-que-buscam-passagens-aereas-revela-pesquisa-22974058.html

Nogueira, M. A. (2009). Development, State and Society: The necessary relations, the possible coalitions to the required institutionality. In *Dialogues for development*. Brasília: Ipea.

Norhafiza, M.S., & Ku Azam, T.L. (2017). Sustaining the Entrepreneurship in Rural Tourism Development. *International Journal of Multicultural and Multireligious Understanding*. . doi:10.18415/ijmmu.v4i6.122

Noronha, G. (2010). *Ceará: O Havaí dos ventos*. Retrieved March 18, 2020, from https://diariodonordeste.verdesmares.com.br/editorias/jogada/ceara-o-havai-dos-ventos-1.503959

Nunes, M., Santos, K., & Azevedo, F. (2014). Turismo e capital social: uma aproximação teórica e conceitual. *Revista de Turismo y Patrimonio Cultural, 12*(2008), 443–452.

Nunkoo, R. (2017). Governance and sustainable tourism: What is the role of trust, power and socialCapital? *Journal of Destination Marketing & Management, 6*(4), 277–285. doi:10.1016/j.jdmm.2017.10.003

Nunkoo, R., & Gursoy, D. (2012). Residents' support for tourism: An identity perspective. *Annals of Tourism Research, 39*(1), 243–268. doi:10.1016/j.annals.2011.05.006

Nunkoo, R., Ramkissoon, H., & Gursoy, D. (2012). Public Trust in Tourism Institutions. *Annals of Tourism Research, 39*(3), 1538–1564. doi:10.1016/j.annals.2012.04.004

O'Connor, A., Stam, E., Sussan, F., & Audretsch, D. (2018). *Entrepreneurial ecosystems. Place-based transformations and transitions*. Springer International Publishing AG.

O'Reilly, C., Chatman, J., & Caldwell, D. (1991). People and organizational culture: A profile comparison approach to assessing person-organization fit. *Academy of Management Journal, 14*(3), 487–516.

OCDE. (2017). *Estudios en Turismo Estudio de la Política Turística de México*. Secretaría de Turismo de México.

OECD. (2005). *Place-Based Policies for Rural Development, Crete, Greece (Case study). Working Party on Territorial Policy in Rural Areas, Public Governance and Territorial Development Directorate*. Territorial Development Policy Committee, OECD.

OECD/Eurostat. (2005). *Oslo manual: Guidelines for collecting and interpreting innovation data* (3rd ed.). Paris: OECD Publishing; doi:10.1787/9789264013100-

OECD/Eurostat. (2018). *Oslo manual: Guidelines for collecting, reporting and using data on innovation* (4th ed.). Paris: OECD Publishing; doi:10.1787/9789264304604-

Ojeda, A., Camacho, C., Hernández, C., & López, R. (2017). Caracterización de las principales tendencias de la gastronomía mexicana en el marco de nuevos escenarios sociales. *Teoría y Praxis, 21*(21), 91–113. doi:10.22403/UQROOMX/TYP21/05

Okumus, B., Okumus, F., & Mckercher, B. (2007). Incorporating Local and İnternational Cuisines in the Marketing of Tourism Destinations: The Cases of Hong Kong and Turkey. *Tourism Management, 28*(1), 253–261. doi:10.1016/j.tourman.2005.12.020

Okumus, F., Köseoglu, M. A., Putra, E. D., Dogan, I. C., & Yildiz, M. (2019). A bibliometric analysis of lodging-context research from 1990 to 2016. *Journal of Hospitality & Tourism Research (Washington, D.C.), 43*(2), 210–225. doi:10.1177/1096348018765321

Oliveira, R. (2019). *Turismo e inovação, 5 questões para pensar a inovação em turismo*. Retrieved December 19, 2019, from http://turismoeinovacao.com/conceito/5-questoes-para-pensar-a-inovacao-em-turismo/

Oliveira, G. B. (2002). A discussion on the concept of development. *FAE Magazine*, *5*(2), 37–48.

Oliveira, P. H. P., Gois, I. P., & Goldstein, M. C. B. (2018). Aplicação do funil de vendas para melhoria de engajamento em um SVA. *Revista da Universidade Vale do Rio Verde*, *16*(2), 1–19. doi:10.5892/ruvrd.v16i2.5020

Ordanini, A., & Parasuraman, A. (2011). Service innovation viewed through a service-dominant logic lens: A conceptual framework and empirical analysis. *Journal of Service Research*, *14*(1), 3–23. doi:10.1177/1094670510385332

Ordanini, A., Parasuraman, A., & Rubera, G. (2014). When the recipe is more important than the ingredients: A qualitative comparative analysis (QCA) of service innovation configurations. *Journal of Service Research*, *17*(2), 134–149. doi:10.1177/1094670513513337

Orfila-Sintes, F., Crespí-Cladera, R., & Martínez-Ros, E. (2005). Innovation activity in the hotel industry: Evidence from Balearic Islands. *Tourism Management*, *26*(6), 851–865. doi:10.1016/j.tourman.2004.05.005

Orfila-Sintes, F., & Mattsson, J. (2009). Innovation behavior in the hotel industry. *Omega*, *37*(2), 380–394. doi:10.1016/j.omega.2007.04.002

Ortiz García, P. (2017). The discourse on entrepreneurship of women from a gender perspective. *Vivat Academia*, *140*(140), 115–129. doi:10.15178/va.2017.140.115-129

Osuna-Rodríguez, M., Rodríguez García, L. & Gómez Parra, M. (2015). *Mujeres, Turismo Rural y Brecha Salarial de Género*. Córdoba: UCOPress, Editorial Universidad de Córdoba.

Ottenbacher, M. C., & Gnoth, J. (2005). How to develop successful hospitality innovation. *The Cornell Hotel and Restaurant Administration Quarterly*, *46*(2), 205–222. doi:10.1177/0010880404271097

Ottenbacher, M. C., & Harrington, R. J. (2010). Strategies for achieving success for innovative versus incremental new services. *Journal of Services Marketing*, *24*(1), 3–15. doi:10.1108/08876041011017853

Ottenbacher, M., & Harrington, R. (2011). A Case Study of a Culinary Tourism Campaign in Germany: Implications for Strategy Making and Successful İmplementation. *Journal of Hospitality & Tourism Research (Washington, D.C.)*, *37*(1), 3–28. doi:10.1177/1096348011413593

Ozen, K.Y. (2017). Competitiveness of Istanbul as A Tourism Destination for Luxury Market. *Journal of Tourismology*, *3*(2).

Öztayşi, B., Baysan, S., & Akpinar, F. (2009). Radio frequency identification (RFID) in hospitality. *Technovation*, *29*(9), 618–624. doi:10.1016/j.technovation.2009.05.014

Pace, L. A. (2016). How do tourism firms innovate for sustainable energy consumption? A capabilities perspective on the adoption of energy efficiency in tourism accommodation establishments. *Journal of Cleaner Production*, *111*, 409–420. doi:10.1016/j.jclepro.2015.01.095

Page, S., & Ateljevic, I. (Eds.). (2009). *Tourism and Entrepreneurship: International Perspectives*. Butterworth - Heinemann.

Palacios-Marqués, D., Merigó, J. M., & Soto-Acosta, P. (2015). Online social networks as an enabler of innovation in organizations. *Management Decision*, *53*(9), 1906–1920. doi:10.1108/MD-06-2014-0406

Pappas, N. (2015a). Achieving competitiveness in Greek accommodation establishments during recession. *International Journal of Tourism Research*, *17*(4), 375–387. doi:10.1002/jtr.1995

Pappas, N. (2015b). Marketing hospitality industry in an era of crisis. *Tourism Planning & Development, 12*(3), 333–349. doi:10.1080/21568316.2014.979226

Patiño Romarís, C. A. (2016). El turismo marinero: un producto diferenciador y emergente de la oferta turística del litoral gallego. In X.M. Santos, P. Taboada, & L. López (Eds.), *Actas del X Congreso Internacional de Turismo Rural y Desarrollo Sostenible* (pp. 401-410). Santiago de Compostela, España: Universidad de Santiago de Compostela.

Pavlovich, K. (2003). The evolution and transformation of a tourism destination network:The Waitomo Caves, New Zealand. *Tourism Management, 24*(2), 203–216. doi:10.1016/S0261-5177(02)00056-0

Pechlaner, H., Smeral, E., & Matzier, K. (2002). Customer value management as a determinant of the competitive position of tourism destinations. *Tourism Review, 57*(4), 15–22. doi:10.1108/eb058390

Pedraza, L., Obispo, K., Vásquez, L., & Gómez, L. (2015). Cultura organizacional desde la teoría de Edgar Schein: Estudio fenomenológico. *Revista Clío América, 9*(17), 17–25.

Peixoto, P. (2016). Os usos sociais dos rios. In P. Peixoto & J. P. Cardielos (Orgs.), *A água como património. Experiências de requalificação das cidades com água e das paisagens fluviais,* (pp. 57-70). Coimbra: IUC.

Perinotto, A. R. C. (2013). *Circulação de imagens turísticas: fotografias de Parnaíba/PI nas mídias* (Doctoral dissertation). Universidade do Vale do Rio dos Sinos, São Leopoldo, Brazil.

Perinotto, A. R. C.; Muniz, R. A.; Muniz Brito, A. S., & Borges, D. M. (2017). Comunicação Turística (...). *Revista Hospitalidade, 1*(1), 1-28.

Perinotto, A. R. C., & Siqueira, R. A. (2018). As novas tendências do marketing digital para o setor turístico. *Turismo: Estudos & Práticas, 7*(2), 86–215.

Peru, E. (2016). *Novoandina – Andean Cuisine's Rich Cultural History.* Eat Peru. Available on: https://www.eatperu.com/novoandina-novo-andean-cuisine/a

Pettersson, K., Ahlb, H., Berglundc, K., & Tillmard, M. (2017). In the name of women? Feminist readings of policies for women's entrepreneurship in Scandinavia. *Scandinavian Journal of Management, 33*(1), 50–63. doi:10.1016/j.scaman.2017.01.002

Pike, S. (2002). Destination image analysis - a review of 142 papers from 1973 to 2000. *Tourism Management, 23*(5), 541–549. doi:10.1016/S0261-5177(02)00005-5

Pikkemaat, B., Peters, M., & Bichler, B. F. (2019). Innovation research in tourism: Research streams and actions for the future. *Journal of Hospitality and Tourism Management, 41*, 184–196. doi:10.1016/j.jhtm.2019.10.007

Pilcher, J. (2017). *Food in World history* (2nd ed.). Routledge Taylor & Francis Group. doi:10.4324/9781315718941

Pine, B., & Gilmore, J. (1999). *The Experience Economy.* Harvard Business School Press.

Piscitelli, A. (2009). Gênero: a história de um conceito. In Igualdades. São Paulo: Berlendis & Vertechia.

Plessis, M. (2007). The role of knowledge management in innovation. *Journal of Knowledge Management, 11*(4), 20–29. doi:10.1108/13673270710762684

Power, S., Di Domenico, M., & Miller, G. (2017). The nature of ethical entrepreneurship in tourism. *Annals of Tourism Research, 65*, 36–48. doi:10.1016/j.annals.2017.05.001

Prezenza, A., & Chiappa, G. (2013). Entrepreneurial strategies in leveraging food as a tourist resource: A cross-regional analysis in Italy? *Journal of Heritage Tourism, 8*(2), 182–192. doi:10.1080/1743873X.2013.767810

Prideaux, B., Tomothy, D. J., & Cooper, M. (2009). Introducing River Tourism – Physical, Ecological, and Human Aspects. In River Tourism (pp. 1-22). CABI Publishing, International.

Proyecto Sagital. (2007). *El papel de la mujer en el sector pesquero: potencialidades en el ámbito del turismo pesquero*. Madrid: Fundación General de la Universidad Politécnica de Madrid. Retrieved from: http://www.cesmuamfar.com/pdf/El_Papel_de_la_Mujer_en_el_Sector_Pesquero._Potencialidades_en_el_%C3%A1mbito_tur%C3%ADstico-pesquero.pdf

Pulido-Fernandez, J., Casado-Montilla, J., & Carrillo-Hidalgo, I. (2019). Olive-oil tourism as a special interest tourism. *Heliyon*, *5*(12), e02975. doi:10.1016/j.heliyon.2019.e02975 PMID:31872132

Rachão, S., Breda, Z., Fernandes, C., & Joukes, V. (2019). Food tourism and regional development: A systematic literature review. *European Journal of Tourism Research*, *21*, 33–49.

Ratten, V., Ferreira, J. J., & Fernandes, C. I. (2017). Innovation management - Current trends and future directions. *International Journal of Innovation and Learning*, *22*(2), 135–155. doi:10.1504/IJIL.2017.085916

Red Española de Mujeres en el Sector Pesquero (REMSP). (2017). *La igualdad en cifras: Diagnóstico sobre la situación de la mujer en el sector pesquero y acuícola*. Ministerio de Agricultura y Pesca, Alimentación y Medio Ambiente. Retrieved from: https://www.mapa.gob.es/es/pesca/temas/red-mujeres/diagnosticoencifrasjunio2017revision_tcm30-425320.pdf

Redação Waves (Brasil). (n.d.). *Marcio Farney promove Projeto Social*. Retrieved March 10 2020 from: https://www.waves.com.br/arquivo/marcio-farney-promove-projeto-social/

Redação. (2009). *Paraíso dos esportes náuticos*. Retrieved March 18, 2020, from https://diariodonordeste.verdesmares.com.br/editorias/verso/paraiso-dos-esportes-nauticos-1.52520

Rede Tucum. (2013). *Caderno de normas e procedimentos internos Turismo Comunitário*.

Rede Tucum. (2020). *Quem somos: Tucum – Rede Cearense de Turismo Comunitário*. Retrieved January 5, 2020, from http://www.redetucum.org.br/rede-tucum/quem-somos/

Reggers, A., Grabowski, S., Wearing, S. L., Chatterton, P., & Schweinsberg, S. (2016). Exploring outcomes of community-based tourism on the kokoda track, Papua New Guinea: A longitudinal study of participatory rural appraisal techniques. *Journal of Sustainable Tourism*, *24*(8–9), 1139–1155. doi:10.1080/09669582.2016.1145229

Reid, R. D., & Sandler, M. (1992). The use of technology to improve service quality. *The Cornell Hotel and Restaurant Administration Quarterly*, *33*(3), 68–73. doi:10.1177/001088049203300319

Reis, A. M. (2017). *Mulheres e viagens: insegurança e medo?* (Undergraduate dissertation). Universidade Federal Fluminense, Rio de Janeiro, Brazil. Retrieved November 15, 2019 from https://app.uff.br/riuff/handle/1/4403

Revista Fácil. (2018). *Tourist Map of Ceará*. Retrieved January 15, 2020, from http://www.revistafacil.net/2017/09/mapa-do-turismo-do-ceara-cresce-e.html?m=1

Reyes Vargas, M.V., Ortega Ocaña, Á.F., & Machado Chaviano, E.L. (2017). Model for integrated community tourism in Ecuador, case study Pastaza. *REVESCO Revista de Estudios Cooperativos*, *123*, 250-275.

Ricco, A. S. (2012). *Tourism as a social and anthropological phenomenon. Tourism.* Space and Local Development Strategies.

Richards, G. (2002). Gastronomy: An essential ingredient in tourism production and consumption? In A. Hjalager & G. Richards (Eds.), *Tourism and Gastronomy* (pp. 3–20). Routledge.

Richards, L., & Dalbey, M. (2006). Creating Great Places: The Role of CitizeParticipation. *Community Development (Columbus, Ohio)*, *37*(4), 18–32. doi:10.1080/15575330609490193

Rico González, M., & Gómez García, M. J. (2009). La participación empresarial de la mujer en las iniciativas de turismo rural en Castilla y León. *Economía Agraria y Recursos Naturales*, *9*(2), 51–77. https://core.ac.uk/download/pdf/6449164.pdf

Rincón-Elías, C., Hermith-Patricia, D., & Bautista-Molina, W. (2018). Innovación social y su importancia en la gestión del conocimiento y la participación ciudadana. trilogía Ciencia Tecnología Sociedad, Volumen 10. *Número*, *18*, 51–61.

Rivera, J. (2002). Assessing a voluntary environmental initiative in the developing world: The Costa Rican certification for sustainable tourism. *Policy Sciences*, *35*(4), 333–360. doi:10.1023/A:1021371011105

Robbins, S. P. (2005). *Comportamiento organizacional*. McGraw Hill.

Rodrigues, L. de A. (2015). *Mulheres no exterior: um estudo sobre medo e gênero* (Undergraduate dissertation). Universidade Federal do Rio Grande do Sul, Porto Alegre, Brazil.

Rodrigues, A., & Rodrigues, A. (2007). *Rede Europeia de Turismo de Aldeia, um Casoe inovação em Turismo rural*. Congresso Internacional de Turismo, Leiria-Oeste.

Roux Fanny. 2013. Turismo comunitario ecuatoriano, conservación ambiental y defensa de los territorios. Federación Plurinacional de Turismo Comunitario del Ecuador (FEPTCE).

Rubin, G. (1993). Tráfico de mulheres: Notas sobre a "Economia Política do Sexo". In SOS Corpo. Recife.

Rueede, D., & Lurtz, K. (2012). *Mapping the Various Meanings of Social Innovation: Towards a Differentiated Understanding of an Emerging Concept*. EBS Business School Research Paper No. 12-03.

Ruhanen, L., Weiler, B., Moyle, B. D., & McLennan, C. L. J. (2015). Trends and patterns in sustainable tourism research: A 25-year bibliometric analysis. *Journal of Sustainable Tourism*, *23*(4), 517–535. doi:10.1080/09669582.2014.978790

Ruiz Ballesteros, E., & Cáceres-Feria, R. (2016). Community-building and amenity migration in community-based tourism development. An approach from southwest Spain. *Tourism Management*, *54*, 513–523. doi:10.1016/j.tourman.2016.01.008

Ruiz Ballesteros, E., & Solís Carrión, D. (2007). *Turismo comunitario en Ecuador. Desarrollo y sostenibilidad social*. Ediciones Abya-Yala.

Ruiz-Ballesteros, E. (2011). Social-ecological resilience and community-based tourism: An approach from Agua Blanca, *Ecuador. Tourism Management*, *32*(3), 655–666. doi:10.1016/j.tourman.2010.05.021

Ruiz, E., Hernández, M., Coca, A., Cantero, P. A., & del Campo, A. (2008). Turismo comunitario en Ecuador. Comprendiendo el community-based tourism desde la comunidad. *PASOS Revista de Turismo y Patrimonio Cultural*, *6*(3), 399–418. doi:10.25145/j.pasos.2008.06.031

Ruschmann, D. (2001). *Turismo e Planejamento Sustentável: a proteção do meio ambiente*. Editora Papirus.

Russell, R., & Faulkner, B. (2004). Entrepreneurship, chaos and the tourism area lifecycle. *Annals of Tourism Research*, *31*(3), 556–579. doi:10.1016/j.annals.2004.01.008

Ryan, T., Mottiar, Z., & Quinn, B. (2012). The dynamic role of entrepreneurs in destination development. *Tourism Planning & Development*, *9*(2), 119–131. doi:10.1080/21568316.2011.630747

Sacramento, P. M., & Teixeira, R. M. (2019). Innovation and Entrepreneur Learning in the Tourism Sector: Cross-Case Analysis in Small and Medium Enterprises in the city of Aracaju, Sergipe. *Revista Brasileira de Pesquisa em Turismo*, *13*(3), 121–139. doi:10.7784/rbtur.v13i3.1586

Sainsbury, B. (2012). *Peru's Gastronomic renaissance*. BBC. Available on: http://www.bbc.com/travel/story/20120803-perus-gastronomic-renaissance

Salavou, H. (2004). The concept of innovativeness: Should we need to focus? *European Journal of Innovation Management*, *7*(1), 33–44. doi:10.1108/14601060410515628

Salem, I. E.-B. (2014). Toward better understanding of knowledge management: Correlation to hotel performance and innovation in five-star chain hotels in Egypt. *Tourism and Hospitality Research*, *14*(4), 176–196. doi:10.1177/1467358414542265

Sampaio, C. A. C. (2007). Turismo como fenômeno humano: Princípios para pensar a ecossocioeconomia do turismo e sua prática sob a denominação turimos comunitario. *Turismo em Analise*, *18*(2), 148–165. doi:10.11606/issn.1984-4867.v18i2p148-165

Sanchez-Gutierrez, J., Gonzalez-Uribe, E., Gutierrez-Govea, A., & Garcia-Jimenez, E. (2012). The effects of intellectual capital and innovation on competitiveness: An analysis of the restaurant industry in Guadalajara, Mexico. *Journal of Competitiveness Studies*, *20*(3/4), 32–46.

Sandvik, I. L., & Sandvik, K. (2003). The impact of market orientation on product innovativeness and business performance. *International Journal of Research in Marketing*, *20*(4), 355–376. doi:10.1016/j.ijresmar.2003.02.002

Sandybayev A. (2016). Strategic Innovation in Tourism. A Conceptual and Review Approach. *International Journal of Research in Tourism and Hospitality,* 2(4), 5-10.

Santich, B. (1996). Looking for Flavour. Kent Town: Wakefield Press.

Santich, B. (2004). The study of gastronomy and its relevance to hospitality education and training. *International Journal of Hospitality Management*, *23*(1), 15–24. doi:10.1016/S0278-4319(03)00069-0

Santos, N. (2012). A gestão das bacias hidrográficas e a valorização turística dos ambientes fluviais: O rio Mondego. *Revista Geonorte*, *4*(4), 210–219.

Santos, N., & Cunha, L. (2008). Novas oportunidades para o espaço rural. Análise exploratória no Centro de Portugal. In *Lazer: Da libertação do tempo à conquista das práticas* (pp. 209–225). IUC.

Sarabia, M., García, L., & Bravo, O. (2019). Responsabilidad Social y Turismo comunitario en la provincia de Santa Elena (Ecuador). *Revista Espacios.*, *40*(30), 2.

Saraniemi, S., & Kylänen, M. (2011). Problematizing the concept of tourism destination: An analysis of different theoretical approaches. *Journal of Travel Research*, *50*(2), 133–143. doi:10.1177/0047287510362775

Sarmah, B., Kamboj, S., & Rahman, Z. (2017). Co-creation in hotel service innovation using smart phone apps: An empirical study. *International Journal of Contemporary Hospitality Management*, *29*(10), 2647–2667. doi:10.1108/IJCHM-12-2015-0681

Sarmiento, I., García, A., Marcial, F., & Hernández, B. (2010). *Panorama de la micro y pequeña empresa en México.* Retrieved June 12, 2019, from https://www.uaeh.edu.mx/investigacion/productos/6520/panorama_de_la_micro_y_pequena_empresa_en_mexico.pdf

Sass, C. (2017). *Precisamos falar de turismo!* Retrieved January 3, 2018, from https://www.ecommercebrasil.com.br/artigos/precisamos-falar-de-turismo/

Scarpato, R. (2002). Gastronomy as a tourist product: the perspective of gastronomy studies. In A. Hjalager & G. Richards (Eds.), *Tourism and Gastronomy* (pp. 51–71). Routledge.

Scavone, L. (2008). Estudos de gênero: Uma sociologia feminista? *Revista de Estudos Feministas*, *16*(1), 288. doi:10.1590/S0104-026X2008000100018

Schein, E. (1988). *Corporate culture and leadership. A dynamic vision.* Plaza y Janes.

Schlüter, R. (2012). Promoting regional cuisine as intangible cultural heritage in Latin America. In *Food and the Tourism Experience: The OECD-Korea Workshop*. OECD Publishing.

Schumpeter, J. (2008). The Theory of Economic Development: An Inquiry into Profits, Capital, Credit, Interest and the Business Cycle. London: Transaction Publishers. (Original publication 1934)

Schumpeter, J. (1947). The creative response in economic history. *The Journal of Economic History*, *7*(2), 149–159. doi:10.1017/S0022050700054279

Schumpeter, J. A. (1934). *The theory of economic development: An inquiry into profits, capital, credit, interest and the business cycle*. Harvard University Press.

Scott, J. (1995). Gênero: Uma categoria útil para análise histórica. *Educação &amp. Realidade*, *20*(2), 71–99.

Secretaría de Economía. (2016). *Responsabilidad Social Empresarial*. Gobierno de México. https://www.gob.mx/se/articulos/responsabilidad-social-empresarial-32705

Seelos, C., & Mair, J. (2005). Social entrepreneurship: Creating new business models to serve the poor. *Business Horizons*, *48*(3), 241–246. doi:10.1016/j.bushor.2004.11.006

Sharif, N. & Ku 'Azam, L. (2017). Sustaining the Entrepreneurship in Rural Tourism Development. *International Journal of Multicultural and Multireligious Understanding*, *4*(6), 31-42.

Shaw, G. (2014). Entrepreneurial Cultures and Small Business Enterprises in Tourism. In A. Lew, C. M. Hall, & A. Williams (Eds.), *The Wiley Blackwell Companion to Tourism* (pp. 120–131). Wiley. doi:10.1002/9781118474648.ch9

Shaw, G., & Williams, A. M. (1998). Entrepreneurship, small business culture and tourism development. In D. Ioannides & K. G. Debbage (Eds.), *The economic geography of the tourist industry* (pp. 235–255). Routledge. doi:10.4324/9780203398425.ch13

Shoham, A., Vigoda-Gadot, E., Ruvio, A., & Schwabsky, N. (2012). Testing an organizational innovativeness integrative model across cultures. *Journal of Engineering and Technology Management*, *29*(2), 226–240. doi:10.1016/j.jengtecman.2012.01.002

Silveira, E. S. (2007). *Por uma Sociologia do Turismo*. Porto Alegre-RG: Zouk.

Simpson, M. C. (2007). *An integrated approach to assess the impacts of tourism community development and sustainable livelihoods. Community Development Journal.*

Sims, R. (2009). Food, place and authenticity: Local food and the sustainable tourism experience. *Journal of Sustainable Tourism*, *17*(3), 321–336. doi:10.1080/09669580802359293

Smallbone, D. (2005). *Improving Access to Business Advices. In Rural Areas: an Evaluation, report commissioned by the Department for Environment, Food and Rural Affairs, Small Business Research Centre*. Kingston University.

Smith, S., & Costello, C. (2009). Culinary tourism: Satisfaction with a culinary event utilizing importance-performance grid analysis. *Journal of Vacation Marketing*, *15*(2), 99–110. doi:10.1177/1356766708100818

Soares, J. R. R., & López, M. P. V. (2015). O uso do etnoconhecimento para o desenvolvimento do turismo sustentável. *Tourism and Hospitality International Journal*, (4), 19-28.

Soares, J. R. R., Gabriel, L. P. M. C., & Sánchez Fernández, M. D. (2017). A influência das fontes de informação na escolha dos turistas acadêmicos: O caso de Santiago de Compostela. *Investigaciones Turísticas*, *14*, 130–153. doi:10.14461/INTURI2017.14.07

Soberanes, M. T. A., Gómez, R. A., & Andrade, M. A. (2018). Opportunity Strategies related to Business Competitiveness in Restaurants in Mexico. Case: Micro-enterprises in the Municipality of Othon P. Blanco, State of Quintana Roo. *Cuadernos Americanos*, *34*(60), 18–29. doi:10.25100/cdea.v34i60.5928

Solvoll, G., Alsos, A., & Bulanova, O. (2015). Tourism entrepreneurship: Review and future directions. *Scandinavian Journal of Hospitality and Tourism*, *15*(sup1), 120–137. doi:10.1080/15022250.2015.1065592

Sosa, A. (2013). *Organizational culture, Japan, the United States, and Mexico: a hypothesis*. Retrieved June 12, 2019, from https://colegios.ingenieria.webunam.mx/sites/default/files/la_cultura_organizacional_japon_estados_unidos_y_mexico_una_hipotesis.pdf?m=1500437373

Soulkite. (2019). *Projeto Vem Voar! Capacidade Técnica-Operacional.* Portfólio Projeto Vem Voar pro-Kitesurf.

Souza, M. L. (2004). Desenvolvimento de comunidade e Participação (3rd ed.). Cortez.

Sparrer, M. (2003). Género y turismo rural. El ejemplo de la Costa coruñesa. *Cuadernos de Turismo*, *7*, 181–197.

Stahel, W. R. (1997). The service economy: 'wealth without resource consumption'? *The Royal Society*, *35*(1728), 1309-1319.

Stoner, J., Freeman, R., & Gilbert, D. (1996). *Administration* (6th ed.). PEARSON - Prentice Hall.

Stueve, A. M., Cooks, S., & Drew, D. (2002). *The Geotourism Study: Phase I. Executive Summary*. Travel Industry Association of America.

Subramanian, A., & Nilakanta, S. (1996). Organizational innovativeness: Exploring the relationship between organizational determinants of innovation, types of innovations, and measures of organizational performance. *Omega*, *24*(6), 631–647. doi:10.1016/S0305-0483(96)00031-X

Summers, G. F., & Wilkinson, K. P. (1992). Review: The community in rural America. *Contemporary Sociology*, *21*(2), 212–213. doi:10.2307/2075444

Sundbo, J., Orfila-Sintes, F., & Sørensen, F. (2007). The innovative behaviour of tourism firms-Comparative studies of Denmark and Spain. *Research Policy*, *36*(1), 88–106. doi:10.1016/j.respol.2006.08.004

Swarbrooke, J., & Horner, S. (2002). *O Comportamento do Consumidor no Turismo*. Aleph.

Tajeddini, K. (2009). The impact of learning orientation on NSD and hotel performance: Evidence from the hotel industry in Iran. *Education, Business and Society*, *2*(4), 262–275. doi:10.1108/17537980911001099

Tajeddini, K. (2010). Effect of customer orientation and entrepreneurial orientation on innovativeness: Evidence from the hotel industry in Switzerland. *Tourism Management*, *31*(2), 221–231. doi:10.1016/j.tourman.2009.02.013

Tajeddini, K., & Trueman, M. (2008). Effect of innovativeness, risk taking, self confidence and the need for achievement on venture performance: Evidence from Switzerland effect of customer orientation and innovativeness on business performance: A study of small-sized service retailers. *International Journal of Entrepreneurship and Small Business*, *6*(2), 280–295. doi:10.1504/IJESB.2008.018633

Tajeddini, K., Trueman, M., & Larsen, G. (2006). Examining the effect of market orientation on innovativeness. *Journal of Marketing Management*, *22*(5–6), 529–551. doi:10.1362/026725706777978640

Tang, T.-W. (2014). Becoming an ambidextrous hotel: The role of customer orientation. *International Journal of Hospitality Management*, *39*, 1–10. doi:10.1016/j.ijhm.2014.01.008

Tang, T.-W. (2016). Making innovation happen through building social capital and scanning environment. *International Journal of Hospitality Management*, *56*, 56–65. doi:10.1016/j.ijhm.2016.04.002

Tang, T.-W., Wang, M. C.-H., & Tang, Y.-Y. (2013). Developing service innovation capability in the hotel industry. *Service Business*, *9*(1), 97–113. doi:10.100711628-013-0220-z

Tardif, C., & Harrisson, D. (2005). Complémentarité, convergence et transversalité: la conceptualisation de l'innovation socialeau CRISES. In CRISES. Centre de Recherche Sur Les Innovations Sociales. Cahiersdu CRISES.

Taylor, J. B. (1970). Introducing Social Innovation. *The Journal of Applied Behavioral Science*, *6*(1), 69–77. doi:10.1177/002188637000600104

Teh, L., & Cabanban, S. (2007). Planning for sustainable tourism in southern pulau Banggi: An assessment of biophysical conditions and their implications for future tourism development. *Journal of Environmental Management*, *85*(4), 999–1008. doi:10.1016/j.jenvman.2006.11.005 PMID:17204361

Theodori, G. (2008). Constraints to the Development of Community. *Community Development (Columbus, Ohio)*, *39*(1), 63–81. doi:10.1080/15575330809489742

Theodori, G. L. (2005). Community and Community Development in Resource-Based Areas: Operational Definitions Rooted in an Interactional Perspective. *Society & Natural Resources*, *18*(7), 661–669. doi:10.1080/08941920590959640

Thériault, G. (2015). Le tourisme communautaire : quand le développement touristique est une affaire de communauté. *Observatoire Québécois du Loisir, 12*(8), 1–4. Retrieved from www.uqtr.ca/oql

Thevenet, M. (1992). *Audit of the corporate culture*. Diaz de Santos.

Thomas, R., Shaw, G., & Page, S. J. (2011). Understanding small firms in tourism: A perspective on research trends and challenges. *Tourism Management*, *3*(5), 963–976. doi:10.1016/j.tourman.2011.02.003

Thomas, R., & Wood, E. (2014). Innovation in tourism: Re-conceptualising and measuring the absorptive capacity of the hotel sector. *Tourism Management*, *45*, 39–48. doi:10.1016/j.tourman.2014.03.012

Thompson, C., Johnson, T., & Hanes, S. (2016). Vulnerability of fishing communities undergoing gentrification. *Journal of Rural Studies*, *45*, 165–174. doi:10.1016/j.jrurstud.2016.03.008

Thompson, V. A. (1965). Bureaucracy and innovation. *Administrative Science Quarterly*, *10*(1), 1–20. doi:10.2307/2391646

Tidd, J., & Bessant, J. R. (2018). *Managing innovation: Integrating technological, market and organizational change* (6th ed.). Willey.

Timothy, D. (2015). *Heritage Cuisines: Traditions, Identities and Tourism*. Routledge. doi:10.4324/9781315752525

Timothy, D., & Ron, A. (2013). Understanding heritage cuisine and tourism: Identity, image, authenticity, and change. *Journal of Heritage Tourism*, *8*(2-3), 99–104. doi:10.1080/1743873X.2013.767818

Tinsley, R., & Lynch, P. A. (2007). Differentiation and tourism destination development: Small business success in a close-knit community. *Tourism and Hospitality Research*, *8*(3), 161–177. doi:10.1057/thr.2008.26

Tocquer, G., & Zins, M. (1999). *Marketing do Turismo*. Instituto Piaget.

Tolkach, D., & King, B. (2015). Strengthening community-based tourism in a new resource-based island nation: Why and how? *Tourism Management*, *48*, 386–398. doi:10.1016/j.tourman.2014.12.013

Tönnies, F. (1988). *Community and Society: Gemeinschaft and Gesellschaft; with a new introduction of John Samples*. Transaction Publishers.

Torres, R. (2003). Cancun's tourism development from a Fordist spectrum of analysis. *Tourist Studies*, *2*(1), 87–116. doi:10.1177/1468797602002001098

Tortosa Conchillo, J. M., Valenciano, J. P., & Uribe Toril, J. (2018). Innovación social como motor del desarrollo local y el emprendimiento social. *Equidad y Desarrollo*, *31*(31), 11–25. doi:10.19052/ed.4416

Tourismusforschung Tirol–Executive Summary. Forschungsbereich 1. Destinations management. MCI Tourismus Destination Research Lab. (2019). https://www.ttr.tirol/sites/default/files/2019-07/2.1%20MCI%20Tourismus%20Destination%20Research%20Lab.pdf

Trejo-Pech, C., Arellano-Sada, R., Coelho, A., & Weldon, R. (2011). Is the Baja California, Mexico, wine industry a cluster? *American Journal of Agricultural Economics*, *94*(2), 569–575. doi:10.1093/ajae/aar080

Tripadvisor. (2015). *Uma a cada quatro brasileiras viaja sozinha*. Retrieved December 01, 2019 from https://tripadvisor.mediaroom.com/2015-03-05-Uma-A-Cada-Quatro-Brasileiras-Viaja-Sozinha

Tseng, C.-Y., Kuo, H.-Y., & Chou, S.-S. (2008). Configuration of innovation and performance in the service industry: Evidence from the Taiwanese hotel industry. *Service Industries Journal*, *28*(7), 1015–1028. doi:10.1080/02642060701882080

Úbeda-García, M., Claver-Cortés, E., Marco-Lajara, B., García-Lillo, F., & Zaragoza-Sáez, P. C. (2018). Continuous innovation in the hotel industry: The development of organizational ambidexterity through human capital and organizational culture in Spanish hotels. *International Journal of Contemporary Hospitality Management*, *30*(12), 3609–3631. doi:10.1108/IJCHM-06-2017-0326

UNESCO. (2015). *Statutes of the International Geoscience and Geoparks Programme and Operational Guidelines for UNESCO Global Geoparks*. UNESCO.

UNESCO. (2019). *Traditional Mexican cuisine - ancestral, ongoing community culture, the Michoacán paradigm*. Available at: https://ich.unesco.org/en/RL/traditional-mexican-cuisine-ancestral-ongoing-community-culture-the-michoacan-paradigm-00400

UNESCO. (2020a). *Cochabamba*. UNESCO. Available on: https://en.unesco.org/creative-cities/cochabamba

UNESCO. (2020b). *Panama City*. UNESCO. Available on: https://en.unesco.org/creative-cities/panama-city

UNESCO. (2020c). *Florianopolis*. UNESCO. Available on: https://en.unesco.org/creative-cities/florian%C3%B3polis

UNESCO. (2020d). *Ensenada*. UNESCO. Available on: https://en.unesco.org/creative-cities/ensenada

United Nations World Travel Organization. (2019). *International Tourism Highlights*. World Tourism Organisation.

United Nations. (1987). *Report of the World Commission on Environment and Development*. United Nations.

UNWTO Measuring Sustainable Tourism. (2017). *A Call for Action, Report of the 6th International Conference on Tourism Statistics*. Manila, Philippines: UNWTO.

UNWTO Statistics and Tourism Satellite Account Programme. (2019). *Measuring Sustainable Tourism: Developing a statistical framework for sustainable tourism*. UNWTO.

UNWTO. (2010)-*International Recommendations for Tourism Statistics 2008: Department of Economic and Social Affairs*. New York: Statistics Division. Retrieved Aug 25 2019 from https://unstats.un.org/unsd/publication/Seriesm/SeriesM_83rev1e.pdf#page=21

UNWTO. (2011). *Global report on women in tourism 2010*. World Tourism Organization.

UNWTO. (2012). *Global Report on Food Tourism*. UNWTO.

UNWTO. (2017). Second Global Report on Gastronomy Tourism. Madrid: World Tourism Organization (UNWTO).

Valencia, C. (2019). *Memoria culinaria de las mayoras en fondas del centro histórico de Puebla: Entrevista etnográfica y cartografía social.* Available at: https://repositorioinstitucional.buap.mx/handle/20.500.12371/4782

Valente, T., & Mielke, E. (2017). Evolução e Potencialidades de Inovação do Turismo de Base Comunitária no Município de Guaraqueçaba – PR (Brasil). *Turismo e Sociedade*, *10*(3), 1–26.

Vallejo, R. (2002). Economy and history of Spanish tourism of the XXI century. *Contemporary History*, *25*, 203–232.

Valls, J. F. (1996). *Las claves del mercado turístico. Cómo competir en el nuevo entorno.* Deusto.

Vargas-Sánchez, A., Plaza-Mejia, M. D. L. Á., & Porras-Bueno, N. (2009). Understanding residents' attitudes toward the development of industrial tourism in a former mining community. *Journal of Travel Research*, *47*(3), 373–387. doi:10.1177/0047287508322783

Vasiliadis, L., Trivellas, P., Belias, D., Meleas, J., Kyriakou, D., & Koustelios, A. (2016). Cultural tourism revisited: the case of Thessaly. In V. Katsoni, & A. Stratigea (Eds.), Tourism and Culture in the Age of Innovation. Springer International Publishing.

Vaz, B. (2008). *Contributos para a Avaliação e Gestão de Praias: A Importância da Percepção dos seus Utilizadores.* Dissertação de Mestrado. Faculdade de Ciências Sociais e Humanas. Universidade Nova de Lisboa. Lisboa.

Vázquez, A., & Rodríguez, J. (2015). La política de desarrollo local: los desafíos de los territorios de desarrollo tardío. *Ciudad y Territorio: Estudios Territoriales*, *186*, 625-638. Retrieved from: https://recyt.fecyt.es/index.php/CyTET/article/view/76436

Victorino, L., Verma, R., Plaschka, G., & Dev, C. S. (2005). Service innovation and customer choices in the hospitality industry. *Managing Service Quality*, *15*(6), 555–576. doi:10.1108/09604520510634023

Vieira, G. (2004). *Geomorfologia dos Planaltos e Altos Vales da Serra da Estrela. Ambientes frios do Plistocénico Superior e Dinâmica Atual. PhD Tesis. Universidade de Lisboa.* CEGOT.

Vieira, J. M. (2011). Inovação no Turismo: Um processo catalisador de desenvolvimento. *Journal of Tourism Studies*, *4*, 133–143.

Vila, M., Enz, C. A., & Costa, G. (2012). Innovative practices in the Spanish hotel industry. *Cornell Hospitality Quarterly*, *53*(1), 75–85. doi:10.1177/1938965511426562

Villavicencio, B. P., Zamora, J. G., & Pardo, G. L. (2016). El turismo comunitario en la Sierra Norte de Oaxaca: Perspectiva desde las instituciones y la gobernanza en territorios indígenas. *El Periplo Sustentable*, *30*, 6–37.

Violier, P. (1993). Le tourisme fluvial dans la région Pays de la Loire et la dynamique des espaces et sociétés rurales. Norois, 159, 371-386. doi:10.3406/noroi.1993.6493

Vitorino. (2019). *O que fazer em Canoa Quebrada.* Retrieved March 19, 2020, from https://vitorinotur.com.br/o-que-fazer-em-canoa-quebrada/

Vivanco, J. S., & Franco, R. (2012). *The types of organizational culture and performance of SMEs in Aguascalientes.* Retrieved June 12, 2019, from http://congreso.investiga.fca.unam.mx/docs/xvii/docs/C28.pdf

Vogel, R., & Güttel, W. H. (2013). The dynamic capability view in strategic management: A bibliometric review. *International Journal of Management Reviews*, *15*(4), 426–446. doi:10.1111/ijmr.12000

Von Schuckmann, J., Barros, L. S., Dias, R. S., & Andrade, E. B. (2018). from slum tourism to smiley selfies: The role of social identity strength in the consumption of morally ambiguous experiences. *Journal of Consumer Psychology, 28*(2), 192–210. doi:10.1002/jcpy.1016

Voopter. (2019). *Viaje, mulher!* Retrieved November 21, 2019, from https://voopter.com.br/viajemulher/

Vuković, A. J., Gagić, S., Terzić, A., Petrović, M. D., & Radovanović, M. (2018). The impact of organisational learning on innovation: Case study of the serbian hotel industry. *Journal for East European Management Studies, 23*(4), 673–692. doi:10.5771/0949-6181-2018-4-673

Wall, G., & Mathieson, A. (2006). *Tourism: change, impacts and opportunities.* Pearson Education Limited.

Walraven, C. I. M. (2009). *Ceará é sede da 18ª BNTM no Nordeste.* Retrieved March 15, 2020, from https://www.skyscrapercity.com/threads/fortaleza-ce-estrutura-para-a-copa-de-2014.818858/page-4#post-37456328

Walters, G., & Mair, J. (2013). Post-Disaster Recovery Marketing Messages for tourism destinations. In Routledge Handbook of Tourism Marketing. Routledge.

Walters, G., & Mair, J. (2012). The effectiveness of post-disaster recovery marketing messages. The case of the 2009 Australian bushfires. *Journal of Travel & Tourism Marketing, 29*(1), 87–103. doi:10.1080/10548408.2012.638565

Wang, C. C., Cater, C., & Low, T. (2006). Political challenges in community-based ecotourism. *Journal of Sustainable Tourism, 24*(11), 1555–1568. doi:10.1080/09669582.2015.1125908

Wang, C. L., & Ahmed, P. K. (2004, December 1). The development and validation of the organisational innovativeness construct using confirmatory factor analysis. *European Journal of Innovation Management, 7*(4), 303–313. Advance online publication. doi:10.1108/14601060410565056

Wang, S., Hung, K., & Huang, W. (2019). Motivations for entrepreneurship in the tourism and hospitality sector: A social cognitive theory perspective. *International Journal of Hospitality Management, 78*, 78–88. doi:10.1016/j.ijhm.2018.11.018

Wang, T.-C., Tang, T.-W., & Cheng, J.-S. (2018). Art-oriented model of hotel service innovation. *International Journal of Contemporary Hospitality Management, 30*(1), 160–177. doi:10.1108/IJCHM-02-2016-0059

WebVenture. (2003). *Rally Mundial de Voo Livre em Sobral.* Ceará. Retreived March 29, 2020 from https://www.webventure.com.br/rally-mundial-de-voo-livre-em-sobral

Weiermair, K., Keller, P., Pechlaner, H., & Go, F. (2010). *Innovation and entrepreneurship: Strategies and processes for success in tourism.* Schmidt.

Welford, R., & Ytterhus, B. (1998). Conditions for the transformation of eco-tourism into sustainable tourism. *European Environment, 8*(6), 193–201. doi:10.1002/(SICI)1099-0976(199811/12)8:63.3.CO;2-W

Wikhamn, W., Armbrecht, J., & Wikhamn, B. R. (2018). Innovation in Swedish hotels. *International Journal of Contemporary Hospitality Management, 30*(6), 2481–2498. doi:10.1108/IJCHM-06-2017-0323

Wilkinson, K. P. (1991). *The community in rural America.* Greenwood Publishing Group.

Williams, H., Williams, R., & Omar, M. (2014). Gastro-tourism as destination branding in emerging markets. *International Journal of Leisure and Tourism Marketing, 4*(1), 1–18. doi:10.1504/IJLTM.2014.059257

Woo, E., Kim, H., & Uysal, M. (2015). Life satisfaction and support for tourism development. *Annals of Tourism Research, 50*, 84–97. doi:10.1016/j.annals.2014.11.001

Woo, E., Uysal, M., & Sirgy, M. J. (2018). Tourism impact and stakeholders' quality of life. *Journal of Hospitality & Tourism Research (Washington, D.C.)*, *42*(2), 260–286. doi:10.1177/1096348016654971

World Bank. (2009). Albania: building competitiveness in Albania. The World Bank.

World Tourism Organization. (2016). *Mekong River-based Tourism Product Development*. UNWTO.

WTTC World Travel & Tourism Council. (2019). *Travel & Tourism Economic Impact 2019*. Retrieved March 29, 2020, from: https://www.wttc.org/-/media/files/reports/economic-impact-research/regions-2019/world2019.pdf

Yepes, V., & Cardona, A. (2000). Proyecto e instalación de infraestructuras higiénicas en las playas turísticas. *Cimbra*, *333*, 56–59.

Yumisaca, J., & Mendoza, E. (2016). La hostelería y el sector intermediario como fortaleza del turismo en Santa Elena: caso Salinas y Montañita. In UPSE CECATERE, Investigaciones Turísticas del Ecuador: fortaleciendo la innovación de los destinos turísticos (pp. 69 -84). La Libertad: Imprenta López.

Zacarias, D., & Loyola, R. (2017). How ecotourism affects human communities. In *D. T Blumstein, B. Geffroy, D. S. Samia. & E. Bessa (Eds.), Ecotourism's promise and peril. A biological evaluation* (pp. 133–151). Springer.

Zahra, S., & George, G. (2002). Absorptive Capacity: A review, reconceptualization, and extension. *Academy of Management Review*, *27*(2), 185–203. doi:10.5465/amr.2002.6587995

Zalpa, G. (2002). Culture in business organizations. *Studies Journal Contemporary Cultures*, *8*(15), 9-33.

Zhang, H., Fu, X., Cai, L. A., & Lu, L. (2014). Destination image and tourist loyalty: A meta-analysis. *Tourism Management*, *40*, 213–223. doi:10.1016/j.tourman.2013.06.006

Zhou, K. Z., Brown, J. R., & Dev, C. S. (2009). Market orientation, competitive advantage, and performance: A demand-based perspective. *Journal of Business Research*, *62*(11), 1063–1070. doi:10.1016/j.jbusres.2008.10.001

Ziegler, R. (2017). Social innovation as a collaborative concept. *Innovation (Abingdon)*, *30*(4), 388–405. doi:10.1080/13511610.2017.1348935

Zouros, N., & Mc Keever, P. (2008). The European Geoparks Network. *Proceedings. 3rd International UNESCO Conference on Geoparks*, 11-12.

Zupic, I., & Čater, T. (2015). Bibliometric methods in management and organization. *Organizational Research Methods*, *18*(3), 429–472. doi:10.1177/1094428114562629

About the Contributors

Jakson Renner Soares is a Professor in the Department of Business Organization and Commercialisation at University of Santiago Compostela (USC) and Professor of the Master in Tourism Business Management at UECE (Brazil). Ph.D. in Tourism Planning. Post-doctoral scholarship (jan.2018-dec.20), granted by the Fundação Cearense de Apoio ao Desenvolvimento Cientifico e Tecnológico (FUNCAP) in collaboration with the Conselho Nacional de Desenvolvimento Científico e Tecnológico (CNPQ) in Brazil. Principal Investigator from the project titled "Innovation in the community tourism network and its contribution to the development of the territory" and I am lead researcher at group GTES: Tourism, Economy and Sustainability. The research lines are related to the impact of tourism as an economic activity in life and regional development: a) economy of sustainable tourism, b) tourism, local development, and sustainability, c) social responsibility and tourism, d) tourist destinations, communication, and image, e) gender studies and tourism, and f) heritage and tourism.

* * *

María Andrade Suárez has a PhD in Sociology at the University of A Coruña. Professor at the University of A Coruña in Degree in Sociology. Lines of work in research on tourism in rural areas, tourism as economic and development agent, rural tourist profiles, images of tourist destinations, tourism consumer behavior and tourism marketing.

Iria Caamaño-Franco is a Doctor for the Universities of A Coruña, Santiago of Compostela and Vigo in planning and management of tourist destinations. Degree in Tourism from the University of A Coruña and University of Murcia. Professor at the University of A Coruña in Degree and master in Tourism. Lines of work in research on Local Development, Sustainability, Heritage, Industrial Tourism, Tourist Intermediation, and Sociology in the Tourism Field.

Gil Célio Cardoso holds a degree in Economics from the Federal University of Ceará (1997), specialization in Strategy and Business Management from the State University of Ceará (2000), a master's degree in Education (Popular Education) from the Federal University of Paraíba (2000); and in Economics (Economic Development and Public Policies) from the Universidad Autónoma de Madrid (2013). He holds a PhD in Social Sciences (Regional Development) from the Federal University of Rio Grande do Norte (2005). Between 2007-2008, he did a postdoctoral internship at the Institut des Hautes Etudes de L'Amérique Latine at Université Paris III (Sorbonne Nouvelle) and, in the period 2008-2009, at the Center for Sustainable Development at the University of Brasilia. He is currently an associate professor at the University of Brasilia, working on a temporary basis at the Federal University of Ceará, working on the undergraduate courses in Public Policy Management and in the Graduate Program in Public Policy Evaluation at UFC. He has experience and interest in the area of Development Socioeconomics, with an emphasis on Regional Development, acting mainly on the following themes: regional economics and public policies; State, Northeast and sustainability; local development, governance and social participation.

Aida Ciro is a professional in the field of communication and tourism development with 10 years of varied experience. Currently a PhD Candidate on place-based sustainable tourism, at the International Doctorate Course in Architecture and Urban Planning, offered by the University of Ferrara and POLIS University. Main fields of expertise include: strategic communication, tourism planning, and promotion, participatory tourism planning, etc. She has and several published works, including: An Assessment of the Policy Framework for the Tourism Sector in Albania, Sustainable Tourism Development in Albania, and Social Innovation and Sustainable Economic Development: Participatory Tourism Destination Management. Member of the Western Balkan Network on Territorial Governance.

Jessica Abigail Cortés González originally from Puebla, is currently in the last year of her degree in Educational Processes from the Benemérita Universidad Autónoma de Puebla. In the fall of 2018 he participated in the project "Good Practices for University Social Responsibility", in the spring of 2019 she collaborated in the design, validation and application of a quantitative instrument to measure University Social Responsibility ". In October 2019 she was a speaker at the V Colloquium organized by the College of Educational Processes on educational evaluation with a critical analysis of the evaluation instruments of the general mathematics and communication programs in the upper middle level of the State of Puebla. Jessica has been a student advisor to the College of Educational Processes to identify and resolve university needs, as well as contributed to the creation of student events that impact the educational and social, cognitive and administrative fields within the university.

Gheorghia Costandachi is Associate Lecturer and Student Success Mentor at the Faculty of Business and Law at the University of Northampton, United Kingdom. She has a Joint Honours degree in Economics and International Development, and Master's in International Tourism Development from University of Northampton.Her research interests include community-based tourism, food tourism and gastronomy and economics of tourism development.

Concepción Nancy de Cristóbal González is studying for a Bachelor's Degree in Educational Processes at the Benemérita Universidad Autónoma de Puebla (BUAP) and is in her 9th semester. In 2018 he carried out an academic mobility of a semester at the Autonomous University of Yucatan where she participated in an investigation on Linguistic Culture and Transculturization itself that was published in the Ensino Magazine of Sciences and Humanities in 2019. She also participated in the city council from the municipality of Cacalchén in the state of Yucatan. In the same year she participated as Organizer of the panel "Transdisciplinarity in Autism Intervention" in the facilities of the Faculty of Philosophy and Letters of BUAP, and participated in the Doing Science Program of the Vice-Rectory for Research and Postgraduate Studies with two Research "The training of young researchers at BUAP" and "Ethics, Responsibility and Social Innovation in organizations" (2019, Puebla).

Maria del Carmen De la Luz Lanzagorta is Doctor in Science, Engineering and Technology Education from the Universidad de las Américas Puebla (UDLAP). She is a member of the National System of Researchers of Mexico (SNI), of University Social Responsibility of Latin America Union (URSULA), the Register of Researchers of the Benemérita Universidad Autónoma de Puebla (BUAP) and the Academic Group of Social Pedagogy and Educational Orientation at BUAP. She was a founding member of the UDLAP research team on University Social Responsibility, awarded by the Inter-American Development Bank (IDB) for its Social Service Program. She is currently part of research groups at UPAEP and BUAP on the topic of Social Responsibility and Innovation and is a Research Professor at the School of Philosophy and Letters of the Benemérita Universidad Autónoma de Puebla.

Maria de Oliveira holds a degree in Mathematics (Licenciatura) from the Universidade Estadual Vale do Acaraú (2010), specialization in Teaching Mathematics from the Instituto Superior de Applied Theology (2011), master's degree in Evaluation in Public Policies from the Federal University of Ceará (2019). She is currently a Technician in Educational Affairs and holds the position of Head of the Department of Education at the Federal Institute of Education of Ceará, campus Itapipoca, having been Coordinator of Education and pedagogy on the same campus. She was a professor in the effective staff of the Government of the State of Ceará, Coordinating professor in the area of Natural Sciences and Mathematics. He has experience in continuing education for high school teachers, educational management, public management and accountability of public resources. He is interested in the development area, with an emphasis on regional development with study on the following themes: tourism, public policies, participatory management, social control.

Catarina Fernandes is an Assistant Professor of Finance and Entrepreneurship at Polytechnic Institute of Bragança, Portugal and a researcher at CEF.UP – Center of Economics and Finance at University of Porto, Portugal. She holds a Licenciatura in Management from ISEG – Lisbon School of Economics and Management, University of Lisbon, Portugal, a Master in Monetary and Financial Economics also from ISEG and a Ph.D in Finance from the Faculty of Economics, UP. Her research interests include Corporate Governance, Corporate Finance and Banking. Over the recent years, her work appeared in academic journals in finance and economics.

Gonçalo José Poeta Fernandes, Coordinator Professor at Polytechnic Institute of Guarda – ESTH/ Scientific Technical Unit of Tourism and Hospitality (Portugal). Education and training: 2008 - PhD in Geography and Regional Planning, specializing in Planning and Territorial Management, Faculty of Social Sciences and Humanities, New University of Lisbon. Research Areas: Tourism and Development, Planning and Territorial Management, Public Policy and Border Areas, Rural Entrepreneurship,Territory and Tourist Destinations, Mountains and geotourism. R&D: Researcher CITUR - Centre for Tourism Research, Development and Innovation. Researcher. CICS.NOVA – Centro Interdisciplinar de Ciencias Sociais & UDI - Research Unit for Inland Development -Instituto Politécnico da Guarda Polytechnic Institute of Guarda. Member of the Executive Board of the National Mountain Research Network of FCT since 2017; Member of the Scientific Council of the Iberian Mountain Research Network (RIIM) since 2017; Member of Working Group 4 of RIS 3 (CCDR Centro) - Territorial innovation: subgroups of Rural innovation and Research and Development and Innovation in Tourism.

Hugo Gomes graduated in Geology since 2007, Master in Geosciences since 2010, from the University of Coimbra and since 2015 PhD in Geology in Quaternary Materials and Cultures, from the University of Trás-os-Montes and Alto Douro. Since 2007, he has been researcher at the Center for Geosciences at F.C.T.U.C, in which he has been involved in several multidisciplinary research projects in Portugal, Spain, Italy, Brazil, Angola and Ethiopia. Since 2016, he has been part of the technical team of Associação Geopark Estrela, performing duties in the area of Science and Research.

Vicente Alejandro Jiménez Olivera has a Doctorate in Mexican Studies from the Center for Higher Studies and Research. Bachelor of Tourism Services Administration, Specialist in Tourist Organizations Directorate and Master in Tourism Competitiveness by the University of Colima. Founding partner of the Research and Business Institute Senryakuteki S.A.P.I. de C.V. and the Iruma Desu A.C. Research Institute He has participated in the elaboration, development and execution of research projects financed by institutions such as: the University of Colima, the Ministry of Culture of Colima, the Federal Tourism Secretariat (SECTUR) and the National Council of Science and Technology (CONACYT), among others. He is the author and co-author of articles, books and book chapters published at the national and international, being also an active participant in Congresses and Seminars of research related to its line of generation and application of knowledge (LGAC): Tourism and Cultural Heritage for productivity, competitiveness and innovation from Mexico.

Eunice Ramos Lopes has an Advanced Training Course in Tourism by the Department of Economics, Management, Industrial Engineering and Tourism from University of Aveiro; a PhD in Antropology (Specialization in anthropology of tourism, politics and images of Culture and museology) awarded by New University of Lisbon; a MSc in Heritage and Museology awarded by New University of Lisbon. She is a Professor at the Polytechnic Institute of Tomar (IPT), at the Management School of Tomar (ESGT). Presently, coordinates the courses of Tourism at the IPT and is a member of the I&D Unit in Technology, Restoration and Arts Enhancement Center. Collaborator at the Network Research Center in Anthropology (CRIA-FCSH – UNL), and research unit Governance, Competitiveness and Public Policies (GOVCOPP) of the University of Aveiro. She is also a member of the editorial and scientific board of a few journals, as well the organizational and scientific committees of international tourism conferences.

Edith Sarai Lozada Sánchez is currently studying the last year of the Degree in Educational Processes at the Benemérita Universidad Autónoma de Puebla. She is a student who has great interesting academic potential and has qualities that distinguish her in an important way: responsible, intelligent, tolerant and hard-working. She has acquired a sufficient amount of theoretical and methodological knowledge to enter basic research. She collaborates in two research stays at the Benemérita Universidad Autónoma de Puebla. The first research entitled Design, validation and application of a quantitative instrument to measure University Social Responsibility. The second research entitled Ethics, Responsibility and Social Innovation in organizations. She showed great capacity for leadership and analysis of information while contributing to the theoretical foundation. In conclusion, she is interested in continuing in the investigation of Social Responsibility.

Irma Magaña Carrillo has a Doctor of Science. International Relations Area Transpacific. Full-time Professor-Researcher, Holder "B", in the Faculty of Tourism of the University of Colima, Mexico. Master in Administration, with Recognition Specialist in Total Quality Management and Quality Control.

Bachelor of Administration in Tourist Companies with a Specialty in Planning and Promotion. Leader of the Academic Body UCOL-CA-60 Tourism and Competitiveness Studies; have as General Knowledge Line: Tourist Competitiveness; Destinations Tourist; Public Tourism Policy; Cultural heritage; Corporate culture; Business gender; Culture and Identity; Total quality; Total Quality Control; Continuous improvement; Culture of 'the Client-Tourist'; Innovation; Asian Studies-Pacific: Japanologist.

Cintia Martins has Ph.D and Master's degrees in Celular and Molecular Biology at the UNESP-Brazil. Graduated in biological Sciences at the UNESP-Brazil. Full Professor of Biological Science course at Universidade Federal do Delta do Parnaíba, Brazil. Founder of Women in Science UFPI extension project.

Karla Melo is studying Professional Master in Tourism Business Management at the State University of Ceará - UECE. Specialist in Organizational Consulting and Human Development at Centro Universitário Estácio (2014). Graduated in Business Administration from Centro Universitário da Grande Fortaleza (2010). She is currently a professor at the Center for Studies and Research in Professional Electronics and Informatics - CEPEP and Manager at the Company TE&A - Educational Tourism & Learning. He has experience in Third Sector Management and Organizational Consulting by the Brazilian Service of Support to Micro and Small Enterprises - SEBRAE.

Arnaldo Mendoza Tarabó, docente de la Carrera Gestión y Desarrollo Turístico de la Universidad Estatal Península de Santa Elena, investigador del Proyecto La artesanía tradicional y su contribución al desarrollo económico y socio-cultural en la parroquia Manglaralto, cantón Santa Elena, Provincia de Santa Elena. Doctorando en Ciencias Económicas, Especialización Turismo en la Universidad de La Habana, Cuba.

Nikola (Nick) Naumov is a Senior Lecturer in Hospitality & Tourism Management at University of Northampton, United Kingdom. Nick has research interests in historical and cultural geography, heritage tourism, World Heritage sites, food tourism and gastronomy. He has a background in food and beverage operations and is currently a Visiting Professor at Meikai University (Japan).

Maria Rita Nunes is a PhD in Tourism from the University of Lisbon, Master in Tourism and Regional Development from the Federal University of Rio Grande do Norte and Bachelor in Tourism from the Federal University of Paraíba. She has experience in the Tourism area mainly in the following themes: criticism of predatory tourism, international real estate investment, crisis of the capitalist system. Interest in studies related to the local community, regional development, social capital, alternative tourism, gender, education and sustainability. At the professional level, she develops activity as a professor at Polytechnic Institute of Tomar and the I&D investigation center Techn&Art-IPT, Portugal.

Antonia Pérez-García is a Doctor for the Universities of A Coruña, Santiago of Compostela and Vigo in Economic Analysis and Business Strategy. Degree in Sociology from the University of A Coruña. Senior Public Relations Technician from the Higher School of Public Relations, attached to the University of Barcelona. Professor accredited as Doctor by the ACSUG (Galician Agency of Quality for the University System). Professor at the University of A Coruña in Degree in Tourism subjects: Sociology of Tourism, Methods and Techniques of Social Research and Human Resources. Lines of work in research on Local Development, Sustainability, Public Relations, Social Responsibility Corporative and Sociology in the Tourism Field.

André Riani Costa Perinotto is a Ph.D. in Communication Sciences (Mediatic Processes) - UNISINOS/RS-Brazil. Master's degree in Geography (Organization of Space) - UNESP/Rio Claro/SP-Brazil; Specialist in Teaching for Higher Education in Tourism and Hospitality at the SENAC/SP-Brazil; and Graduated in Tourism at the UNIMEP/SP-Brazil. Full Professor of the Bachelor's Degree in Tourism at the Universidade Federal do Delta do Parnaíba, Brazil. Professor of the Master's degree in Tourism Business Management at the UECE (State University of Ceará). Former Head of the Bachelor's Degree in Tourism at the UFDPar (Head of Department and Coordination of Course).

Rui Alexandre R. Pires is lecturer of Management Accounting and Pricing at the Escola Superior de Comunicação, Administração e Turismo at the Instituto Politécnico de Bragança (EsACT-IPB), Portugal. His research interests focus on strategic management accounting, management accountant roles, strategic decision making process, and strategic resources and capabilities.

Raquel Santiago Romo, Master in Planning and Management of destinations and products tourism, member of the GTES Research Group: Tourism, economics and sustainability, of the Universidad Estadual do Ceará (UECE)and doctorate in the Doctoral Programme in Regional Development and Economic Integration of the Universidade de Santiago de Compostela (USC).The main lines of research has been related to the tourism sector since gender perspective and sustainability

Eliana Silva is a PhD student at the Post-Graduate Program in Sociology at the Federal University of Ceará. Bachelor of Social Sciences from the Federal University of Ceará (2010). Master in Sociology by the Graduate Program in Sociology at the Federal University of Ceará (2014). She was substitute professor at the State University Vale do Acaraú (Sobral - CE), 2015-2017. He has experience with Coordination of Extension Projects, in the area of Territorial Development (NEDET-UVA) and University Entrepreneurship (OSESP - UVA).

João Tomaz Simões began his studies in Communications (2003) and later attended a degree in Tourism (2008). Before starting his teaching career in Social and Human Sciences (Portuguese Language, History and Geography of Portugal, Administration and Tourism), he performed as a Director of F&B(2010), while finishing in 2012 the Master in Development of Cultural Tourism Products, in the School of Management in Tomar (ESGT) of the Polytechnic Institute of Tomar (IPT). He is a certified Specialist in Tourism and Leisure and is currently completing his PhD in Tourism at the Institute of Geography and Spatial Planning (IGOT) of the University of Lisbon, in partnership with the Higher School of Hotel Management and Tourism of Estoril (ESHTE). At the professional level, he develops activity as a professor at Polytechnic Institute of Tomar and the I&D investigation center Techn&Art-IPT, Portugal.

Merita Toska, PhD, is an Expert for Economic Development and Finances, with over 13 years of experience in the field. 2014, she was awarded a Doctoral Degree (PhD) in "Financial Markets and Intermediaries" from the University of Bologna, Italy, having obtained the second level degree on "Economics and Business Administration" in 2007 and bachelor degree in "Business and Economics" in 2004 from the same institution. She has authored and co-authored various research articles including:

"Regional Economic Cooperation: the State of Play in the Western Balkans and Future Scenarios", 2016; "Balkan Countries: the Path to Nominal (Economic) Convergence', 2016; Towards Regional Operational Programing, 2016; "Are inflation expectations in Albania rational? – Empirical evidence from surveys data", 2015; "Informal Economy in Albania", 2013; "Modeling the quarterly GDP - Role of economic and survey indicators", 2010 etc. Research interests include: economic and financial development, financial sector development, alternative financing sources, informal economy, local finances management and budgeting, natural resources management, entrepreneurship and business planning.

Index

A

B

C

D

E

F

G

H

I

J

L

M

O

P

R

S

T

V

W

www.ingramcontent.com/pod-product-compliance
Lightning Source LLC
LaVergne TN
LVHW060410300926
849162LV00006B/1394